Artillery in the Era of the Crusades

History of Warfare

VOLUME 122

The titles published in this series are listed at *brill.com/hw*

Artillery in the Era of the Crusades

Siege Warfare and the Development of Trebuchet Technology

By

Michael S. Fulton

BRILL

LEIDEN | BOSTON

Cover illustration: Subayba castle, looking east from tower 9 (AUTHOR'S PHOTO).

The Library of Congress Cataloging-in-Publication Data

Names: Fulton, Michael S., author.
Title: Artillery in the era of the Crusades : siege warfare and the development of trebuchet technology / by Michael S. Fulton.
Other titles: Siege warfare and the development of trebuchet technology
Description: Leiden ; Boston : Brill, [2018] | Series: History of warfare ; volume 122 | Includes bibliographical references and index.
Identifiers: LCCN 2018025935 (print) | LCCN 2018033523 (ebook) | ISBN 9789004376922 (ebook) | ISBN 9789004349452 (hardback : alk. paper)
Subjects: LCSH: Catapult. | Siege artillery--History--To 1500. | Siege warfare--History--To 1500. | Crusades.
Classification: LCC U875 (ebook) | LCC U875 .F85 2018 (print) | DDC 355.4/40902--dc23
LC record available at https://lccn.loc.gov/2018025935

Typeface for the Latin, Greek, and Cyrillic scripts: "Brill". See and download: brill.com/brill-typeface.

ISSN 1385-7827
ISBN 978-90-04-34945-2 (hardback)
ISBN 978-90-04-37692-2 (e-book)

This book is printed on acid-free paper and produced in a sustainable manner.

For Laura

Contents

Foreword

In 2011, I had the privilege of digging in southern Italy as part of a group coordinated by Fabio Colivicchi through Queen's University, Canada. On 16 July, we visited Pompeii; there, Professor Colivicchi made a point of showing us (I like to think me in particular) the north wall of the city, where the scarring left by Sulla's artillery in 89 BC can still be seen. I had completed my MA a couple years earlier but had been struggling to come up with an original line of doctoral research, looking beyond the traditional limits of history to archaeology and the burgeoning field of photogrammetry in hopes of finding inspiration. By the end of that hot July day I had a clear thesis topic. My plan was to try to determine the strength of medieval stone-throwing engines by studying the datable impact signatures discernable at castles in the Levant. After a few months of research I determined that this was not only a feasible project but that there was a significant gap in the scholarship relating to medieval mechanical artillery. Unfortunately I began my doctoral programme, with the fantastic crusades scholars at Cardiff, just as the Syrian civil war was beginning. Perhaps the least significant consequence of this terrible conflict was that it prevented me from visiting certain sites in Syria, forcing me to shift my approach from an archaeological study to a historical one and place far more emphasis on the surviving literary evidence than I had originally planned.

This book is the ultimate outcome of that epiphany back at Pompeii in 2011; however, it would not have been possible without the help and support of a number of people between then and now. In Wales, thanks are due first to Denys Pringle, whose help, guidance and input over the past few years has been instrumental. Thanks also to Helen Nicholson and John France, who also read and commented on what was an early version of this book, and to Peter Edbury who graciously read parts and has generously shared many thoughts over the years. In Canada, W.T.S. Tarver has enthusiastically shared his experiences building and operating trebuchets and Mark Denny has been incredibly generous in running numerous theoretical mathematical scenarios through his trebuchet algorithm. Denys Pringle, Michael Eisenberg and others have kindly shared some of the photos that are included and CGIAR have allowed me to include the background topographical data of the small-scale regional maps, composed in QGIS, found below. Numerous archaeologists based around the world have generously shared information relating to their past and ongoing excavations and to all of them I am deeply grateful. Finally, thanks are due to the Zinman Institute of Archaeology, which provided the postdoctoral fellowship that allowed for the final preparation of this book, to Rabei Khamisy, my sponsor, and to his incredible family in Mi'ilya. To everyone involved, my most sincere thanks.

M.S.F.
Mi'ilya, Galilee, 2018

List of Illustrations

Maps

Plans

Figures

Images

Plates

Abbreviations

BL	British Library
BNF	Bibliothèque nationale de France
ar.	arabe
fr.	français
lat.	latin
pers.	persan
CCCM	Corpus Christianorum, continuatio mediaeualis
CSCO, Scr. Syri	Corpus scriptorum Christianorum orientalium, scriptores Syri
DRHC	Documents relatifs à l'histoire des croisades
FRA	Fontes rerum austriacarum
Loeb	Loeb Classical Library Series
MGH	Monumenta Germaniae historica, Societas aperiendis fontibus rerum Germanicarum medii aevi
SS	Scriptores
SrG	Scriptores rerum Germanicarum
SrG NS	Scriptores rerum Germanicarum Nova Series
PO	Patrologia orientalis
PPTS	Palestine Pilgrims' Text Society Library
RHC	*Recueil des historiens des croisades*, Académie des inscriptions et des belles-lettres
Ar	Documents Arméniens
Oc	Historiens occidentaux
Or	Historiens orientaux
RHGF	Recueil des historiens des Gaules et de la France
RS	Rerum Britannicarum medii aeui scriptores, or Chronicles and Memorials of Great Britain and Ireland in the Middle Ages [Rolls Series]
SI	Istituto/fonti per la storia d'Italia
SHF	Société de l'histoire de France

Names

Place names used in this study	Alternatives
‘Ajlun	al-Rabad
‘Akkar	Gibelacar
Amida	Amid / Diyar Bakr
Apamea	Afamiya / al-Mudiq
‘Arqa	Archas
Arsuf	Arsur
‘Atlit	Castrum Peregrinorum / Chastiau Pelerin
Baghras	Gaston
Beaufort	Belfort / Shaqif Arnun
Belmont	Suba
Belvoir	Kawkab
Bethgibelin	Beit Gibelin / Bayt Jibrin
al-Bira	Birecik
Bourzey	Qal‘at Barza / Rochefort
Qaqun	Caco
Castrum Arnoldi	Yalu
Chastel Blanc	Safita
Chastel Neuf	Hunin
Chastel Rouge	Castrum Rubrum / Qal‘at Yahmur
Crac des Chevaliers	Hisn al-Akrad
Cursat	al-Qusayr
Darum	Dayr al-Balah
al-Habis	Petra
Harim	Harrenc
Jabala	Gibellum
Jacob’s Ford	Chastellet / al-‘Atra
Jubayl	Giblet / Byblos
al-Jundi	Qal‘at Sadr
Kafr Lam	Cafarlet
Kerak	Petra Deserti / al-Karak
Latrun	Toron de les Cabelleros / Toron des Chevaliers
Li Vaux Moise	al-Wu’ayra
Maraclea	Maraqiyya
Margat	Marqab

Minat al-Qal'a	Ashdod-Yam
Mirabel	Majdal Yaba
Montferrand	Ba'rin
Montfort	al-Qurain
Montreal	Shawbak
Mount Tabor	Qal'at al-Tur
Ruad	Arwad
Safed	Saphet
Saone	Sahyun / Qal'at Salah al-Din
Sidon	Sagette / Saida
Sidon Land Castle	al-Mu'azza
Sidon Sea Castle	al-Bahr
Subayba	Nimrod
Toron	Tibnin
Tortosa	Tortouse / Tartus / Antartus
Trapessac	Darbsak
Turbessel	Tell Bashir
Vetula	Qal'at Bani Qahtan

Where possible, the names most often used by historians and archaeologists of the crusader period have been used and Arabic personal and place names have been transcribed using basic Roman characters for convenience. Personal names have been spelled according to their most popular form in Western scholarship, i.e. 'Alexius Comnenus' is given in the Latin form whereas 'Niketas Choniates' is given in the Greek; similarly, 'Zanki' and 'Tughtakin' will be used rather than 'Zangi' and 'Tughtagin', while 'Qaraqush' will be used rather than 'Karakush'. European names have been anglicised where possible.

Weights and Measures*

dirham	Egypt	3.125 g
	Damascus	3.09 g
	Iran	3.2 g
ratl (pl. *artal*)	Iraq	0.40625 kg
	Egypt	0.4375 kg
	Damascus	1.85 kg
	Aleppo	2.27-2.28 kg
	Hama	2.062 kg
	Homs	2.7 kg
	Sahyzar	2.137 kg
	Tripoli	1.968 kg
	Palestine	2.2-2.5 kg
qintar	100 *artal*	[see above]
	Cyprus (100 *rotuli*)	226.4 kg
	Rum (100 *lodra*)	56.44 kg
	of gold	42.33 kg
cubit	Egyptian/ordinary/*al-sauda'* (black)	0.540 m
	al-yad (hand)	0.499 m
	bi'l-najjari/al-mi'mariyyah (carpenter's/building)	0.789 m
	hashimi/al-'amal/al-malik (practical/royal)	0.665 m
ba'	Egypt	1.995 m

* The Islamic measures are taken from Hinz, *Islamische Masse und Gewichte*, trans. Marcinkowski, *Measures and Weights in the Islamic World*. The Cypriot *qintar* has been taken from Coureas, *The Assizes of the Lusignan Kingdom of Cyprus*, p. 58.

Regional Maps

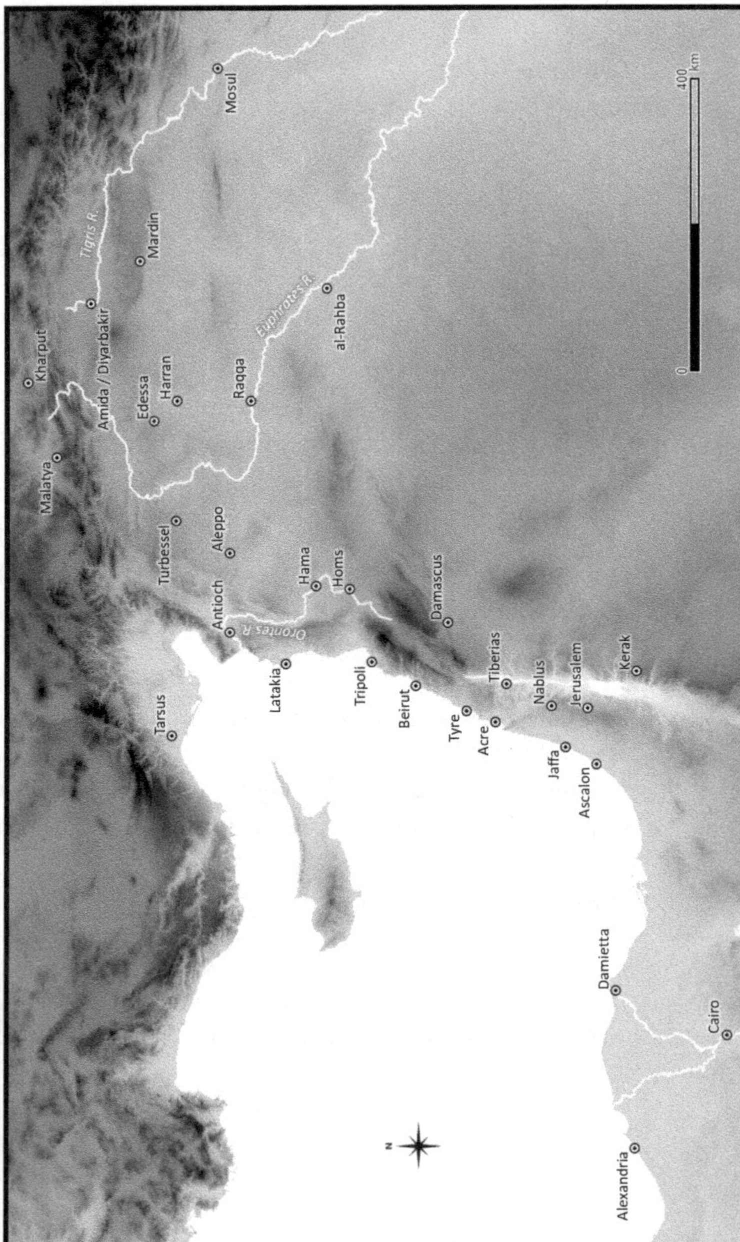

MAP 1 *General theatre of the crusades*

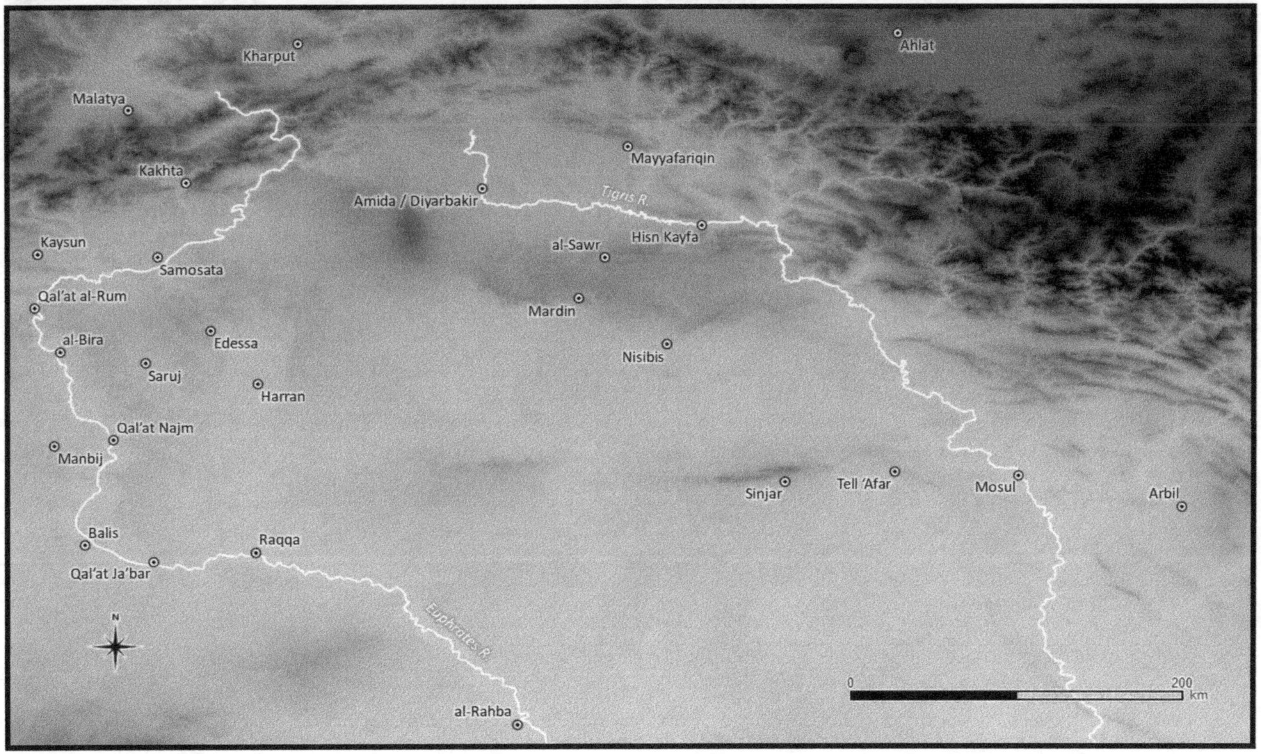

MAP 2 *The Jazira*

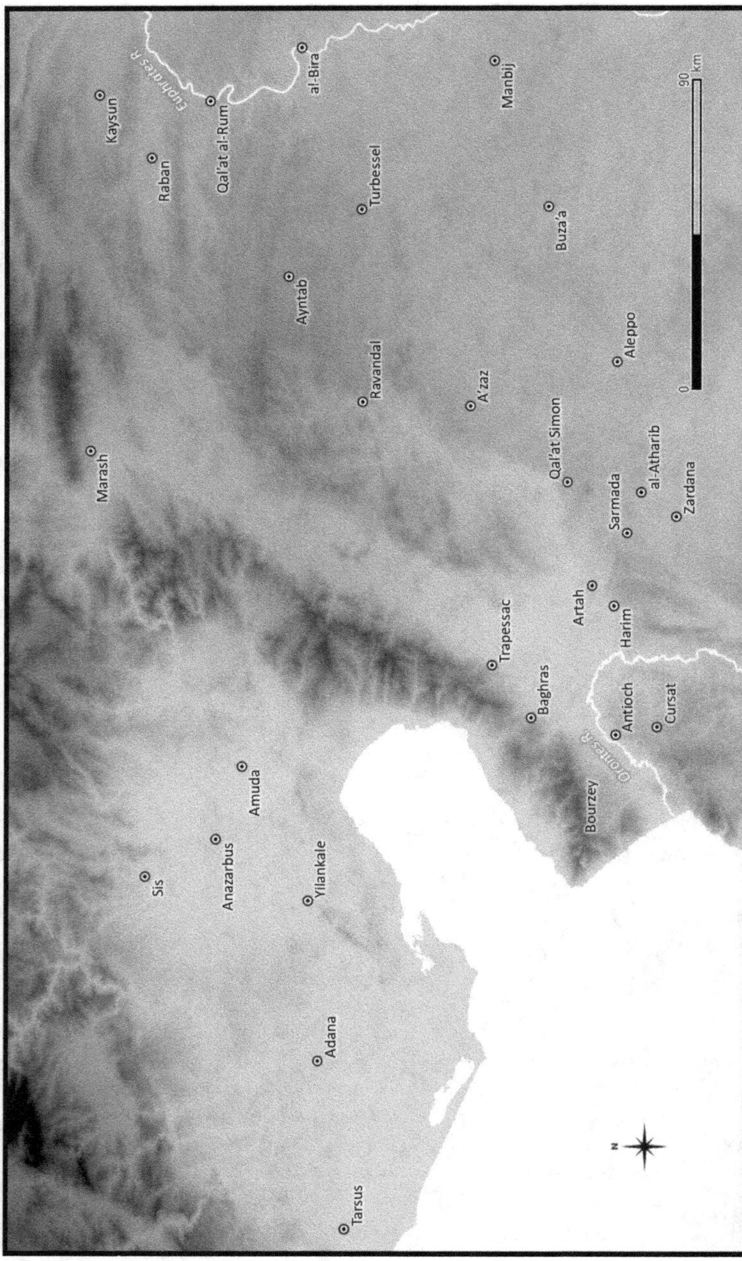

MAP 3 *Cilicia and north-western Syria*

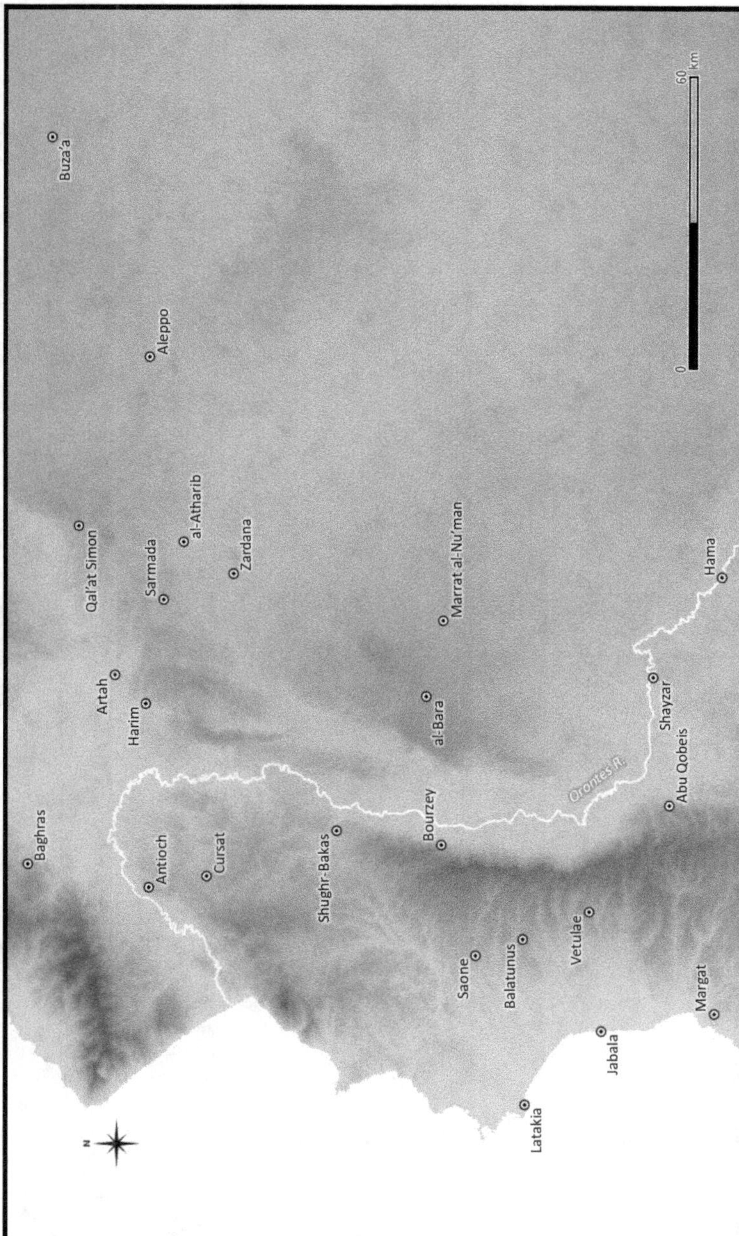

MAP 4 *Western Syria*

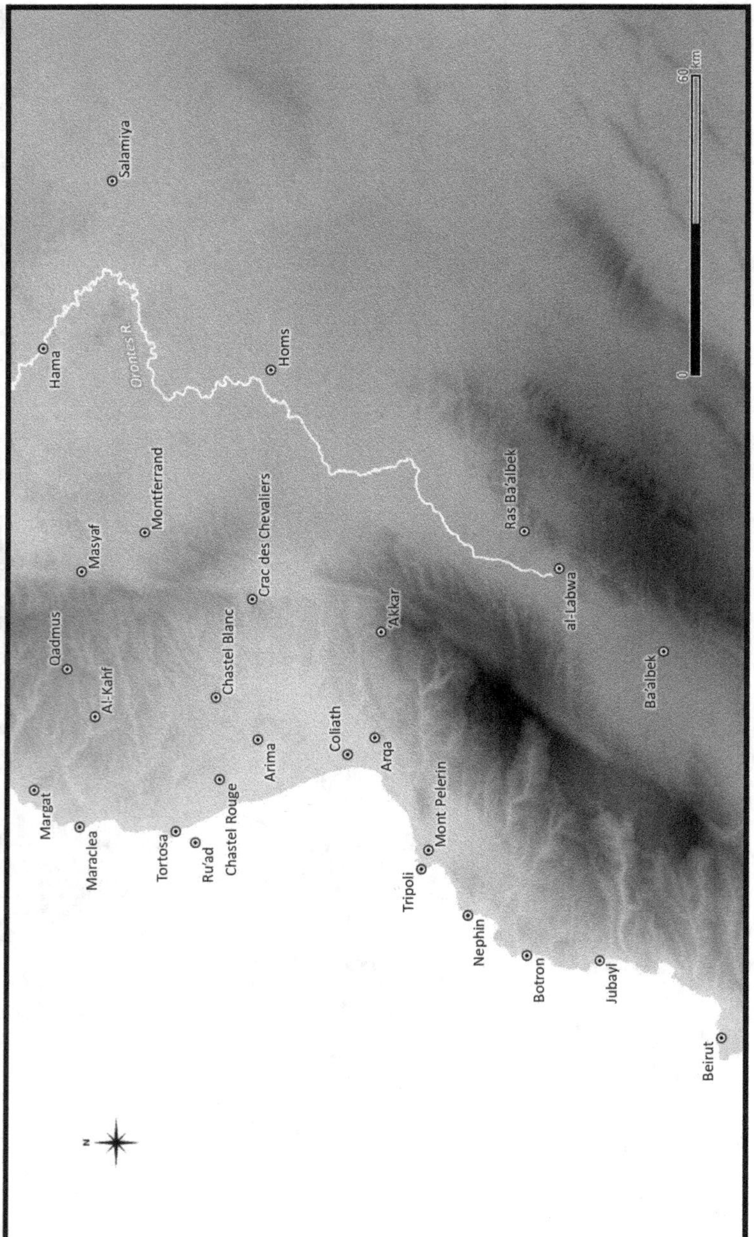

MAP 5 *South-western Syria and northern Lebanon*

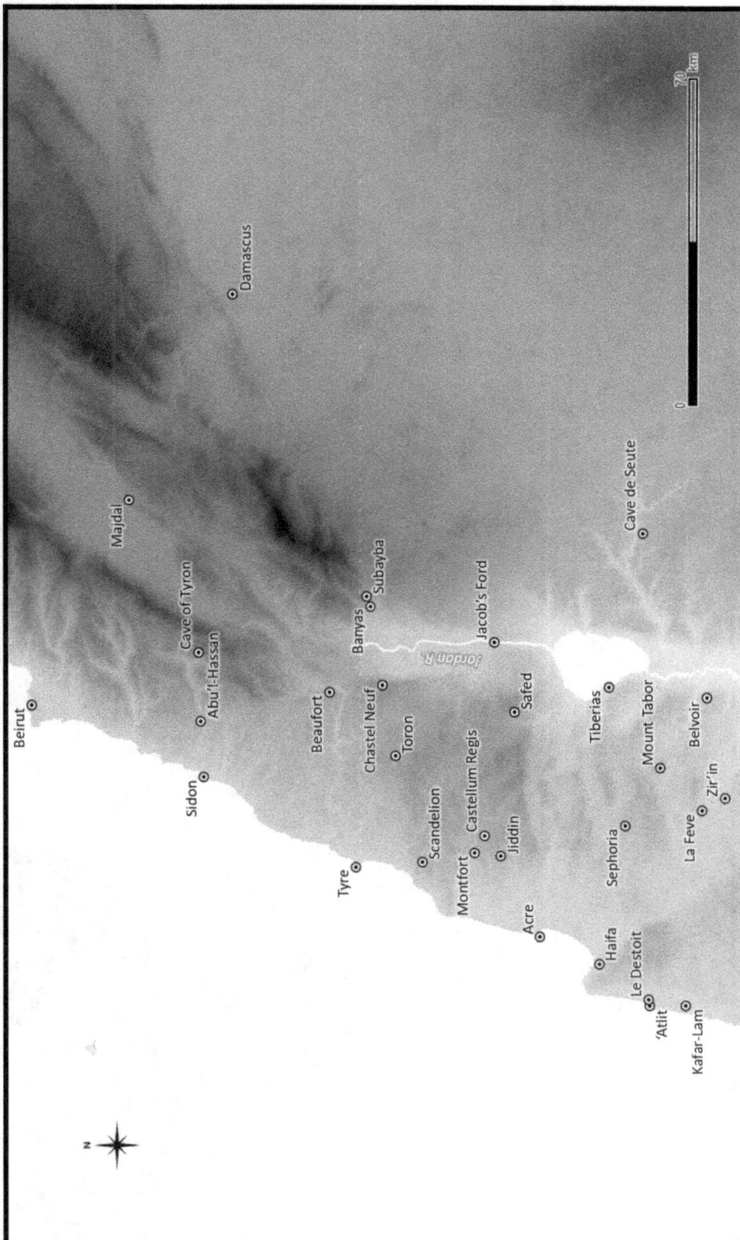

MAP 6 *Southern Lebanon and northern Palestine*

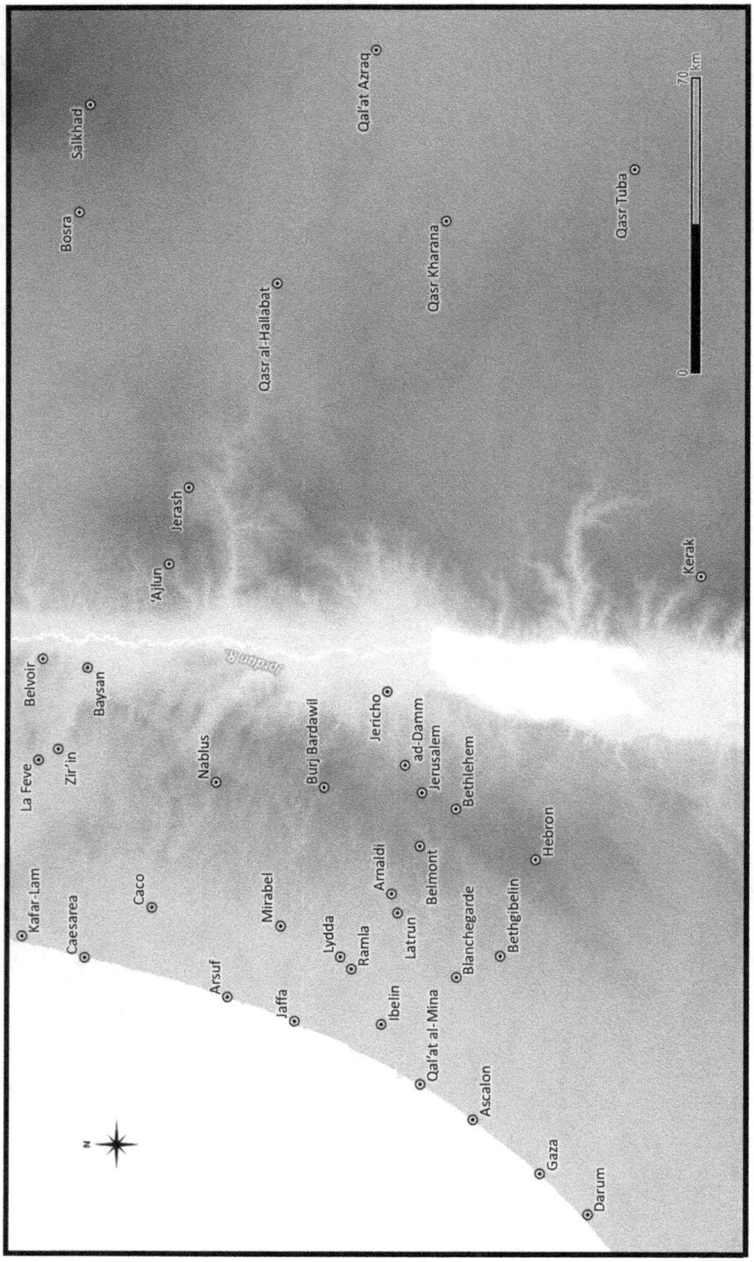

MAP 7 *Southern Palestine and Transjordan*

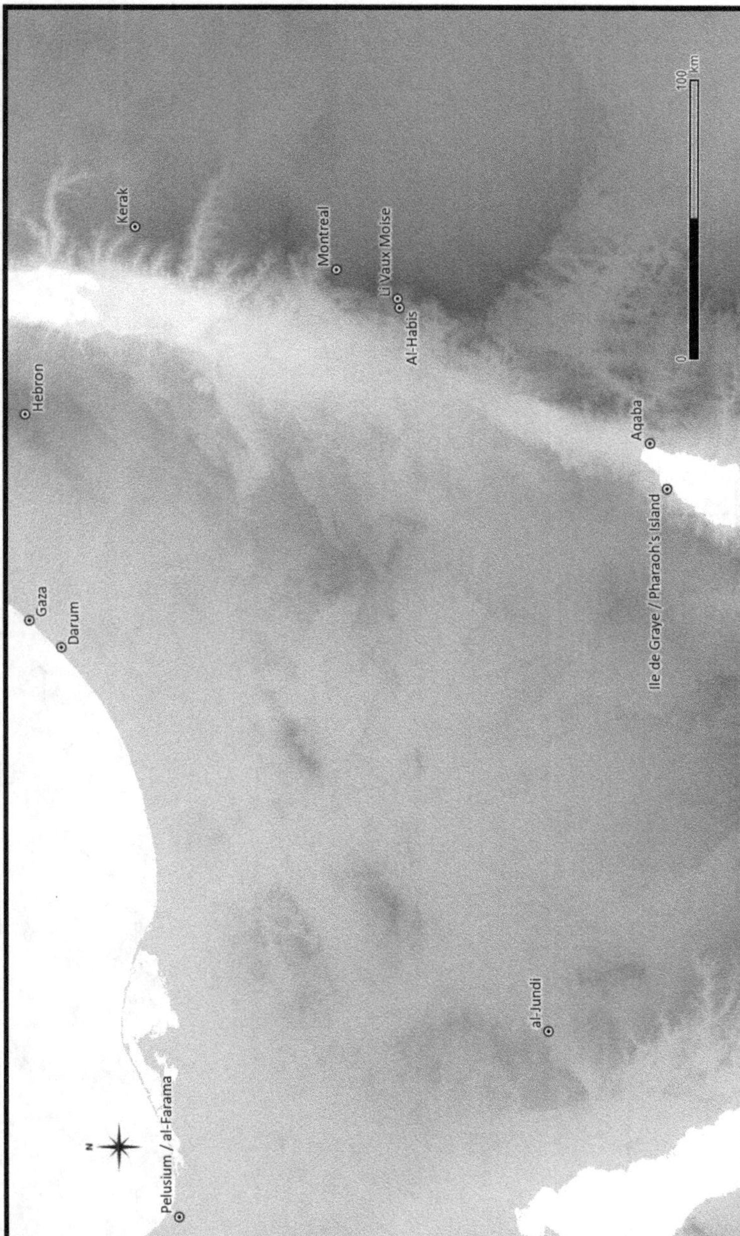

MAP 8 *Sinai and Wadi 'Araba*

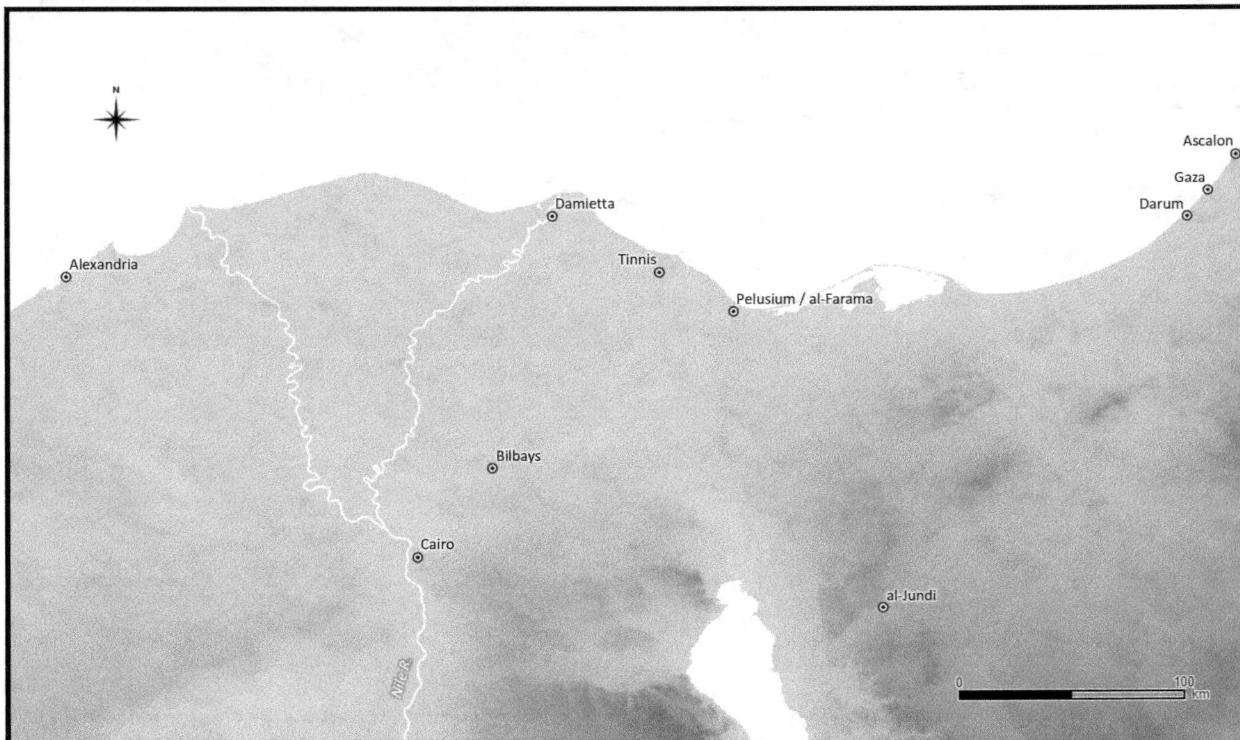

MAP 9 *Northern Egypt*

Introduction

Medieval mechanical artillery has interested historians for centuries, yet our understanding of these engines remains fairly poor. Whereas technical treatises, detailed contemporary descriptions and illustrations survive from the classical and Renaissance periods, evidence dating to the Middle Ages is relatively scarce. In the absence of lucid contemporary descriptions, medieval artillery (mechanical stone-throwing machines) tends to be dealt with quite generally or studied in a very broad context. This has allowed historians, often relying on the most sensational anecdotal scenarios, to put forward various theories relating to the development and capabilities of these engines.

In order to avoid the generalisations that have accompanied previous studies, the focus here will be on a specific region and a relatively short period of time: the Levant, from 1097 to 1291. From the arrival of the First Crusade in Anatolia until the expulsion of the Franks from Palestine and Syria, this was a dynamic period of conflict, coexistence and cooperation between Latin, Greek, Armenian and Syrian Christians, as well as Sunni and Shiite Arabs, Turks and local Syrian Muslims. This nexus of cultures left a rich collection of literary sources, composed by a range of individuals from different cultural backgrounds writing from different perspectives. Although these contemporary sources present an informative window into the use and steady development of artillery technology during this period, their testimony must be reconciled with examinations of available archaeological evidence, the topography that influenced the use of artillery in each scenario and an understanding of the mechanics that governed the function of these engines.

Unfortunately, most previous studies have dealt with only the most sensational descriptions of artillery and have neglected the wealth of less detailed references that can help to construct a more complete picture of these machines and how they were used. Likewise, few have attempted to incorporate archaeological evidence or a sound appreciation of the practical physics of trebuchet technology. Such shortfalls have allowed scholars to accept at face value certain reports and figures, often provided by medieval authors with limited understanding of contemporary siege technology, leading to assumptions that these machines were far more powerful than appears to have been the reality.

Over the past two centuries or so, the views of many scholars have been influenced by the significance of modern artillery. At times, historians have even drawn direct parallels between modern howitzers and the mechanical engines

 | DOI 10.1163/9789004376922_002

regarded as their medieval predecessors.[1] In his seminal work on twelfth-century crusader warfare, R.C. Smail highlighted the inability of crusader castles to halt advancing armies, pointing out that these were not the forts of the eighteenth century with guns that could effectively close roads and obstruct the movements of large bodies of troops.[2] While few people associate the power of medieval defensive artillery with that of the modern era, there remains a pervasive tendency to characterise offensive engines as proto-bombards, the exclusive function of which was to smash holes in walls. Much as Smail forced us to reconsider the role of castles, it is time to re-evaluate the development of artillery during this period and the ways in which these engines were employed.

Among the best studies of medieval siege warfare in the Levant since that of Smail are those of Randall Rogers, Christopher Marshall and John France. Rogers and Marshall focused on the twelfth and thirteenth centuries respectively while France examined sieges in an around the crusader principalities in a broader study of medieval warfare in the Latin world.[3] All three works operate primarily from a Frankish perspective and only Rogers examines artillery in any detail. Contrastingly, Paul Chevedden has operated primarily from a Muslim point of view and has perhaps devoted more attention to medieval artillery than any other modern scholar since Kalervo Huuri. His theories, however, tend to be built around a selection of choice anecdotes.[4]

The present study is, in part, a response to the ideas of these men: an attempt to improve our understanding of the development and role of mechanical artillery during this period and to integrate this into our conception of contemporary siege warfare; and, to address the exaggerated notions of power that have become popular in recent decades. In order to do this, it is necessary

---

1 Chevedden's statement that "The Big Bertha of the pre-modern era was the trebuchet" is perhaps the most direct, Chevedden, "King James I," p. 313. As will be shown, such a comparison is quite inaccurate, or at the very least misleading.

2 Smail, *Crusading Warfare*, esp. pp. 204-5.

3 Rogers, *Latin Siege Warfare*; Marshall, *Warfare in the Latin East*; France, *Victory in the East*; France, *Western Warfare*. Rogers devotes the most attention to artillery; however, his primary concern is siege towers when dealing with Frankish sieges in the Levant during the twelfth century. He claims that the use of artillery was better suited to the Normans of South Italy and examines it in this context, despite its more frequent appearance in sources dealing with the Latin East.

4 See Chevedden, *The Citadel of Damascus*; Chevedden, "Artillery of King James I," pp. 47-94; Chevedden, "The Hybrid Trebuchet," pp. 179-222; Chevedden, "Fortifications and the Development of Defensive Planning," pp. 33-43; Chevedden, "Invention," pp. 71-116; Chevedden, "Black Camels," pp. 227-77; Chevedden, "King James I," pp. 313-39.

to look at the development of these engines in the lead up to the twelfth century in the various cultural contexts that would play a role in the struggle for control of the Levant during the period of the crusades. Furthermore, it is important to appreciate the fundamental mechanics of these machines and to possess an understanding of them as energy systems. The wealth of evidence, textual and archaeological, can then be critically examined to determine not only how these engines developed, but how widely this technology was known and employed, if the technology of certain groups was superior to that of others, and how these engines were used by military forces with different siege traditions and tactical preferences.

When all of the evidence is considered, it will become apparent that mechanical artillery was an important part of siege warfare during the era of the crusades; however, these engines did not develop into super-weapons that drastically changed the nature of war in the region or the design of fortifications. Light traction trebuchets, familiar to all parties at the end of the eleventh century, remained important weapons through to the end of the thirteenth century. These machines were complimented, probably from around the late twelfth century, by the development of the counterweight trebuchet, a more powerful variety that was employed in an increasingly different way. Despite the increasing use and scale of these latter engines, they do not appear to have caused any kind of revolution in the nature of siege warfare.

CHAPTER 1

Background

As the First Crusade made its way across Europe at the end of the eleventh century, the trebuchet seems to have been the dominant type of stone-throwing artillery. The torsion-powered stone-throwers of the classical era appear to have fallen from use at some point during the Early Middle Ages and were replaced, directly or after a period of delay, by the simpler swing-beam or levered family of artillery, which we now label broadly as 'trebuchets'. The matter of when, or even if, torsion artillery was replaced by trebuchets is a contested one. Although there appears to be insufficient evidence at this time to conclusively prove or disprove that the use of torsion stone-throwers continued through the Middle Ages, it can be observed that trebuchets were the most common, if not only, mechanical stone-throwers employed by Latin, Greek and Muslim forces on the eve of the First Crusade.

The traction trebuchet was the earliest and simplest variety, powered by the pulling force of a team of operators. These were relatively light engines best suited to targeting people and improvised defences rather than fortified masonry. At some point around the late twelfth century, these machines came to be supplemented by an increasingly heavy variety that was powered by the force of a falling counterweight, rather than a team of pullers. Whereas the traction trebuchet remained a fairly light engine, counterweight trebuchets grew larger and more powerful through the thirteenth century, allowing them to threaten battlements and eventually the much thicker walls below.

Classical Artillery

The first stone-throwing engines used in Europe and the Near East were adapted from earlier arrow-shooting machines. The *gastraphetes*, or *katapeltikon*, which was invented around the start of the fourth century BC, is typically judged to have been the earliest of these arrow-shooters. It was little more than a composite bow mounted on a wooden stock and operated by a mechanised draw and release. This defensive weapon quickly developed to incorporate a winch, increasing the potential tension energy that could be drawn from the bow.[1]

1 For this early engine and its successors, see Marsden, *Historical Development*, pp. 48-85, see also pp. 1-47; Soedel and Foley, "Ancient Catapults," pp. 150-60; Chevedden, "Artillery in Late

 | DOI 10.1163/9789004376922_003

Around the middle of the fourth century BC, the more powerful two-armed torsion catapult was invented. Rather than employing a bow, a fairly stiff horizontal arm was inserted into each of two vertically mounted coils of hair or sinew, giving the appearance of a horizontal bow without the central section. The two arms were pulled back against the force of the coils as the string running between their outward ends was drawn back. Initially these engines shot arrows or bolts, which were placed in the grooved stock that extended backwards from between the two coils, but the far greater potential torsion energy stored in the coils, relative to the energy stored in a tension bow of comparable size, allowed larger engines of this type to shoot stone projectiles.

Vitruvius (d. c. 15 BC) divided torsion bows into two categories: bolt-throwing *catapultae* (or *scorpiones*) and stone-throwing *ballistae*.[2] Some historians suggest that by this point some of these torsion stone-throwers were capable of breaching fortifications.[3] Although Vitruvius provides figures for a machine capable of throwing a stone projectile in excess of 150 kg, he himself concedes that such an engine may never have been (previously) built. His figures instead represent an exercise in which he demonstrates a supposed ideal proportional relationship between the various mechanical components of an engine and the corresponding size of the projectiles that it was meant to throw. While it is tempting to fixate on certain accounts and even examples of quite large projectiles, it is important to keep in mind that most Hellenistic projectiles found in the Near East weigh less than 15 kg and most dating to the Roman period weigh less than 5 kg.[4]

Torsion bows were employed throughout the Roman period, noted most often in a defensive capacity but also commonly employed by attackers.[5] These

---

Antiquity," pp. 134-36; DeVries and Smith, *Medieval Military Technology*, pp. 118-19; Campbell, *Greek and Roman Artillery*, pp. 3-15; Campbell, "Ancient Catapults," pp. 678-82.

2 For a full description, see Vitruvius, *De Architectura Libri Decem* 10.10-11, ed. F. Krohn, pp. 245-50, trans. Morgan, pp. 303-8.

3 For example, Soedel and Foley, "Ancient Catapults," pp. 150-60; DeVries and Smith, *Medieval Military Technology*, pp. 118-19.

4 For surveys of classical projectiles at various Near Eastern sites, see Marsden, *Historical Development*, pp. 79-83; Shatzman, "Stone-Balls from Tel Dor," pp. 52-72; Campbell, *Greek and Roman Artillery*, pp. 19-22; Nossov, *Ancient and Medieval Siege Weapons*, pp. 137-42; Holley, "*Ballista* Balls," pp. 349-65; Holley, "Stone Projectiles," pp. 35-55; Syon and Yavor, "Gamla," pp. 40, 42-44; Adan-Bayewitz and Aviam, "Iotapata," pp. 131-65; Aviam, "The Archaeological Illumination," pp. 376-77; Wilkins, Barnard and Rose, "Roman Artillery Balls," pp. 66-80; Savage and Keller, "Archaeology in Jordan," p. 478.

5 Famous examples of their use include the siege of Massalia (Marseilles) during Caesar's Civil War in the first century BC and Titus's siege of Jerusalem in the first century AD, Caesar, *De bello civili* 2.1-2, ed. and trans. Peskett, pp. 124-27; Josephus, *The Jewish Wars* 5.6, trans. Whiston, pp. 810-11.

weapons became such an intrinsic part of Roman poliorcetics that Frontinus (d. c. 103 AD) deemed their development to have reached its peak long before his own time.[6] Centuries later, these same weapons can be found in the works of Ammianus (d. c. 395 AD) and Vegetius (fl. fourth century AD).[7] Ammianus clearly states that the torsion bow remained in use through the late Roman imperial period, a point confirmed by archaeological evidence dating to the late fourth century.[8] By the fifth century, however, these engines appear to have been used almost exclusively as bolt-shooters and were complimented by the appearance of metal tension bows around this time.[9] Through this cyclical course of development, the heavy tension bow reappeared and replaced the two-armed catapult as precision seems to have been preferred to power.

Around the fourth century AD, the terminology used to identify the torsion bow changed: *catapulta*, traditionally denoting an arrow-shooting machine, was generally replaced by *ballista*, formerly used to identify a stone-throwing engine.[10] This transition serves as an important warning, revealing that the meaning of traditional artillery terms could change over time. Furthermore, it indicates that changes in terminology did not always reflect technological advancements, nor were such advancements always apparent in the vocabulary used by contemporary authors.[11] When dealing with later medieval sources, especially those composed prior to the twelfth century, it can be difficult to discern whether the terminology, often classical in origin, was meant to identify a stone-throwing engine or one that shot bolts.

The other type of torsion engine described by Vegetius is the *onager* or 'wild ass', so labelled on account of its kick. In the fourth century, this was the heaviest ballistic weapon in the Roman arsenal, used to target personnel as well as other engines.[12] This one-armed stone-thrower used the power of a single large

6 Frontinus, *Strategmata* 3.preface, ed. McElwain and trans. Bennett, p. 204.

7 Ammianus, *Res gestae* 23.4.1-3, ed. Eyssenhardt, p. 270, trans. Yonge, p. 322; Vegetius, *Epitoma rei militaris* 4.9, 22, ed. Lang, pp. 134-35, 143-44.

8 For brief summaries of the archaeological remains of Roman artillery, see Bishop and Coulston, *Roman Military Equipment*, pp. 58-61, 88-90, 168-70, 206-8; Baatz, "Recent Finds in Ancient Artillery," pp. 1-17.

9 The tension-powered *arcuballista* was possibly the predecessor of the medieval siege crossbow. Such an engine may have been used at the siege of Paris in the eighth century, see Abbo, *Bella Parisiacae urbis*, ed. Pertz, pp. 9-10.

10 For interpretations of this shift in terminology, see Marsden, *Historical Development*, pp. 188-89; Marsden, *Technical Treatises*, pp. 25-51; Campbell, "Ancient Catapults," p. 690; Chevedden, "Artillery in Late Antiquity," pp. 137-42.

11 For the changing nature of the term *machina* between antiquity and the Middle Ages, see Popplow, "The Concept of *Machina*," pp. 83-90.

12 Vegetius, *Epitoma rei militaris* 4.22, ed. Lang, pp. 143-44.

horizontal coil, against which the perpendicularly extending throwing arm was drawn back. Upon release from its relatively horizontal loading position, the arm was abruptly stopped at a desired angle by a sturdy crossbeam, allowing the projectile, held in a sling at the end of the arm, to fly free along a desired parabolic path. Although not as accurate as the two-armed torsion bolt-shooter, which shot its projectiles along a fairly flat trajectory, the advantage of the one-armed engine was its power and comparative simplicity.[13] These factors, as well as the deterioration of the administrative infrastructure and technical expertise in the late Roman world, probably contributed to the decline of the stone-throwing torsion bow and the eventual abandonment of all torsion weapons in western Europe sometime around the sixth century or perhaps the seventh.

The range of the *onager* was made possible by the sling at the end of the throwing arm. This essentially elongated the arm while adding negligible mass, forcing the projectile to travel farther in the same amount of time before release. This increased the projectile's acceleration and release velocity without retarding the angular velocity of the throwing arm or increasing the potential energy in the coil (by using a larger one), which would have required the whole structure of the engine to be strengthened. This simple component made the whole system far more efficient.

Fundamental to the question of what type of artillery was employed by the earliest crusaders and their Muslim counterparts, is the debate concerning the extent to which torsion-powered machines remained in use through the Early and High Middle Ages and at what point, if any, they were replaced by engines of a new type unknown in classical Europe and the Middle East. At the root of the debate is the lack of clear source descriptions, illustrations or material evidence to support either the continued use of torsion engines or their replacement.

At least some medieval figures were acquainted with the technology of torsion artillery because classical texts that describe these engines are known to have survived and circulated during the High Middle Ages. For instance, Geoffrey Plantagenet appears to have had access to a copy of Vegetius's work at the siege of Montreuil-Bellay in 1147.[14] Although it is tempting to see him looking to the ancients for inspiration during a tough siege, there is little to suggest that he, or any of his contemporaries, ever constructed a torsion stone-thrower.

13 Two-armed catapults required constant adjustment to balance the resistance of the two spring coils.

14 John of Marmoutier, *Historia Gaufredi ducis Normanorum*, ed. Halphen and Poupardin, pp. 215-19.

At Montreuil-Bellay, Geoffrey's solution was the use of some kind of incendiary, which was thrown from what seems, from its description, to have been a traction trebuchet, a machine-type unknown to Vegetius and his contemporaries.

When analysing the use of any siege engine, it is important to bear in mind the broader context: Geoffrey, like his predecessors a millennium earlier, was a seasoned campaigner, familiar with poliorcetics and the construction of siege engines. If there is any truth in this story, it seems he was simply interested in exploring his options during what was apparently a difficult siege. Geoffrey's ingenuity and experience are evident; he is said to have used both towers and rams, as well as a number of types of artillery, in his effort to take Montreuil-Bellay.

The survival of such classical texts, the use of unclear terminology by medieval sources and the reappearance of the *onager* in fifteenth-century illustrations have fuelled over two hundred years of debate concerning the medieval use of classical engines.[15] The primary issue is the lack of clear descriptions and the inconsistent use of terminology. Considering the general uncertainty that characterises the Early Middle Ages, and broad spectrum of interpretations regarding the technological sophistication of its cultures, three general theories can be put forward. First, that the knowledge, institutions and administrative network that allowed for the manufacturing and maintaining of torsion engines endured the retreat of Roman influence, leading to the continued use of these engines well into the Early Middle Ages. Second, that these factors were not present, leading to a period without mechanical artillery until the introduction of the traction trebuchet. Third, that while it remained possible to employ torsion engines, this technology was abandoned in favour of the simpler trebuchet upon its arrival. Each theory has received support at one time or another.[16]

---

15 For summaries of the historiography, see Rogers, *Latin Siege Warfare*, Appendix 3, pp. 254-73; Chevedden, "Artillery in Late Antiquity," pp. 131-34. For the most recent study, see Purton, "The Myth of the Mangonel," pp. 79-90.

16 For arguments supporting the continued use of classical torsion stone-throwers, see Grose, *Military Antiquities*, 1:380-84; Köhler, *Die Entwickelung des Kriegswesens*, 1:139-211; Oman, *A History of the Art of War*, p. 543, for his discussion of classical machines, pp. 131-49; King, "The Trebuchet," p. 459; Bradbury, *The Medieval Siege*, pp. 253, 259; D. Bachrach, "English Artillery," pp. 1,408-30. See also D. Bachrach, *Warfare in Tenth-Century Germany*, pp. 160-64. For arguments supporting the disappearance of torsion stone-throwers, for different reasons, see Dufour, *Mémoire sur l'artillerie des anciens*, p. 97; Bonaparte, *Études sur le passé et l'avenir de l'artillerie*, 2:26-54; Payne-Gallwey, *A Summary of the History*, pp. 7-8; Schneider, *Die Artillerie des Mittelalters*, pp. 1-60, esp. pp. 1-26; Huuri, *Zur Geschichte*

The best arguments for the use of torsion stone-throwers in Europe after the sixth century are the continued use of classical terms and the lack of conclusive evidence that they were not used; but neither of these arguments is particularly strong. Such engines were no more powerful, much more complicated, and far more dangerous to operate than traction trebuchets, given the pent-up stresses within the coil and then violent stop of the arm against the crossbeam when operated. Traction trebuchets, by comparison, were capable of a similar range, a much higher shooting rate, and were far simpler to construct, use and maintain. Even the heavier counterweight models that were developed later, although typically much larger, were quite mechanically simple, requiring only a moderate level of skill to construct. Perhaps the best argument for the discontinued use of torsion stone-throwers, is the complete lack of illustrative evidence. From the late eleventh century, when European manuscript illustrations become more numerous, dozens of traction trebuchets are portrayed, but the earliest attempt to depict what appears to be an *onager* dates to the 1320s, by which time numerous images of counterweight trebuchets had appeared.[17] Given the classical descriptions of these engines, which many illuminators would have read, it is hard to explain why none were depicted if they had remained in use.

For present purposes, it will be assumed that all stone-throwers were of the trebuchet type during the period of the crusades, which appears a near certainty when examining the sources. Even historians who contend that torsion stone-throwers remained in use during the High Middle Ages tend to agree that trebuchets were the dominant form of artillery in Europe and the Near East by the twelfth century. However, there is less agreement concerning when and where the traction trebuchet was first used by European and Middle Eastern armies. The terminology, once more, poses the greatest challenge to a consensus of opinion.

Terminological Issues

Most references to pre-fourteenth-century artillery in Europe are found in the chronicles, annals and letters of churchmen who, in most cases, had little if

des Mittelalterlichen Geschutzwesens, pp. 51-65, 212-17; Chevedden, "Invention," pp. 73-74; Chevedden, "King James I," pp. 315-16; Kennedy, *Crusader Castles*, p. 107; France, *Western Warfare*, p. 119.

17 For a brief discussion of the surviving visual evidence of medieval artillery, see Appendix 2.

any expertise in siege warfare. The diverse and inconsistent pool of terminology that they employed appears to reflect this. Until the term 'trebuchet' and its variants begin to appear near the end of the twelfth century, most vocabulary used to identify ballistic siege machines was drawn from antiquity. The terms *funda, catapulta, manganum, balista* and *tormentum*, all found in classical sources, were used by various medieval authors to identify mechanical stone-throwers prior to the First Crusade. Of the few new terms that appeared in the Early Middle Ages, *petraria* (pl. *petrariae*) was the only one that found widespread usage.

Some sources appear to have favoured the older terms in order to evoke a sense of imagery familiar to an audience of fellow clerics acquainted with the same classical texts, or as part of a more general effort to present their work in a classical style to give it greater authority.[18] The terminology of many appears to reflect regional and personal preferences, while for others ignorance of the 'correct' vocabulary probably led them to employ familiar terms. Matters are further complicated when a single term is used to identify a siege engine that appears to take on a combined nature, such as a siege tower that incorporates a ram or some kind of ballistic machine.[19] In general, it seems that many medieval authors in Europe did not consider the clear, consistent and accurate use of artillery terminology to be a high priority prior to the thirteenth century.

These challenges have not stopped historians from attempting to provide terminological guidelines. In the eighteenth century, Francis Grose ambiguously grouped together an array of terms that he found in certain sources and was content to conclude that some threw darts and others stones, and that many threw both.[20] A century later, Gustav Köhler interpreted *petraria* as a traction trebuchet, suggesting that it had Islamic origins. He believed that *mangana* referred to a one-armed torsion engine and that *manganella* was used to identify a smaller two-armed rock-throwing torsion machine, which was rediscovered through the course of the First Crusade.[21] In the early twentieth century, Rudolf Schneider, who argued that there was a period relatively devoid of artillery following the collapse of the Western Roman Empire,

---

18 For example, Rahewin's account of Emperor Frederick I's wars is heavily influenced by ancient Greek and Latin authors, for an example see, Otto of Freising and Rahewin, *Gesta Friderici I* 4.46, ed. Waitz, pp. 285-87, trans. Mierow, pp. 278-79. See also Rogers, *Latin Siege Warfare*, pp. 127-28; Fino, "Machines de jet médiévales," p. 25.

19 One such example is the Norman siege tower used at Brevol, Orderic Vitalis, *Historiae ecclesiasticae* 8.24, ed. and trans. Chibnall, 4:288-91. See also France, *Western Warfare*, p. 118.

20 Grose, *Military Antiquities*, 1:380-84.

21 Köhler, *Die Entwickelung des Kriegswesens*, 1:139-211.

suggested that the Vikings invented the traction trebuchet in the ninth century, from which point all terms identifying artillery referred to such engines. Although he postulated that certain terms might correspond with particular varieties of this technology, he recognised the complexity and possible futility of trying to establish a strict terminological framework.[22]

Kalervo Huuri, who greatly expanded the study of medieval artillery in the mid-twentieth century by examining Eastern sources, defined *mangonella* as an engine that threw quite light projectiles, up to about 5 kg, while *petrariae* and *mangana* threw stones 50-70 kg. He judged the heavier engines to be trebuchets and suggested the weaker machines might have been light traction trebuchets or one-armed torsion catapults. A third, heavier, category was added with the development of the counterweight trebuchet.[23] Ada Bruhn Hoffmeyer and D.J. Cathcart King similarly believed that the *onager* remained in use through the Middle Ages, although it came to be known as the *manganellus* or *mangonel*. Hoffmeyer also linked the Arabic term *'arrada* with the *onager*. King suggested that *petraria* was originally a generic term for artillery but came to refer specifically to the traction trebuchet.[24] More recently, Christopher Marshall, admittedly influenced by the nineteenth-century work of Charles Oman, has suggested that 'petrary' and 'mangonel' were both torsion weapons, while John France has approached the issue more cautiously, proposing that the *petraria* and *mangana* may have referred to larger machines while *mangonella* and *tormentum* denoted smaller ones.[25]

Any effort to determine strict definitions for these terms must address the changing use of the vocabulary. From about the ninth century, *catapultae* and *tormenta* appear less frequently, while *petrariae*, *mangana* and *mangonella* are found more often. This, however, does not necessarily imply that there was a technological change at this point. These were newer terms: *petraria* dates to the Early Middle Ages and *manganon*, used by Greek authors from the fourth century, gradually permeated into Arabic as *manjaniq* (pl. *majaniq*) and Latin as *manganum* (pl. *mangana*).[26] The Arabic term *'arrada* (pl. *'arradat*) seems to have been derived from the Aramaic translation of the Greek *onagros*, or Latin *onager*.[27] Similar to the way in which *petraria* and *manganum* were at times

22 Schneider, *Die Artillerie des Mittelalters*, pp. 1-60, esp. pp. 1-26.

23 Huuri, *Zur Geschichte des Mittelalterlichen Geschutzwesens*, pp. 57-65.

24 Hoffmeyer, "Military Equipment," pp. 131-34; King, "The Trebuchet," pp. 458-62.

25 Marshall, *Warfare in the Latin East*, p. 213; France, *Western Warfare*, p. 119. Cf. Oman, *A History of the Art of War*, pp. 136-40, 543-45.

26 Gillmor, "Introduction of the Traction Trebuchet," p. 7; Bradbury, *The Medieval Siege*, p. 253.

27 Chevedden, "Artillery of King James I," pp. 58-59.

used as synonyms in Latin Europe, some Arabic sources seem to have used *manjaniq* and *'arrada* interchangeably to describe the same type of ballistic machine, most likely a traction trebuchet. Like their Latin counterparts, most Arabic sources were content to mention the presence of one or more engines without providing a description or many details. In his influential study of trebuchets from a technological perspective, Donald Hill suggested that the term *manjaniq* was a general one applied to the artillery of Arab forces.[28]

Paul Chevedden has disputed the notion that such terms were often used generally or loosely as synonyms. In a Byzantine context, he has argued that some authors used *manganon* and *manganikon* to identify the lightest trebuchets, which employed a single vertical pole to support the axle, although others used these to refer to trebuchets more broadly. He postulates that the *lambdarea*, or *labdarea*, was a trebuchet with a *lambda*-shaped trestle frame, while *helepolis*, *petrobolos*, *petrarea* and *sphendone* were all used to refer to trebuchets with triangular-shaped trestle frames. Chevedden concludes that Anna Comnena thus identifies separate machine types when noting the use of *petrobolos* and smaller *lithobolos* in the twelfth century, equating these with trestle and pole varieties of trebuchets respectively.[29] In an Arabic context, he equates *'arrada* with the simple pole trebuchet and *manjaniq* with those with a trestle framework.[30]

Neither the search for etymological roots nor the picking and choosing of terms, which the sources fail to demonstrate were applied consistently to singular mechanical forms, has proven to be an adequate basis upon which to establish a terminological typology. Latin sources frequently equate various terms and existing vocabulary was clearly applied to new technologies. For example, Otto of Freising mentions a stone-throwing *balista*, which he also calls a *manga*, during his account of the siege of Tortona in 1155.[31] *Ballistae*, which originally shot stones but became arrow-shooters in late antiquity, are again described as throwing stones during the siege of Ascalon in 1153.[32] It is hard to identify any intended difference between the *mangana* and *predeirae* mentioned in Caffaro's account of the siege of Almeria in 1146 or the *petraria* and *mangonella* in Roger of Wendover's account of the siege of Lincoln in 1217.[33]

28 Hill, "Trebuchets," p. 100.

29 Chevedden, "Invention," p. 79.

30 Chevedden, "King James I," pp. 321-22.

31 Otto of Freising, *Gesta Friderici I* 2.21, ed. Waitz, p. 124, trans. Mierow, p. 135.

32 *Auctarium Aquicinense*, ed. Bethmann, p. 396. See also *Auctarium Affligemense*, ed. Bethmann, pp. 401-2.

33 Caffaro, *Annales*, ed. Belgrano, 1:34-35, trans. Hall and Phillips, pp. 69-70; Roger of Wendover, *Flores Historiarum*, ed. Hewlett, 2:214-15, trans. Giles, 2:393-94.

However, this does not rule out the likelihood that some sources used certain or different terms to point out mechanical distinctions, only that this was not done uniformly. William of Tyre identifies the artillery at the siege of Jerusalem in 1099 as *mangana* and *petrariae*, grouping them as if synonyms, but later notes that *mangana* threw smaller stones than certain other engines, which he does not label.[34]

Like *manganum*, the Latin term *petraria* also has Greek roots. While Byzantine artillery vocabulary remained fairly consistent through the sixth century, *petrarea* appears in the seventh-century and seems to have become synonymous with *manganon* sometime before the end of the ninth century. Carroll Gillmor has suggested that the term could then have infiltrated Latin from the eighth century through Frankish interaction with the Greeks of Italy.[35] Paul the Deacon, writing in the eighth century, records the Byzantines' use of a siege engine, which he claims was commonly called a *petraria*, to throw the head of a certain Lombard leader over the walls of a besieged city in 663.[36] If those favouring the theory that classical torsion engines fell from use in the sixth or even seventh century are correct, *petraria* may have referred exclusively to trebuchet-type engines in a Latin context. Regardless, *petraria* (literally 'stoner' or 'stone thrower') became the most common term used exclusively to identify artillery in Latin sources by the end of the twelfth century.

While the use of more specific vocabulary became more common with time, ambiguous terms appear at least as frequently up to and throughout much of the twelfth century. Whereas William of Tyre mentions the use of *machinae* at one point in his account of the siege of Jerusalem in 1099, the later *Eracles* translation replaces this with the terms *perriers* and *mangoniaux*. Roger of Wendover, in his version of William's account, calls these engines *petrariae* and *trebuculi*, the latter being a term that would not have been understood by those taking part in the siege.[37] While terminological specificity may help clarify the use and types of later artillery, it remains unclear at what point trebuchet-type engines were first used in Europe and the Near East. Although many cultural interactions have been put forward as the possible catalyst for

---

34 *machinas iaculatorias, quas mangana et petrarias vocant*, William of Tyre, *Chronicon* 8.6, 13, ed. Huygens, 1:392-93, 403, trans. Babcock and Krey, 1:350-51, 362.

35 Gillmor, "Introduction of the Traction Trebuchet," pp. 7-8.

36 Paul the Deacon, *Historia Langobardum* 5.8, ed. Waitz, p. 189, trans. Foulke, p. 221.

37 William of Tyre, *Chronicon* 8.6, ed. Huygens, 1:392-93; *Eracles* 8.6, RHC Oc 1, p. 563; Roger of Wendover, *Flores Historiarum*, ed. Coxe, 2:137-38, trans. Giles, 1:430.

the westward transference of this technology, there is little solid evidence to attribute credit to any of them.

Early Traction Trebuchets in the Near and Middle East

The traction trebuchet appears to have been developed in Asia centuries before the first of such engines were employed in the Near East or Europe. Links between trebuchet technology and China were drawn in the mid-nineteenth century by Louis-Napoleon, and have since been developed by scholars such as Huuri, Needham and Yates.[38] In his assessment of these Chinese engines, Donald Hill estimated that a typical beam, whether consisting of a single mass or several spars lashed together, was between 5.6 m and 8.4 m long, tapering from a diameter of 12.5 cm to 7 cm at the tip of the long arm. His sound understanding of the mechanical physics led him to suggest that light engines probably had a beam ratio of 5:1 or 6:1 and heavier engines about 2:1 or 3:1. Such engines, according to Hill, probably shot projectiles weighing 1-60 kg about 75-120 m.[39] These Chinese machines appear to have been developed between the third and fifth centuries BC. Textual evidence places them in the Byzantine sphere of influence by the end of the sixth century and the Arab world during the seventh, while illustrative evidence places them in Transoxania by the eighth century.[40] By the ninth century, they appear to have been employed in northwestern Europe.[41]

At its simplest, a traction trebuchet could be little more than two beams and a sling. These 'pole trebuchets' made use of a single vertical support topped with a fork or some other kind of simple axle that acted as the fulcrum point for the throwing arm, which was fitted with the sling.[42] Although relatively light, these could throw heavier projectiles than a staff sling, which was little more than a hand-held stick with a sling at the end, to greater distances and similarly required only a single operator. These pole trebuchets could easily yaw around their y-axis and their minimalist structure allowed them to be

38 Bonaparte, *Études sur le passé et l'avenir de l'artillerie*, 2:35; Huuri, *Zur Geschichte des Mittelalterlichen Geschutzwesens*, pp. 56-65, 212-27; Needham, "China's Trebuchets," pp. 107-45; Yates, "Siege Engines," pp. 414-19; Needham and Yates, *Science and Civilization in China*, pp. 203-40.

39 Hill, "Trebuchets," p. 102.

40 See Appendix 2, **no. 1**.

41 Gillmor, "Introduction of the Traction Trebuchet," p. 1; Chevedden, "Invention," p. 74; Chevedden, "King James I," pp. 315-16.

42 See Appendix 2, **no. 2**.

moved with relative ease, making them the most flexible type of trebuchet. There is general consensus that these were antipersonnel weapons; however, as soon as the single supporting pole is replaced by a trestle frame and additional pullers are added, some scholars rush to suggest that these sturdier engines were meant to damage masonry.[43]

Evidence from Asia confirms that traction trebuchets had reached a complete level of development centuries before the technology arrived in the West. Despite sensationalised accounts of extremely large and destructive traction trebuchets, these engines appear to have remained quite light. Ahead of the First Crusade, the traction trebuchet was clearly embraced by Latin, Greek and Muslim armies, but when and through whom the technology was transmitted from Asia remains unclear.

There would have been few obstacles to the spread of this technology. While these engines would have had little value to nomadic peoples and those unaccustomed to siege warfare, those with more developed siege traditions would have found them easy to replicate. Unlike torsion engines, which required a much higher level of skill to ensure that the coils would perform effectively and the framework could endure the necessarily high stresses, the mechanics of a trebuchet are incredibly simple. Little more than seeing such an engine from a distance, noticing the rotation of the beam and function of the sling, would have allowed someone with basic carpentry experience to construct a functional prototype that could be further refined through use or experimentation. Although sizable counterweight engines would have required more skill to construct, it would have been the manoeuvring of the large components, which required rigging, rather than the design of these engines that would have benefited from greater experience.[44]

Early Byzantine Trebuchets

The Byzantines are most commonly nominated as the first to have used swing-beam artillery technology in the region surrounding the Mediterranean. Artillery seems to have been an important weapon in the Byzantine arsenal, both for attack and defence, but it remains unclear when torsion engines fell from use and traction trebuchets were adopted. Most references from the sixth

43 For example, Gillmor, "Introduction of the Traction Trebuchet," p. 4.

44 This author built his first trebuchet at the age of fifteen with no more than hand tools. The small counterweight engine had a beam about 3.5 m long and could throw a golf ball about 80 m or a rock the size of a grapefruit or softball about 50 m. These were not very complicated engines and required less expertise to construct than some have suggested, for example, Purton, *Early Medieval Siege*, esp. p. 298.

century give few details that might allow for a determination of what types of machines were used, such as the apparently powerful stone-thrower employed by the defenders of Amida in 502-3.[45]

An account of the siege of Thessalonica by the Avaro-Slavs in the late sixth century appears to provide the earliest definitive description of traction trebuchets in the region.[46] According to the first-hand account found in the *Miracles of St Demetrius*, the besiegers spent the first and second days of the siege constructing stone-throwers, as well as rams and a shelter for mining. The artillery was ready on the third day and was then used to throw stones until the siege was lifted on the seventh day, after the ram had been disabled and the efforts of the miners had been thwarted by the defenders.[47]

The *ballistrae* of the Avaro-Slavs are described as tetragonal, the frames tapering as they rose. To these were affixed beams, clad in iron at their ends,[48] and to them were fashioned very long framing timbers. Slings were hung at the back ends of these beams and ropes from the front, which when pulled, released the sling and propelled a stone into the air with a loud noise. The framework was covered on its three exposed sides with boards to protect the crew within from arrows, indicating they were within range of the city's archers.[49] Apart from the 'loud noise' at release,[50] all factors suggest that these were traction trebuchets. As for their size, when the protective panelling of one of the engines was set on fire by a flaming arrow, the machine was simply carried away by its operators, repaired and then carried back into place to resume shooting the following day. To be moved with such ease suggests that these were light machines of an antipersonnel scale.

The account states that the garrison hung cloth curtains and mattresses in front of their battlements to lessen the impact of the Slavic artillery and that a large section of wall was destroyed by these engines. While this suggests that these were quite powerful machines, this may not have been the case, as the

45 Joshua the Stylite, *Chronicle* 53, trans. Wright, pp. 41-42.

46 For the defences of Thessalonica, see A.W. Lawrence, "A Skeletal History of Byzantine Fortification," p. 185.

47 *Miracles of St Demetrius*, ed. Lemerle, pp. 148-54.

48 Six centuries later, al-Tarsusi similarly describes cladding certain components of a traction trebuchet with iron.

49 *Miracles of St Demetrius*, ed. Lemerle, p. 154. While Vryonis argues for a literal reading of this description, Lemerle suggests otherwise in his edition, *Miracles of St Demetrius*, ed. Lemerle, p. 154 n. 65; Vryonis, "The Evolution of Slavic Society," p. 384 n. 23.

50 For the 'loud noise', cf. Seneca, *Qaestiones naturales* 2.16, ed. Monfrè, p. 558; trans. Clarke, p. 67. *Nam balistae quoque et scorpiones tela cum sono expellunt.* See also Marsden, *Historical Development*, p. 97.

practice of padding battlements to protect them from siege engines was nothing new. Precedent for this can be found at a number of classical sieges and later defenders would employ such a technique at Mayet (1098) and Jerusalem (1099).[51] At Thessalonica, the padding is described as only cloth hung from thin rods, suggesting that this was not a significant barrier. Furthermore, the author's assertions that fifty engines were arrayed against the eastern wall, that they were constructed within two days, and that they threw 'mountains' are conflicting. Even with modern equipment, to build a machine of this type with dimensions sufficient to threaten classical walls would take a number of days; to compound this by fifty is unimaginable even if all of the Slavs were skilled craftsmen. The destruction that is noted appears to have been added to illustrate the perceived 'mountainous' size of the projectiles that were thrown: an emotive expression of the experience rather than an accurate description of the reality. Consider that it is unlikely that early modern siege guns could have breached these impressive walls in so short a period. While much has been made of this account and its significance, all of these observations rely on the accuracy of a single source.[52]

Speros Vryonis Jr. has concluded that the Slavs only recently gained the ability to build siege engines, employing them for the first time in 587 at the siege of Appiareia in Moesia Inferior.[53] During this siege, according to the account of Theophylact Simocatta, a captured Byzantine soldier betrayed the methods of building a siege engine, a *helepolis*, to the Avars – in a single day, the Slavs jumped from constructing ladders to building machines capable of reducing cities.[54] For this theory to work, however, the siege of Thessalonica must have taken place in 597 rather than 586.[55]

---

51 Orderic Vitalis, *Historiae ecclesiasticae* 10.10, ed. and trans. Chibnall, 5:258-59; Ralph of Caen 124-25, RHC Oc 3, pp. 692-93, trans. Bachrach and Bachrach, pp. 140-42; Albert of Aachen 6.17.19, pp. 424-29; Fulcher of Chartres 1.27.8, ed. Hagenmeyer, pp. 297-98, trans. Ryan, pp. 120-21.

52 Although he appears to have been an eyewitness, the author may have written out his account decades after the events that he witnessed, Purton, *Early Medieval Siege*, pp. 29-32.

53 Vryonis, "The Evolution of Slavic Society," pp. 385-87. He forms this judgment primarily on the descriptions of Procopius, but at no point does Procopius ever explicitly state that the Slavs lacked the knowledge or ability to construct siege equipment. The presumption is thus based on the absence of evidence.

54 Theophylact Simocatta, *History* 2.16.1-11, trans. Whitby and Whitby, pp. 65-66.

55 The siege is reported to have been carried out on Sunday, 22 September, during the reign of Maurice, a combination appearing only in 586 and 597. For the broader context and

The anecdote about the traitor at Appiareia appears to be a convenient, rather than accurate, explanation. The mechanics of a traction trebuchet are so straightforward that it would not have been necessary to compel someone to betray the secrets of this technology. The simplicity of these engines allows crews to become quickly familiar with the shooting process and designers to adjust the beam and sling lengths to achieve a desired range through little more than a brief period of trial and error.[56] The machines built at Thessalonica seem to reflect this tradition rather than a secret and devastating weapon bequeathed by a traitor. Furthermore, there is little to suggest that the *helepolis* mentioned by Theophylact Simocatta was a ballistic engine; the term was more often used by Greek sources to denote a siege tower. In this light, the story seems little more than a racially motivated explanation of how a supposedly 'barbaric' people were able to replicate and incorporate a piece of 'civilised' technology, comparable to Henry of Livonia's explanation of how the Estonians acquired trebuchet technology in the early thirteenth century.[57]

It is possible that the Avars were responsible for importing the traction trebuchet from Asia;[58] however, it is at least as likely that it entered the Byzantine sphere of influence through some other cultural conduit, allowing for the possibility that the Byzantines exposed the Avaro-Slavs to this technology, rather than the other way around.[59] Tracy Rihll has even expressed scepticism towards the theory that Byzantine trebuchet technology originated in Asia, suggesting it may have been developed directly from the staff sling.[60] Regardless, the traction trebuchets employed by Byzantine forces remained fairly light and were used most effectively in supporting roles.[61]

---

a discussion of the dating, see P. Yannopoulos, "La pénétration Slave à Argos," pp. 339, 359, 364.

56 For trials demonstrating the limited practice required to effectively operate a traction trebuchet, see Tarver, "The Traction Trebuchet," pp. 158-61.

57 See Fulton, "The Diffusion of Artillery Terminology," pp. 16-17.

58 Purton, among others, has suggested that the Avars may have been responsible for bringing the traction trebuchet to Europe, Purton, *Early Medieval Siege*, pp. 29, 33.

59 It is predominantly historians of western Europe who have credited the Byzantines with developing trebuchet technology once it reached the Mediterranean. Cf. A.W. Lawrence, "A Skeletal History of Byzantine Fortification," p. 222 and n. 137.

60 Rihll, *The Catapult*, pp. 263-64, esp. n. 71. This theory has received little endorsement.

61 For example, Ocbas (opposite Martyropolis/Mayyafriqin) (591) and Tiflis (Tbilisi) (627/28), Evagrius, *Ecclesiastical History* 6.15, ed. p. 176, trans. p. 303; Movses Dasxuranci, *Patmut'iwn Aluanic* 2.11, trans. Dowsett, p. 85.

Early Muslim Trebuchets

As trebuchet technology appears to have been present in the Balkans by the end of the sixth century, so too are there indications that it had reached the fringes of the Arab world by this time or shortly thereafter. The earliest reference to artillery in Islam is found in 630, when a *manjaniq* was used during Muhammad's siege of Ta'if, a town in Arabia. The engine is not described but it was probably a traction trebuchet.[62] Where the Arabs might have first been exposed to this technology is far from clear, but Ibn Ishaq notes that certain figures had been sent to Yemen around this time to learn about siege engines, including artillery.[63] Whether this claim is accurate, or if the Arabs learned about traction trebuchets from the Byzantines, or even the Persians, is hard to prove.

There is little evidence that the Arabs employed any advanced siege equipment during their conquest of Byzantine Palestine and Syria in the first half of the seventh century. The tactics employed at the sieges of Damascus, Homs, Tikrit and Mosul seem to have been restricted to blockade and guile, staples of Muslim siege warfare throughout the centuries to come. Al-Tabari, writing around the start of the tenth century but using earlier sources, states that the Muslims used artillery to bombard Persian Ctesiphon in 637, making use of at least twenty engines of two different sizes according to one of his sources.[64] These engines, however, may have been constructed by Persian deserters.[65] In the early 640s, al-Baladhuri, writing in the ninth century but similarly making use of earlier sources, notes the use of artillery (*'arradat*) by the defenders of Ra's al-'Ayn in the Jazira and by the Greek forces that briefly retook Alexandria.[66] In the account of Bar Hebraeus, seventy-two stone-throwing engines, almost certainly an exaggerated figure, were used during the Muslim siege of Caesarea in Palestine.[67] In 653, according to the seventh-century account of Sebeos, the Arabs launched a great fleet from Alexandria against Chalcedon, which carried mangonels (*mangganon*) and other throwing machines (*mek'enayk'*).[68] This suggests that by this point certain Muslim armies pos-

62 Hill judged this to have been a classical torsion engine, Hill, "Trebuchets," p. 100; Hill, "Military Technology," p. 262. Setting terminology aside, a traction trebuchet would be better suited to the tribal structure of the Prophet's early army than a more complicated and higher-maintenance torsion engine.

63 Ibn Ishaq, *Sirat Rasul Allah*, trans. Guillaume, pp. 587-91.

64 Al-Tabari, *Tarikh*, trans. Juynboll, 13:9, 11.

65 Purton, *Early Medieval Siege*, pp. 37-41.

66 Al-Baladhuri, *Kitab Futuh*, trans. Hitti, 1:276, 348.

67 Bar Hebraeus, *Makhtebhanuth Zabhne*, trans. Budge, 1:97.

68 Sebeos, *History* 36 (50), trans. Thomson, 1:144-45, cf. trans. Bedrosian, pp. 169-70.

sessed the technology to build and operate artillery, without foreign assistance, a lesser feat compared with launching the associated flotilla.

Like the story about the Slavs at Appiareia, the Syriac *Maronite Chronicle* relates how, in about 664, a master carpenter from Paphlagonia offered to build ʿAbd al-Rahman ibn Khalid a *manganike*, with which to take Synnada (Suhut), in exchange for a guarantee of his safety and that of his family. In this case, however, the story serves to illustrate the disparity in expertise and experience between the Arabs and the Byzantines. Said to have been built from long logs, the Arabs erected the engine in front of a gate and proceeded to discharge three shots towards the fortress: the first hit the gate, the second fell short and the third fell even shorter. The engine portrayed in this account is a traction trebuchet; after their failed shooting experience, the defenders taunted the crew by yelling out "Pull your weight, Khalid's men, you are drawing badly." The defenders proceeded to target the besiegers' engine with one of their own, destroying it with a single shot.[69] While this might be evidence that not all Muslim armies were experienced with the use of traction trebuchets, the episode might also have been invented to point out the perceived inferiority of the Arabs, portrayed as ignorant of trebuchet technology. For Christians of minority sects, such as the author, Byzantine rule was preferred to that of the Muslims, who had occupied greater Syria and were at that time pushing west into Anatolia.[70]

In 682, during the siege of Mecca, a certain *manjaniq* was named *umm farwa* (mother of the hair), suggesting this was a traction trebuchet named after the pulling ropes that hung from its beam. To the east, Muhammad ibn Qasim employed an engine called *ʿarus* (the bride), a traction trebuchet supposedly operated by 500 men, against Debal, in Sind.[71] Because the mechanics of a traction trebuchet make it impossible for 500 pullers to operate an engine of this type simultaneously,[72] this may be an exaggeration to present the superiority of the Muslims over the Buddhists. The engine was apparently used for nothing more than knocking down the flagpole on top of a Buddhist temple in advance of a general assault.

The lack of clear and detailed contemporary accounts prevents ruling out the possibility that some early Muslim armies may have employed torsion artillery. While many historians assume that the Muslims obtained their knowledge of trebuchet technology from the Byzantines, Hill suggested that

---

69 *Maronite Chronicle*, trans. Palmer, pp. 34-35.

70 Cf. Purton, *Early Medieval Siege*, p. 47.

71 Al-Baladhuri, *Kitab Futuh*, trans. Murgotten, 2:217-18.

72 Fulton, "Myth of the Hybrid Trebuchet."

they may have acquired this through the course of their campaigns in central Asia during the Umayyad offensives from around 660, implying all earlier mentions referred instead to torsion engines.[73] Despite its increasing use by Muslim armies, regardless of its power source, artillery rarely had a direct influence on the outcome of Near Eastern sieges in the Early Middle Ages. Treachery, blockades and frontal attacks were most often responsible for successful sieges, while most breaches were opened through mining.

Later Byzantine and Muslim Artillery

Few episodes shed further light on what form Near Eastern artillery took during the eighth and ninth centuries. Most references provide no indications of power and little about how these engines were used.[74] In most instances where artillery is portrayed as particularly powerful or destructive, there appears to be at least a degree of exaggeration, as these sieges were typically concluded through mining or a frontal attack. In 766, at the siege of Kamacha (Kamkh), the Byzantine defenders employed covers of Cyprus wood to protect themselves from the stones of the Muslims' artillery, while inflicting numerous casualties with their own stone-throwers before the stronghold was taken.[75] Michael the Syrian, writing in the twelfth century, similarly notes that it was the defenders of Balis, rather than its fortifications, that suffered when it was bombarded by Abdallah ibn Tahir around 823. The defenders, loyal to Nasr ibn Shabath, forced the Christians up to the parapet to act as human shields, leading Abdallah to turn to mining. Not long after, Abdallah invested Kaysum and used his artillery to damage the houses within the town as winter approached, throwing stones that reportedly weighed the load of an ass. In response, Christian women and children were again made to stand on top of the walls and act as human shields.[76] Despite the suggested size of the projectiles, there are no indications that these engines threatened the town's fortifications; rather, it is again the danger they posed to individuals that is emphasised.

Al-Tabari's account of Caliph al-Mu'tasim's siege of Amorium ('Ammuriyya) in 838 reinforces the sense that artillery was an increasingly important siege weapon. Stone-throwers were apparently transported with the army on this campaign, contributing to the slow movement of the caliph's baggage in the

73 Hill, "Trebuchets," pp. 100-3; Hill, "Military Technology," pp. 262-63.

74 The nondescript mentions of artillery (*majaniq*) used by Constantine V at Malatya and by Muslim forces at Dvin (Dabil) are fairly typical, al-Baladhuri, *Kitab Futuh*, trans. Hitti, 1:291, 314.

75 Al-Baladhuri, *Kitab Futuh*, trans. Hitti, 1:288.

76 Michael the Syrian, *Chronique* 12.12, trans. Chabot, 3:52-53.

lead up to the siege. These "powerful mangonels" were solidly built yet powered by only four operators. They seem to have been used to support a siege tower, initially throwing sheepskins filled with earth to fill the ditch between the tower and wall ahead of it. Although the tower became stuck and had to be abandoned, artillery continued to play a role in the siege. It proved capable of destroying improvised wooden defences and helped drive defenders away from a breach that was created when a section of wall collapsed.[77] In 852/53, Byzantine forces used artillery to throw incendiaries into Ushtum, burning the town during a raid along the Egyptian coast, and the tenth-century *Chronicon Salernitanum* describes Muslim forces employing "a machine that we call a *petraria*" during the siege of Salerno in 871.[78]

References to the use of *majaniq* and *'arradat* in defence increase with time, suggesting similar engines to those deployed by attacking forces became a more intricate part of defensive planning. At the siege of Marand (848/49), artillery was employed by both the attackers and defenders, reportedly killing 100 and wounding 400 on each side during the eight-month siege.[79] When Baghdad was refortified in the 860s, the outer portion of the Shammasiyyah Gate was provisioned with five large mangonels, the largest was named *al-ghadban* (the angry), while an *'arrada* was set up on the inner portion of the gate and six more were positioned nearby in the direction of Raqqat al-Shammasiyyah. Four *'arradat* were positioned on each side of the Baradan Gate and similar arrangements were made for the other gates. Each engine had a team of pullers assigned to it as well as an operator, who would have been the one loading and holding each stone in place before it was pulled out of his hands by the force of the pulling crew.[80]

When Baghdad was attacked in 865, its defensive artillery targeted people who approached the gates, claiming responsibility for most of the casualties sustained by the besiegers on the first day. The attackers also employed artillery, but a sally by a party of defenders debilitated one engine that had been brought up to attack the Khurasan Gate. A push against the Shammasiyyah Gate involved the use of an engine to throw naphtha and incendiaries against the engines defending the gate, but the assault was pushed back. In another attempt against the Shammasiyyah Gate, eight *'arradat* were placed on boats.

---

77 Al-Tabari, *Tarikh*, trans. Bosworth, 33:100, 109-12. See also Michael the Syrian, *Chronique* 12.20, trans. Chabot, 3:97-100.

78 Al-Tabari, *Tarikh*, trans. Kraemer, 34:126-27; *Chronicon Salernitanum* 113, ed. Westerbergh, p. 127.

79 Al-Tabari, *Tarikh*, trans. Kraemer, 34:79-80.

80 Al-Tabari, *Tarikh*, trans. Saliba, 35:40-41.

These reportedly killed a hundred defenders while assault forces attempted to scale the gate. The attack ultimately failed and the head of the only besieger who had managed to reach the top of the gate was thrown back to his comrades by one of the defenders' engines. Defensive artillery was also responsible for repulsing another attack launched from the east, during which al-Durghuman, a notable figure among the besiegers, was killed when a stone struck him in the chest. Al-Tabari makes a point of mentioning the defenders engines one last time before the final battle and end of the siege.[81] Although most other descriptions contain far fewer details, the use of artillery seems to have been similarly important elsewhere, such as at the defence of al-Mukhtarah in 881 and Amida in 899, although success varied.[82] When Tarsus was taken by the Byzantines in 965, its towers, said to number 100, were reportedly defended by *'arradat* and *majaniq* as well as archers.[83]

Artillery remained an important weapon as frontier warfare between the Byzantines and Muslims continued through the tenth and eleventh centuries. While these engines typically threw stones, on a number of occasions they were also used to throw the heads of slain opponents, a practise that would continue through the Middle Ages. During the siege of Chandax (Heraklion) by Nicephorus II Phocas in 960-61, artillery was used by both attackers and defenders. In one account, a live ass was thrown over the town walls to the starving defenders.[84] Historians have interpreted this event in different ways, but it seems very unlikely that any engine of the day would have been able to throw the mass of a donkey, let alone deal with its irregular shape and potential movement.[85] Heads are a different matter and will be addressed in Chapter 3.

Eric McGeer has analysed the siege weapons available to the Byzantines between the tenth and eleventh centuries on the basis of four surviving texts.[86] His work highlights the pre-eminent use of mining and even siege towers

81 Al-Tabari, *Tarikh*, trans. Saliba, 35:46-47, 59-60, 63, 95.

82 Al-Tabari, *Tarikh*, trans. Fields, 37:43, 46; trans. Rosenthal, 38:75.

83 Bosworth, "The City of Tarsus," p. 281.

84 Theodosius the Deacon, *Expugnatio Cretae* 4, ed. and trans. Migne, pp. 1,040-41. Heads were also thrown during this siege, Michael Attaleiates, *History* 28.4.6, ed. and trans. Kaldellis and Krallis, pp. 410-17.

85 For various interpretations of this episode, see Huuri, *Zur Geschichte des Mittelalterlichen Geschutzwesens*, p. 150; Chevedden, "The Hybrid Trebuchet," p. 188; Fulton, "Myth of the Hybrid Trebuchet."

86 McGeer, "Byzantine Siege Warfare," pp. 123-29. The sources are Hero of Byzantium, the anonymous *De obsidione toleranda*, the *Taktika* of Nikephoros Ouranos and the *Strategikon* of Kekavmenos. The first two are from the classical *poliorketika* tradition, the latter two were composed by contemporary soldiers.

rather than artillery, which was generally employed defensively or grouped with besieging archers and slingers in a supportive position. McGeer concludes that despite the wealth of theoretical works on siege warfare, technologically and tactically it was a fairly simple art, especially concerning artillery.[87] Attrition, guile and mining were the most popular siege techniques of both Byzantine and Muslim forces.

Aside from the 'mountains' thrown at Thessalonica, the asses' loads tossed at Kaysum and the live ass hurled into Chandax, most descriptions of artillery present these engines as light supportive weapons. The accounts of Tugril's siege of Byzantine Manzikert in 1054 provide another exception and one within living memory of the First Crusade. Aristakes Lastivertc'i describes how the Turks' initial trebuchet was countered by a similar defensive engine, which would shoot the stones of the besieging machine right out of the sky. In response, the assailants prepared another trebuchet, a *baban*, that required four hundred men to pull certain ropes and threw stones weighing 60 *litrs* (about 20 kg) from its sling. From a position fortified against the defenders' artillery with cotton loads, the *baban* opened a breach with its first shot, but was later burnt in a ruse by the Byzantines.[88] In his slightly later account, Matthew of Edessa emphasises that the attackers focused on mining; it was only after the garrison had successfully countermined that Tugril brought up a large engine, said to have been built by Emperor Basil. Matthew claims that this machine weighed 15 *adil* (about 3,400 kg) and inflicted a number of casualties among the town's sentinels. In this account, the engine was temporarily put out of action by a defensive machine, before it was burnt in a sally.[89]

This seems to be an excellent example of source exaggeration. Although it has been suggested that these accounts describe some kind of super traction, or hybrid, trebuchet that really was powered by 400 pullers, the mechanics simply do not allow for this.[90] Instead, Aristakes seems to have presented a sensationalised rendition of events, which Matthew was then left to interpret when composing his later history. By comparison, the latter's account is fairly reasonable: Matthew emphasises the besiegers' mining activities and describes the engine as targeting the defenders along the parapet rather than the wall itself, as is also found in Aristakes' account. When taken together, both authors agree that a notable stone-thrower was used by the Turks and that the

87 McGeer, "Byzantine Siege Warfare," p. 129.

88 Aristakes Lastivertc'i, *Patmut'iwn*, trans. Bedrosian, pp. 100-5.

89 Mathew of Edessa, *Patmut'iwn* 2.3, trans. Dostourian, pp. 86-88 (p. 87 n. 5 for the weight conversion).

90 Fulton, "Myth of the Hybrid Trebuchet."

defenders also made use of artillery. Matthew inflates the scale of the engine to better fit with the sensational account of his predecessor; writing almost a century after the siege, it seems unlikely that Matthew would have been able to provide an accurate measurement of the engine's weight. Rather than indicating a significant step forward in trebuchet technology, this event actually appears to support McGeer's theory that artillery, while known to the Byzantines, was generally used as a supportive weapon during the eleventh century. If this was indeed a relatively large engine, it might explain why it attracted so much attention and its scale was so exaggerated.

Similar exaggerated accounts can be found in other, usually later, accounts of events. For instance, Ibn al-'Adim describes how Sayf al-Dawla Khalaf ibn Mula'ib al-Ashhabi, having captured Salamiyya, used a mangonel to throw Ibrahim al-Hashimi from one of its towers in 1089/90.[91] For the most part, however, artillery is presented as a light antipersonnel weapon, as is described by Matthew of Edessa at the siege of Edessa by Alp Arslan around 1070/71.[92] Sources writing from either the attacking or defending perspective of a siege had reason to exaggerate the scale or power of besieging artillery, adding to the prestige of the attacking force or resilience of the defenders, but such examples remain exceptions rather than the norm.

Early Traction Trebuchets in Europe

It remains unclear where and when the traction trebuchet was first employed in north-western Europe. The apparent use of trebuchets by the Slavs, Byzantines and Muslims by the seventh century, seems to offer three possible routes of transmission: overland from the Balkans; through Byzantine Italy; or via the Muslims through their conquest of the Iberian Peninsula and forays into southern France. Regardless of how these engines first came to Europe, they became as much a part of European siege warfare by the late eleventh century as they were in the Near East.

The paucity of surviving contemporary accounts from early medieval Europe has contributed to much of the disagreement and debate over what form artillery took during this period and how widely it was used. If the traction trebuchet was adopted by the Byzantines in the sixth century, it stands to reason that it would have been employed in Italy not long after. Paul the Deacon's

91 Ibn al-'Adim, *Bughyat al-Talab fi tarikh*, trans. Lewis, in *Islam*, 1:64. Cf. the reported throwing of a spy at the siege of Jerusalem (1099) in Chapter 3.

92 Matthew of Edessa, *Patmut'iwn* 2.56, trans. Dostourian, pp. 131-32.

use of the term *petraria*, equating it with *mangonella*, marks one of the earliest appearances of this term in its Latin form.[93] This might suggest that a relatively new type of engine had come into use in Italy in the mid-seventh century, requiring him to explain it, but there is little to support this. Even if the traction trebuchet arrived at a later point, it was almost certainly employed in this theatre by the Carolingian conquest of the Lombards in the eighth century. It is quite possible, however, that the Franks encountered these engines before their intervention in Italy.

Carroll Gillmor has suggested that the Franks may have acquired knowledge of these engines through their encounters with Muslim Spain in the eighth and ninth centuries and Peter Purton has postulated that this exchange might have resulted from the Muslim siege of Toulouse in 721.[94] Although Gillmor notes that it was during the ninth century that there was a noticeable shift in terminology and the *manganum* appears in the Carolingian siege arsenal, the notion that Charlemagne and Louis the Pius sometimes travelled with a siege train earlier in the eighth century is not direct evidence that trebuchets were employed.[95] Taking another step back, there is little to confirm what type of artillery Muslim forces were using in the early seventh century as they swept across Spain and into France.

The south of France hosted many of Europe's most significant eighth-century sieges; but successful or otherwise, there is very little evidence for the actual use of artillery by the Franks. Although simple siege engines were used, notably rams, there is little evidence to directly refute Schneider's theory that there was a period in the history of north-western Europe when tension bows were the most powerful ballistic weapons.[96] Like the Arabs, the late Merovingians and Carolingians appear to have found the blockade to be their most effective siege tactic; however, the limited source material and sparse descriptions of sieges are enough to cast a degree of doubt on any interpretation. To the north, the Saxons are recorded using *petrariae* against Frankish Eresburg in 776.[97] Similar to the Muslims at Synnada, they struggled to use these engines effectively. While this might suggest this technology was still new to them, the

---

93 Paul the Deacon, *Historia Langobardum* 5.8, ed. Waitz, p. 189, trans. Foulke, p. 221.

94 Gillmor, "Introduction of the Traction Trebuchet," pp. 6-8; Purton, *Early Medieval Siege*, pp. 57-59.

95 Gillmor, "Introduction of the Traction Trebuchet," pp. 6-7. The ease with which traction trebuchets can be constructed exposes this to question if nothing else.

96 Purton has concluded that torsion siege weapons probably fell from use during the Merovingian period and that if trebuchet technology was known in the West, the Franks seem to have initially lacked the skill to employ it, Purton, *Early Medieval Siege*, pp. 66-68.

97 *Annales regni Francorum*, ed. Kurze, p. 44.

reliability of the account is questionable. It seems most likely that the Saxons would have learned about artillery from the Franks, but this is hard to prove.

Most of the surviving evidence suggests that artillery technology had been transmitted through Italy and southern France; however, we know little about the interactions between the Germanic groups that came to rule western Europe and their neighbours to the southeast. As the Bulgars are said to have employed artillery against Adrianople (Edirne) in 813,[98] it is possible that this technology remained in use in the Balkans since the siege of Thessalonica by the Avaro-Slavs. Accordingly, the Germanic rulers of central and northern Europe may have been exposed to this technology from multiple directions. The absence of sensational accounts and lack of attention given to the first use of this technology as it spread reinforces the suggestion that these were quite simple machines, of little use unless employed effectively.

During the relative stability that accompanied the peaceful years of Carolingian rule, certain high-ranking clergymen took advantage of the readily available supply of stone tied up in town walls to build new churches and cathedrals. Unknowingly, this exposed their flocks to new dangers in the ninth century with the onset of Viking raids up many of Europe's rivers. Although the sources remain extremely poor for the most part, the Scandinavians seem to have employed some form of artillery at times. In turn, it is recorded that Charles the Bald used "new and sought-out types of machines" against a Norse army pinned in Angers in 873.[99] It is tempting to jump to the conclusion that this indicates the use of a new type of stone-thrower; however, this is the extent of the description and the machines could just as easily have been a new type of ram or even a siege tower. Casting caution aside, historians such as Ferdinand Lot have argued that this represents the first definitive use of trebuchets in Europe, influenced in part by the reported presence of Byzantine engineers, who might have built or brought knowledge of these engines.[100] This obscure but enticing reference gains further significance when considered in the context of the siege of Paris twelve years later.

98 *Historia de Leone*, ed. Bekker, p. 347.

99 *Igitur ex omnibus partibus urbe obsidione circumdata, multis diebus undique summa virtute dimicatur, nova et exquisita machinamentorum genera applicantur; sed conatus regis prosperitatis effectum non optinuit, quia et loci facies non facilem prebebat accessum et paganorum valida manus, quia pro vita res erat, summo resistebat conamine.* Regino of Prum, *Chronicon*, ed. Kurze, p. 106. Cf. *Annals of St Bertin and St Vaast*, ed. Dehaisnes, p. 235.

100 Lot, *L'Art Militaire*, pp. 221-22. Cf. Purton, *Early Medieval Siege*, p. 83.

In the autumn of 885, a composite Viking force laid siege to Frankish Paris, which controlled access along the Seine.[101] Abbo's eyewitness account of the siege, which lasted through the summer of 886, provides a small moment of clarity amidst the narrative fog of the Early Middle Ages. More than any other document, this has become a battleground for scholars arguing for and against the existence of trebuchet and torsion technology in Europe prior to the period of the crusades. A number of siege engines, including *catapultae*, *mangana* and *ballistae*, are repeatedly mentioned but at no point are they clearly described, thus allowing for various interpretations.[102]

Although this is a rare example of a detailed account of an early medieval siege, the degree to which it has been used as evidence for and against the use of classical torsion artillery is entirely unjustified. While Oman claimed that the arsenal of Vegetius was clearly employed by the Vikings, having acquired this (torsion) technology through a half-century of raiding against the Franks, Schneider saw sufficient evidence to conclude that the Vikings had invented the traction trebuchet, fostered by their nautical traditions. Like Schneider, Gillmor regards the use of the term *mangana* as indicative of a new type of siege weapon, but suggests that torsion engines were also used, associating these with the *catapultae* because of their supposedly slower shooting rate.[103]

These arguments and many like them hinge primarily on one paragraph of text. Abbo relates,

> Our men prepared heavy beams, each with an iron tooth at its end, so as to damage the machines of the Danes more quickly. With upright frames of the same length, they built what are commonly called *mangana*, from which they threw vast stones, which could crush the low race of savages.[104]

There is nothing conclusive in this statement to prove what type of engines these were. The iron prong affixed to the end of the throwing arm strongly suggests that this was used to hold a sling, a feature of both the traction trebuchet and the *onager*. However, the suggested framework of the engine, the use of

---

101 Abbo, *Bella Parisiaca urbis*, ed. Pertz, pp. 6-46. Cf. *Annals of St Bertin and St Vaast*, ed. Dehaisnes, pp. 322-27.

102 Abbo, *Bella Parisiacae urbis*, ed. Pertz, pp. 6-46.

103 Oman, *A History of the Art of War*, pp. 141-48; Schneider, *Die Artillerie des Mittelalters*, pp. 53-54, 60-61; Gillmor, "Introduction of the Traction Trebuchet," p. 5.

104 *Magno cum pondere nostri / Tigna parant, quorum calibis dens summa peragrat, / Machina qup citius Danum Quisset terebrari; / Conficiunt longis aeque lingis geminatis / Mangana quae proprio vulgi libitu vocitantur, / Saxa quibus iaciunt ingentia, seo iaculando / Allidunt humiles scaenas gentis truculentae*. Abbo, *Bella Parisiacae urbis* 1 ll. 360-66, ed. Pertz, p. 18.

traction trebuchets by both Muslim and Byzantine forces well before this point, and the absence of evidence to indicate the continued use of the *onager*, all suggest this was more likely a traction trebuchet.

It is important to view Abbo's account, like others of the period, with caution. At one point during the early winter, he claims that the defensive engines in the city were used to support the bridgehead tower on the north bank of the river; however, he later describes the tower as being beyond their effective range following its capture. Such inconsistencies, along with the absence of a clear description of the framework or shooting process of one of these engines, and the single mention of the term *mangana*, compared to six mentions of *catapulta* and two of *balista*, unfortunately allow for few definite conclusions regarding the form of these engines.

Following the siege of Paris the muted nature of the sources resumes and references to artillery are rare through the tenth century and into the eleventh. *Tormenta* are mentioned at the siege of Heimburg (1073),[105] but for the most part artillery tends to be identified with more general terms, such as *machina*.[106] Little more can be discerned regarding the nature of trebuchet technology in Europe up to the departure of the First Crusade.

There is little evidence to suggest that stone-throwing torsion engines were employed beyond the sixth century. Such engines had few advantages to support their continued use following the introduction of the traction trebuchet, wherever and whenever that might have occurred. While relying on isolated events in particular sources, like the sieges of Thessalonica and Paris, is hardly a good historical practice, they unfortunately provide the clearest evidence available. Suffice it to say that when the sources become more numerous and more detailed at the end of the eleventh century, the lack of any evidence to support the use of torsion artillery by Latin armies implies that this type of artillery had already fallen from popular use. Likewise, the clearer descriptions of traction trebuchets that eventually appear, similar to those found in the Near East prior to this, give no indication that these engines were of a new mechanical type. This strongly suggests that these machines had been in use for quite some time previously.

In the second half of the twelfth century, suggestions of a new type of engine begin to appear. In 1174, the first shot loosed by William the Lion of Scotland's *periere* at Wark was discharged early and flew backwards hitting a

105 *Carmen de bello Saxonico* l. 111, ed. Holder-Egger, p. 4.

106 For an examination of this in an Anglo-Norman context, see Fulton, "Anglo-Norman Artillery," pp. 3-8.

Scottish knight.[107] A similar event took place in 1206 when a Russian engine misfired at the siege of Holm.[108] The descriptions of these accidents seem to imply that the authors of the accounts were familiar with the machines that were employed, but that the operators may not have properly understood how to use them. Both engines were almost certainly trebuchets. Their slings were probably of an incorrect length, leading to an early or extremely late release, or the projectiles simply fell out of their slings early in the shooting sequence. Although these events may speak to the inexperience of the operators, the nature of these accidents is perhaps more suited to a counterweight trebuchet than a traction variety. Regardless, the victim at Wark was apparently unscathed and his counterpart at Holm was wounded rather than killed, suggesting that the responsible engines were not large. While such instances hint at the development of the counterweight trebuchet, they are far from conclusive evidence. Ammianus describes a similar accident in which a Roman soldier was killed when a torsion engine misfired backwards at the siege of Maogamalcha (in Mesopotamia) in 363.[109]

Appearance of the Counterweight Trebuchet

The counterweight trebuchet seems to have been developed during the period of the crusades. The earliest known surviving illustration of such an engine is found in al-Tarsusi's treatise on weaponry, composed in Egypt around 1180.[110] Although al-Tarsusi's work indicates that this technology was known and almost certainly employed in the Near East before the Third Crusade, there is little consensus among scholars as to when and where the counterweight trebuchet was first invented.

Using al-Tarsusi's illustration and its accompanying description as a reference point, most scholars assume that this technology was developed at some point in the second half of the twelfth century, most likely somewhere around the Mediterranean. By the early thirteenth century, such engines appear to have been employed across much of Europe and by 1218 *trabuculi* are noted at the crusaders' siege of Damietta.[111]

107 Jordan Fantosme, *Chronique* ll. 1,245-55, ed. and trans. Hewlett, pp. 306-8.

108 Henry of Livonia, *Chronicon Lyvoniae* 10.12, ed. Arbusow and Bauer, p. 42, trans. Brundage, p. 63.

109 Ammianus, *Res gestae* 24.4.28, ed. Eyssenhardt, p. 308, trans. Yonge, p. 362.

110 See Appendix 2, **no. 34**.

111 See Fulton, "The Diffusion of Artillery Terminology," pp. 1-10.

Origin of the Counterweight Trebuchet

Although most scholars date the invention of the counterweight trebuchet to the twelfth century, opinions vary on where it was first used geographically. According to some, the earliest description of a counterweight trebuchet can be found in Niketas Choniates' account of the Byzantine siege of Zevgminon in 1165.[112] Andronicus Comnenus is said to have employed engines (*petrobolous mekhanas*) that threw stones weighing a *talent* (about 30 kg), each making use of a sling, winch and beam.[113] Using these engines in conjunction with mining efforts, the attackers were able to compromise the targeted stretch of defences, causing the fortifications to begin to crumble. Niketas describes Andronicus's use of a similar engine at the siege of Nicaea in 1184.[114]

It is not the power of the engines described by Niketas that has drawn attention but rather the mention of winches, a component not found on a traction trebuchet. It is possible that Niketas leaned on classical sources when composing his history and that his descriptions reflect earlier accounts of torsion-powered engines; however, it seems at least as plausible that they were counterweight trebuchets that he describes. As he was writing somewhat after these events, it is also possible that the descriptions provided by Niketas more accurately reflect Byzantine artillery used at the very end of the twelfth century or start of the thirteenth. The prominent position that he gives to Andronicus at both Zevgminon and Nicaea, claiming he personally adjusted and aimed the engines, an honour which Henry of Livonia similarly bestowed on the duke of Saxony at Mesoten in 1220,[115] is also cause for suspicion. It is not uncommon to find Byzantine commanders, especially emperors, showered with praise and placed in focal positions in Niketas's work and those of contemporary Byzantine sources, suggesting aspects of certain anecdotes were exaggerated in order to enhance the commanders' prestige. Accordingly, it is possible that Niketas associated Andronicus with the latest siege engines of his own day, regardless of whether or not they were employed at these particular sieges.

---

112 Niketas Choniates, *Historia*, trans. Magoulias, p. 76. For suggestions that this is the earliest evidence of the use of a counterweight trebuchet, see, for example, DeVries and Smith, *Medieval Military Technology*, p. 127; Kennedy, *Crusader Castles*, p. 107. Cf. Chevedden, "Invention," pp. 86-87.

113 For the Greek terms, see Dennis, "Byzantine Heavy Artillery," p, 113; Chevedden, "Invention," pp. 86-87.

114 Niketas Choniates, *Historia*, trans. Magoulias, p. 156.

115 Henry of Livonia, *Chronicon Lyvoniae* 23.8, ed. Arbusow and Bauer, p. 163, trans. Brundage, p. 181.

Paul Chevedden has suggested that the origins of the counterweight trebuchet are to be found earlier in Byzantine history. According to Chevedden, it was Emperor Alexius Comnenus who invented the counterweight trebuchet, postulating that Alexius's eastward campaigns against the Seljuks might have been the impetus. The first of these new engines were then provided to the crusaders, who used them against Nicaea in 1097. This theory, however, rests entirely on the account of events given by the emperor's daughter decades later and her remark that most of the engines were constructed "to an unorthodox design of his [Alexius's] own which surprised everyone."[116] As will later become apparent, there is nothing in the contemporary accounts of the siege of Nicaea to suggest that any new or particularly powerful engines were used by the crusaders.

Many have hypothesised that counterweight artillery was used at Tyre in 1124.[117] Using William of Tyre's account of events, Chevedden has suggested that Havedic, the Armenian summoned by the Franks to improve the performance of their artillery, brought with him Byzantine knowledge of counterweight trebuchet technology, which according to his theory had been developed by Alexius less than thirty years earlier.[118] Ronnie Ellenblum has rejected the idea that counterweight engines were first used at Nicaea, pointing to the lack of evidence to support this. However, similarly citing no more than William of Tyre's account, composed at least half a century after the fact, Ellenblum confidently asserts that a counterweight trebuchet was used at Tyre.[119] Searching in other directions, some scholars have looked to the Islamic world for the origins of the counterweight trebuchet.

David Nicolle favours a theory that counterweight artillery was developed under Islam during the tenth century, although he concedes there is little sound evidence of such. This forces him, like Chevedden, to rely on a selection of exceptional anecdotes, many of which appear to contain clear exaggerations when evaluated alongside similar contemporary accounts.[120] Randall Rogers has taken a more cautious approach, acknowledging the lack of clear

116 Chevedden, "Invention," pp. 86-87; Chevedden, "King James I," p. 316. Anna Comnena, *Alexiad* 11.2.1, trans. Sewter, pp. 335-36. The context of this description bears resemblance to the engines supposedly invented by Alfred of England in the ninth century, Florence of Worcester, *Chronicon ex chronicis*, ed. Thorpe, 1:89, trans. Forester, p. 66.

117 For a more detailed discussion, see Chapter 4.

118 Chevedden, "Invention," p. 92. Cf. William of Tyre, *Chronicon* 13.10, ed. Huygens, 1:597-98, trans. Babcock and Krey, 2:15-16.

119 Ellenblum, *Crusader Castles*, pp. 193-94, 208-10. Chevedden has since refuted Ellenblum's criticism, Chevedden, "King James I," p. 317 and n. 9.

120 Nicolle, "Early Trebuchet," pp. 270-72.

evidence. He has postulated that the increasing professionalism of the experts employed by Roger II of Sicily and his successors might have contributed to the development of the counterweight power system in south Italy, though he does not rule out the possibility that counterweight trebuchets might also have developed in the realm of Islam.[121]

Those arguing that the genesis of the counterweight trebuchet was to be found in Europe typically point to the appearance of new terminology, roughly contemporary with al-Tarsusi's treatise. However, like discerning when or whether torsion stone-throwers fell from use during the Middle Ages and when traction trebuchets came into use, terminology offers little to pin down exactly when the first counterweight trebuchets were employed.

Frankish Terminology

The earliest surviving mentions of the term 'trebuchet' date to the late twelfth century. *Trabuchellus* is found alongside *manganum* and *prederia* in an agreement of fealty made in Vicenza on 6 April 1189.[122] A decade later, *trabucha* can be found with *predariae* at the siege of Castelnuovo Bocca d'Adda in the account of Iohannes Codagnellus.[123] Despite the eagerness of some scholars to view this as clear evidence of the presence of counterweight trebuchets in northern Italy, John France has astutely pointed out that Codagnellus does not describe the engine that he identifies with this term and even seems to indicate that they are fairly light in subsequent references.[124] D.J. Cathcart King went so far as to discard this reference in his study of trebuchets because it is listed along with other machines in an effort to secure a ditch.[125]

The origin of the term trebuchet is not clear. King suggested that it came from *trebucher*, to rock or tilt,[126] while William Sayers has provided a more thorough etymological study.[127] The variety of forms that appear in the early

121 Rogers, *Latin Siege Warfare*, pp. 247-48.

122 *Nec cum mangano, nec Trabuchello, aut cum Prederia, vel Balista vel Archu traham, nec aliquem trahere permittam, nec faciam, si vetare potero usque ad finem sue consularie*, in Verci, *Storia degli Ecelini*, no. 52, 3:97.

123 Iohannes Codagnellus, *Annales Placentini*, ed. Pertz, p. 420, ed. Holder-Egger, p. 25. See also France, *Western Warfare*, pp. 121-23; Huuri, *Zur Geschichte des Mittelalterlichen Geschutzwesens*, p. 171.

124 France, *Western Warfare*, p. 122, p. 273 n. 33.

125 King, "The Trebuchet," p. 461.

126 King, "The Trebuchet," p. 461.

127 Sayers, "The Name of the Siege Engine *trebuchet*," pp. 189-96.

thirteenth century, including *trabocco*, *tribok*, *tribuclietta*, *trubechetum*, and numerous other vernacular renderings, appear to obscure any certainty.[128]

From about the late 1210s, variations of 'trebuchet' appear more frequently in the sources and seem to identify more definitely counterweight trebuchets. In the *Chanson de la Croisade Albigeoise*, Simon de Montfort's northern crusaders as well as the defenders of Toulouse are said to have employed *trabuquetz*. A French engine of this type apparently required only one shot to smash a tower of Castelnaudary and one more to destroy its great hall in 1211, obvious exaggerations.[129] But the power of the Toulousians' trebuchets is also emphasised during their defence of the city and concurrent siege of the citadel in 1217-19:

> Inside Toulouse so many carpenters were busy building strong fast-firing double trebuchets that no tower or hall, rampart or merlon was left undamaged in the Narbonnais castle confronting them.[130]

However, other engines are also characterised in this way: Simon de Montfort's *calabres* and *peirers* are credited with breaching one of the towers, which granted the crusaders temporary access to Toulouse.[131]

Rather than the notion of increased power, it is the mention of certain components and operational procedures that suggest something new. In one instance it is recorded that "more than ten thousand tallied on the ropes" of these engines as they assailed the Narbonnais castle, possibly suggesting that these were traction engines. Soon afterwards, however, it is noted that the defenders "ran to the ropes and wound the trebuchets",[132] indicating that the ropes were instead required to haul or winch back the main beams of counterweight trebuchets. To shoot one of these engines, the defenders "then released their ropes",[133] either the means or a symbolic reference to allowing the counterweight to fall. Although there are few indications that these engines were significantly stronger than the traction trebuchets positioned next to them, they appear to have been quite accurate. Their slings were loaded with

128 For more on the appearance of these terms, see Huuri, *Zur Geschichte des Mittelalterlichen Geschutzwesens*, pp. 63-64.

129 *Chanson de la croisade Albigeoise* 92, ed. Meyer, 1:93-94, trans. Shirley, pp. 51.

130 *E lains en Toloza ac aitans carpentiers / Que fan trabuquetz dobles e firens e marvers, / Qu'el castel Narbones que lor es frontaliers / No i remas tor ni sala, dentelh ni murs entiers*, *Chanson de la croisade Albigeoise* 192, ed. Meyer, 1:285, trans. Shirley, p. 141.

131 *Chanson de la croisade Albigeoise* 198, ed. Meyer, 1:311-13, trans. Shirley, pp. 156-57.

132 *Chanson de la croisade Albigeoise* 203, ed. Meyer, 1:332, trans. Shirley, pp. 155, 167.

133 *Chanson de la croisade Albigeoise* 204, ed. Meyer, 1:333, trans. Shirley, p. 168.

well-dressed stones, which on one occasion destroyed an assailing penthouse.[134] Because each shot of a counterweight trebuchet is powered by the same amount of force, these can be extremely accurate machines when loaded with regularly shaped ammunition of a consistent mass. This is a less significant consideration when operating traction engines, as the force applied by the pullers naturally varies from shot to shot.

The defenders of Toulouse appear to have enjoyed the help of two experts, Bernard Parayre and Master Garnier. It was the task of these two men to "manage the trebuchets as they were accustomed to this work".[135] Critically, their charge does not appear to have extended to the *calabres*, *engenhs* and *peirers* that are mentioned only one line above. A rapid demand for experienced artillerists may have developed if the spread of the terminology through the region represents the diffusion of counterweight technology. In 1226, the defenders of Avignon are noted to have made use of *trabucheta* as well as *petrarie*, and *mangonella*.[136]

This new terminology appears elsewhere in Europe around the same time. In Germany, Otto IV is reported to have used a *tribok*, or *tribracho*, in 1212 against Weißensee.[137] In Britain, the earliest use of the term 'trebuchet' accompanies the invasion of Prince Louis (later Louis VIII of France) and his siege of Dover in 1216-17. The *Historia des Ducs de Normandie* mentions the use of a *trebuket* at this siege and Richard of Morins similarly notes the use of a *tribuclietta* along with many other engines.[138] Roger of Wendover appears to refer to the same engine but uses the traditional term *petraria*.[139] Although Roger uses the term *trebuculus* to identify engines at the sieges of Jerusalem (1099) and Damietta (1218-19), he at no point uses this term in a European context.

---

134 *Chanson de la croisade Albigeoise* 204, ed. Meyer, 1:333, trans. Shirley, p. 168.

135 *Chanson de la croisade Albigeoise* 213, ed. Meyer, 1:377, trans. Shirley, p. 191, see also 198, ed. 1:310, trans. p. 155.

136 *Chronico Sancti Martini Turonensi*, ed. Holder-Egger, p. 73.

137 *Et inde progrediens, obsedit oppidum Wizense, quod similiter expugnavit usque ad arcem. Ibi tunc primum cepit haberi usus instrumenti bellici, quod vulgo tribok appelleri solet. Annales Marbacenses*, ed. Wilmans, p. 172. *Movens igitur castra contra miniciunculam in Salcza tribracho illo, cognomento tribock, muris inminet. Cronica Reinhardsbrunnensis*, ed. Holder-Egger, p. 580. *Tribracho* is given in the *Chronica S Petri Erfordensis*, ed. Holder-Egger, p. 383. See also Schneider, *Die Artillerie des Mittelalters*, p. 28.

138 *Histoire des ducs de Normandie et des rois d'Angleterre*, ed. Michel, pp. 188, 192-96; Richard of Morins, *Annales Prioratus de Dunstaplia*, ed. Luard, pp. 48-49.

139 Fulton, "Anglo-Norman Artillery," pp. 21-22; Roger of Wendover, *Flores Historiarum*, ed. Coxe, 2:135, Hewlett, 2:191, 229, trans. Giles, 1:429, 2:374-75, 406-7, cf. ed. Hewlett, 2:252, 310, trans. Giles, 2:426, 478-79.

The relatively sudden appearance and spread of this new terminology in the early thirteenth century does not rule out the likelihood that counterweight artillery was employed in Europe before large contingents of crusaders associated with the Third Crusade began arriving in the Levant. Likewise, the earliest appearances of these new terms is not conclusive evidence that they were initially used exclusively to refer to counterweight trebuchets nor that they were the only terms used to identify such engines. After returning home from the Third Crusade, Philip II Augustus may have used counterweight trebuchets at Verneuil in 1194[140] and Château Gaillard in 1203-4,[141] though the sources do not use any new terminology to identify these engines. Similarly, the prefabricated engines that Richard I made use of at Nottingham in 1194 might also have been counterweight trebuchets, although the sources do not distinguish them with new terminology.[142] Henry of Livonia, writing in the Baltic region in the 1220s, does not employ any new terminology, but it is likely that he, like Roger of Wendover, used traditional terms to identify the use of counterweight trebuchets.[143] When considering that certain sources used traditional vocabulary to identify counterweight trebuchets after the appearance of the new terminology, it is quite likely that traditional terms were applied to this new technology before the first recorded use of the term 'trebuchet'.

It seems very unlikely that either the crusaders or Muslims first learned about the counterweight trebuchet during the Third Crusade. Al-Tarsusi's work confirms that the Muslims appreciated this technology beforehand, while the appearance of the term 'trebuchet' in Europe prior to the departure, let alone return, of most crusaders suggests prior knowledge of this technology. As will be discussed in Chapter 5, it seems both sides were well acquainted with the technology of the other when the crusaders arrived and that the stone-throwing engines of each side appear to have been remarkably similar.

First appearing in the early to mid-thirteenth century, variants of the terms *bricola* and *blide* were also associated with the counterweight trebuchet. The term *bricola* is most often used today to refer to a type of counterweight trebuchet with a single vertical support and two counterweights, hung from the two ends of a forked short arm. *Blide*, found most often in German contexts, is

140 Roger of Wendover, *Flores Historiarum*, ed. Hewlett, 2:123, trans. Giles, 2:135; Ralph of Diceto, *Ymagines histriarum*, ed. Stubbs, 2:114-15.

141 William the Breton, *Philippide* 7, ed. Delaborde, pp. 201-4; William the Breton, *Gesta Philippi Augusti* 128-129, ed. Delaborde, pp. 218-19.

142 Roger of Howden, *Chronica*, ed. Stubbs, 3:238-40, trans. Riley, 2:314-16; *Pipe Roll 40* (*1194*), ed. Stenton, p. 43.

143 Fulton, "The Diffusion of Artillery Terminology," pp. 13-21. See also Fulton, "Anglo-Norman Artillery," pp. 22-24.

typically synonymous with 'counterweight trebuchet'.[144] Frederick II is said to have dispatched *bricolae*, along with *prederiae* and *machinae*, to the Holy Land in 1242 and *biblieta* are mentioned in the *Excidium* during the siege of Acre in 1291, but neither mention is particularly descriptive nor corroborated by other sources.[145] Unlike variants of *trebuchetum*, which appear relatively frequently in thirteenth-century Frankish sources, these other terms were not widely embraced in the context of the crusades. Although this might suggest that one or more different types of engines were employed in Europe but not by crusaders or the Frankish population living in the Near East, this seems unlikely given the close links between Europe and the Latin principalities.

Muslim Terminology

Muslim sources also began to use a series of new Arabic terms to classify the *majaniq* employed in the Levant during the thirteenth century. The *shaytani* and *qarabugha* varieties appear to have been traction trebuchets, often found in the greatest numbers and seldom praised. The *manjaniq maghribi* seems to have been a counterweight trebuchet, later replaced by the *ifranji* type.[146] These labels do not correspond with those given by al-Tarsusi in the late twelfth century. According to al-Tarsusi, the 'Frankish' trebuchet was a light traction engine; however, the same term came to refer to the most powerful counterweight trebuchets in the late thirteenth century.[147] None of al-Tarsusi's other designators were carried forward. Even more perplexing, the use of these thirteenth-century terms was essentially limited to the late Ayyubid and early Mamluk periods, almost disappearing from use at the start of the fourteenth century.

Having examined the Arabic terminology used before the appearance of these new terms, Chevedden has argued that augmenting adjectives such as *kabir* (big), *'azim* (great), and *ha'il* (frightful or huge) were used to specify moments of technological change.[148] This essentially allowed him to interpret the

---

144 Etymologically, Chevedden has interpreted *bricola* to come from the Latin *bi-coleus* ('double testicle') in reference to its design, Chevedden, "Invention," p. 109. David Nicolle has suggested that the term simply refers to the divided 'arms' of the short arm of the beam, while the variant *biblieta* might refer to the division or opening of the short arm like a book, Nicolle, "The Vocabulary of Medieval Warfare," p. 173.

145 Caffaro (cont.), *Annales*, ed. Imperiale, 3:128; *Excidium Aconis* 2.3, ll. 122-48, ed. Huygens, pp. 67-68.

146 For an initial study of these terms, see Khamisy and Fulton, "Manjaniq Qarabugha," pp. 179-201.

147 See Chapter 7 and Appendix 2.

148 Chevedden, "Invention," pp. 90-91.

meaning of these terms as he wanted. Putting forward a three-stage theory of development (consisting of the traction trebuchet, hybrid trebuchet and counterweight trebuchet), he has argued that the use of such adjectives from the twelfth century onwards was meant to indicate the use of counterweight trebuchets, such as at the sieges of Baghdad (1157), Alexandria (1174), Masyaf (1176), and Kerak (1184).[149] The theory seems overly convenient and relies on the mechanically impractical 'hybrid' trebuchet having been a historical reality.[150] Although it appears probable that counterweight trebuchets were coming into use by the late twelfth century, there is little solid evidence to confirm their use at any particular siege before the start of the thirteenth century.

Considerable uncertainty remains when dealing with the development of medieval artillery. Issues such as how long torsion engines remained in use and when trebuchet-type engines were adopted in various regions remain open for debate. With such fundamental questions left unresolved, it is hard to discern the power of the engines used between the sixth and eleventh centuries. From the end of the eleventh century, however, relatively detailed descriptions of artillery emerge from the theatre of the crusades, allowing for a comparatively good understanding of the machines employed in the Levant during the twelfth and thirteenth centuries.

149 Chevedden, "Invention," p. 93.

150 See Fulton, "Myth of the Hybrid Trebuchet."

CHAPTER 2

Mechanics

The term 'trebuchet' first appears in the late twelfth century. Because it was never used to refer to earlier torsion engines, modern historians have embraced the word as a broad, but clear, technical term to identify the whole family of swing-beam, or levered, stone-throwing siege engines. This family of machines can be divided into two subcategories based on their power source: traction trebuchets, which relied on the strength of people pulling on ropes; and counterweight trebuchets, which made use of a deadweight and gravity.

The primary component of a trebuchet, regardless of its power source, was a beam that rotated around a horizontal axle, elevated and supported by a framework. By offsetting the axle along the length of the beam, the engine made use of mechanical advantage: when a downward force was applied to the end of the short arm of the beam, the end of the long arm rose at a quicker rate due to its greater distance from the axle. The simplistic but essential addition that made this pivoting beam a practical weapon was the sling. This magnified the mechanical advantage, forcing the projectile to travel even farther, and thus faster, before it was released. One end of the sling was fixed near the end of the long arm of the beam while the other was fitted with a loop that slid over a hook or 'finger' that extended from the end of the long arm. Ideally the finger would be made of metal, reducing friction as the loop slipped free and unleashed the couched projectile at the point of release. By altering the length of the sling, and even the angle of the finger if possible, the point at which the sling released the projectile, and thus the angle at which it was thrown, could be adjusted. This was the easiest way to adjust the range of the engine, avoiding the complicated task of changing the mass of the counterweight or the force with which the pullers were tugging. The value of the sling seems to have been well appreciated as the handheld staff-sling, Vegetius's *fustibalus*, which was little more than a stick with a similar sling and finger arrangement at its end, can be seen in a number of medieval illustrations.[1]

The nature of the forces that powered these engines led to two profound differences between traction and counterweight trebuchets. The first was

1 Vegetius, *Epitoma Rei Militaris* 3.14, ed. Lang, pp. 98-99. For illustrations of the staff-sling, see *Biblia Sancti Petri Rodensis* (Paris: BNF, MS lat. 6 (3)), fol. 134r; Peter of Eboli, *Liber ad honorem Augusti* (Bern: Burgerbibliothek, BBB Cod. 120.II), fol. 111r; Matthew Paris, *Chronica maiora* (Cambridge: Corpus Christi College Lib., MS 16, pt. 2), fols. 56r, 59v.

 | DOI 10.1163/9789004376922_004

scale. The size of a traction trebuchet was restricted by the finite distance that a person can pull a rope in a single tug, ultimately limiting the amount of energy that these engines could harness. Free from this restriction, a counterweight trebuchet could be built as large as the strength of its components and skill of its builders allowed. This permitted them to throw larger projectiles to greater distances and inflict considerably more damage. The second was shooting rate: the necessarily light components of a traction trebuchet allowed for a rapid shooting rate, while counterweight trebuchets required significantly more time between shots. By making use of pullers, the force that powered a traction trebuchet could be applied, and removed, almost instantaneously. This allowed the long arm, the heavier end of the beam, to return naturally to a loading position after each shot. Making use of the constant pull of gravity, the beam of a counterweight trebuchet was subject to a period of pendular swinging after each shot, as energy stored in the counterweight that had not been transferred to the projectile slowly dissipated. There was then the laborious task of pulling or winching back the long arm of the beam, by far the lighter end of the beam due to the counterweight attached to the end of the short arm. Only after this was done could the sling be loaded for the next shot. These fundamental differences made traction trebuchets effective antipersonnel weapons, more powerful than slings and archers but capable of a similar shooting rate, while counterweight trebuchets could be used to target structures and defences.

Traction

The traction trebuchet had the benefit of being a fairly light engine. It was simple to assemble and shift if necessary, and could be operated by crews with little experience. When worked by trained teams in relay, they were capable of maintaining high shooting rates over extended periods of time. At the siege of Lisbon in 1147, traction trebuchets employed by the crusaders achieved a reported shooting rate of four shots per minute, a rate that was achieved by W.T.S. Tarver in trials using groups of relatively untrained volunteers.[2] Although the force with which each crew member pulled invariably fluctuated, Tarver's trials have revealed that these engines were more accurate than is often believed. So long as projectiles of about the same mass were used, the slightly different power that went into each shot made the use of finely shaped

2 See Chapter 4.

stone balls unnecessary. This, along with the rapid rate at which they could shoot, meant that roughly tooled stones or even fieldstones were suitable ammunition.

The scale of traction trebuchets remained relatively constant throughout the Middle Ages. This was not due to a lack of development, but rather the finite distance that a person can effectively pull a rope in a downward direction (less than 1.5 m). This limits the potential length of the short arm: the point at which the ropes are attached to the end of the arm cannot travel farther than the distance that the crew can pull. If the short arm is too long, the crew will stop pulling before the beam has rotated far enough to release the projectile; if it is too short, the projectile will be released while the crew is still pulling, failing to capture all of the energy provided by the pullers.

The length of the short arm then influences the proportionate dimensions of other components, such as the length of the long arm, which in turn determines the ideal length of the sling. While these other components are also influenced by the building materials, their strength and weight, and the mass of the intended projectiles, the ideal length of the short arm remains relatively constant. Accordingly, there is a theoretical ideal set of dimensions for any traction trebuchet, taking into consideration the strength of the pulling crew and mass of the various components and projectiles. Yet these engines did not require precise engineering because the interrelation between all of the components provided plenty of room for adjustments. For example, if a long arm is shorter than might be theoretically ideal, the length of the sling can be adjusted to compensate for this, allowing for an effective, if slightly less efficient, shooting sequence. So long as the length of the short arm is appropriate, there is a forgiving operational window for the dimensions of the other components.[3]

The limited distance that the end of the short arm can travel has further consequences on the shooting sequence. Illustrations often show traction trebuchets with their beams in a roughly horizontal starting, or loaded, position. This would allow for a slightly longer short arm than might be expected, ensuring that the crew is able to accelerate each end of the beam to a sufficient speed before the projectile is released. This, however, limits the potential length of the sling because the slightly decreased rotation of the beam gives it less time to swing around before release. This arrangement appears to have been the result of practical experimentation and may in part have been inspired to better incorporate an additional operator: the loader. This individual

3 See Fulton, "Myth of the Hybrid Trebuchet."

would place a projectile in the sling and then continue to hold the couched projectile during the early stages of pulling, allowing all slack in the ropes to be taken up, before the sling and projectile were pulled loose.[4] This would maximise the efficiency of the crew's pulling motion and potential shooting rate. Al-Tarsusi hints that this may have been a way of extracting further energy from the resistance of the beam;[5] however, this alluring suggestion has not yet been demonstrated through trials and seems somewhat fanciful when the physics are examined closely. Sling length would vary by machine and would need to be adjusted if a significantly larger or smaller projectile was used. When considering the mechanics, it seems unlikely that many slings would have been longer than half the length of the long arm of the beam.

The restricted size of the traction trebuchet also limits the size of the pulling crew. As there is limited room directly below the short arm, the direction of force spreads outwards as more pullers are added. Power can thus be increased only to a limited extent by utilising more pullers. If the pullers are positioned behind the axle, rather than below the short arm, there is near limitless room; however, this requires an even smaller short arm as the distance that an individual can pull horizontally across their body is less than they can pull vertically downwards. Although there are a few historical illustrations of crews standing behind the axle of an engine, these are relatively rare and do not appear to reflect a common practice.[6] The physics also prevents crews from pulling in stages or relays during a single shooting sequence. Even with hugely exaggerated components, neither the rate of acceleration nor true velocity would be sufficient to allow for a practical shooting scenario.[7] Despite the numerous surviving medieval images of traction trebuchets, none appear to depict a supersized engine of this type. Likewise, descriptions tend to exaggerate the number of pullers or the size of the stones that were thrown rather than the scale of certain mechanical components.[8] These often indicate that the author had an appreciation of how these engines worked but lacked a sound understanding of their mechanics. Accounts of teams composed of hundreds

4 For depictions of this operator, see Appendix 2, **nos.** 3, 4, 13, 19, 20, 25 and 28.

5 For a translation of this passage of al-Tarsusi's treatise, see Appendix 2.

6 For an example, see Appendix 2, **no.** 25.

7 Perhaps the closest mechanical solution might have been to dispense with the short arm and exaggerate the axle, as can be seen in a drawing of a counterweight engine by Leonardo da Vinci, Leonardo da Vinci, *Codex Atlanticus* (Milan, Biblioteca Ambrosiana), fol. 57r, cf. fol. 50v. There is no evidence that such an engine was ever built.

8 For exaggerated pulling crews, see al-Baladhuri, *Kitab Futuh*, trans. Murgotten, 2:217-18; Aristakes Lastivertc'i, *Patmut'iwn*, trans. Bedrosian, pp. 103-4; *History of the Patriarchs*, trans. Khs-Burmester et al., 3.2:218.

of pullers, which some modern historians have taken literally, can be dispelled as pure hyperbole.[9]

Because of the fundamentally limited length of the short arm, and the knock-on effect that this has for the proportional dimensions of other components, the average crew size would appear to have been around ten pullers. This number allows everyone to stand almost directly under the short arm, the ideal position to maximise the transfer of energy. If the pullers stand farther back, towards the axle, it is possible to rotate the beam farther but resistance is increased as the direction of force is brought more in line with the axle; if they stand farther ahead, initial resistance is lowered but potential rotation is decreased. Experimental trials using reconstructed engines have confirmed that the ideal position for the pulling crew was directly below the short arm as well as the redundancy of any more than about ten to twelve pullers.[10]

Hybrid

In an attempt to validate textual references to excessively large teams of pullers, Paul Chevedden has developed the theory of a 'hybrid trebuchet'. Suggestions that pullers worked in conjunction with a counterweight can be traced back to Viollet-le-Duc, who seems to have believed that all trebuchets had a counterweight and that the pulling crews, clearly described in certain accounts, helped to propel the counterweight downward.[11] This, however, is an impractical design, it being easier to simply increase the mass of the counterweight so as to render the extra pullers redundant. Furthermore, any added energy provided by the pullers would lead to an uneven application of force, decreasing accuracy.

Whereas Viollet-le-Duc envisioned a counterweight trebuchet aided by pullers, Chevedden's proposed engine was a traction trebuchet aided by a counterweight. According to the latter theory, the counterpoise would offset the weight of a significantly heavier beam, which was needed in order to throw larger, wall-breaking projectiles.[12] On the surface this appears like a practical

9 Fulton, "Myth of the Hybrid Trebuchet."

10 For a documented case study, see Tarver, "The Traction Trebuchet," pp. 157-58. These findings have been confirmed by the crews operating the traction trebuchets both at Caerphilly castle and the Middelaldercentret. My thanks to W.T.S. Tarver who, over the past few years, has kindly shared his previous and ongoing experiences building and operating traction trebuchets.

11 Villet-le-Duc, *Dictionnaire*, 5:236, 239 (and fig. 14), for his broader discussion of artillery see 5:219-42.

12 Chevedden has argued that the hybrid trebuchet was developed under Islamic rule in the early eighth century and by the ninth it had spread through the Byzantine Empire to

innovation, as adding mass to the short arm of a heavy beam would assist in raising the long arm; however, it does not take into account the limiting factors that dictate the function of a traction trebuchet nor some basic physics.

Because the short end of the beam cannot be lengthened, due to the finite distance that a person can pull, an exceptionally large counterpoise is required, as it must be positioned quite near the axle. Although this would help balance the beam, the force provided by the pulling crew must then be enhanced simply to overcome the inertia of the overall heavier beam. While this might be seen to explain references to extremely large pulling crews, it is necessary to consider that the restricted length of the short arm demands that the energy provided by the pullers be sufficiently efficient to accelerate it to an effective velocity across a space of about 1 m. This is not like a tug-of-war that comes down to the net application of energy, but rather a rapid jerk, a sudden burst of power, to be completed in less than a second. As the size of the crew increases, its ability to effectively complete the necessary motion decreases as the logistical challenges mount. One solution to overcoming the slow and ineffective acceleration of the short arm would be to increase considerably the length of the long arm; however, this would in turn increase the mass of the beam, the number of required pullers and the organisational challenges. Finally, there remain the same issues of space that limit the scale and number of pullers that can operate a typical traction trebuchet. Nevertheless, intrigued by the potential of such an engine, the Danish Middelaldercentret attempted to reconstruct such a machine. Although of relatively modest scale, it proved to be a complete failure and was quickly retired.

Traction trebuchets worked best when the beam was as light as possible, allowing for a swift transfer of energy. Some engines may have benefitted from a very small counterpoise manifest in additional structural support added to the short arm. Many traction trebuchets appear to have made use of a fanning 'rake', which effectively extended the end of the short arm along a horizontal two-dimensional plane. This allowed more pullers to stand more directly below the point at which their ropes were attached to the beam.[13] The additional structural supports needed to contend with the uneven lateral forces would have naturally helped to balance the long arm of the beam without adding significant mass.[14]

---

northern Europe. He contends that such a machine was used by the Crusaders against Lisbon (1147) as well as Damietta (1218), Chevedden, "The Hybrid Trebuchet," pp. 183-98; Chevedden, "Invention," p. 74; Chevedden, "King James I," pp. 324-25.

13 For clear illustrative examples, see Appendix 2, **nos. 3-6**, 11, 12, 17 and 18-22.

14 See also Tarver, "The Traction Trebuchet," p. 156.

For these reasons it is extremely unlikely that traction trebuchets ever threw significantly heavy projectiles or were at any point considered breaching weapons. Although the concept of a transition stage in the development of the counterweight trebuchet is appealing, it seems impractical.[15] Simply put, the traction trebuchet was most effective as an antipersonnel weapon. It was reliable, affordable and required minimal skill and time to construct and operate. It was relatively mobile and could be grouped easily with others to concentrate fire. It was able to take advantage of ammunition that could be gathered locally and prepared in little time. Although these engines could be used to destroy improvised wooden defences, such as hoardings, and might even threaten the integrity of a crenellated parapet over an extended barrage, they were most effective against the defenders seeking refuge behind such obstacles.[16]

Counterweight

Although more powerful and visually impressive, the mechanics of a counterweight trebuchet are perhaps even simpler than those of its traction cousin. By using a counterweight rather than pullers to propel the short arm downward, the amount of force that could be applied was limited only by the strength of the beam, axle and framing. This allowed designers to scale these engines up or down according to the building materials and skill of the craftsmen that they had at their disposal.

Taking any component as a starting point, the ideal scale of all other parts can be calculated in proportion. Although medieval craftsmen could not mathematically model these engines, a complete engine system requires an incredibly complicated algebraic expression even by today's standards, they grasped the proportional relationships between the various components. Like the traction trebuchet, if some components are slightly longer or shorter, heavier or lighter, than might be ideal, this can typically be mitigated by an adjustment to a different component to yield an effective shooting sequence.

Historical sources of the Late Middle Ages suggest a variety of ideal ratios between the long and short arms of the beam. Marino Sanudo advised a ratio

15 The assumed existence of hybrid variants has so permeated scholarship that Kostick notes the use of such at the siege of Jerusalem in 1099 without any reference to Chevedden in his bibliography. He, however, also suggests that torsion engines were present at this siege, Kostick, *The Siege of Jerusalem*, pp. 82-83.

16 For a more complete discussion of the suggested 'hybrid trebuchet', see Fulton, "Myth of the Hybrid Trebuchet."

of 6:1 in the early fourteenth century, a proportion fairly close to ideal if maximum range is desired.[17] The theoretical mathematics suggest that the ratio should be slightly less than this, especially when throwing larger weights. Personal experience, field testing, and historical illustrations suggest that a ratio of 5:1 is a safe and effective arrangement when the counterweight outweighs the projectile by at least 100:1. If range is most desirable, a disparity of 1000:1 is generally more appropriate.

To maximise the strength of the frame, and minimise the materials required, the axle was typically supported by a trestle: two vertical triangles, often with a central vertical support bisecting them, set parallel to each other. These were joined to a horizontal base structure that added further support while the tops were connected by the main axle.[18] The long arm of a counterweight trebuchet would typically have been drawn back about 135° from its vertical position of rest. From a mathematical point of view, a greater (steeper) angle would generate more energy, as the counterweight would fall a greater distance and reach a higher velocity. From a carpentry standpoint, however, the higher the axle is placed the more difficult it is to assemble the machine and the sturdier the accompanying structure must be to withstand the additional forces. While increasing the strength and stability of the structure would allow for a higher axle and the beam to be pulled back farther, the added strength would also permit the use of a longer beam and a heavier counterweight at the original loading angle. Because the latter would generally transfer the most energy, it would appear that the length of the long arm of most counterweight trebuchets was around 50% greater than the height of the axle, creating a ratio of 3:2.

Practical carpentry issues ultimately restricted the scale of the largest counterweight trebuchets. While adding mass to the counterweight of an engine with a sturdy frame will increase power, the corresponding amount of energy added to the system decreases exponentially: for each unit of mass that is added, a slightly less amount of energy is gained each time. This is because the system becomes less efficient as less resistance is encountered, allowing the counterweight to fall at an increasingly natural rate. To significantly increase power, a new engine is required, one with a higher axel and proportionally larger components. The construction of ever larger engines, however, sees building challenges increase geometrically, while power enjoys only linear

17 Marino Sanudo, *Liber secretorum* 2.4.22, ed. Bongars, pp. 79-80, trans. Lock, pp. 135-36.

18 For illustrative evidence in European sources, see Appendix 2, **nos. 35, 39, 40, 42, 57, 70-74, 76-79, 91, 92** and **95**. For Muslim examples, see **nos. 103-107, 109, 113, 115, 116, 118, 120, 123, 125** and **128**.

growth. Accordingly, practical construction costs, including materials, labour and expertise, ultimately restricted the size, and thus power, of the largest historical counterweight trebuchets.

The sling length of a counterweight trebuchet was the easiest variable to adjust. Whereas the mathematics show that a sling of equal or near-equal length to the long arm of the beam is ideal, a slightly shorter length is often more effective in the field, due to various mechanical inefficiencies. A long sling allows for maximum mechanical advantage, while adding only negligible mass to the system. By extending the sling horizontally between the struts of the engine, when in the starting or loaded position, the projectile it carries is forced to travel as far as possible through the shooting sequence, maximising release velocity. Furthermore, as the projectile slides horizontally along a trough between the struts of the engine, it exerts little retarding resistance before it is lifted off the ground.

Rather than replace traction trebuchets, counterweight trebuchets supplemented them in a different role. Their slower shooting rate and greater mass made them more difficult to reposition, or even yaw, leaving few incentives to employ a small counterweight engine rather than a comparable traction type. Although less accurate, traction trebuchets might be expected to achieve the same result, albeit with more shots, in a similar amount of time. Accordingly, it was only profitable to employ counterweight trebuchets if they were capable of harnessing noticeably more energy, allowing them to throw significantly larger stones or similarly sized stones greater distances.

The counterweight trebuchet's slower shooting rate was a result of its power source. At the end of each shooting sequence, any energy that had not been transferred to the projectile was dissipated through a subsequent swinging process. This period of post-release pendular swing was longer for those engines that made use of a fixed counterweight (one that was an intricate part of the short arm rather than hung from it). Counterweights attached with a hinge fell straighter, thus allowing for a more efficient transfer of energy, while those firmly fixed to the short arm fell in a wider arc, remaining a constant distance from the main axle. Only once the beam and counterweight were still could the engine be cocked. To pull the long arm back from its vertical position of rest, thus lifting the short arm and counterweight, treadwheels or winches were required for the largest engines while the beams of smaller engines could be pulled into place manually with ropes. The amount of time it took to cock each engine and the number of men required to do so varied according to the size of the machine. A small engine, which could be cocked manually, might be capable of dispelling two shots within 5 minutes. Larger engines, requiring pulleys and winches, might be expected to take upwards of 30 minutes to achieve the same, although they could be expected to throw significantly larger projectiles.

Ammunition for counterweight trebuchets was typically specialised. Because of the power source's consistent energy input, these were exceptionally accurate engines provided projectiles of equal mass and shape were used. Accordingly, projectiles were often finely tooled into spheres, giving them a uniform cross-section along any axis. While this provided a reasonably aerodynamic form, it also ensured that the performance of each projectile remained constant and predictable, allowing them to travel along any axis or even spin in flight. This left only environmental factors and the effects of small mechanical variances, such as the stretch in the ropes, to influence the accuracy of each shot. Although this allowed these engines to constantly hit a specific target, additional investments of money and time were required as each stone had to be shaped by a mason or stonecutter.[19]

Counterweight trebuchets were the most powerful ballistic siege weapons to be used in Europe and the Near East until the bombard was introduced. Although their ability to breach curtain walls and pulverise towers is debatable, the fact that they could remove battlements and expose defenders along parapets to antipersonnel weapons, without the need to construct expensive, complicated and vulnerable siege towers, was significant. Furthermore, the psychological effects that these engines could have on besieged defenders should not be underestimated.

At first glance it can seem perplexing why the developmental leap that led to the invention of the counterweight trebuchet took so long. This, however, oversimplifies the matter with the benefit of hindsight, even though the mechanics had been understood for centuries. The ancient *shaduf*, which had been used for millennia in Egypt to bring up water from the Nile, was based on a similar design, while the *tollenno*, described by Vegetius, made use of a 4:1 fulcrum and counterweight to elevate men to a besieged parapet.[20] Adding knowledge of the Roman *onager*, the continued use of the staff sling and established use of traction trebuchets to the mix, and it seems all of the fundamental components had been appreciated for centuries. Arranging these components is also quite straightforward: children demonstrate an appreciation of

---

19 On 30 June 1224 Henry III summoned an additional 30 quarrymen and stonecutters to shape projectiles to be used against Bedford, Amt, "Besieging Bedford," p. 111. There is evidence that standardised ammunition was stored in Northumberland during the Middle Ages, Hill, "Trebuchets," p. 104.

20 Vegetius, *Epitoma Rei Militaris* 4.21, ed. Lang, pp. 142-43. For a thirteenth-century illustration of a similar contraption, depicted elevating dead bodies, see *Morgan Bible* (New York: Pierpont Morgan Library, MS M.638), fols. 10v, 35v.

the basic mechanics when they lay a spoon across some kind of fulcrum and use it to catapult peas across the dinner table. However, Chevedden seems quite justified in citing the apocryphal story of Columbus's egg in relation to the development of trebuchet technology.[21] Historians tend to fixate on why it took so long to invent the counterweight trebuchet, yet it seems equally reasonable to question why the traction trebuchet was not independently developed in the West, as all of the fundamental components were present, before this technology arrived from the Far East.

The Mathematics

It is relatively simple to express each mathematical component of a counterweight trebuchet: the system is little more than three masses and three axles (a fulcrum and two pendulums), powered by the force of gravity. However, it is extremely complicated to represent a complete shooting sequence as each component complicates the equation geometrically. For purposes here, the function of a counterweight trebuchet will be analysed; traction trebuchets operate on the same essential principles but the lack of a uniform and consistent downward force makes it difficult to accurately express a shooting sequence mathematically.

Following Hill's first attempt to express the function of a counterweight trebuchet in mathematical terms, more detailed studies have been put forward by mathematicians and mechanical engineers. Although the essentials are of a high-school level, the full process is extremely difficult to express without computer software. Today the foremost mathematical representations of trebuchet dynamics are offered by Mark Denny and Donald B. Siano. Siano employs complex numerical models using Mathematica programming, plotted using Cartesian coordinates. Denny's range-based study is more approachable but still relies on a numeric rather than algebraic evaluation of a complete counterweight trebuchet.[22] For clarity, the basic components will be outlined, loosely

21 Chevedden introduces this in his suggestion of a 'hybrid trebuchet', Chevedden, "The Hybrid Trebuchet," p. 184. Hill postulated the initial failure to utilise a long sling may have been responsible for the delayed development of the counterweight trebuchet, Hill, "Trebuchets," pp. 108-9.

22 Denny, "Siege Engine Dynamics," 561-77; Siano, *Trebuchet Mechanics*. See also Denny, *Ingenium*, pp. 61-91. Siano's work has served as the exclusive mathematical basis for studies such as Saimre, "Trebuchet," 61-80. I am deeply indebted to Mark Denny for the innumerable scenarios and setups he has run through his algorithms to test the viability of various theories that I have come up with, which stretch far beyond the parameters of his

following Denny's model, to the limits that basic algebra will allow. For convenience and clarity, four sets of data will be used to provide examples (Fig. 1), and the following abbreviations will be employed:

Mp	Mass of the projectile (kg)
Mcw	Mass of the counterweight (kg)
Mb	Mass of the Beam (kg)
Mla	Mass of the long arm (m)
Msa	Mass of the short arm (m)
Lb	Length of the beam (m)
Lla	Length of the long arm (m)
Lsa	Length of the short arm (m)
Ls	Length of the sling (m)
Lax	Height of the main axle (m)
Lcw	Length of the counterweight's centre of mass from the beam (m)
Hcw	Displacement of the falling hinged counterweight (m)
r	Radius of the projectile (m)
Ap	Cross-sectional area of the projectile (m)
Rmax	Maximum theoretical horizontal range (m)
Rh	Horizontal range (m)
Rv	Vertical range (m)
V	True velocity of the projectile at release (m/s)
Vmax	Maximum theoretical release velocity (m/s)
Vh	Horizontal velocity (m/s)
Vv	Vertical velocity (m/s)
t	Shooting sequence time to release (s)
Tf	Flight time (s)
Ek	Kinetic energy (J)
Ep	Potential energy (J)
Eff	Efficiency (%)
Ib	Moment of inertia ($kg{\cdot}m^2$)
g	Acceleration due to gravity (9.8 m/s^2)
r	Density of air (J/kg·K)
C	Coefficient of drag (shape-dependent)
θ	Angle of the short arm to the vertical below the beam
ϕ	Angle of the short arm to the vertical hanging counterweight

article. For a simpler introduction to the physics of this mechanical process, see Chevedden, et al., "The Trebuchet: Recent Reconstructions and Computer Simulations," pp. 66-71.

ψ	Angle of the long arm to the loaded sling
α	Angle of release to the horizontal

* Note, all angles will be represented in degrees but calculated in radians.

FIGURE 1 *Arrangements A, B, C and D*[a]

	Mp (kg)	Mcw (kg)	Lb (m)	Lla (m)	Lsa (m)	Ls (m)	Lcw (m)
Arrangement A	100.00	10,000.00	12.00	10.00	2.00	10.00	2.00
Arrangement B	10.00	1,000.00	5.00	4.00	1.00	4.00	1.00
Arrangement C	0.45	45.36	1.52	1.22	0.30	0.99	0.30
Arrangement D	20.00	5,000.00	10.00	8.00	2.00	6.00	2.00

a *Arrangements A* and *B* are based on values used by Donald Hill, *Arrangement C* is a roughly optimal set of values provided by Donald Siano and *Arrangement D* is an original data set.

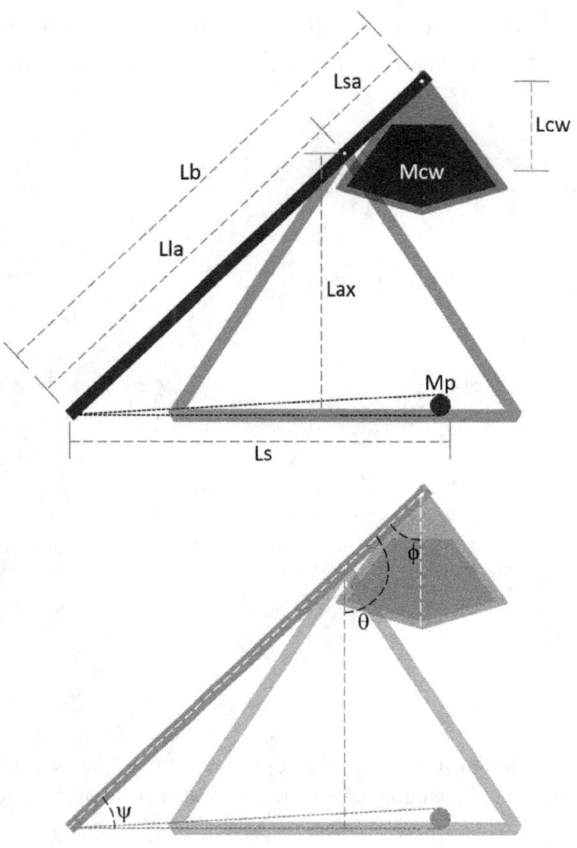

FIGURE 2
Components of a counter-weight trebuchet

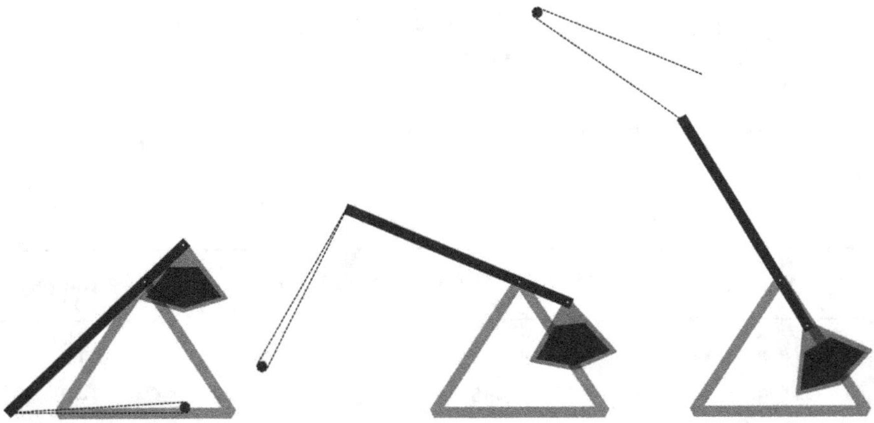

FIGURE 3 *Shooting sequence of a hinged counterweight trebuchet*

Energy System

At its core the shooting sequence of a counterweight trebuchet is simple: potential energy stored in the suspended counterweight is transferred to the projectile couched in the sling.[23] Thus a simplified expression for maximum range can be expressed if the masses and measurements of the basic components are known:

$$\text{Rmax} = 2\left(\frac{\text{Mcw}}{\text{Mp}}\right)\text{Hcw}\cdot\text{Eff}$$

Or, if the velocity and angle of release are known:

$$\text{if Vh} = \text{V}\cdot\cos\alpha$$

$$\text{and Vv} = \text{V}\cdot\sin\alpha$$

$$\therefore\ \text{Rh} = 2\,\text{V}^2\left(\frac{\sin\alpha\cdot\cos\alpha}{\text{g}}\right)$$

$$\text{and Tf} = 2\left(\frac{\text{Vv}}{\text{g}}\right)$$

23 For working through the basic trigonometry and helping provide the essential platform from which I built my numerical representation of the shooting sequence, I would like to thank Ross Underhill.

The maximum potential velocity of a perfectly efficient system (Eff = 1) can also be shown or, alternatively, the efficiency of a system with a known velocity:

$$\text{if Rmax} = 2\left(\frac{\text{Mcw}}{\text{Mp}}\right)\text{Hcw}$$

$$\text{where Hcw} = \text{Lsa}\ (1 - \cos\theta)$$

$$\therefore\ \text{Vmax} = \sqrt{2\left(\frac{\text{Ep}}{\text{Mp}}\right)\text{Eff}}$$

$$\text{and Eff} = \text{V}^2\left(\frac{\text{Mp}}{2\ \text{Ep}}\right)$$

$$\text{where Ep} = \text{Mcw}\cdot\text{Hcw}\cdot\text{g}$$

The energy within the projectile can also be shown:

$$\text{Ek} = \text{V}^2\left(\frac{\text{Mp}}{2}\right)$$

$$\text{or Ek} = \frac{1}{2}\text{Mp}\cdot\text{V}^2$$

Although convenient, these representations are drastic simplifications and as representations of an actual engine they fail to take into account such basic factors as the mass of the beam. For the sake of simplicity, certain modelling limitations will continue to be observed, notably that this remains a frictionless system (unless represented by Eff), the sling is considered to have no mass, the beam has a uniform cross-section, and every component is rigidly defined so that there is no flex or stretch at any point through the shooting sequence.

Mechanical System

To model the mechanics of a trebuchet, it is helpful to start with the beam itself. Devoid of a sling or hinged counterweight, a maximum release velocity is achieved at a release angle of 0° (vertical):

$$V = Lla \sqrt{2\left(\frac{v}{I}\right)(\sin\theta + \cos\alpha)}$$

$$\text{where } v = (Mcw \cdot g \cdot Lsa) - (Mp \cdot g \cdot Lla) - \left[Mb \cdot g\left(\frac{Lla - Lsa}{2}\right)\right]$$

$$\text{where } I = (Mcw \cdot Lsa^2) + (Mp \cdot Lla^2) + Ib$$

$$\text{where } Ib = \frac{1}{3} Mb \left(\frac{Lla^3 + Lsa^3}{Lla + Lsa}\right)$$

The impracticality of this arrangement is obvious. If *Arrangement A* is used and the beam is loaded at 135°, a projectile released at 45° will travel 26.57 m/s or 29.19 m/s at 0°. Only 10% of the kinetic energy within the machine is transferred to the projectile and only about 4% of the potential energy stored within the counterweight is communicated to the projectile.[24] Although this beam-only arrangement is entirely impractical, what should be highlighted is that when the mass of the projectile is reduced the launch speed intuitively increases but the efficiency of the system decreases, a trend that continues as additional elements of complexity are added.

The next mathematical component to be added is a hinged counterweight. In a scenario where it is desirable to throw an extremely large projectile, relative to the corresponding counterweight, only a short distance, the absence of a sling is theoretically beneficial. Such an arrangement can even be seen in certain historical images.[25] In reality, without the added arcing movement of a sling, it would have been necessary to stop the rotation of the beam with something like a crossbeam while it was still accelerating in order to achieve an effective release angle. Despite the impracticalities of such a scenario, it can be expressed in a simplified form to reflect an ideal throwing situation, where the centre of mass of the counterweight is in line with the long axis of the beam:

$$V \approx \sqrt{\frac{v[\sin(\theta - 90°) + \cos\alpha] - Mcw \cdot Lcw \cdot g(1 - \cos\alpha)}{\frac{1}{2}(Mp \cdot Lla^2 + Ib)}} \cdot Lla$$

Arrangement A yields a release velocity of 31 m/s at a release angle of 45°, or 37 m/s if release occurs at 0°.

24 $EK = \frac{Mp \cdot Lla^2}{I}$. With so little energy conveyed to the projectile, most of the remainder would cause a long period of beam swing, further decreasing the potential shooting rate.

25 See Appendix 2, nos. 63-65.

The addition of a sling to the rotating beam and hinged counterweight significantly complicates the mathematics. Furthermore, there arises the need to place a restriction on the initial path of the projectile, representing its horizontal slide before it is lifted into the air; without this, the projectile's natural arc would travel below its initial elevation. Rather than digging a trench to allow for this, using a low-friction trough increases the efficiency of the system. But as this represents a complete counterweight trebuchet, so too is it unfeasible to express this scenario algebraically and little more can be added to what has already been provided by Siano and Denny.[26] The last component to add is the extension over which the loose end of the sling is looped: the hook or low-friction 'finger'.[27] Although some analysts regard the angling of the finger as a significant design component,[28] personal experience has revealed that this is more practically a point of fine tuning.[29] Finally, theories surrounding the 'propping' of the counterweight, as seen in a number of fifteenth-century manuscripts,[30] will not be discussed; there is little to suggest that such a design feature was used during the period of the crusades, if ever it was effectively employed.

Using Mathematica, Siano sums up the comparative results between the various scenarios discussed above using *Arrangement C*, each provided with an optimal release angle for maximum range (Fig. 4).[31]

The addition of a hinged counterweight and sling significantly increase both range and efficiency; however, these values decrease notably when the beam is given mass. This reinforces how each mechanical component contributes to the overall efficiency of a trebuchet, although certain retarding elements, such as friction, have still been omitted.

At this point the decelerating effect of air resistance can be incorporated, determined by a shape-specific coefficient of drag, about 0.45 to 0.50 for a sphere:

$$\text{Horizontal deceleration} = \text{Vh}^2 \left(\frac{\text{C} \cdot \text{Ap} \cdot \rho}{2\text{Mp}} \right)$$

* Note, deceleration is expressed here as a positive value (rather than negative acceleration).

---

26 See Siano, *Trebuchet Mechanics*, pp. 15-20; Denny, "Siege Engine Dynamics," pp. 572-74.

27 For a mathematical representation of the 'finger', see Siano, *Trebuchet Mechanics*, pp. 25-31.

28 For example, Saimre, "Trebuchet," p. 71.

29 These sentiments are also shared by Peter Vemming, see Vemming Hansen, "Experimental Reconstruction," p. 202.

30 For examples of this, see Appendix 2, **nos.** 77, 79, 86 and 91.

31 Siano, *Trebuchet Mechanics*, p. 22.

FIGURE 4 *Arrangement C: varying levels of complexity (Siano)*

	Angle of Release (α)	Horizontal Range (Rh)	Time to Release (t)	Efficiency (Eff)
Beam-only	38.0°	11.70 m	0.40 s	11%
Hinged CW, no sling	22.6°	61.87 m	0.36 s	59%
Hinged CW, sling	18.4°	86.56 m	0.40 s	83%
Hinged CW, sling (with 2.27 kg beam)	19.3°	69.49 m	0.46 s	67%

Using a drag coefficient of 0.5, reflecting a near spherical projectile, the graph below demonstrates the rate of horizontal deceleration of projectiles with masses 5-60 kg (with a consistent density of 2,500 kg/m^3), shot at 55 m/s and 35 m/s at angles of 30° and 45°. The air is about 15°C, with a density of 1.2 kg/m^3; the air pressure is about 101.325 kPa and generally dry (Fig. 5).

The geometric relationship between a projectile's mass and its deceleration is significant: although much lighter projectiles are often selected to maximise release velocity, they will decelerate more rapidly than heavier projectiles. Accordingly, many modern reconstructed counterweight trebuchets shoot projectiles that weigh about 10 kg, which provide a balance between release velocity and in-flight performance.

Ideal Proportions

Hill's expression of the shooting sequence of a trebuchet with a hinged counterweight is somewhat simpler in approach than those of Denny and Siano. All three, however, are constructed around an attempt to maximise potential range. Assuming a 135° loading angle and 45° release angle, Hill bases his calculations on a 50-grade timber beam weighing 35 lbs/ft^3 (561 kg/m^3) and working fibre stress of 3,000 lbs/in^3 (83,039,700 kg/m^3). With this he constructed a compound pendulum to yield a theoretical range (using $Rh = \frac{V^2}{g}$).[32] When comparing his results with those achieved by the engine reconstructed by Louis-Napoleon and Favé in the nineteenth century, Hill determined that the efficiency of their reconstructed engine was about 56%. If neglecting air resistance, he believed a maximum efficiency of around 70% is achievable. Hill then used his calculations to model two postulated machines (Fig. 6).

32 Hill, "Trebuchets," pp. 112-13.

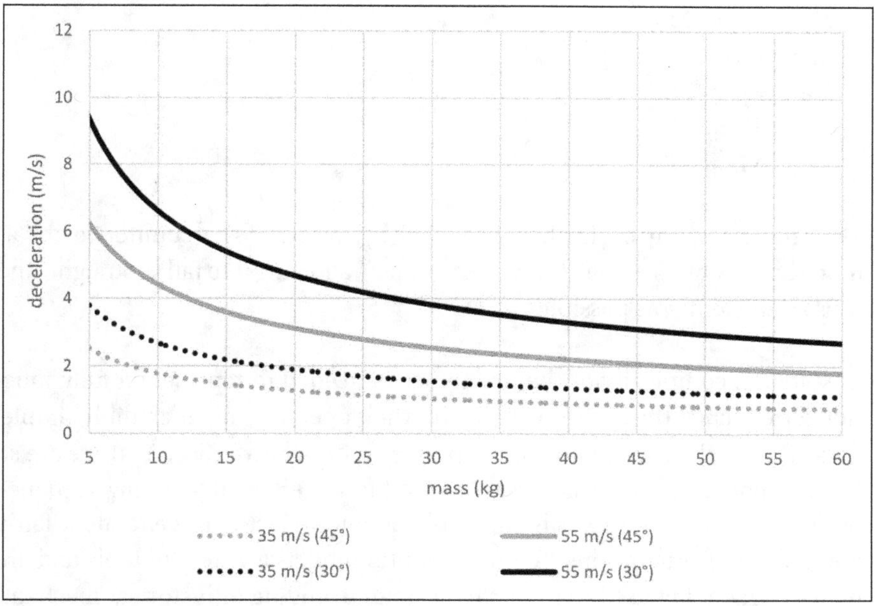

FIGURE 5 *Horizontal deceleration due to air resistance*

FIGURE 6 *Hill's proposed engines*

	Mp	Mcw	Lla	Ls	Rh
Machine 1	45.36 kg	4,535.92 kg	9.14 m	4.57 m	224.03 m
Machine 2	226.80 kg	9,071.85 kg	14.63 m	9.14 m	260.60 m

By adjusting the length of the short arm, preserving a ratio of between 6:1 and 5:1, these are the maximum theoretical ranges that Hill was able to achieve after factoring in an efficiency of 70%.[33]

Siano's conclusions agree, placing the maximum mechanical efficiency of a counterweight trebuchet at just over 70%. Summing up his findings, Siano provides a set of approximately ideal proportions that are based on the mass of the beam and length of the short arm:[34]

33 Marino Sanudo similarly recommended a ratio of 5:1 or 6:1 in the early fourteenth century, Marino Sanudo, *Liber secretorum* 2.4.22, ed. Bongars, pp. 79-80, trans. Lock, pp. 135-36.

34 Siano, *Trebuchet Mechanics*, p. 41.

$$
\begin{aligned}
Mp &= 0.2 \times Mb \\
Mcw &= 2.0 \times Mb \\
Lla &= 3.75 \times Lsa \\
Ls &= 3.75 \times Lsa \\
Hcw &= Lsa
\end{aligned}
$$

This, however, can be further improved by lowering the counterweight as much as possible, so long as Lax > (Lsa + Lcw), allowing it to fall as straight, and thus as efficiently, as possible.

Despite the complexity of the mathematics needed to represent a fully functional trebuchet, the basic mechanics of their operation are incredibly simple. This allowed them to be built and operated with relative ease. As there are so few components, each is necessary to produce an efficient shooting sequence but so too can they be easily manipulated and adjusted to generate a fairly optimal set of proportions. Today, computer models are helpful tools and are increasingly being used to improve our understanding of historical mechanical processes.[35] However helpful, these are no substitute for real-world recreations, which not only express actual forces that are near impossible to compute, such as friction, flex and wear, but convey the functional issues and challenges associated with construction, operation and maintenance.[36] Although an understanding of the mathematics is not necessary to operate a trebuchet, an appreciation of the governing physics will inevitably lead to a better understanding of the historical context and allow for a better evaluation of the sources.

35 For example, Kooi and Bergman, "An Approach to the Study of Ancient Archery," pp. 124-134.

36 For a selection of reconstructed trebuchets, see Appendix 4.

CHAPTER 3

First Crusade

No archaeological evidence survives to indicate the scale or form of the early stone-throwing siege engines used by the Franks or their Muslim counterparts during the First Crusade. We are therefore dependent on the surviving literary sources, which fortunately include a number of eyewitness accounts, even if the most detailed of them are all written from a Frankish point of view. Among the participants of the First Crusade, the anonymous author of the *Gesta Francorum* and Raymond of Aguilers composed detailed and influential accounts of their travels and experiences as they made their way to Jerusalem. Another participant, Fulcher of Chartres, was among those who broke off from the main army and accompanied Baldwin of Boulogne to Edessa prior to the siege of Antioch. Although not an eyewitness of the crusaders' activities once they left Anatolia, Fulcher, as well as others who were not present, including Albert of Aachen, who was back in Europe, and Ralph of Caen, who would make his way to Syria about a decade later, provide original details that they seem to have learned directly from participants who survived to recount their adventures.

Nicaea: 1097

Having crossed Europe, the various regional units that would make up the main wave of the First Crusade assembled at Constantinople. These forces were represented by a number of significant nobles: Bohemond of Taranto led the Italian Normans; the Alsatians were led by Godfrey of Bouillon; Count Raymond of St Gilles of Toulouse and Adhémar of Le Puy, a papal legate, headed the Provençal contingent. Hugh of Vermandois, Stephen of Blois, Robert of Normandy and Robert of Flanders each commanded another significant force.

Setting out from Constantinople, the first challenge that faced this army in Anatolia was the city of Nicaea. A stronghold of the Eastern portion of the Roman Empire since the fourth century, it had been acquired by the Seljuks less than two decades before the crusaders arrived. According to Raymond of Aguilers, Nicaea was surrounded "with such lofty walls that the city feared neither the attack of enemies nor the force of any machine."[1] Considerable stretches of

1 *Praeterea muris ita eminentibus cingitur, ut nullorum hominum assultus, nullis machinae impetus vereatur*, Raymond of Aguilers, *Historia Francorum* 3, RHC Oc 3, p. 239, trans. Hill and

 | DOI 10.1163/9789004376922_005

the city's classical defences can still be viewed today, although the walls have not survived to their original height.

The bulk of the Frankish army arrived at Nicaea on 6 May 1097, followed by Raymond of St Gilles and his Provençal forces a week later and a final contingent, under Robert of Normandy and Stephen of Blois, arrived in early June, just in time to see the city surrender.[2] The sources make it clear that artillery was used throughout the siege. The anonymous author of the *Gesta Francorum* and Guibert of Nogent relate that, soon after arriving, the main body of Franks began to construct machines to aid their early attacks.[3] Raymond of Aguilers and Albert of Aachen note the construction of further stone-throwers by the Provençal forces when they took up a position to the south of Nicaea.[4] This contingent arrived just in time to defeat a relief force sent by Kilij Arslan and wasted little time using their engines to hurl the heads of fallen Turks into the city.[5] Fulcher of Chartres, traveling with the forces of Stephen of Blois and Robert of Normandy, notes that *petrariae* and *tormenta* were also built by this group, the last to arrive.[6]

The use of artillery is widely noted by the sources, but so too is its limited power. Raymond of Aguilers clearly states that the stones thrown by the Provençal engines had minimal effect against the city's defences. Albert of Aachen confirms the initial ineffectiveness of the crusaders' artillery but adds that when the number of engines was increased, from just two, some cracks were

---

Hill, p. 25. For a detailed description of Nicaea's defences, see France, *Victory in the East*, pp. 143-44.

2 *Gesta Francorum* 2.7-8, ed. and trans. Hill, pp. 13-17; Raymond of Aguilers, *Historia Francorum* 3, RHC Oc 3, p. 239, trans. Hill and Hill, pp. 25-26; Fulcher of Chartres, *Historia Hierosolymitana* 1.9.2-10.1, ed. Hagenmeyer, pp. 177-81, trans. Ryan, pp. 79-81. The *Gesta* places Robert of Flanders at the siege before the arrival of the Provençals while Fulcher states he arrived with Stephen of Blois only days before the siege ended.

3 *Gesta Francorum* 2.8, ed. and trans. Hill, p. 14; Guibert of Nogent, *Die gesta per Francos* 3.6-10, ed. Huygens, pp. 145-55, trans. Levine, pp. 62-66.

4 Raymond of Aguilers, *Historia Francorum* 3, RHC Oc 3, p. 239, trans. Hill and Hill, p. 26; Albert of Aachen, *Historia Ierosolimitana* 2.32, ed. and trans. Edgington, pp. 114-15, cf. 2.29, 32, pp. 110-11, 116-17.

5 *Gesta Francorum* 2.8, ed. and trans. Hill, p. 15; Guibert of Nogent, *Die gesta per Francos* 3.6, ed. Huygens, pp. 145-47, trans. Levine, pp. 62-63; Roger of Wendover, *Flores historiarum*, ed. Coxe, 2:83, trans. Giles, 1:394. Cf. *Chanson d'Antioche* 74, trans. Edgington and Sweetenham, p. 146. This force had been sent by Kilij Arslan, who was then busy besieging Malatya. Matthew of Edessa notes Kilij Arslan's use of artillery against Malatya, Matthew of Edessa, *Patmut'iwn*, trans. Dostourian, pp. 163-64.

6 Fulcher of Chartres, *Historia Hierosolymitana* 1.10.6, ed. Hagenmeyer, pp. 185-87, trans. Ryan, p. 82.

opened and a few stones fell free of the wall, a point reiterated and embellished by William of Tyre in his later account.[7] A hexameter written by Robert the Monk states that the Franks threw stones and other blunt projectiles into the city, suggesting either that these were aimed at the parapet but missed or that targets inside the city were deliberately selected.[8]

The damage inflicted by the time that Stephen of Blois and Robert of Normandy arrived was caused by mining rather than artillery. Penthouses had been used to shelter sappers while nearby artillery provided additional cover, suppressing the garrison's ability to interrupt the work of the miners.[9] The use of artillery to support mining efforts will remain a recurring tactic through the following two centuries. The extent of the damage that the Franks were able to inflict is unclear as they were denied the opportunity to assault the sections of the city's walls that they had weakened. About seven weeks into the siege, to the surprise of the crusaders, the garrison capitulated to a Byzantine force that had made contact across the Ascanian Lake (Lake Iznik), the shores of which lapped the city's western walls at that time.[10]

What is perhaps most striking in this, the opening siege of the crusades, is the universal notice given to artillery in the contemporary accounts.[11] Most eyewitness sources note that the contingent of the army with which they were then associated constructed such engines almost as soon as it arrived. This implies that the various Frankish contingents had experience with artillery of the sort constructed at Nicaea before they had left for the East and that the use of such engines was nothing exceptional, suggesting knowledge of this technology was widespread in western Europe at the time. These appear to have been

7 William of Tyre, *Chronicon* 3.8-9 (9-10), ed. Huygens, 1:205-8, trans. Babcock and Krey 1:161-63.

8 Robert the Monk, *Historia Iherosolimitana* 3.3, RHC Oc 3, p. 756, trans. Sweetenham, p. 104.

9 Raymond of Aguilers, *Historia Francorum* 3, RHC Oc 3, p. 239, trans. Hill and Hill, p. 26; *Gesta Francorum* 2.8, ed. and trans. Hill, p. 15; Guibert of Nogent, *Die gesta per Francos* 3.6, ed. Huygens, pp. 145-47, trans. Levine, pp. 62-63; Ralph of Caen, *Gesta Tancredi* 17, RHC Oc 3, p. 615, trans. Bachrach and Bachrach, p. 40; Anna Comnena, *Alexiad* 11.1, trans. Sewter, pp. 333-35; Albert of Aachen, *Historia Ierosolimitana* 2.30, 34-36, ed. and trans. Edgington, pp. 112-13, 118-25.

10 Anna Comnena, *Alexiad* 11.2, trans. Sewter, pp. 337-38; *Gesta Francorum* 2.8, ed. and trans. Hill, pp. 16-17; Raymond of Aguilers, *Historia Francorum* 3, RHC Oc 3, pp. 239-40, trans. Hill and Hill, pp. 26-27. For an excellent analysis of the siege of Nicaea, see France, *Victory in the East*, pp. 143-69.

11 Later sources also tend to stress the use of artillery during this siege. For example, William of Tyre, *Chronicon* 3.6-9 (7-10), ed. Huygens, 1:203-8, trans. Babcock and Krey, 1:158-63; Roger of Wendover, *Flores historiarum*, ed. Coxe, 2:84, trans. Giles, 1:394.

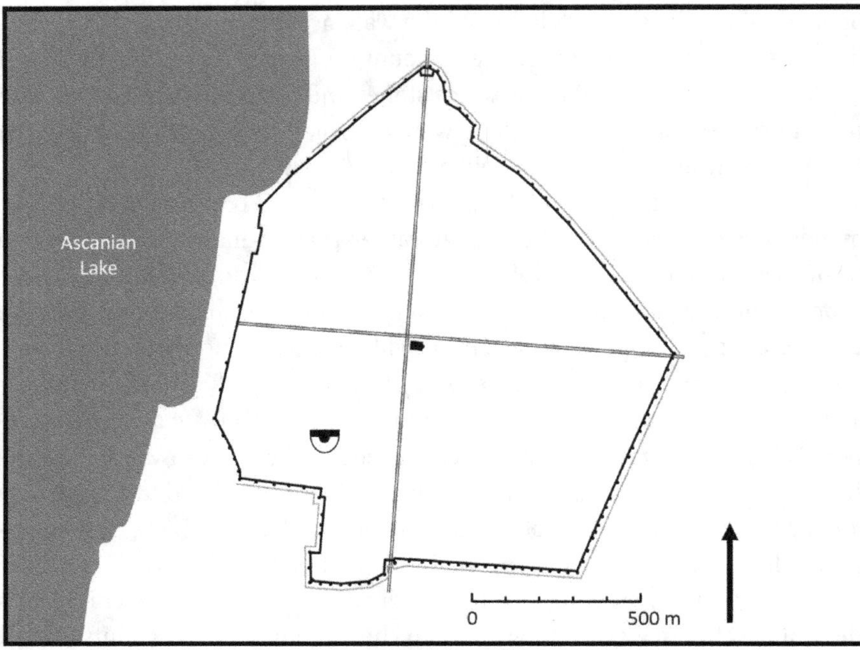

PLAN 1 *Nicaea: surviving town defences with later thirteenth-century forewall*

antipersonnel engines, used to suppress the activities of the defenders; however, after a month and a half, it seems reasonable to expect that a number of stones and bricks might have been knocked free from the battlements.

Although unable to threaten Nicaea's fortifications, the crusaders' artillery was powerful enough to throw human heads over the city's walls. If the walls were originally 10 m high, the heads would require a release velocity of at least 20 m/s to clear the battlements if thrown at an angle of 45°. This, however, would place an engine only slightly more than 20 m from the base of the wall ahead of it. Working from these figures, a rough scale of the trebuchets in question can be postulated. For example, a light beam 4.5 m long with a ratio of 5:1, fitted with a 1 m sling, a counterweight of 600 kg (representing approximately the pulling force of eight to ten men) and a projectile of 4.5 kg (representing a human head), when loaded at a horizontal angle, with the sling extending vertically, is theoretically capable of releasing a projectile at about 25 m/s at 45°, or 26 m/s at a slightly lower trajectory of 35° (Fig. 7).[12] This would allow the

12 This rough model has been calculated using the masses and angles involved, which can be expressed as the product of the projectile's acceleration (Ap) and time of the shooting sequence (t) if the acceleration of the short arm (Acw) is known:

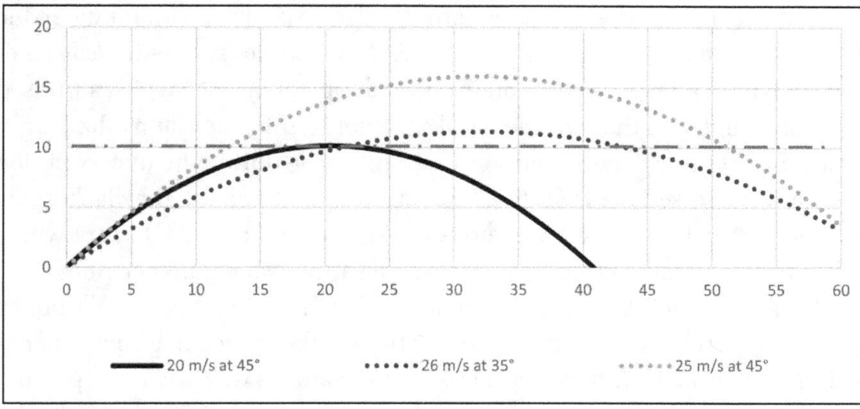

FIGURE 7 *Throwing heads*

Mass of projectile (Mp)	4.5 kg	4.5 kg	4.5 kg
Release velocity (V)	20 m/s	26 m/s	25 m/s
Release angle (α)	45°	35°	45°
Max. horizontal range (Rh)	41 m	65 m	64 m
Horizontal range at 10 m elevation (Rh when Rv = 10)	23 m	44 m	51 m
Kinetic energy (Ek)	900 J	1,521 J	1,406 J

engine to be placed at a slightly safer distance around 50 m from the wall. This is probably a fairly conservative model, as contemporary engines were almost certainly capable of release velocities exceeding 30 m/s.

Although the accounts of the siege appear to confirm that the Franks had prior knowledge of the artillery technology employed at Nicaea and that the power of these machines was modest, it has been argued that the Franks received counterweight trebuchets from the Byzantines before they left Constantinople. The basis of this suggestion is found in the testimony of Anna Comnena, who, despite being well-informed of events, composed her account many decades later and was not an eyewitness of the siege of Nicaea. Accord-

$$V \approx Ap \cdot t$$

$$\text{where } Ap = \frac{4\pi \cdot Ls\left(\frac{180-\psi}{360}\right)}{t^2} + \left(Acw \cdot \frac{Lla}{Lsa}\right)$$

$$t = \sqrt{4\frac{\left(Lsa \cdot \sin\left[\frac{1}{2}(\theta-\alpha)\right]\right)}{Acw}}$$

$$Acw = \frac{(Mcw \cdot g)-\left(Mp \cdot g\frac{Lla}{Lsa}\right)}{Mcw+Mp}$$

See Chapter 2 for a more detailed discussion of the mathematics, Chapter 5 for the use of this equation to model a counterweight engine, and Appendix 4 for trials with reconstructed traction engines.

ing to Anna, Emperor Alexius, her father, judged that even with innumerable forces the Franks would not be able to take Nicaea so "constructed *helepoleis* of several types, but mostly to an unorthodox design of his own which surprised everyone" and sent these to the Franks.[13] Prior to this, she claims the Franks made a brief attack on the walls of Constantinople during the uneasy period before they crossed the Hellespont but, at this point at least, they lacked any *helepoleis*.[14] This contrasts numerous examples earlier in her history where Byzantine commanders are found employing artillery and other engines.[15]

Using Anna's account, Chevedden has argued that Alexius personally built a number of trebuchet-type machines for the Franks, associating such engines with Anna's use of the term *helepoleis*.[16] This conflicts with the more popular interpretation of events, which is supported by Randall Rogers and John France. The latter historians suggest that Alexius might have helped the crusaders by providing materials and possibly even some advice, but that knowledge of the machines built at Nicaea was not foreign to the Franks and that it was almost certainly Latin craftsmen who built them.[17] Using Anna's praise for her father and his commission of 'various' and 'new' designs, Chevedden suggests this was the point at which the first counterweight trebuchet was invented, designed by no less a figure than Emperor Alexius himself.[18] Besides Anna's account, however, there is no evidence to justify such a leap. Thus although Greek aid should not be underemphasised in accounting for the early successes of the First Crusade, particularly with regard to provisioning, there is no reason to assume that the engines built by the Franks in Asia Minor were significantly different or any more powerful than contemporary engines built in southern Italy, greater France or western Germany.

As Rogers has pointed out, there are three essential ingredients necessary to build siege engines: timber, fasteners and manpower.[19] Timber was available in the environs of Nicaea and it seems fasteners, nails and rope, would have

13 Anna Comnena, *Alexiad* 11.2, trans. Sewter, p. 336.

14 Anna Comnena, *Alexiad* 10.9, trans. Sewter, pp. 319-20.

15 For example, Constantine Dalassenus's siege of Chios, Anna Comnena, *Alexiad* 7.8, trans. Sewter, p. 234.

16 Chevedden, "Invention," pp. 76-78. The term *helepolis*, 'city taker', appears to have been used as a fairly general term. Although most often associated with a siege tower, it was used by various sources at various times to identify a variety of engines. For suggestions that the term referred specifically to a counterweight trebuchet at this point, see Chevedden, "King James I," p. 316; Dennis, "Byzantine Heavy Artillery," pp. 99-115.

17 See Rogers, *Latin Siege Warfare*, pp. 21-22; France, *Victory in the East*, p. 165.

18 Chevedden, "Invention," p. 86.

19 Rogers, *Latin Siege Warfare*, pp. 21-22.

been more abundant at this point than at many later phases of the crusade. Latin labour was also in abundance. There is no evidence that a Byzantine engineering expert was present in the Frankish siege camp and the fact that the machines do not appear to have been disassembled and transported with the crusaders as they pushed east suggests that their designs were quite simple and unexceptional.

Construction of the Frankish artillery was most likely directed by members of certain baronial households. Sometimes these individuals were retainers with a demonstrated skill in this field; however, in many instances, construction was probably overseen by a major lord himself.[20] While more elaborate engines would require significant carpentry skills, the artillery and other engines built at Nicaea appear to have been small enough and sufficiently simple that only a moderate level of skill would have been required. The anonymous Lombard, who reportedly built a sturdy penthouse for the Lotharingians during the siege, was probably one of a number of freelance profiteers with transferable construction skills who often accompanied such large armies. The presence of such figures ensured that skilled labour was always available. Although it lasted seven weeks and saw the construction of a number of engines, the siege of Nicaea would prove to be a relatively easy success compared to the Franks' next challenge.

Antioch: 1097-98

Antioch had long been a strongly defended city. Geographically, the city is the first major urban centre south of the Nur Mountains (Amanus Mountains), which divide Anatolia from greater Syria. It is also midway between the Mediterranean and the Syrian Coastal Mountains (Jabal al-Nusayriyya), the northernmost extension of the mountain range marking the Great Rift in the Near East. Antioch had been under Muslim rule for much of the past four and a half centuries, but it remained a largely Christian city. It was often an important stronghold on the Muslim-Byzantine frontier and had been under continuous Turkish rule for little more than a decade when the crusaders arrived in October 1097.

Antioch had been captured a number of times over the preceding centuries. In most instances, the city had been taken by surprise attack or treachery –

20 For the case of Robert of Bellême and Robert of Normandy in the decade leading up to the First Crusade, see Orderic Vitalis, *Historiae ecclesiasticae* 8.16, 24, ed. and trans. Chibnall, 4:232-37, 288-95.

most damage sustained by the city's fortifications was the result of earthquakes rather than siege activities.[21] The walls were well provisioned with large towers, spaced less than 50 m apart across exposed sections of the enceinte. The formidable strength of this city's now dilapidated defences still impresses, as it certainly did when the Franks first viewed them. Raymond of Aguilers remarked that Antioch feared no besieging army, regardless of its size, a sentiment echoed by Fulcher of Chartres and Stephen of Blois.[22]

The siege began on 21 October 1097.[23] The Franks adopted a strategy of close blockade, employing siege forts to secure their positions and help interrupt the flow of men and materials in and out of the city.[24] In much the same way as it had fallen to the Turks, and the Byzantines before them, the city was betrayed to the Franks. After the siege had gone on for about eight months, the Franks, led by Bohemond, were able to enter the city one night in early June 1098. By the end of the following day, the only remaining point of resistance was the citadel on top of Mt Silpius. This final bastion surrendered a few days later,

---

21 The Sassanids took Antioch by storm, scaling its Roman walls, in the 540s. The city burnt and depopulated before the attackers withdrew but its defences were left intact. The city walls were subsequently developed in the reigns of Justinian, Zimisces and Basil II. The city fell under Muslim influence during the Arab expansion but was retaken by Nicephorus II Phocas in 968. About 1085, forces of the Seljuk sultan, Suleiman, took the city by surprise, those who had fled to the citadel were granted favourable terms after a short blockade, A.W. Lawrence, "A Skeletal History of Byzantine Fortification," pp. 196-97; Oman, *A History of the Art of War*, pp. 527-28; Matthew of Edessa, *Patmut'iwn*, trans. Dostourian, pp. 147-48. Oman discounts any notion that the defences of Antioch were altered or improved in any way while the city was under Muslim control.

22 Raymond of Aguilers, *Historia Francorum* 5, RHC Oc 3, p. 242, trans. Hill and Hill, p. 31. Fulcher of Chartres, *Historia Hierosolymitana* 1.15.2, ed. Hagenmeyer, p. 217, trans. Ryan, p. 92. *Antiochiam vero urbem maximam, ultra quam credi potest firmissimam atque inexpugnabilem*, Stephen of Blois, *Letters*, RHC Oc 3, p. 888. Although Fulcher was not at the siege of Antioch, he visited the city only a couple years later.

23 Fulcher of Chartres, *Historia Hierosolymitana* 1.15.1, ed. Hagenmeyer, pp. 215-17, trans. Ryan, p. 92; Albert of Aachen, *Historia Ierosolimitana* 3.36-38, ed. and trans. Edgington, pp. 196-201.

24 *Gesta Francorum* 5.12-13, 8.18-19, ed. and trans. Hill, pp. 29-30, 39-43; Raymond of Aguilers, *Historia Francorum* 7-8, RHC Oc 3, pp. 247-50, trans. Hill and Hill, pp. 41-46; Stephen of Blois, *Letters*, RHC Oc 3, pp. 889-90; Anselm of Ribemont, *Letters*, RHC Oc 3, pp. 890-93; letter of the people of Lucca on crusade to all faithful Christians, trans. Krey, in *The First Crusade*, pp. 161-62; Robert the Monk, *Historia Iherosolimitana* 5.4, RHC Oc 3, pp. 793-94, trans. Sweetenham, p. 138. Robert the Monk states that the third fort was an old castle, rather than a monastery, which Tancred repaired. The cloistered layout of a monastery, intended to shut out the world, was a naturally defensible type of structure and monasteries were used as counterforts before and after the time of the First Crusade.

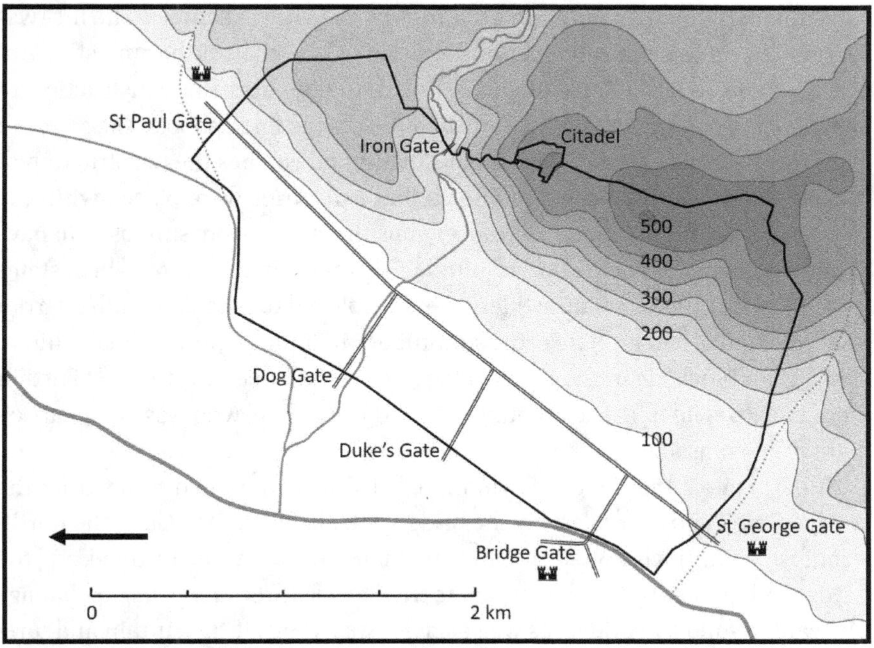

PLAN 2 *Antioch: town defences and siege forts of the First Crusade*

following the Franks' desperate victory over the relief army led by Kerbogha of Mosul.[25]

Unlike at Nicaea, siege machines figured very little into the crusaders' siege of Antioch. According to Guibert of Nogent, all sorts of siege engines were of no use and, had the city not been betrayed, the famine and other difficulties that the Franks had endured would have been for nothing.[26] Although descriptions of the siege of Antioch are far more substantial and detailed than those of Nicaea, references to artillery, or any siege machinery, are few. This, however, reflects the passive strategy that the Franks adopted given the size of the city and strength of its defences rather than the effectiveness or sophistication of their siege technology. The Franks had enough trouble containing the garrison's raids without having to defend numerous siege engines, which were often expensive to build and vulnerable to attack.

25 For the siege of Antioch and the battle that followed, see France, *Victory in the East*, pp. 197-296.

26 *Huic obsidendae quarumcumque machinarum balistarumve tormenta nil poterant*, Guibert of Nogent, *Die gesta per Francos* 6.16, ed. Huygens, p. 250, trans. Levine, p. 116.

Artillery is noted on only two occasions by eyewitness sources. The first was during efforts to stem sallies from the Bridge Gate, the southernmost of Antioch's three western gates. Raymond of Aguilers notes the construction of *machinae* in connection with efforts to build a siege fort opposite the gate, but calls the engines useless.[27] Anselm of Ribemont describes the construction of this fort in one of his letters and notes that *balistariae* were placed within it; however, it is unclear what type of weapons these men were supposed to have been employing.[28] The second is found in a passage in the *Gesta*, which states that *machinae*, presumably artillery, were employed to help defend the barrier that the Franks had built near the summit of Mt Silpius. This had been built to assist their blockade of the citadel after the city had been captured.[29] Further details are provided in the account of Albert of Aachen, who was not a participant of the crusade.

Albert claims that before Raymond of St Gilles attempted to obstruct the Bridge Gate, he tried to destroy the bridge outside of the Dog Gate, the northernmost of Antioch's western gates. In their third attempt to break up the bridge, Albert states that the crusaders tried to use three *mangenae* to damage the gate beyond the bridge, as well as nearby sections of the curtain and forewall. The engines, however, failed to make any impression upon the defences. Although Albert suggests that they were meant to destroy fortifications, these stone-throwers might instead have been used against the defenders opposing the crusaders, who employed their own *mangenae* and shot down on the Franks with arrows from the parapet. In any case, the Franks failed to destroy the bridge and resorted to obstructing it with felled trees and large stones.[30] It is noteworthy that in all three instances where artillery is mentioned it was used to help stem the egress of hostile defending forces.

---

27 Raymond of Aguilers, *Historia Francorum* 7, RHC Oc 3, p. 248, trans. Hill and Hill, p. 41.

28 Anselm of Ribemont, *Letters*, RHC Oc 3, pp. 890-93. It is possible that *balistaria* is meant to identify a machine rather than an operator; however, this is of no more help identifying the type of machine. The term was often associated with mechanical shooting devices, although these could range from traction trebuchets to crossbows depending on the source.

29 *Gesta Francorum* 9.26, ed. and trans. Hill, p. 62.

30 Albert of Aachen, *Historia Ierosolimitana* 3.40-42, ed. and trans. Edgington, pp. 202-5. Cf. *Chanson d'Antioche* 147, trans. Edgington and Sweetenham, p. 182. Peter Tudebode describes the use of a penthouse, similar to that used in Raymond's second effort against the Dog Gate, to damage the Fortified Bridge outside of the Bridge Gate in spring 1098, although he mentions no artillery supporting it, Peter Tudebode, *Historia de Hierosolymitano Itinere* 8.1, RHC Oc 3, pp. 50-51. Neither of these episodes are found in the *Gesta*.

For the defenders' use of artillery, it is again the writers who were absent who provide the most useful insights. In addition to the engines that were used against Raymond's efforts to destroy the bridge outside of the Dog Gate, Albert of Aachen claims that *mangenae* were used to dispatch the heads of Adalbero of Metz and a female companion, following their capture and execution.[31] Fulcher of Chartres also claims that the Muslims, using *petrariae* and *fundabula*, threw the heads of fallen Franks back to their comrades, while Robert the Monk ascribes this practice to the Franks, who employed *balistae* to do so.[32] Collectively, these accounts of throwing heads over the town walls suggest that the artillery used at Antioch was comparable to that employed at Nicaea.

In contrast to the accounts of the siege of Nicaea, it is the infrequency with which artillery is mentioned that is notable at Antioch. Neither the *Gesta* nor Ralph of Caen make any mention of artillery being used outside of the walls of Antioch and Fulcher of Chartres, who notes its use in defence, gives no indication that artillery was used against the city by the Franks. Thomas Asbridge has suggested that the Franks lacked the materials or craftsmen to construct siege engines at Antioch;[33] however, such an argument seems improbable. The Franks built notable engines at the preceding siege of Nicaea and subsequent sieges of Ma'arrat al-Nu'man, 'Arqa and Jerusalem, but at none of these would they seem to have had access to a body of skilled labour greater than that which passed through the nearby ports of Latakia and St Symeon during the siege of Antioch. Furthermore, the material and skill to build three siege forts was evidently present. The apparently modest scale and simplicity of contemporary artillery would have posed few building challenges to any who had seen, much less operated, such an engine previously. Rather than a shortage of skill or materials, artillery was simply less valuable or necessary given the tactics chosen at Antioch. At Nicaea, artillery had been used to support sappers, frontal attacks and an overall more aggressive siege strategy; it was far less useful during a fairly passive blockade.

31 Albert of Aachen, *Historia Ierosolimitana* 3.41, 46, ed. and trans. Edgington, pp. 202-3, 208-11.

32 Fulcher of Chartres, *Historia Hierosolymitana* 1.15.10, ed. Hagenmeyer, p. 221, trans. Ryan, p. 94; Robert the Monk, *Historia Iherosolimitana* 4.3, RHC Oc 3, p. 777, trans. Sweetenham, p. 123. Similar to Fulcher's account, William of Malmesbury claims that the garrison used *ballistae* and *petrariae* to throw the heads of Christian Syrians and Armenians, slain within the city, into the Frankish camp, William of Malmesbury, *De gestis rerum Anglorum* 4.361, ed. Stubbs, 2:417, trans. Giles, pp. 379-80. Cf. *Chanson d'Antioche* 187, trans. Edgington and Sweetenham, p. 209; Michael Attaleiates, *History* 28.4.6, ed. and trans. Kaldellis and Krallis, pp. 414-15.

33 Asbridge, *The Crusades*, p. 66.

Ma'arrat al-Nu'man and 'Arqa: 1098

The next significant siege of the First Crusade was that of Ma'arrat al-Nu'man.[34] On 27 November 1098, Raymond of St Gilles and Robert of Flanders encamped with a contingent of crusaders near the town's walls. An unsuccessful attempt was made to storm the town using ladders the following day and a second attempt failed a few days later, by which time Bohemond had arrived with a rival force.[35] A third assault was carried out on 11 December. This time, making use of a mobile siege tower in addition to ladders, the Franks appear to have gained a foothold on the parapet by nightfall and the town was pillaged the next day.[36]

Ralph of Caen, who was not present, appears to be the only source who claims that the Franks employed artillery at this siege.[37] The engines are portrayed as more powerful than in previous references but this seems to be a result of his poetic writing style. Although the Franks may not have employed artillery at Ma'arrat al-Nu'man, its defenders did. Raymond of Aguilers, the *Gesta*, Guibert of Nogent, Ralph of Caen and Robert the Monk all note the defenders' use of at least one stone-thrower to target the siege tower built by the Provençal forces.[38]

34 The siege of Albara has been omitted on the basis that the sources offer little insight into the techniques that were employed and the general lack of details.

35 Raymond of Aguilers, *Historia Francorum* 14, RHC Oc 3, p. 268, trans. Hill and Hill, pp. 75-76.

36 *Gesta Francorum* 10.33, ed. and trans. Hill, pp. 78-79; Raymond of Aguilers, *Historia Francorum* 14, RHC Oc 3, pp. 268-70, trans. Hill and Hill, pp. 76-79; Guibert of Nogent, *Die gesta per Francos* 6.18, ed. Huygens, pp. 251-54, trans. Levine, pp. 116-17; Albert of Aachen, *Historia Ierosolimitana* 5.30, ed. and trans. Edgington, pp. 374-77; Ralph of Caen, *Gesta Tancredi* 104, RHC Oc 3, p. 679, trans. Bachrach and Bachrach, pp. 121-22; Fulcher of Chartres, *Historia Hierosolymitana* 1.24.3, ed. Hagenmeyer, p. 261, trans. Ryan, p. 113; Robert the Monk, *Historia Iherosolimitana* 8.5-7, RHC Oc 3, pp. 846-49, trans. Sweetenham, pp. 183-86; Ibn al-Qalanisi, *Dhail*, trans. Gibb, p. 46; Ibn al-'Adim, *Zubdad*, RHC Or 3, pp. 586-87. Robert the Monk adds a battering ram to the equipment used by the Franks and Albert of Aachen specifies that the wood for the engines came from the castle of Talamria, in the Syrian mountains.

37 Ralph of Caen, *Gesta Tancredi* 96, 103-4, RHC Oc 3, pp. 674, 678-79, trans. Bachrach and Bachrach, pp. 115, 121-22. Cf. William of Tyre, *Chronicon* 7.9, ed. Huygens, 1:353-55, trans. Babcock and Krey, 1:310-12.

38 Raymond of Aguilers, *Historia Francorum* 14, RHC Oc 3, p. 269, trans. Hill and Hill, p. 78; *Gesta Francorum* 10.33, ed. and trans. Hill, p. 78; Guibert of Nogent, *Die gesta per Francos* 6.18, ed. Huygens, pp. 251-54, trans. Levine, pp. 116-17; Ralph of Caen, *Gesta Tancredi* 104, RHC Oc 3, p. 679, trans. Bachrach and Bachrach, pp. 121-22; Robert the Monk, *Historia Iherosolimitana* 8.5, RHC Oc 3, p. 846, trans. Sweetenham, p. 183.

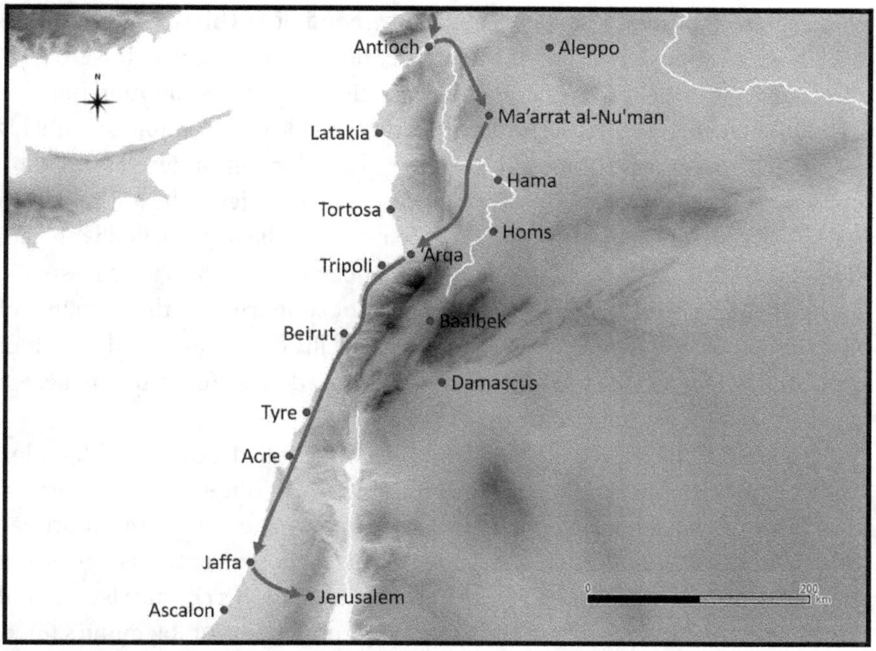

MAP 10 *First Crusade's movement through Syria and Palestine, 1097-99*

Following the capture of Ma'arrat al-Nu'man, Raymond of St Gilles, joined by Robert of Normandy and Tancred, headed south in January 1099. They took the interior road to the east of the Coastal Mountains, securing free passage past Shayzar and then Homs, where they turned west towards 'Arqa, Tripoli and the coast.[39] 'Arqa was an ancient town positioned on a tell at the north-western edge of the Lebanon, about 5 km from the coast and 25 km from Tripoli to the southwest. Unlike the urban strongholds hitherto besieged by the Franks, the topography surrounding 'Arqa offered no readily accessible line of approach. This made frontal attacks more difficult and denied the option of using a siege tower.

The siege began on 14 February 1099 and an artillery duel quickly developed, encouraged when Frankish mining efforts proved ineffective. Anslem of Ribemont was among those who fell amid this exchange.[40] Ralph of Caen states

39 Ibn al-Athir, *al-Kamil*, trans. Richards, 1:18.

40 Raymond of Aguilers, *Historia Francorum* 15-16, RHC Oc 3, pp. 275-78, trans. Hill and Hill, pp. 87-92; Ralph of Caen, *Gesta Tancredi* 105-10, RHC Oc 3, pp. 680-83, trans. Bachrach and Bachrach, pp. 123-27; Albert of Aachen, *Historia Ierosolimitana* 5.31-36, ed. and trans. Edgington, pp. 376-87; Fulcher of Chartres, *Historia Hierosolymitana* 1.25.1-3, ed. Hagenmeyer,

that each of the three Frankish leaders, Raymond of St Gilles, Robert of Normandy and Tancred, oversaw the construction of *tormenta*, which were then directed against the town.[41] His further assertion that it was common practice to build such engines and the implication that each leader was able to build his own machines, as was the case at Nicaea, appears to confirm that this technology was well known across western Europe by the late eleventh century. Given Ralph's free writing style, however, his account must be approached with caution. As at Ma'arrat al-Nu'man, the degree of destruction that he suggests the Frankish artillery was capable of causing is not supported by the accounts of his contemporaries. The Frankish engines posed little real threat to the walls of 'Arqa and, when mining proved difficult, the crusaders were content to accept a payoff when it was offered by the emir of Tripoli.

Meanwhile, the stalled siege of 'Arqa had compelled Godfrey of Bouillon and Robert of Flanders to join their comrades. Having left Antioch for Latakia, they besieged Jabala during the first half of March. When false reports arrived that their fellow crusaders at 'Arqa were in danger, they accepted a large payoff and headed south. Unfortunately we know little about this coastal siege, nor of the attacks made on Maraclea and Tortosa by contingents of Raymond's force. In May 1099, following the arrival of Godfrey and Robert of Flanders, the crusaders resumed their march to Jerusalem. 'Arqa remained under Muslim rule for another ten years.

Jerusalem: 1099

The conquest of Jerusalem appears to have been the accepted goal of the First Crusade by the time the various contingents set out across Europe; however, others also coveted the holy city. With the distraction caused by the Frankish presence in Syria, al-Afdal, vizier of Fatimid Egypt, besieged Jerusalem in July 1098, at that time ruled by a Seljuk emir.[42] Both Ibn al-Qalanisi and the later account of Ibn al-Athir emphasise the use of artillery and claim that the city's

---

pp. 265-67, trans. Ryan, p. 113; Guibert of Nogent, *Die gesta per Francos* 6.23, ed. Huygens, pp. 264-65, trans. Levine, p. 122; *Gesta Francorum* 10.34-35, ed. and trans. Hill, pp. 83-84. Cf. Ralph of Caen, *Gesta Tancredi* 106, RHC Oc 3, pp. 680-81, trans. Bachrach and Bachrach, p. 123.

41 Ralph of Caen, *Gesta Tancredi* 105, RHC Oc 3, p. 680, trans. Bachrach and Bachrach, p. 123.

42 It may be presumptuous to tie this attack directly to the fall of Antioch. The amount of time required to assemble the Fatimid army meant that this process likely started before the city fell, although news of this blow to the 'Abbasids no doubt encouraged their preparations and strengthened their resolve. Fatimid forces appear to have required about two

fall was a direct result of the damage that it inflicted.[43] Although it is tempting to see these Muslim engines as superior to their Frankish counterparts, subsequent Muslim sieges suggest that this was not necessarily so.

Situated in the hills of Judea, between the coastal plain and the Jordan Valley, the trace of Jerusalem's walls had changed over the centuries. By the time that the Franks arrived, the walls stretched 4 km along a line that roughly corresponds with the present Ottoman trace. The northern front of the city extended for about 1.2 km and was naturally approachable, standing at a slightly lower elevation than the remainder of the plateau which extends farther north. Under Muslim rule, the Roman defences along this line had been strengthened with the addition of a forewall and a rock-cut fosse.[44] The eastern side of the city, facing the Kidron Valley, was inaccessible to a sizeable force given the gradient of the topography. The southern half of the western face was also difficult to approach, while the northern half was anchored by two strong towers: the city's citadel, known as the Tower of David, in the centre of Jerusalem's western defences, and the Quadrangular Tower (Tancred's Tower or Tower of Goliath), which secured the northwest corner of the city.[45] To the south, the city could be approached along a constricted path that led northeast to the Tanners' Gate, just west of al-Haram al-Sharif/the Temple Mount, or from the plateau that extended south from the western end of the southern defences. This plateau, about 150 m square, was the southern extension of Mt Zion. Both areas had been left exposed when the line of the southern city walls was redrawn under Muslim rule.[46] From the Frankish sources it appears that a ditch had been added along at least the Mount Zion portion of this front by 1099.

---

months to assemble and then march to Ascalon around the start of the twelfth century, Hamblin, *The Fatimid Army*, pp. 180-85.

43 Ibn al-Qalanisi, *Dhail*, trans. Gibb, p. 45; Ibn al-Athir, *al-Kamil*, trans. Richards, 1:21. Even Albert of Aachen notes the use of artillery during the Fatimid siege of Jerusalem, though his source for this information is unknown, Albert of Aachen, *Historia Ierosolimitana* 6.31, ed. and trans. Edgington, pp. 442-43.

44 For descriptions of the medieval defences of Jerusalem and excavation findings, see Boas, *Jerusalem*, pp. 46-49; Pringle, "Town Defences," pp. 80-81; Weksler-Bdolah, "The Fortification System"; Mazor, "The Northern City Wall of Jerusalem"; Reich and Shukron, "Excavations in the Mamillah Area"; Broschi and Gibson, "Excavations along the Western and Southern Walls"; Broschi, "Mount Zion," p. 285.

45 For a contemporary description of the citadel, see Fulcher of Chartres, *Historia Hierosolymitana* 1.26.4, ed. Hagenmeyer, pp. 284-85, trans. Ryan, p. 117. Fulcher claims that the stones were fused with lead, a practice described by Vitruvius, Vitruvius, *De architectura*, 2.8.4, ed. Krohn, trans. Morgan, pp. 51-52. Cf. Boas, *Jerusalem*, pp. 66-75.

46 Prawer has suggested this may have occurred during the latter half of the tenth century or following an earthquake around 1033, Prawer, "The Jerusalem the Crusaders Captured,"

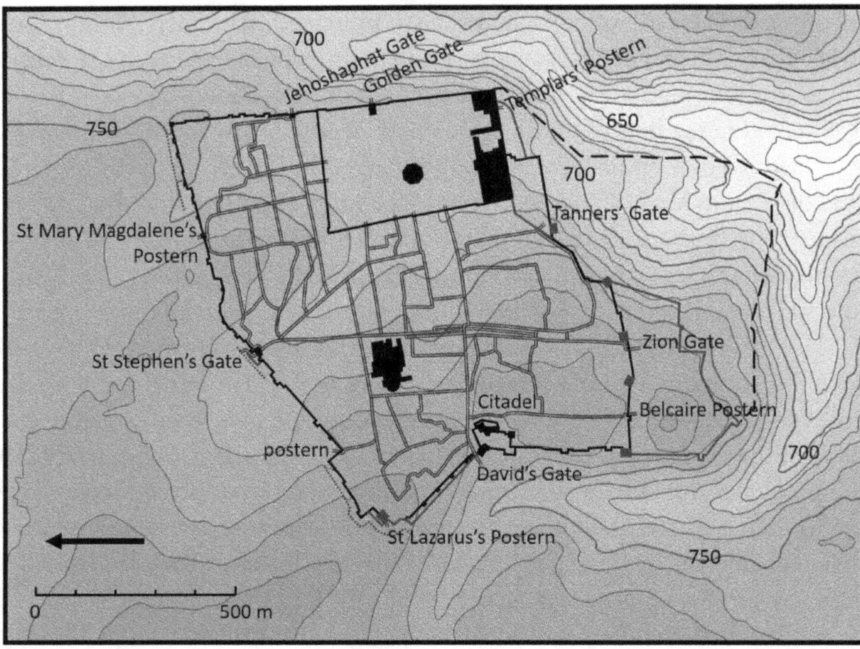

PLAN 3 *Jerusalem: present Ottoman walls with earlier discernable defences, including the Ayyubid extension around Mount Zion and the ancient wall around the lower city*

As the Franks approached Jerusalem, the population of the surrounding region sought refuge within the city. The garrison gathered supplies, equipment and whatever timber they could from the surrounding area, burning the rest and poisoning the region's wells.[47] The Christian population of Jerusalem may also have been expelled to eliminate the risk of a fifth column, as might have been done at Antioch two years earlier.

When the crusaders arrived on 7 June 1099, they immediately set about establishing a siege, dividing themselves into two groups. The first deployed along the western half of the northern wall: Robert of Normandy positioned himself near the church of St Stephen, opposing the St Stephen's Gate (Damascus Gate) with Robert of Flanders to his right and Godfrey and Tancred stationed around the Quadrangular tower. Raymond of St Gilles initially opposed

p. 2. Boas and Pringle believe that the redrawing of the southern line took place in the eleventh century, Boas, *Jerusalem*, pp. 43-44; Pringle, "Town Defences", p. 79. Later, during the Ayyubid period, Mount Zion was completely enclosed in an effort to strengthen this section of the southern front.

47 Benvenisti estimates the population of Jerusalem may have swelled to 40,000 at this point, Benvenisti, *The Crusaders*, pp. 35-36.

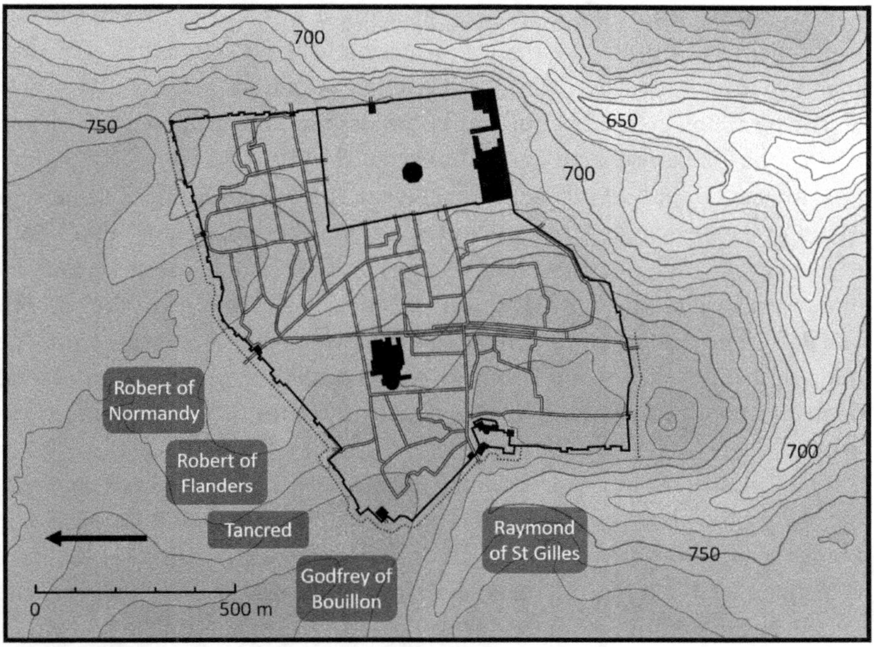

PLAN 4 *Jerusalem: initial deployment of the crusaders in 1099*

David's Gate (Jaffa Gate) but, given the restrictions of the terrain in this area, he soon repositioned his forces to the south, establishing them on Mt Zion near the church of St Mary of Mt Zion.[48] Although a more suitable position from which to assail the city, the restrictions of the plateau forced this group to encamp fairly close to the city's walls and may have kept them within range of the defending archers.

On 13 June 1099, the Franks made their first assault. Although the northern force was able to gain the forewall, their efforts to take the main wall failed.[49]

48 *Gesta Francorum* 10.37, ed. and trans. Hill, p. 87; Peter Tudebode, *Historia de Hierosolymitano Itinere* 14.1, RHC Oc 3, p. 102; Raymond of Aguilers, *Historia Francorum* 20, RHC Oc 3, pp. 292-93, trans. Hill and Hill, pp. 116-17; Albert of Aachen, *Historia Ierosolimitana* 5.46, ed. and trans. Edgington, pp. 402-5.

49 *Gesta Francorum* 10.37, ed. and trans. Hill, p. 88; Raymond of Aguilers, *Historia Francorum* 20, RHC Oc 3, p. 293, trans. Hill and Hill, p. 117; Fulcher of Chartres, *Historia Hierosolymitana* 1.27.2, ed. Hagenmeyer, pp. 293-94, trans. Ryan, p. 119; Albert of Aachen, *Historia Ierosolimitana* 6.1, ed. and trans. Edgington, pp. 406-7; Ralph of Caen, *Gesta Tancredi* 118-119, RHC Oc 3, pp. 688-89, trans. Bachrach and Bachrach, pp. 134-36. Ralph of Caen gives a mystical account of Tancred receiving direction from a monk to attack the city on this day and his discovery of a cache of hidden lumber, which proved to be just enough for the single siege ladder. It is more likely that the Franks had multiple ladders, allowing them to

Following this test of the defenders' strength, it was agreed that machinery would be necessary to take the city, resulting in the construction of no less than two towers, artillery, a ram and a number of ladders.

The terminology used by the sources to describe the artillery used by and against the Franks is characteristically vague. The terms *machinae* and *mangenae* are used by Albert of Aachen to refer to the Franks' artillery while Raymond of Aguilers adds *tormenta* and *petrariae* when identifying that of the garrison.[50] Writing decades later, William of Tyre uses the terms *mangana* and *petrariae* and Roger of Wendover calls these machines *petrariae* and *trubuculi*.[51] None of the sources provide a detailed mechanical description. The main difficulty faced by the Franks when constructing their artillery was the need for wood, a far greater challenge when it came to erecting their siege towers.

Finding tall and straight trees in the Judean hills would have been challenging even if the area was more fertile in the Middle Ages than it is now.[52] The Franks scoured the area as far as Nablus to find wood, possibly aided by local Christians who had recently been evicted from Jerusalem.[53] To this end, the Franks benefited from the timely arrival of a Genoese fleet at Jaffa on 17 June. Trapped in the harbour by Fatimid vessels, the Genoese party, under William Embriaco, was escorted to Jerusalem by a detachment of crusaders.[54] Critically, the force also brought back timber that had been stripped from the Genoese

---

take the forewall, but not enough of sufficient scale to take the main wall. The region's local timber is predominantly short and rarely straight – perhaps only one of the Frankish ladders was of sufficient height and strength to provide access to the parapet of the primary curtain.

50 Albert of Aachen, *Historia Ierosolimitana* 6.2, ed. and trans. Edgington, pp. 406-7; Raymond of Aguilers, *Historia Francorum* 20, RHC Oc 3, pp. 298-99, trans. Hill and Hill, pp. 125-26.

51 William of Tyre, *Chronicon* 8.6, ed. Huygens, 1:392-93, trans. Babcock and Krey, 1:351; Roger of Wendover, *Flores historiarum*, ed. Coxe, 2:135, trans. Giles, 1:429.

52 Adrian Boas has suggested that this may have been exacerbated by the series of sieges which had repeatedly consumed the surrounding timber over the preceding millennium, Boas, *Jerusalem*, pp. 5-7.

53 Ralph of Caen, *Gesta Tancredi* 123, RHC Oc 3, p. 691, trans. Bachrach and Bachrach, p. 139; Peter Tudebode, *Historia de Hierosolymitano Itinere* 15.1-3, RHC Oc 3, pp. 106-7; Albert of Aachen, *Historia Ierosolimitana* 6.2-4, ed. and trans. Edgington, pp. 406-9; *Gesta Francorum* 10.38, ed. and trans. Hill, p. 90.

54 Raymond of Aguilers, *Historia Francorum* 20, RHC Oc 3, pp. 294-95, trans. Hill and Hill, pp. 119-20; Caffaro, *De liberatione*, ed. Belgrano, pp. 110-11, trans. Hall and Phillips, pp. 116-17; Peter Tudebode, *Historia de Hierosolymitano Itinere* 14.1-2, RHC Oc 3, pp. 103-4; *Gesta Francorum* 10.37, ed. and trans. Hill, pp. 88-89. Raymond adds that the tools which the Genoese

ships. The long beams, oars and even the ships' fasteners may have been regarded as divinely delivered, while Provençal carpenters benefitted from the skill of the sailors who helped them construct their siege machines.

As at Ma'arrat al-Nu'man, the Franks oriented their strategy around siege towers. Godfrey oversaw the construction of one tower on the northern side of the city, charging Gaston of Béarn with this task, while Raymond of St Gilles saw to the erection of another, under William Ricau, on Mount Zion.[55] Constructed primarily from the wood taken from the Genoese vessels, Raymond's tower would have been as sturdy as the conditions allowed.[56] The defenders appear to have viewed this tower as the more threatening, as Albert of Aachen states that nine *mangenae* were arrayed against the Provençal tower, while just five targeted Godfrey's tower.[57] The defenders' engines threw both stones and Greek fire; the shortage of water that afflicted the crusader army also posed a serious threat to their machines.[58]

When the Franks had completed their siege towers and supporting artillery, as well as a ram, mantlets and ladders, they set about coordinating an attack. The defenders of Jerusalem had strengthened their fortifications across from the northern siege tower as best they could. They added timber defences and raised the height of their towers with wooden turrets, mitigating the elevation advantage that an opposing siege tower would enjoy. This compelled the northern assailants to attempt a daring manoeuvre: moving their tower from its original position, near the Quadrangular Tower, about a kilometre to the

---

brought proved invaluable in the construction of the southern siege tower, Raymond of Aguilers, *Historia Francorum* 20, RHC Oc 3, p. 298, trans. Hill and Hill, p.125.

55 Raymond of Aguilers, *Historia Francorum* 20, RHC Oc 3, p. 297, trans. Hill and Hill, pp. 123-24; *Gesta Francorum* 10.37, ed. and trans. Hill, p. 88. Large parties of captured and coerced Muslim labourers may have taken part in handling the large beams, salvaged from the Genoese ships, which were used to construct the Provençal tower.

56 Apparently using William of Tyre's account, some have added the construction of a second northern tower, bringing the total to three siege towers, Bradbury, *The Medieval Siege*, p. 116; Benvenisti, *The Crusaders*, pp. 37-38; William of Tyre, *Chronicon* 8.12, ed. Huygens, 1:401-2, trans. Babcock and Krey, 1:560-61.

57 Albert of Aachen, *Historia Ierosolimitana* 6.15, ed. and trans. Edgington, pp. 422-23. Ralph of Caen and Fulcher of Chartres also make it clear that a number of defensive engines were used to target the northern tower, Ralph of Caen, *Gesta Tancredi* 123, RHC Oc 3, p. 691, trans. Bachrach and Bachrach, p. 139; Fulcher of Chartres, *Historia Hierosolymitana* 1.27.6, ed. Hagenmeyer, p. 296, trans. Ryan, p. 120.

58 Raymond of Aguilers, *Historia Francorum* 20, RHC Oc 3, pp. 293-94, trans. Hill and Hill, pp. 118-19; *Gesta Francorum* 10.37, ed. and trans. Hill, p. 89.

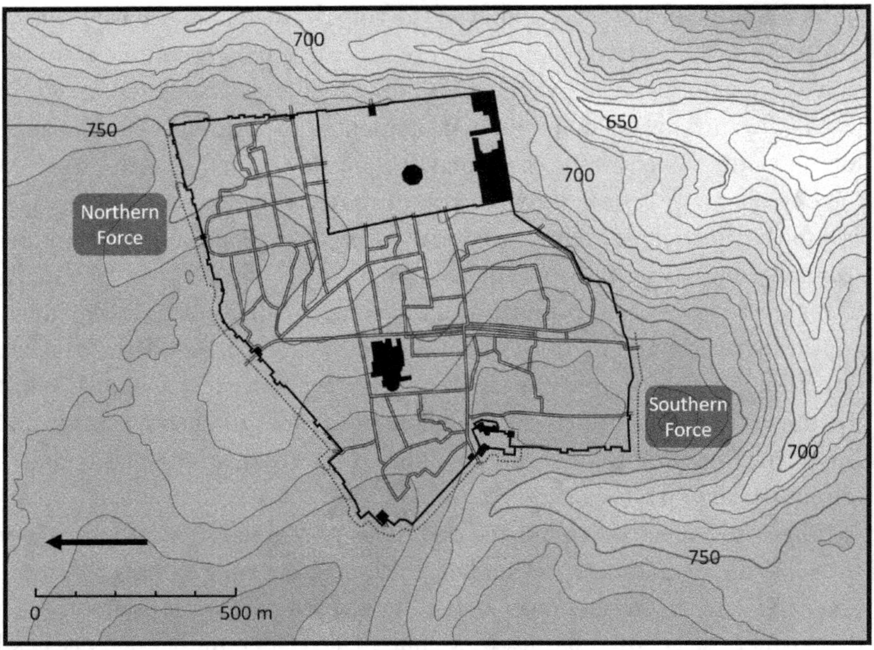

PLAN 5 *Jerusalem: final deployment of the crusaders in 1099*

east during the night of 9-10 July.[59] As this tower had not received any of the sturdy beams brought by the Genoese, the northerners were able to disassemble the tower and move it to its new location in sections.[60] Although not an eyewitness to these events, Albert of Aachen states that the Franks moved their *mangenae*, and accompanying piles of ammunition, along with the

59 *Gesta Francorum* 10.38, ed. and trans. Hill, p. 90; Raymond of Aguilers, *Historia Francorum* 20, RHC Oc 3, p. 298, trans. Hill and Hill, pp. 124-25; Fulcher of Chartres, *Historia Hierosolymitana* 1.27.5, ed. Hagenmeyer, pp. 295-96, trans. Ryan, p. 120; Peter Tudebode, *Historia de Hierosolymitano Itinere* 15.2, RHC Oc 3, pp. 107-8; Guibert of Nogent, *Die gesta per Francos* 7.6, ed. Huygens, pp. 275-78, trans. Levine, pp. 128-29; Ralph of Caen, *Gesta Tancredi* 123, RHC Oc 3, p. 691, trans. Bachrach and Bachrach, p. 139; Albert of Aachen, *Historia Ierosolimitana* 6.9, ed. and trans. Edgington, pp. 414-15.

60 It seems unlikely that the tower was designed so that it could be disassembled, as it took three days just to reassemble it once it was moved. Ralph of Caen claims that this manoeuvre was planned all along and the theory has since been supported by Thomas Asbridge, who suggests that the move took place on the night of July 13-14, rejecting the dating found in the *Gesta* and related sources, Asbridge, *The Crusades*, pp. 96-99. The ambiguous timing surrounding the construction of Robert of Normandy's tower at Courcy in January 1091, might suggest that such a tactic had been attempted before, Orderic Vitalis, *Historiae ecclesiasticae* 8.16, ed. Chibnall, 4:232-35.

tower. Three of these stone-throwers were then erected in front of the tower's new position.[61]

Supported by their artillery, the northern force of Franks used their ram to batter the forewall on 13-14 July.[62] The shaky siege tower was pushed up behind the ram and on Friday 15 July it was brought up against the main wall. In addition to denying the defenders along the rampart the protection of their battlements, the tower allowed the Franks to bridge the gap and gain access to the parapet. As the first Franks began stepping onto the wall, others were encouraged to bring up ladders and the northern defences were successfully stormed.[63] To the south, the Provençal contingent launched a corresponding attack.

The southern force met stiff opposition from the defenders' artillery as they attempted to finish filling a section of the fosse ahead of their siege tower on 14 and 15 July. Unlike north of the city, there was no room to manoeuvre the southern tower, leaving the defenders free to concentrate and then calibrate their artillery. The barrage of stones thrown against the sturdy tower had taken a toll by the time it was pushed forward on 15 July. In addition to stones, the defenders used their artillery to throw incendiaries as the tower drew closer. It seems these combustibles proved effective and the crippled tower was on the verge of being withdrawn when news arrived around noon that elements of the northern force had broken into the city. Encouraged once more and with chaos consuming Jerusalem and its defenders, the Provençal besiegers abandoned their tower and instead used ladders to force their way up to the parapet.[64] From about midday on 15 July 1099, Jerusalem was subject to a thorough and brutal sack.

---

61 Albert of Aachen, *Historia Ierosolimitana* 6.9, ed. and trans. Edgington, pp. 414-15.

62 Rams create fractures in the masonry ahead of them, crushing rather than puncturing. The fractures form seems, so most of the mass that is removed is extracted by men with picks and prybars, rather than being smashed out of place by the machine itself. Because such engines are extremely cumbersome, France has argued, building on Albert of Aachen's account, that the ram used by the northern Franks was burnt *in situ* to clear its bulk from the path of the approaching tower, France, *Victory in the East*, pp. 349-50.

63 Albert of Aachen, *Historia Ierosolimitana* 6.9-12, 16-19, ed. and trans. Edgington, pp. 414-19, 422-29; Fulcher of Chartres, *Historia Hierosolymitana* 1.27.5-10, ed. Hagenmeyer, pp. 295-99, trans. Ryan, pp. 120-21; Ralph of Caen, *Gesta Tancredi* 124-27, RHC Oc 3, pp. 691-94, trans. Bachrach and Bachrach, pp. 139-43; Raymond of Aguilers, *Historia Francorum* 20, RHC Oc 3, pp. 299-300, trans. Hill and Hill, p. 127; *Gesta Francorum* 10.38, ed. and trans. Hill, pp. 90-92.

64 Raymond of Aguilers, *Historia Francorum* 20, RHC Oc 3, pp. 298-99, trans. Hill and Hill, pp. 125-27; *Gesta Francorum* 10.38, ed. and trans. Hill, pp. 90-91; Peter Tudebode, *Historia de Hierosolymitano Itinere* 15.3, RHC Oc 3, pp. 108-9; Fulcher of Chartres, *Historia Hierosolymitana* 1.27.7, 11, ed. Hagenmeyer, pp. 296-97, 299-300, trans. Ryan, pp. 120-21; Albert of

The Franks' Artillery

Artillery was an important tool used by both the crusaders and Jerusalem's defenders. As with previous sieges of the First Crusade, the best indications of its power are somewhat morbid. The most illustrative descriptions of the capabilities of these engines are associated with an episode related by Peter Tudebode and Albert of Aachen concerning the treatment of a Muslim spy.

> *Peter Tudebode*:
> They took him, with bound hands and feet, and put him in the sling of a machine called a *petrera* and intending with all their strength to hurl him into the city, they were unable to do so. For with such force it happened that his bonds broke before he arrived at the wall of the city and he was dismembered.[65]

> *Albert of Aachen*:
> After this and other reports, the messenger was handed back to the soldiers, and he was put into the *tormentum* of one of the mangonels with hands and feet bound, so that he would be thrown over the walls after the first and second charge. But the skin of the mangonel was too heavily weighed down by the weight of his body and did not throw the wretch far. He soon fell onto sharp stones near the walls, broke his neck, his nerves and bones, and is reported to have died instantly.[66]

Aachen, *Historia Ierosolimitana* 6.16-17, ed. and trans. Edgington, pp. 422-25. Raymond of Aguilers' account of the southern push is the most detailed but contains some conflicting information. For example, he notes the existence of a forewall, which none of the other sources mention and for which there is no archaeological evidence. He also appears to suggest that the Provençal force built more than one siege tower; although possible, this is not corroborated by the other sources.

65 This author's translation, *atque eum acceptum, ligatis manibus ac pedibus, posuerent eum in funda cuzusdam ingenii, quod petrera vocatur, atque cum omnibus viribus suis cogitantes eum projicere infra civitatem, nequiverunt. Nam cum tanto impetu venit, quod, ruptis vinculis, antequam ad murum pervenisset civitatis, dilaceratus est.* Peter Tudebode, *Historia de Hierosolymitano Itinere* 15.2, RHC Oc 3, p. 107.

66 Adapted from Edgington's translation, *Post hanc et ceteras relationes militibus restitutus, tormento cuiusdam mangene ligatis manibus et pedibus est inmissus ut sic post primam et secundam inundationem trans muros iataretur. Sed nimio corporis eius pondere pellis mangene grauata, non longe miserum proiecit. Qui mox iuxta muros corruens super asperos silices, fractis ceruicibus, neruis et ossibus, in momento extinctus fuisse refertur.* Albert of Aachen, *Historia Ierosolimitana* 6.14, ed. and trans. Edgington, pp. 422-23.

The machine described here is a traction trebuchet. The unfortunate spy was apparently placed in the sling, which was attached to the end of the long arm of the beam, and the men pulling were hauling on ropes attached to the end of the short arm. If the machine here was of the same dimensions as that which has been postulated may have been used against Nicaea, but given a larger pulling crew, equivalent to 1,000 kg, the prisoner could have been theoretically accelerated to about 20 m/s. This would subject the wretch to a flight of around 40 m, or slightly farther if he landed at a lower elevation, as might have happened if he were thrown into the town ditch.

In reality, however, there would have been a number of challenges facing any attempt to throw a person with a traction trebuchet. First, there is the restricted length of the sling to consider: most slings were no longer than half the length of the long arm, which would cause problems when attempting to load a person more than 1.5 m tall, even if he were curled up as much as possible. Second, there is the size and irregular shape of the person. Slings were typically designed to throw projectiles about the size of large citrus fruit and, even if a sufficiently large sling was employed, the irregular shape and mass of the individual would cause issues through the shooting sequence. Finally, there is the considerable mass of the individual. The model used above, and earlier in the context of the siege of Nicaea, assumes that the sling will have no trouble rotating, despite the mass of the projectile that it carries. A more complex model would reveal more accurately the effect of the person's mass as it inhibited the rotation of the sling, most likely resulting in a failed shooting sequence.

Even if the anecdote is fictitious, the sources' descriptions of the difficulties associated with throwing this person are well founded, as are their remarks of the violence to which the spy was subjected even before he landed. Informal testing carried out with a reconstructed counterweight trebuchet, which was used to throw conglomerate masses of masonry, demonstrated the magnitude of the reciprocal centrifugal and centripetal forces to which projectiles are exposed as they rapidly accelerate before release. The chunks of masonry that were loaded into the sling of this engine were often torn apart through the course of the shooting sequence, ripping stone from mortar before the point of release.[67]

There is little to suggest that the engines used at Jerusalem would have been sufficiently powerful to effectively throw a projectile as heavy as a person.[68]

67 My thanks to the operators of the trebuchets at Caerphilly castle for sharing the results of this experiment with me.

68 Centuries later, John Froissart provides an account of a trebuchet being used successfully to throw an emissary back into Auberoche during the Hundred Years War, Froissart, *Chroniques* 1.1.218, ed. Luce, 3:65.

There are no indications that the Franks employed counterweight trebuchets at this siege and, given the limitations of traction power, it seems unlikely, or at least impractical, that a mass this great, let alone of an irregular size and shape, was successfully thrown towards the walls of the city. Considering the limited availability of suitable timber, and lack of any reference to wood brought from the Genoese ships being used to construct artillery, the Frankish engines were unlikely any stronger than those that had been used against Nicaea or Antioch. As at these earlier sieges, the Franks used their artillery to target Jerusalem's defenders, rather than its defences.

Albert of Aachen and Ralph of Caen claim that sacks were hung from the walls to lessen the impact of the Frankish artillery. While this may have been a reality, Albert, and Ralph following him, may have been inspired by the more common practice of using padding to protect siege towers from artillery and masonry fortifications from battering rams. In the account of Fulcher of Chartres, who like Albert and Ralph was not present at the siege, the Muslims tied two beams across the battlements as additional protection. It was these beams, according to Fulcher, that later served as the bridge from Godfrey's tower to the wall.[69] Although these beams might have fulfilled a similar role to the sacks of padding found in other accounts, it is at least as likely that Fulcher was attempting to describe a type of improvised hoarding: rather than shielding the walls, the beams provided additional temporary cover for the defenders along the battlements.[70] By securing the beams over and in front of the merlons, the Muslims would have been able to protect themselves better from the Frankish archers and artillery, as well as those shooting at them from the top of the northern siege tower.[71] Decreasing the size of the crenels in this way would have left enough space for the defenders to shoot through, but the added protection would have restricted their vertical field of view. This handicap is less dramatic than it may initially seem: most arrows and stones shot or slung at

69 Ralph of Caen, *Gesta Tancredi* 124-25, RHC Oc 3, pp. 692-93, trans. Bachrach and Bachrach, pp. 140-42; Albert of Aachen, *Historia Ierosolimitana* 6.17-19, ed. and trans. Edgington, pp. 424-29; Fulcher of Chartres, *Historia Hierosolymitana* 1.27.8, ed. Hagenmeyer, pp. 297-98, trans. Ryan, pp. 120-21.

70 Unlike more developed hoarding or machicolations, these improvised measures would not likely have provided the means to assault those at the base of the walls, directly below the defenders.

71 This is all postulated. Fulcher did not arrive in Jerusalem until the following year so was not an eyewitness of the siege, although neither was Albert of Aachen, who never visited the East. The latter's account states that this timber was a large tree trunk, covered in combustibles, which the defenders attempted to use to burn Godfrey's tower, Albert of Aachen, *Historia Ierosolimitana* 6.18-19, ed. and trans. Edgington, pp. 426-29.

the besiegers approaching the base of a wall would have come from flanking towers rather than the battlements above. Instead of leaning over the parapet, completely exposing themselves in order to shoot at attackers at the base of the wall below them, defenders directly above were more likely to drop large stones, requiring minimal visibility to do so.

The Muslims' Artillery

Despite the emphasis that has so far been placed on the crusaders' artillery, the sources suggest that the Muslims' stone-throwers were at least as effective and almost certainly more numerous. In his later account, William of Tyre judged the defenders' machines to have been superior to those of the Franks because they were constructed from materials brought into Jerusalem ahead of the siege, supplemented with structural beams harvested from buildings within the city.[72] The Muslims' *tormenta* and *petrariae* outnumbered those of the crusaders by nine or ten to one, according to Raymond of Aguilers, while Albert of Aachen gives the definitive figure of fourteen *mangenae*, nine of which were directed against the Provençal siege tower.[73] In addition to stones, these engines were used to hurl incendiaries. These conglomerates of pitch, sulphur and other components were placed in containers that were either provided with iron spikes or coated with a sticky substance that would help them adhere to the Frankish siege towers.[74]

The defenders' decision to array more engines against the Provençal siege tower probably reflects the strength of this tower. If the crusaders were compelled to build it within range of these engines, due to the limits of the plateau, the defenders might also have hoped to destroy the tower before it could come into use, concentrating their artillery to the south in order to achieve this. From prepared positions, the defending engines were able to seriously cripple the siege tower by the final stages of the siege.

To the north, the crusaders were free to construct their tower well out of range. This contingent of the Frankish army appears to have made up for the comparative weakness of its siege tower with their decision to suddenly move it to face a portion of the northern wall that the defenders had not spent as

72 William of Tyre, *Chronicon* 8.8, 13, ed. Huygens, 1:395-96, 403-4, trans. Babcock and Krey, 1:354, 361-63.

73 Raymond of Aguilers, *Historia Francorum* 20, RHC Oc 3, p. 298, trans. Hill and Hill, p. 125; Albert of Aachen, *Historia Ierosolimitana* 6.15, ed. and trans. Edgington, pp. 422-23.

74 Fulcher of Chartres, *Historia Hierosolymitana* 1.27.6, 9, ed. Hagenmeyer, pp. 296, 298-99, trans. Ryan, pp. 120-21; Raymond of Aguilers, *Historia Francorum* 20, RHC Oc 3, pp. 298-99, trans. Hill and Hill, pp. 125-26.

much time preparing. The Muslims were forced to reposition their artillery in response to this manoeuvre, but the urban environment behind the walls posed challenges to relocating and then shooting these engines.[75] According to Albert of Aachen, the Muslims' artillery struggled to hit the northern siege tower as it advanced because there was insufficient open space between the buildings to allow the engines to shoot effectively. Besides its own footprint, each traction trebuchet would have required an area of space sufficient to accommodate its pulling crew in front and the free movement of its beam and sling behind, an area generally at least 8 m by 4 m.

Albert's account clearly places these engines on the ground, rather than on top of towers. Although they would have enjoyed considerable shelter in the shadow of the curtain wall, out of sight from the Frankish besiegers, the wall would have become an obstacle as the siege tower drew closer. If, for example, a defensive engine, which threw projectiles at 30 m/s, was positioned 15 m inside a curtain wall that was 10 m high, it might enjoy considerable success against a siege tower around 80 m away (65 m beyond the curtain wall). Once the tower was within 70 m, however, the engine would find it increasingly challenging to find its target. If the crew opted to decrease the amount of force applied to each shot, accuracy would suffer as each puller tempered the force of his pull to a different degree, leading to inconsistent shooting sequences. Alternatively, if a constant amount of pulling force was maintained but the release angle was increased, lobbing stones high into the air, slight differences in the amount of energy applied to each shot would be magnified, increasingly so as the shooting sequence became shorter and the flight time of each projectile was extended. Not until the tower was close enough that the angle of release could be lowered, allowing a projectile to strike its target before reaching the zenith of its natural flight path, would accuracy increase once more (Fig. 8).[76]

Despite these challenges, enough shots were brought to bear against the northern tower that its structural weakness became apparent. Lacking the more sturdy beams salvaged from the Genoese ships, the tower had begun to lean to one side by the time it reached the northern wall. Probably subject to a longer period of bombardment, the southern tower was less fortunate despite its sturdier construction.

---

75 Albert of Aachen, *Historia Ierosolimitana* 6.17, ed. and trans. Edgington, pp. 424-25. Ralph of Caen also notes this redeployment, Ralph of Caen, *Gesta Tancredi* 123, RHC Oc 3, p. 691, trans. Bachrach and Bachrach, p. 139.

76 For another examination of this practice, see Chapter 5. For evidence of the use of trebuchets on top of towers, see Chapter 9.

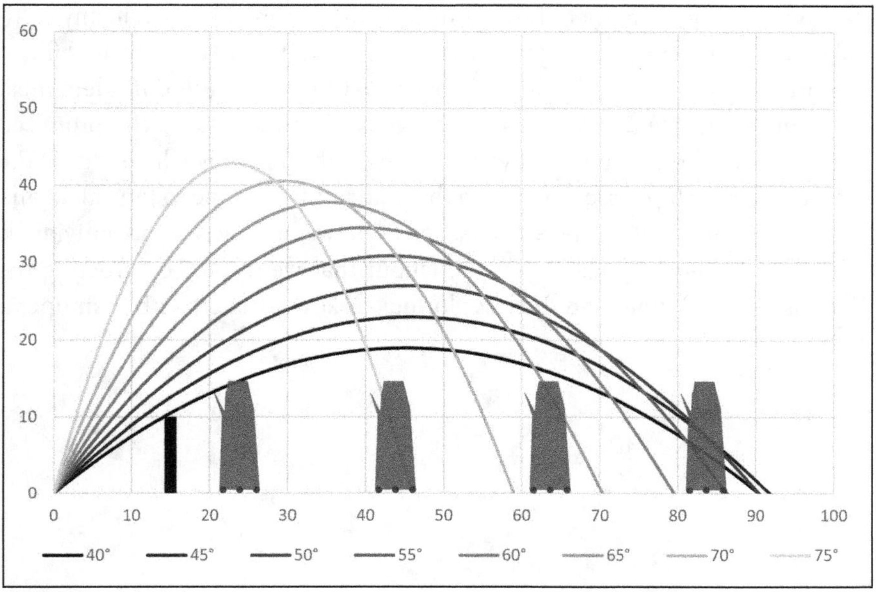

FIGURE 8 *Targeting a siege tower at 30 m/s*

The stone-throwers employed throughout the First Crusade were almost certainly traction trebuchets.[77] Although the conduct of the Franks at Jerusalem is often portrayed as the result of a two-year learning experience, this is a drastic over simplification.[78] The individuals involved would have sharpened their martial skills through the course of the crusade, but this is true of any extended campaign. While some of its leaders gained more experience than they had previously gathered in Europe, many others were seasoned veterans who had undertaken siege operations previously in the West. The tactics employed by the Franks at each siege were the product of pre-existing European traditions and not new methods learned during their travels through Anatolia and the Levant. The artillery that they employed was of a known and familiar type, used, as it was in Europe, as an antipersonnel weapon.[79]

The Franks possessed neither a superior tradition in arms nor technology. The artillery used by the crusaders appears to have been fairly similar in both

77 Most of scholars who suggest otherwise are not well acquainted with the broader study of medieval artillery, this oversight is easy to excuse as their focus lies elsewhere, for example, Asbridge, *The First Crusade*, pp. 55, 127; Kostick, *The Siege of Jerusalem*, pp. 82-83.

78 For example, Rogers, *Latin Siege Warfare*, p. 87. Cf. France, *Victory in the East*, p. 370.

79 For the comparative use of artillery in the Anglo-Norman sphere, see Fulton, "Anglo-Norman Artillery," pp. 5-8. See also Purton, *Early Medieval Siege*, pp. 155-208.

capability and form to that of their Muslim adversaries. Rather than any tactical or technological advantage, it was the determination of the rank and file of the army and the unity of their purpose that benefited the crusaders most through the course of the crusade. The goal of Jerusalem and the promised grace that its liberation offered was enough to inspire many who endured the incredible hardships associated with the march from Europe to Jerusalem and the months of gruelling siege conditions in between. It was these individuals who were largely responsible for the labour that went into constructing the Frankish siege engines and the less glorious tasks associated with their operation.

CHAPTER 4

Twelfth Century

Following the sieges of the First Crusade, artillery was used both offensively and defensively throughout the twelfth century. However, our understanding of how, when and where these engines were used in the early decades of this century is probably skewed due to the nature of the sources. Whereas a number of surviving first-hand accounts provide a comprehensive picture of events in Palestine and southern Syria, our understanding of affairs farther north is less complete. Although some detailed contemporary histories were composed in north-western Syria and south-eastern Anatolia, there was a greater number of competing interests in this area. Accordingly, this probably allowed for a number of smaller conflicts around the frontiers of the county of Edessa to go unrecorded.

Although there is anecdotal evidence that artillery was used at sieges conducted by the Frankish lords of Antioch and Tripoli, and against their early possessions by neighbouring Muslim powers, it is more frequently mentioned during sieges in and around the kingdom of Jerusalem. While it is possible that the gentle topography of the coastal plain of Palestine encouraged the use of artillery, this is not a satisfactory explanation given the apparent use of these engines by both attackers and defenders in western Syria during the First Crusade and later periods. It is possibly no more than a coincidence that the authors based in the south were more disposed to mention the use of artillery. Unfortunately there is little archaeological evidence of the engines used during this period that might be used to evaluate the accuracy of these early sources, upon whom our understanding of this period is largely reliant.

The closest Frankish source to events in Palestine during the early twelfth century was Fulcher of Chartres, chaplain of Baldwin of Boulogne since his journey to Edessa during the First Crusade. When Baldwin travelled south to succeed his brother Godfrey as ruler of Jerusalem in 1100, becoming the first to take the title of king, Fulcher accompanied his lord and from Jaffa and Jerusalem composed and reworked his chronicle. Fulcher repeatedly states that his work was written to give a true and accurate account of events surrounding the Franks in the Holy Land; however, it is not as complete as might be desired and what has survived appears to be his third redaction. Like the similar narratives composed by his contemporaries, it is in part a work of propaganda. Although he was based in Palestine from late 1100, Fulcher maintained an interest in affairs to the north throughout the remainder of his life. While he appears to

 | DOI 10.1163/9789004376922_006

have made use of the *Gesta* and Raymond of Aguilers' account before composing his own history of the First Crusade, the subsequent portion of his work is original, relying on his appreciation of affairs around him and information obtained from others.[1]

The other critical Frankish account of this period is that of Albert of Aachen. In the words of Susan Edgington, it is "the most complete, most detailed, and most colourful of the contemporary narratives of the First Crusade and the careers of the first generation of Latin settlers in Outremer (1095-1119)."[2] Although Albert never visited the East himself, his account was written independent of any knowledge of the narratives composed by his contemporaries. He appears to have relied primarily on the oral testimony of those who had been on crusade. The wealth of detail provided in this history makes it an invaluable source, even if the author's distance from events encourages extra caution.

William of Tyre, who would pen the most influential history of the Franks in the Levant, made use of the accounts of both Fulcher and Albert when composing his narrative of events much later in the twelfth century. William's rise through the ranks of the Church, eventually becoming archbishop of Tyre and chancellor of the kingdom of Jerusalem, probably gave him access to many other works, allowing him to use these now lost sources to supplement the works of his predecessors.[3] In later decades, William's account forms the definitive version of events from the Frankish perspective, although composed in a deliberately classical style. While all three authors wrote with a slightly different audience in mind, each probably hoped that their work would inspire Europeans to take up the cross and help defend Latin control of the Holy Land.

From a Muslim point of view, the work of Ibn al-Qalanisi, a high-ranking Damascene civil servant, is of paramount importance up to the author's death in 1160. The nature of his position provided an ideal perspective from which to compose a chronicle of events from first-hand knowledge of pertinent documents and oral testimony.[4] In the same year that Ibn al-Qalanisi died, another great Muslim historian, Ibn al-Athir, was born. The latter made extensive use of the histories written before his own time, including that of Ibn al-Qalanisi, to compile a more complete, if more distant, history of events, written

---

1 For a discussion of Fulcher of Chartres, his sources and work, see Fulcher of Chartres, *Historia Hierosolymitana*, trans. Ryan, pp. 3-56, esp. pp. 42-46.

2 Albert of Aachen, *Historia Ierosolimitana*, ed. and trans. Edgington, p. i, for a discussion of Albert, his sources and work, see pp. xxi-xxxvii.

3 For a discussion of William of Tyre, his sources and work, see Edbury and Rowe, *William of Tyre*, pp. 13-58.

4 For a discussion of Ibn al-Qalanisi, his sources and work, see Ibn al-Qalanisi, *Dhail*, trans. Gibb, pp. 8-14; Christie, "Ibn al-Qalanisi," pp. 7-28.

mainly from Mosul. Ibn al-Athir does not appear to have held a high administrative or bureaucratic post, but was patronised by the Zankids, towards whom his work is slightly partial.[5]

A first-hand Muslim account, based in the north, has not survived from the early twelfth century. However, Ibn al-'Adim's mid-thirteenth-century history of Aleppo, and the works of other later Muslim sources, provide a reasonable window into the events of this region. A number of quite detailed accounts were also composed by Armenian and Syriac sources.[6] For the early twelfth century, Matthew of Edessa's history is chief among these.[7]

Testimony provided by each author must be considered independently as well as alongside contemporary accounts. From the end of the First Crusade until later in the twelfth century, few of our sources were eyewitnesses to many of the sieges found in their accounts and most would have had little direct knowledge of contemporary mechanical artillery; however, each would have been familiar with the general concept. The descriptions of many appear to reveal influences from both earlier writings and oral descriptions of contemporary engines provided by those more acquainted with the art and practice of siege warfare. Given this imprecision, it will be highlighted when certain sources were eyewitnesses to events. Even when considering certain eyewitness accounts, such as those of Usama ibn Munqidh and the later poetic work of 'Imad al-Din al-Isfahani, issues of reliability, exaggeration and bias still need to be taken into consideration. Until sources become more numerous later in the period, allowing for accounts and descriptions to be compared more closely, it can often be more helpful to consider how each source treats artillery in a general sense and why they might emphasised these engines at some sieges but not at others.

The Opening Decades

The Palestinian Coast

Building outward from the bases developed around Jerusalem and Jaffa, the initial thrusts of Frankish expansion in Palestine were across Galilee and along the Mediterranean coast. Contemporary accounts of the sieges of these coastal

5 For a discussion of Ibn al-Athir, his sources and work, see Ibn al-Athir, *al-Kamil*, trans. Richards, 1:1-5; Micheau, "Ibn al-Athir," pp. 52-83.

6 For a discussion of Ibn al-'Adim, his sources and work, see Eddé, "Ibn al-'Adim," pp. 109-35.

7 For a discussion of Mathew of Edessa, his sources and work, see Matthew of Edessa, *Patmut'iwn*, trans. Dostourian, pp. 1-16.

towns are the most descriptive and provide the clearest picture of Frankish artillery during the early period of Latin rule in the Levant. At the siege of Arsuf in 1099, Albert of Aachen states that Godfrey of Bouillon spent six weeks constructing *mangenae* and other siege engines. As had been common practice during the First Crusade, these stone-throwers were used to target Arsuf's defenders, rather than its defences.[8] Despite the support provided by these engines, and the ultimate use of two siege towers, the siege failed.[9] The following year, Albert asserts that seven *mangenae* were used to support a siege tower built during the Veneto-Frankish assault on Haifa led by Tancred.[10] In 1101, Baldwin I renewed efforts against Arsuf, this time with Genoese assistance. Despite the apparent collapse of a Frankish siege tower, which buckled under the weight of its occupants, the town quickly surrendered. Given the established trend of setting up artillery to support siege operations, especially when a siege tower was used, it is perhaps peculiar that artillery is not even summarily mentioned during this siege.[11] After Arsuf had been garrisoned, the Franks turned on Caesarea, 35 km to the north.

At Caesarea, efforts once more focused on the construction of a siege tower. This was built from the masts and oars of ships. It is tempting to suggest that prepared timbers were also used to construct the *petrariae* that were used to protect the tower, but this is not confirmed by any of the sources. After fifteen days of bombardment, and with their siege tower still incomplete, a general assault was launched with ladders. The zeal of the Franks and apparent effectiveness of their artillery, which Fulcher of Chartres claims was able to damage the parapet, allowed the Franks to storm the town.[12] The location of this dam-

8 Albert of Aachen, *Historia Ierosolimitana* 7.1-5, ed. and trans. Edgington, pp. 486-93.

9 This failure is often attributed to the absence of Italian naval assistance, a point emphasised by William of Tyre the better part of a century later, William of Tyre, *Chronicon* 9.19, ed. Huygens, 1:446, trans. Babcock and Krey, 1:409-10. For an analysis of the Fatimid fleet at this time, see Hamblin, "The Fatimid Navy," pp. 77-83.

10 Albert of Aachen, *Historia Ierosolimitana* 7.22-25, ed. and trans. Edgington, pp. 516-21.

11 See Albert of Aachen, *Historia Ierosolimitana* 7.54, ed. and trans. Edgington, pp. 562-63; Fulcher of Chartres, *Historia Hierosolymitana* 2.8, ed. Hagenmeyer, pp. 395-400, trans. Ryan, pp. 151-53; Caffaro, *Annales*, ed. Belgrano, p. 9, trans. Hall and Phillips, p. 53; William of Tyre, *Chronicon* 10.13 (14), ed. Huygens, 1:468-69, trans. Babcock and Krey, 1:433-35.

12 Fulcher of Chartres, *Historia Hierosolymitana* 2.9, ed. Hagenmeyer, pp. 400-4, trans. Ryan, pp. 153-55; Albert of Aachen, *Historia Ierosolimitana* 7.55-56, ed. and trans. Edgington, pp. 564-67; Caffaro, *Annales*, ed. Belgrano, pp. 9-12, trans. Hall and Phillips, pp. 53-56; William of Malmesbury, *De gestis rerum Anglorum* 4.380, ed. Stubbs, 2:444-45, trans. Giles, pp. 405-6; William of Tyre, *Chronicon* 10.14 (15), ed. Huygens, trans. Babcock and Krey, 1:435-37. For less detailed Muslim perspectives, see Ibn al-Qalanisi, *Dhail*, trans. Gibb, p. 51; Ibn al-Athir, *al-Kamil*, trans. Richards, 1:47-48.

age suggests that the Frankish artillery was targeting the battlements, the defenders behind them, or softer targets inside the town. This corresponds with Albert of Aachen's description of these engines, in conjunction with the Frankish archers, wearing down the defenders. The Frankish artillery probably posed little threat to the city's fortifications; however, these engines most likely left impact signatures in the soft *kurkar* sandstone of the city walls.[13] The amount of energy capable of producing such an effect would also have been sufficient to damage most improvised defences that might have been set up along the parapet, similar to those used at Jerusalem in 1099. It is likely to have been reports of this superficial damage that so impressed Fulcher.

Baldwin I next turned his attention toward Acre. The city had been besieged unsuccessfully by his brother Godfrey in 1100 and Baldwin himself failed before its walls in 1103.[14] Fulcher of Chartres, followed by William of Tyre, notes that the latter attempt was a brief siege, confined mainly to burning the surrounding gardens. Albert of Aachen, however, gives a more sensational account of a five-week siege in which the Franks made use of a siege tower and *mangenae* and were on the verge of success before a Muslim relief force arrived by sea.

When Baldwin moved against Acre in May 1104, he did so with the aid of a Genoese fleet. After the city had been assaulted for around twenty days, again with the assistance of siege engines, the defenders were compelled to seek terms of surrender.[15] William of Tyre stresses the ingenuity of the Frankish artillery that was placed around the city. He also seems to imply that artillery was mounted on top of the Franks' siege towers when he states that large stones were thrown from them. If stone-throwing engines were positioned on top of these siege towers, their size would have been restricted. Other accounts place artillery on top of siege towers at sieges such as Breval (1092) and Du-

---

13 Cf. the siege of Arsuf (1265) in Chapter 7.

14 For the siege of 1100, see Ibn al-Qalanisi, *Dhail*, trans. Gibb, p. 51; Ibn al-Athir, *al-Kamil*, trans. Richards, 1:47. For the siege of 1103, see Fulcher of Chartres, *Historia Hierosolymitana* 2.22, ed. Hagenmeyer, pp. 456-57, trans. Ryan, pp. 174-75; William of Tyre, *Chronicon* 10.25 (26), ed. Huygens, 1:485, trans. Babcock and Krey, 1:453; Albert of Aachen, *Historia Ierosolimitana* 9.19, ed. and trans. Edgington, pp. 660-61. William of Tyre asserts that the failure of the siege of Acre in 1103 was due to a lack of naval support.

15 Fulcher of Chartres, *Historia Hierosolymitana* 2.25.1, ed. Hagenmeyer, pp. 462-63, trans. Ryan, p. 176; Albert of Aachen, *Historia Ierosolimitana* 9.27, ed. and trans. Edgington, pp. 670-71; Caffaro, *Annales*, ed. Belgrano, p. 14, trans. Hall and Phillips, p. 57; Caffaro, *De liberatione*, ed. Belgrano, pp. 121-22, trans. Hall and Phillips, p. 123; William of Tyre, *Chronicon* 10.27 (28), ed. Huygens, 1:486-87, trans. Babcock and Krey, 1:454-56; Ibn al-Qalanisi, *Dhail*, trans. Gibb, pp. 61-62. Ibn al-Athir does not mention any siege equipment at this siege, Ibn al-Athir, *al-Kamil*, trans. Richards, 1:78-79.

razzo (1081);[16] however, reports of large stones being thrown from the tops of siege towers are common and in most cases the apparent trajectory of these stones suggests that they were tossed by hand.

Although the details found in William of Tyre's later description may add certain embellishments to the earlier accounts upon which this part of his chronicle is based, they are nevertheless important in confirming how these engines were employed. His descriptions of damage, inflicted along the parapet and suffered by buildings behind the defensive curtain, reinforce the notion that the Frankish artillery targeted Acre's battlements. Accordingly, literal readings of more colourful statements, which have led some historians to conclude that these engines "destroyed the city walls and buildings within", are entirely unfounded.[17] Even if the artillery of the day was much less impressive than William might suggest, the frequency with which these engines were employed is significant.

In 1106, Baldwin and the Patriarch, according to Albert of Aachen, spent forty days constructing artillery (referred to as both *mangenae* and *tormenta*) at Acre. This was done in preparation for an attack they planned against Sidon, counting on the assistance of naval forces that had arrived from England, Denmark and the Low Countries.[18] Although the attack never materialised, the episode reveals the degree of importance that Albert appears to attribute to these engines. Perhaps this reflects the emphasis his sources placed on artillery after returning from the Holy Land, or simply an inferred significance given the frequency with which it was mentioned.

Artillery remained an important weapon as the Franks entered the second decade of their tenure in the East. Albert of Aachen notes the use of artillery by Baldwin I at Sidon in 1108 and by Bertrand of Toulouse at Tripoli in 1109. Ibn al-Qalanisi corroborates the use of artillery at Tripoli but not at Sidon and Fulcher of Chartres makes no mention of such engines at either siege.[19] Al-

16 Orderic Vitalis, *Historiae ecclesiasticae* 8.24, ed. and trans. Chibnall, 4:288-89; Anna Comnena, *Alexiad* 4.1, trans. Sewter, p. 135.

17 This quote is from Benvenisti, *The Crusaders*, p. 81; however, the practice of taking such enthusiastic testimony literally is endemic.

18 Albert of Aachen, *Historia Ierosolimitana* 10.3-4, ed. and trans. Edgington, pp. 718-23.

19 Albert of Aachen, *Historia Ierosolimitana* 10.46, 11.11, ed. and trans. Edgington, pp. 760-61, 782-83; Ibn al-Qalanisi, *Dhail*, trans. Gibb, pp. 87, 88-90. For the siege of Tripoli, see also Fulcher of Chartres, *Historia Hierosolymitana* 2.40-41, ed. Hagenmeyer, pp. 526-33, trans. Ryan, pp. 193-96; Caffaro, *De liberatione*, ed. Belgrano, p. 123, trans. Hall and Phillips, pp. 123-24; Caffaro, *Annales*, ed. Belgrano, p. 14, trans. Hall and Phillips, p. 57. Tripoli had been blockaded to some degree since 1102, originally by Raymond of St Gilles and then his nephew William Jordan, before Bertrand arrived.

bert states that the Franks also used artillery at Beirut and once more against Sidon in 1110, as well as at the drawn-out, and eventually unsuccessful, siege of Tyre in 1111-12. William of Tyre includes stone-throwing machines in his later account of the siege of Beirut and Ibn al-Qalanisi, followed by Ibn al-Athir, notes their use against Tyre. Artillery was also used defensively at Beirut, according to Ibn al-Qalanisi, and similarly employed by the defenders of Sidon and Tyre, according to Albert of Aachen.[20]

Although artillery was frequently used by the Franks against the cities of coastal Palestine during the first decade of the twelfth century, the siege tower was the paramount or focal engine employed at many of these engagements. Artillery was typically used in a secondary role, supporting these towers. Accordingly, it was frequently mentioned but given only nominal notice on most occasions. The repeated use of such engines, however, is notable: indicating that the value of these machines justified the time, cost and manpower required to construct and operate them.

The North

Evidence of the use of artillery in northern Syria is more obscure; for instance, no such machines are mentioned in Ralph of Caen's otherwise detailed account of Tancred's siege of Latakia.[21] As in Palestine, Albert of Aachen notes the presence of artillery at more sieges in the first quarter of the twelfth century than any other contemporary source. He mentions the use of such engines at Apamea in 1106, al-Atharib in 1110 and specifies that twelve *mangenae* were used against Vetula in 1111.[22] At the latter two sieges in particular, Albert emphasises the damage caused by these engines. He asserts that the engines built

---

20 For Beirut, Albert of Aachen, *Historia Ierosolimitana* 11.15-16, ed. and trans. Edgington, pp. 788-89; William of Tyre, *Chronicon* 11.13, ed. Huygens, 1:515, trans. Babcock and Krey, 1:484-86; Ibn al-Qalanisi, *Dhail*, trans. Gibb, pp. 99-100. For Sidon, Albert of Aachen, *Historia Ierosolimitana* 11.31-34, ed. and trans. Edgington, pp. 804-9. For Tyre, Albert of Aachen, *Historia Ierosolimitana* 12.6, ed. and trans. Edgington, pp. 830-33; Ibn al-Qalanisi, *Dhail*, trans. Gibb, pp. 122-23; Ibn al-Athir, *al-Kamil*, trans. Richards, 1:157. Only Albert of Aachen mentions artillery at the siege of Sidon in 1108, Albert of Aachen, *Historia Ierosolimitana* 10.46-51, ed. and trans. Edgington, pp. 760-67.

21 Ralph of Caen, *Gesta Tancredi* 144, RHC Oc 3, pp. 706-9, trans. Bachrach and Bachrach, pp. 159-63.

22 Albert of Aachen, *Historia Ierosolimitana* 10.20-22, 11.43-47, ed. and trans. Edgington, pp. 738-41, 818-25. For Apamea, cf. Ibn al-Qalanisi, *Dhail*, trans. Gibb, pp. 69-70; Ibn al-Athir, *al-Kamil*, trans. Richards, 1:92-93. For al-Atharib, cf. Matthew of Edessa, *Patmut'iwn*, trans. Dostourian, p. 206; Ibn al-Qalanisi, *Dhail*, trans. Gibb, pp. 105-6; Ibn al-Athir, *al-Kamil*, trans. Richards, 1:153.

at al-Atharib required a number of days to construct, they were then brought forward to a fortified position from which they played against the stronghold for more than two months. This ultimately led to the collapse of the citadel (or perhaps a section of one of its significant towers), which damaged two adjoining towers. Tellingly, this 'breach' remained defensible, suggesting the actual damage was far less severe than what is described. At Tancred's three-month siege of Vetula, artillery was able to damage a portion of the barbican, probably a stretch of battlements along the outer wall or perhaps a weak outer gate, allowing the besiegers to storm the compromised section of defences.

Albert's descriptions of dramatic destruction in the north are checked by Walter the Chancellor, another contemporary who, unlike Albert, was based in Antioch and was an eyewitness of certain events in the region. According to Walter, when Ilghazi besieged al-Athrib in 1119, he used his artillery three or four times a day to deprive the besieged towers of their defences and kill their defenders. The effects of this bombardment were augmented by rows of archers and men throwing javelins who added to the hail of projectiles.[23] It is clear from Walter's description that these efforts were part of a broader strategy to suppress the stronghold's defenders while his sappers worked to undermine the fortress – these men were the real threat to the defensibility of the stronghold.

From a Muslim perspective, Ibn al-'Adim confirms Tancred's use of artillery at al-Atharib in 1110 but emphasises instead his use of a ram and makes no mention of a breach.[24] He later notes Ilghazi's use of four mangonels against Zardana in 1122. These engines might have played a supporting role in the capture of the stronghold's outer defences a few days after the siege began.[25] Unfortunately, few details are provided concerning Baldwin II's use of artillery at Balis in 1122, at which point the king was acting as regent of Antioch.[26]

Although it is possible that artillery was used at many more sieges, despite not being mentioned by most sources, the rough terrain of eastern Anatolia and northern Syria may have favoured a more mobile form of warfare. Siege tactics had to be more flexible when confronting the generally smaller and more topographically isolated strongholds of these regions. However, it is hard to extend this suggestion to the Aleppo plateau of western Syria, between the Orontes and the Euphrates, where the topography is relatively gentle. In this

23 Walter the Chancellor, *Bella Antiochena* 2.10, ed. Hagenmeyer, p. 99, trans. Asbridge and Edgington, p. 146.

24 Ibn al-'Adim, *Zubdat al-Halab*, RHC Or 3, pp. 597-98.

25 Ibn al-'Adim, *Zubdat al-Halab*, RHC Or 3, pp. 632-33.

26 Ibn al-'Adim, *Zubdat al-Halab*, RHC Or 3, p. 634.

region, contested intensely by the powers of Antioch and Aleppo, the relative proximity of hostile forces may have discouraged the use of more elaborate siege engines, given the likelihood that any investment in machinery would be lost if hostile forces arrived to break the siege. Besides the main urban centres, the defences of this region were also more modest and made greater use of the topography when compared with the fortifications of the coastal towns of Palestine, providing additional challenges to the use of siege engines in some cases and making frontal attacks a more attractive option in others.

Unlike the set-piece sieges undertaken in the early twelfth century against the urban centres of the Palestinian littoral, most of the successful sieges in Syria at this time were concluded by a rapid frontal assault and escalade, or a quick surrender in the face of the imminent threat of such. By comparison, the towns of coastal Palestine were relatively far from potential relief forces, more approachable due to the gentle topography surrounding them, and the greater circumference of their town walls meant that they were reliant on much larger garrisons and bodies of defenders. These conditions, in addition to the regular availability of Italian maritime support along the coast, encouraged the Franks to employ siege towers and supporting artillery. The smaller strongholds of Syria were often positioned in strong topographical positions and defended by smaller garrisons, making frontal assaults an attractive option, while those that were more geographically isolated and farther from potential relief were susceptible to strangulating blockades. The major urban centres of this region, including Aleppo, Hama and Homs, had more extensive and approachable defences, as well as larger bodies of defenders, but were not subject to earnest siege efforts during this period. Thus the apparent absence of artillery at many sieges north of Palestine at this time may reflect geographical and political factors as much as the disparity of source material.

Jaffa: 1123

Although Frankish siege efforts have been emphasised up to this point, it is important to acknowledge the similar use of artillery by Muslim forces. The sieges of the First Crusade confirm that Muslims in the region were as familiar with these engines as were their Frankish counterparts. Likewise, as noted above, artillery was often mentioned in the defence of Muslim strongholds during the opening decades of the twelfth century. There are few examples, however, of the offensive use of these engines by Muslim forces, in no small part owing to the limited number of Muslim sieges undertaken in Palestine during this period. In the first decade of the twelfth century, the only Latin reference to Muslims employing artillery offensively is found in Albert of Aachen's account of an attack on Castellum Arnaldi, a small castle commanding the main approach to Jerusalem from the coast through the Judean hills.[27] In an equally rare Arabic reference, Ibn al-Qalanisi notes the use of artillery by Tughtakin of Damascus during his siege of rebellious Baalbek in March-April 1110.[28] Despite his proximity, Ibn al-Qalanisi provides few details concerning the Damascene siege engines or how they were used before an amicable surrender was arranged thirty-five days after the siege had begun.

In May 1123, Fatimid forces from Ascalon took advantage of Baldwin I's captivity to make an assault on Jaffa. What is particularly interesting about this episode is that Fulcher of Chartres claims that the attackers brought artillery with them. The implication that these machines were prefabricated, along with Fulcher's claim that they threw sizable stones, appears to suggest that these were significant siege engines. However, Fulcher's established tendency to stress the power of artillery and the relatively small size of even 'large' projectiles at this time should be taken into consideration. A close reading of Fulcher's account reveals that these *tormenta* were used in a familiar manner: to target the battlements of Jaffa and more generally to apply pressure, both physically and psychologically.[29] Fulcher checks any notion that these were exceptionally large engines when he states that after five days of bombardment the walls had sustained minimal damage and that most was inflicted along the parapet. His claim that these machines could shoot from beyond the range of the garrison's archers may also be an exaggeration, meant more to emphasise their impressive range than to give an accurate definition of it. Testimony from other sieges, such as the siege of Acre in 1189-91, directly contradicts this notion, as - so it appears - does surviving illustrative evidence.[30] Trials with reconstructed traction trebuchets suggest these engines had a range slightly less than 200 m, the approximate range of a medieval archer shooting for maximum distance.[31]

It is only when Fulcher notes the approach of a Frankish relief force that the real thrust of the Egyptian assault on Jaffa becomes clear. Under partial cover of their artillery, Fatimid sappers were undermining the town walls. Unable to

---

27 Albert of Aachen, *Historia Ierosolimitana* 10.14, ed. and trans. Edgington, pp. 730-33. This castle was rebuilt in 1132-33 by the patriarch of Jerusalem, William of Tyre, *Chronicon* 14.8, ed. Huygens, 2:639-40, trans. Babcock and Krey, 2:58. See also Pringle, *Secular Buildings*, no. 231, pp. 106-7; Pringle, "Templar Castles between Jaffa and Jerusalem," pp. 103-8.

28 Ibn al-Qalanisi, *Dhail*, trans. Gibb, pp. 96-98.

29 Fulcher of Chartres, *Historia Hierosolymitana* 3.17, ed. Hagenmeyer, pp. 661-63, trans. Ryan, pp. 240-41. See also *Historia Nicaena* 80, RHC Oc 5, p. 184. Cf. Ibn Muyassar, *Akhbar Misr*, RHC Or 3, p. 469.

30 See Appendix 2, **nos. 3, 4, 12** and **19**. For the siege of Acre, see Chapter 5.

31 See Appendix 4.

open a breach, which would have allowed them to overwhelm the relatively small garrison before relief arrived, the Egyptians were compelled to dismantle their engines and evacuate them with the army to their ships. Despite Fulcher's apparent confirmation of the value of these engines, having been withdrawn rather than abandoned, it is the threat posed by the Fatimid miners that William of Tyre emphasises in his later account of this siege, omitting any mention of artillery.[32]

Tyre: 1124

In 1124 the Franks launched what was probably their largest siege operation since the First Crusade. Apart from Ascalon, Tyre was at this point the only remaining Muslim-held outpost on the Palestinian coast. The ancient Phoenician stronghold was surrounded by the sea and joined to the mainland by a shallow tombolo, which had been enlarged and developed into an isthmus during Alexander the Great's famous siege in 332 BC.[33] A millennium and a half later, the proverbial strength of Tyre was still emphasised by those who visited.[34] The impetus for the siege in 1124 was the arrival of a Venetian fleet. The Venetians obviously intended to undertake a siege while in the Holy Land as Fulcher of Chartres states that they had brought long timbers with them, specifically those that were well suited for building siege engines.[35] This is the only cargo that Fulcher notes and, apart from the number of ships that made up the fleet, the only real detail that he highlights about this force upon its arrival.

Under the leadership of the patriarch of Jerusalem and the doge of Venice, Tyre was invested from mid-February 1124. The Franks attempted to establish a blockade, digging two lines of trenches around the east end of the isthmus,

32 William of Tyre, *Chronicon* 12.21, ed. Huygens, 1:572-73, trans. Babcock and Krey, 1:545-47. The assault is also found in the account of Ibn Muyassar who states that the Fatimids had hoped for 'Abbasid support, they withdrew when it was not forthcoming, Ibn Muyassar, *Akhbar Misr*, RHC Or 3, p. 469.

33 Marriner, Morhange and Meulé, "Holocene morphogenesis," pp. 9,218-23.

34 For example, William of Tyre, *Chronicon* 22.30 (29), ed. Huygens, 2:1057, trans. Babcock and Krey, 2:501; Ibn Jubayr, *Rihla*, trans. Broadhurst, pp. 319-20. See also Anna Comnena, *Alexiad* 14.2, trans. Sewter, p. 442; Ibn al-Athir, *al-Kamil*, trans. Richards, 2:328. For the medieval defences of Tyre, see Chapter 5.

35 Fulcher of Chartres, *Historia Hierosolymitana* 3.14, ed. Hagenmeyer, pp. 655-57, trans. Ryan, pp. 238-39; William of Tyre, *Chronicon* 13.6, ed. Huygens, 1:593, trans. Babcock and Krey, 2:10. See also *Historia Nicaena* 80, RHC Oc 5, p. 185.

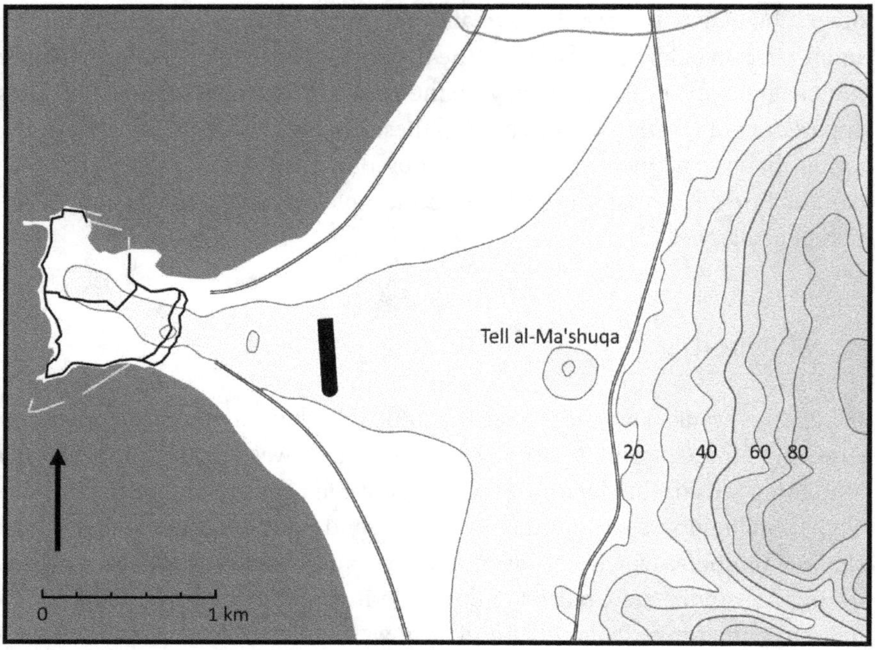

PLAN 6 *Tyre: traces of discernable and probable town defences*

cutting off the besieged and providing protection from any field forces that might try to break the siege from the landward side. Meanwhile, all but one of the Venetian ships were pulled up on shore.[36] Carpenters then used the materials brought by the Venetians, and probably harvested additional timber from one or more of the beached vessels, to build siege towers and other engines.[37]

The sources use various terms to identify the Frankish artillery. Fulcher of Chartres ambiguously notes the construction of *machinae* at the start of the siege.[38] A half-century later, William of Tyre specifies that some of these were *machinae iaculatoriae*, which could throw large stones against the city's walls and towers and terrorise the townspeople, while these are rendered as *perrieres* and *mangoniaux* in the subsequent *Eracles* translation.[39] The added em-

36 William of Tyre, *Chronicon* 13.4, 6, ed. Huygens, 1:590, 593-94, trans. Babcock and Krey, 2:7, 10; *Anonymous Syriac Chronicle*, trans. Tritton, pp. 94-95.

37 The rough February seas would have made beaching the fleet necessary; however, a significant part of it was launched later in the siege, restricting the amount of timber that the Franks could have harvested from these vessels.

38 Fulcher of Chartres, *Historia Hierosolymitana* 3.31.1, ed. Hagenmeyer, pp. 721-23, trans. Ryan, p. 262.

39 William of Tyre, *Chronicon* 13.6, ed. Huygens, 1:593, trans. Babcock and Krey, 2:10; *Eracles* 13.6, RHC Oc 1, p. 563.

phasis on artillery found in William's account is absent from Fulcher's. Although this may suggest a degree of creative licence was taken by the later author, many Syriac sources also specify that artillery was built by the Franks to support their siege towers.[40]

The garrison also employed artillery in their defence of the city. William of Tyre claims that the hail of projectiles unleashed by the defenders, from both artillery within the city and archers along the battlements, placed those guarding the Frankish siege towers and others moving to join or relieve the attackers within them in great peril.[41] Although there were still dangers faced by those inside, it would appear that the defenders' artillery was unable to threaten the immediate structural integrity of the siege towers. Accordingly, the strength of the defending and attacking engines would appear to have been roughly comparable.

Although William of Tyre asserts that the Frankish artillery threw 'huge' projectiles and that the walls and towers of Tyre were nearly demolished by the barrage, his reference to numerous shots passing over the walls and damaging buildings within the city suggest that it was the parapet that was targeted. Similar allusions to impressive power but also the concentration of fire along the battlements can be found in Fulcher's account. Had these engines posed a serious threat to Tyre's defences, there would have been little need to target the parapet. A more productive breach could be opened if shots were focused on lower sections of wall, causing the unsupported masonry above to collapse under its own weight. Despite William of Tyre's colourful imagery of dust clouds created by the ferocity of the Frankish barrage, there is little evidence to suggest that engines as implicitly powerful as this were employed even a half-century later when William was writing.

At one point in the siege, Fulcher of Chartres notes that the defenders mounted a raid against the most formidable of the Franks' machines, successfully burning it. Fulcher describes this engine as having been used to "shatter the towers and the city walls by hurling rocks and riddling the defences with holes."[42] Although no medieval trebuchet was ever capable of throwing projectiles cleanly through defensive masonry, this suggests that the projectiles were

40 Matthew of Edessa, *Patmut'iwn*, trans. Dostourian, pp. 233-34; *Anonymous Syriac Chronicle*, trans. Tritton, p. 95. Cf. Ibn al-Qalanisi, *Dhail*, trans. Gibb, pp. 170-72; Ibn al-Athir, *al-Kamil*, trans. Richards, 1:252-53.

41 William of Tyre, *Chronicon* 13.6, ed. Huygens, 1:593-94, trans. Babcock and Krey, 2:10-11.

42 ... *qua solebant turres antea iactis lapidibus conquassari et vehementissime perforari.* Fulcher of Chartres, *Historia Hierosolymitana* 3.32.1-2, ed. Hagenmeyer, pp. 728-30, trans. Ryan, pp. 264-65.

thrown hard enough to leave impressions, as seems to have been the case at Caesarea in 1101. However, this episode becomes more confusing when compared with William of Tyre's account of a very similar sally, which again led to the burning of a notable Frankish engine. William does not state what type of engine was ultimately destroyed, but his implication that the defenders targeted the siege towers suggests it was more likely one of these that was set alight than a notable trebuchet.[43] Unlike Fulcher, William states that the fire was extinguished and that the group responsible was apprehended and killed, leaving it entirely possible that the two sources are describing separate events.

Havedic

The climax of the siege in William of Tyre's account involves a certain character known as Havedic. William states that a particular stone-thrower of the defenders was wreaking havoc on the Frankish siege towers. Despite noting earlier that experts were manning the Frankish artillery, William claims that no one in the besiegers' camp was sufficiently skilled to destroy this menacing engine. Accordingly, an Armenian expert, Havedic, was summoned from Antioch and then paid from the public treasury upon his arrival. The Franks successfully targeted the defensive engine with Havedic's help, though William provides no further details.[44]

Havedic and the events surrounding him at Tyre are found only in William of Tyre's account of the siege.[45] Fulcher's silence is particularly notable as he was most likely living and writing in Jerusalem at this point in time.[46] Nevertheless, many historians have lent their interpretations to this episode, paying little consideration to the absence of a corroborative account.[47] Among them, Ronnie Ellenblum has argued that this may represent the first use of a counterweight trebuchet by Frankish forces.[48] Similarly, Paul Chevedden suggests that the Armenian introduced Byzantine gravity-powered artillery technology to the siege.[49] However, there is little evidence to suggest that any one of the ma-

43 William of Tyre, *Chronicon* 13.10, ed. Huygens, 1:597, trans. Babcock and Krey, 2:15.

44 William of Tyre, *Chronicon* 13.10, ed. Huygens, 1:597-98, trans. Babcock and Krey, 2:15-16.

45 Babcock and Krey suggested that William's specificity alone makes his account of this episode credible, William of Tyre, trans. Babcock and Krey, 2:16 n. 23.

46 Fulcher of Chartres, *Historia Hierosolymitana* 3.34.5, ed. Hagenmeyer, p. 736, trans. Ryan, p. 267.

47 For the figure of Havedic (Awétik) and the context, see Dédéyan, *Les Arméniens*, pp. 462-63, 513, 536, 861-63.

48 Ellenblum, *Crusader Castles*, p. 210.

49 Chevedden, "Invention," p. 92. This appears to clash with Chevedden's own theory that such engines had already been used by the Franks at Nicaea in 1097, Chevedden, "Invention," pp. 76-79.

chines employed at Tyre was more significant, powerful, or sizable than those used elsewhere by the Franks previously or that any Frankish engine was significantly more impressive than those of the defenders. Although it is possible that there was experimentation with small counterweight trebuchets as early as 1124, it cannot be concluded with any kind of certainty that Havedic was brought to Tyre to construct such an engine.

Those who suggest that Havedic was brought to build or operate a counterweight trebuchet appear to overlook certain fundamentals related to this technology. Firstly, any suggestion that a small counterweight trebuchet is more difficult to construct than a traction trebuchet of similar size is baseless: additional skill is only required when the components become too large to be moved by hand. Secondly, there are few benefits to employing a small counterweight trebuchet. Although such an engine might produce slightly more power than a traction trebuchet of comparable size, it will suffer from a much lower shooting rate. It is this trade-off that ensured that the traction trebuchet remained in use long after large counterweight trebuchets were developed. Finally, William of Tyre appears to imply that it was a need for greater accuracy that led to the summoning of Havedic, but it is hard to argue that an expert would have been needed to improve the accuracy of a counterweight trebuchet. Regardless of scale, the counterweight trebuchet was an extremely accurate engine. Provided the components, masses and loading angles were not altered between shots, projectiles of equal shape and mass would strike almost the same position shot after shot. Given the fixed alignment of the axle, shots would always fly along a fairly straight plane, with little deviation to the right or left. The engine's range could be altered quite predictably by making any number of small adjustments, such as changing the length of the sling or angle of the release finger. Such changes allowed subsequent shots to be walked in or out relatively easily. If accuracy was the problem that required an expert to address, it is more likely that the machines were traction trebuchets. Traction models require teams to pull in a single synchronised motion to maintain accuracy. Range can be adjusted by altering the amount of force applied by the team of pullers or by attempting to maintain a constant amount of force and altering the length of the sling, the angle of the release finger or the mass of the projectile. Regardless of which method is chosen, honing in on and maintaining a specific range is much more difficult.

William of Tyre provides no hint that Havedic constructed any kind of engine upon his arrival, let alone one with a new design or type of power source. If there is any truth in this anecdote, the artillery expert was probably summoned to improve the aim of the besiegers' traction trebuchets and to manage the crews operating them. As the targeted engine is described as being inside the city, thus almost certainly hidden from view, accuracy would have been an understandable issue. Although the Franks' artillery has attracted significant attention, it was the blockade that compelled the garrison of Tyre to sue for peace. On 7 July, in an anti-climactic end to a siege that had lasted five months, the garrison of Tyre marched out and Frankish forces took control of the city.[50]

Offensives of Zanki and John Comnenus

To the north, most references to artillery during the 1120s are found in Matthew of Edessa's chronicle. Among them, Matthew includes the employment of stone-throwers at Balak's siege of Kharput in 1122, after its Frankish prisoners had seized control of the castle, as well as his siege of Muslim-held Manbij in 1124. Although the Syriac sources, notably Michael the Syrian, Bar Hebraeus and the *Anonymous Syriac Chronicle*, claim that it was artillery rather than mining that compelled the Franks to surrender Kharput, Ibn al-ʿAdim appears to be the only one among the Muslim and Frankish sources to place artillery at either siege, noting its use at Kharput and against the citadel of Manbij.[51] Matthew also places artillery at al-Bursuqi's siege of ʿAzaz in 1125. This is corroborated by Fulcher of Chartres, William of Tyre and the *Anonymous Syriac Chronicle*; however, no mention of artillery is made by the Muslim sources.[52]

50 Fulcher of Chartres, *Historia Hierosolymitana* 3.34.1-3, ed. Hagenmeyer, pp. 733-35, trans. Ryan, p. 266; William of Tyre, *Chronicon* 13.13-14, ed. Huygens, 1:600-2, trans. Babcock and Krey, 2:18-21; Ibn al-Qalanisi, *Dhail*, trans. Gibb, pp. 170-72; Ibn al-Athir, *al-Kamil*, trans. Richards, 1:252-53.

51 For Kharput, Matthew of Edessa, *Patmut'iwn*, trans. Dostourian, pp. 229-30; Michael the Syrian, *Chronicon* 15.13, trans. Chabot, 3:211-12; Bar Hebraeus, *Makhtebhanuth Zabhne*, trans. Budge, 1:251; *Anonymous Syriac Chronicle*, trans. Tritton, p. 93. See also Fulcher of Chartres, *Historia Hierosolymitana* 3.23-26.2, ed. Hagenmeyer, pp. 676-91, trans. Ryan, pp. 246-54; Ibn al-Qalanisi, *Dhail*, trans. Gibb, p. 169; Ibn al-ʿAdim, *Zubdat al-Halab*, RHC Or 3, p. 637; Ibn al-Athir, *al-Kamil*, trans. Richards, 1:244; William of Tyre, *Chronicon* 12.18, ed. Huygens, 1:568, trans. Babcock and Krey, 1:541-44. Cf. Orderic Vitalis, *Historiae ecclesiasticae* 11.26, ed. and trans. Chibnall, 6:110-25. For Manbij, Matthew of Edessa, *Patmut'iwn*, trans. Dostourian, pp. 231-32; Ibn al-ʿAdim, *Zubdat al-Halab*, RHC Or 3, pp. 641-42. See also Fulcher of Chartres, *Historia Hierosolymitana* 3.31, ed. Hagenmeyer, pp. 721-29, trans. Ryan, pp. 262-64; Ibn al-Athir, *al-Kamil*, trans. Richards, 1:251.

52 Matthew of Edessa, *Patmut'iwn*, trans. Dostourian, pp. 234-36; Fulcher of Chartres, *Historia Hierosolymitana* 3.42.5-11, ed. Hagenmeyer, pp. 763-67, trans. Ryan, pp. 277-79; William of Tyre, *Chronicon* 13.16, ed. Huygens, 1:604-6, trans. Babcock and Krey, 2:25; *Anonymous Syriac Chronicle*, trans. Tritton, p. 97. See also Ibn al-Athir, *al-Kamil*, trans. Richards, 1:258.

Matthew specifies that twelve engines were used against ʿAzaz but that the breaches that were opened in the walls, also mentioned by Ibn al-Qalanisi, were brought about by mining. Despite the number of sources that take notice of artillery at these sieges, none of them provides any clear indication of the strength or power source of the machines. When considered in relation to instances of Fatimid-employed artillery, such as at the defences of Jerusalem (1099), Beirut (1110), Sidon (1110) and Tyre (1111-12, 1124), and assaults on Castellum Arnaldi (1106/7) and Jaffa (1123), as well as the apparent use of artillery by Tughtakin of Damascus against Baalbek (1110), the rulers of Mosul were far from the only Muslims in greater Syria making use of mechanical stone-throwers in the first quarter of the twelfth century.[53]

Al-Sawr: 1134

One of the most revealing descriptions of the potential power of Muslim artillery in the early twelfth century can be found in an eyewitness account of Zanki's siege of al-Sawr (Savur) composed by the warrior poet, Usama ibn Munqidh.

> Zanki then set up the mangonels against the citadel, which took down one side of it. But not enough of it was brought down for the men to use the breach to get up into the citadel. However, one of the atabeg's bodyguards, a man from Aleppo called Ibn al-ʿAriq, climbed up through the breach and set to striking the enemy with his sword. But they injured him with a number of wounds and threw him down from the tower into the moat. By then, our men had overwhelmed them at that breach and we took possession of the citadel. The representatives of the atabeg climbed up to the citadel and took possession of its keys, sending them to Timurtash [ibn Ilghazi of Mardin], and granting Zanki the citadel.[54]

It seems reasonable to assume that the breach mentioned was made in the battlements rather than the bulk of the wall; this would explain why it was still

---

Ibn al-Qalanisi appears mistaken when placing the assault on ʿAzaz with the events of AH 517 (1123-24), an understandable error given the similarities in circumstance between Balak's assault on Manbij in 1124 and al-Bursuqi's assault on ʿAzaz in 1125, Ibn al-Qalanisi, *Dhail*, trans. Gibb, p. 170. Chevedden has argued that counterweight trebuchets were used at ʿAzaz, pointing to the mention of *manganiqe rawrbe* (great mangonels) by the later *Anonymous Syriac Chronicle* as evidence of such, Chevedden, "Invention," p. 92.

53 See Appendix 1.

54 Usama ibn Munqidh, *Kitab al-Iʿtibar*, trans. Cobb, p. 168. Cf. Ibn al-Athir, *al-Kamil*, trans. Richards, 1:306.

difficult to exploit, hardly assailable by more than one man at a time as a degree of climbing was still necessary, but sufficient to provide access to the rampart. Although the citadel fell to a general assault, using the breach to access the parapet, Usama's indication that some form of surrender took place, evident with the transference of keys, implies that the assault was perhaps less overwhelming than his description implies.[55] An entirely different set of events a few years later provides an opportunity to gain further insight.

Montferrand: 1137

Zanki besieged Homs in May 1137. Failing to make progress, he cut his losses after a few weeks and moved on to invest the Frankish castle of Montferrand from early June.[56] On a rise overlooking Raffaniya, Montferrand was a bastion of Frankish influence near Homs and Hama and was described by Ibn al-Athir as "one of the Franks' strongest and most impregnable castles."[57] Zanki had previously besieged Pons of Tripoli in the castle unsuccessfully in 1133;[58] four years later, the castle's defence was led by Fulk of Jerusalem and Raymond II of Tripoli, after an attempt to break the siege failed and the relief party was forced to take refuge with the besieged.

Because of the rugged strength of the castle's position, Zanki encamped his army in the plains below Montferrand, assigning the task of assaulting the castle to the emirs of his army in turn.[59] Usama describes the siege as a fairly lax endeavour, pressed without vigour by the emirs who faced it.[60] The later accounts of Ibn al-'Adim, Ibn 'Abd al-Zahir and Ibn al-Furat, however, present

---

55 It is interesting that Zanki opted for an attack focussed on escalade, supported by artillery, as Usama mentions the presence of forces from Khursan, a virtual pseudonym for miners. For this convention see Usama's accounts of the sieges of Kafartab in 1115 and Masurra in 1132, and Ibn al-Qalanisi's accounts of the sieges of Banyas in 1132 and Edessa in 1144, Usama ibn Munqidh, *Kitab al-I'tibar*, trans. Cobb, pp. 85-87, 171; Ibn al-Qalanisi, *Dhail*, trans. Gibb, pp. 216-17, 267. See also Prouteau, "Beneath the Battle," pp. 105-17.

56 Homs was then an ally of his opponents in Damascus, Ibn al-Qalanisi, *Dhail*, trans. Gibb, pp. 237-38; Ibn al-Athir, *al-Kamil*, trans. Richards, 1:325-25.

57 Ibn al-Athir, *al-Kamil*, trans. Richards, 1:336.

58 William of Tyre, *Chronicon* 14.6, ed. Huygens, 2:637-38, trans. Babcock and Krey, 2:56; Ibn al-Qalanisi, *Dhail*, trans. Gibb, pp. 221-22; Ibn al-Athir, *al-Kamil*, trans. Richards, 1:303.

59 Usama ibn Munqidh, *Kitab al-I'tibar*, trans. Cobb, p. 169.

60 The rotating emirs possibly judged their individual prospects of taking the stronghold on the day that they opposed it, and the plunder which would accompany its fall, as too slight to risk the commitment of resources a real attempt would require. The castle was also well defended: when the relief force joined the garrison, the castle's provisions would have been consumed at a much more rapid rate but the combined defending force would have been capable of launching powerful sorties.

a more active siege, emphasising Zanki's use of artillery – ten mangonels according to the former historian and fourteen according to the latter Mamluk authors.[61] William of Tyre seems to support this latter interpretation:

> Zangi continued his vigorous attacks upon the besieged with unremitting zeal. The very walls shook under the impulse of his mighty engines. Millstones [or large stones] and huge rocks hurled from the machines fell into the midst of the citadel, shattered the houses within, and caused intense fear to the refugees there.[62]

The phrases that William uses to describe the destruction are part of an established trend: they emphasise the destruction inflicted within the enceinte and the psychological impact of the barrage as much as the damage sustained by the defences. Although it might be suggested that sources unaccustomed to taking part in military actions, such as William of Tyre, might overemphasise the damage sustained by non-military structures more familiar to them, these unfortified buildings, and even non-combatants, could have been legitimate targets. By bombarding the interior of an enceinte fairly indiscriminately, besiegers may have hoped to increase the hardships faced by defenders, eroding their sense of security and weakening morale. While the Franks inside Montferrat endured Zanki's siege, Antioch faced the wrath of the Byzantines.

Antioch: 1137

At the culmination of a fairly rapid campaign, which saw John Comnenus take Tarsus and many of the other fortified towns and castles in Cilicia that had fallen under Armenian or Frankish rule, the Byzantine army pushed into Syria and besieged Antioch.[63] Accounts of the siege differ dramatically. William of Tyre characterises this as an aggressively prosecuted siege, claiming that

61 Ibn al-'Adim, *Zubdat al-Halab*, RHC Or 3, p. 673; Ibn 'Abd al-Zahir, *al-Rawd*, trans. al-Khowayter, 2:652; Ibn al-Furat, *Tarikh*, trans. Lyons and Lyons, 2:115.

62 *Sanguinus interea, obsessos continuis urgens molestiis, menia tormentis quatiens, machinis molares et saxa ingentia iaculatoriis in medium contorquens presidium, domos prosternit interius non sine multa inclusorum formidine.* William of Tyre, *Chronicon* 14.28, ed. Huygens, 2:667, trans. Babcock and Krey, 2:89.

63 This accumulation of territory in Cilicia was the primary objective of the Byzantine campaign, Vahram, *Chronicle*, trans. Neumann, pp. 31-32; Matthew of Edessa, *Patmut'iwn*, trans. Dostourian, pp. 241-42; Niketas Choniates, *Historia*, trans. Magoulias, pp. 12-18. Antioch, which had not been returned to the Byzantines following its capture by the Franks in 1098, passed to Raymond of Poitiers, son of William IX of Aquitaine, in 1136 when he married Constance, daughter of the late Bohemond II and Alice of Jerusalem.

"Mighty machines and engines were placed in strategic positions around the city and ever-increasing pressure was exerted upon the place."[64] Ibn al-Qalanisi instead characterises the siege as a fairly passive blockade.[65] According to John Kinnamos, who presents the broader campaign as a response to a scandalous conspiracy hatched by the prince of Antioch to resist Byzantine influence in Cilicia, the siege began as a blockade but attacks became more frequent as time went on. Although he places artillery at the siege of Anazarbos weeks earlier, he makes reference to no such engines at Antioch.[66] Niketas Choniates appears to confirm that John Comnenus was travelling with artillery but omits any mention of a siege, asserting that the emperor was welcomed into Antioch, and Vahram does not describe the Byzantine army as continuing beyond Cilicia.[67]

Besides the variation presented by the different accounts, circumstances surrounding the siege bring into question William of Tyre's characterisation of it. Fulk's choice to continue to Montferrand rather than move to relieve Antioch and the decision, let alone ability, of Raymond of Poitiers, prince-regent of Antioch, to leave his besieged city to assist Fulk and Raymond II at Montferrand and then slip back into Antioch afterwards are telling. It may be that William misunderstood or slightly exaggerated things. John Kinnamos claims that Raymond and Baldwin of Marash left Antioch to rescue Fulk while the Byzantines were still in Cilicia and were able to return, fearing their approach, before the Greek army arrived at Antioch.[68]

Even if William misjudged the timing of events and overstated the tempo of the siege, his emphasis on the Byzantines' use of artillery should not be dismissed without further consideration.

> The emperor, for his part, caused immensely heavy rocks to be hurled from the mighty machines and engines. In this way, he sought to weaken and break down the defences of the city and to shatter the walls and

---

64 *Ordinatis itaque per girum machinis et ingentibus tormentis congruis stationibus collocatis, urbem vehementius cepit artare.* William of Tyre, *Chronicon* 14.24, ed. Huygens, 2:663, trans. Babcock and Krey, 2:85.

65 Ibn al-Qalanisi, *Dhail*, trans. Gibb, pp. 240-41.

66 John Kinnamos, *Deeds* 1.7-8, trans. Brand, pp. 21-24.

67 Niketas Choniates, *Historia*, trans. Magoulias, pp. 14-16, 18; Vahram, *Chronicle*, trans. Neumann, pp. 31-32.

68 John Kinnamos, *Deeds* 1.8, trans. Brand, p. 23. Caution must also be taken with John's account, as he freely admits that he did not have the best understanding of the events of this campaign, see 1.8, trans. Brand, p. 25.

> towers of the gate of the Bridge. Aided by a strong band of slingers, they sought at long range to prevent the townspeople from defending the walls and were ever on the watch for an opportunity to approach and undermine the fortifications.[69]

Like many of William's descriptions of artillery, this passage is in itself inconsistent. If the Byzantine artillery was sufficient to break down the defences of Antioch, why did these remain effectively defended, let alone require sappers to undermine them? When considering the topography, William's account may have been influenced by those of the First Crusade: the Bridge Gate was one of only a few areas where artillery may have been used in 1097-98.[70] Any hopes of discerning an accurate sense of this siege are dashed if an attempt is made to reconcile this with the less detailed but drastically differing accounts of other sources.[71]

Shayzar: 1138

After peace was established between Raymond of Poitiers and John Comnenus, there seems little doubt that the Byzantines unleashed the full might of their artillery at the subsequent siege of Shayzar in 1138. The citadel of Shayzar sits on a river-cut spur that rises from the left bank of the Orontes, overlooking the town to the west. Following their arrival, the Greeks positioned their camp on the west side of the river and surrounded the town.[72]

Ibn al-Athir states that the Byzantines employed eighteen trebuchets against Shayzar.[73] Ibn al-'Adim gives the same number and adds that four *la'ab* (lighter

---

69 ... *imperator quoque versa vice tormentis ingentibus et machinis iaculatoriis cautes inmanissimos et inmensi ponderis contorquendo, a Porta Pontis menia cedendo et turres, urbis claustra debilitare et effringere nitebatur. Dispositis quoque per girum legionibus, sagittis, et omni missilium genere necnon et fundibulariorum manu proterva cives a muri propugnatione arcebant eminus et ad subfodienda menia aditum et oportunitatem nitebantur vendicare.* William of Tyre, *Chronicon* 14.30, ed. Huygens, 2:670, trans. Babcock and Krey, 2:92.

70 See Chapter 3.

71 In addition to the accounts noted above, cf. Michael the Syrian, *Chronicon* 16.8, trans. Charbot, 3:245; Bar Hebraeus, *Makhtebhanuth Zabhne*, trans. Budge, 1:264.

72 For the defences of Shayzar, see Tonghini and Montevecchi, "The Castle of Shayzar: Access System," pp. 201-24; Tonghini and Montevecchi, "The Castle of Shayzar: Recent Archaeological Investigations," pp. 137-50; Tonghini, "Die Burg Saizar," pp. 234-40; Kennedy, *Crusader Castles*, p. 181; Müller-Wiener, *Castles of the Crusaders*, *p.* 55. See also Ibn al-Athir, *al-Kamil*, trans. Richards, 2:87-89.

73 Ibn al-Athir, *al-Kamil*, trans. Richards, 1:340.

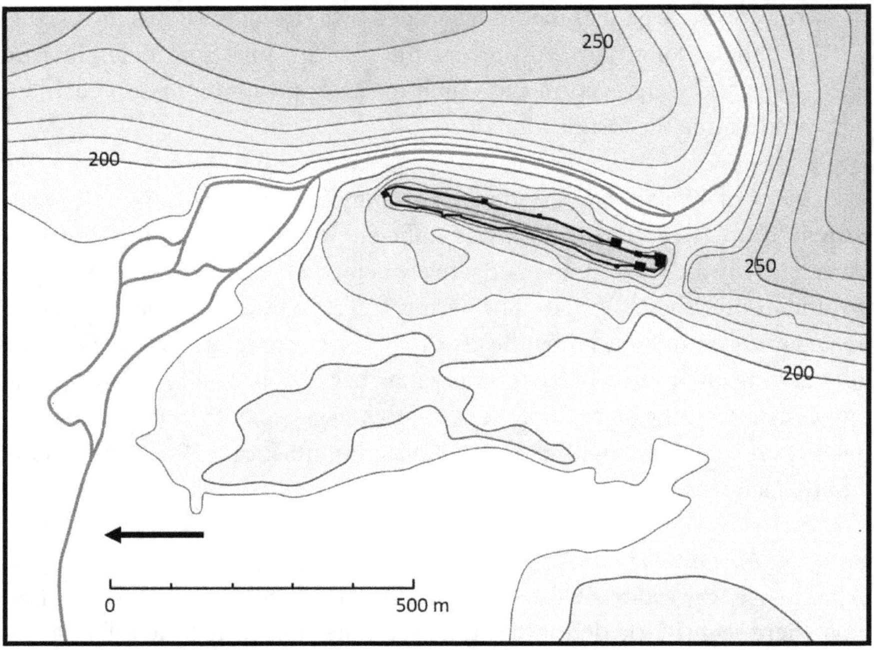

PLAN 7 *Shayzar: citadel with the area of the town to the west*

stone-throwers)[74] were also erected.[75] From the description provided by William of Tyre, it must have seemed as though hell was unleashed upon the town:

> From the machines, set up in strategic positions, poured forth constant volleys of heavy stones which shook the towers and walls and even the houses of the people within. Under the repeated blows of these enormous missiles, the fortifications, on which the inhabitants had relied as their greatest defence, were utterly overthrown and in their fall wrought dreadful havoc among the townspeople.[76]

74 For this term, see the description provided by al-Tarsusi, translated in Chevedden et al., "The Traction Trebuchet," p. 461.

75 Ibn al-ʿAdim, *Zubdat al-Halab*, RHC Or 3, pp. 677-78. This source states that the Byzantines besieged Kafartab, bombarding it with mangonels, and compelled the garrison to surrender before moving against Shayzar. Unfortunately, his descriptions provide few details that might allow for a better understanding of these engines.

76 *Hic demum machinis congrua provisione dispositis, turres ac menia et infra muros civium domicilia gravium immissione molarium incessanter concutiunt et crebris ictibus et vicaria immissorum cautium repetitione non sine multa inhabitantium strage funditus deiciunt, in*

This rhetoric suggests a greater degree of destruction than most of William's earlier descriptions of artillery damage. It is possible that the Byzantines made use of more powerful engines than most of those employed by their Frankish and Muslim neighbours; alternatively, William may have been inspired to characterise them as such given the prestige of their leader and Roman heritage. It is important to bear in mind that William was a boy at this time and the source that he later used when composing his history is unknown. From a Greek perspective, Niketas Choniates, born the better part of two decades after the events in question, states that although the Byzantine artillery was able to open breaches in the defences of Shayzar, its fortifications remained entirely defensible.[77] It is possible that the 'breaches' were rather sections of fortifications where the battlements had been damaged, as seems to have been the case at al-Sawr three years earlier.

William of Tyre states that John Comnenus walked among his troops, urging those who worked the engines to take better aim and increase their shooting rate.[78] Although the evidence is far from definitive, this kind of encouragement would have been more useful to teams working traction trebuchets than larger counterweight engines. It is almost impossible for experienced operators to speed up any component of the loading and shooting process of the latter, from waiting for the beam to come to a natural state of rest to the steady hauling or winching back of the arm, while the consistent shooting sequence of such engines makes them naturally accurate. On the other hand, the effective operation of a traction trebuchet depends on close teamwork: synchronicity and consistency are both required to maintain a close measure of accuracy and rapid shooting rate. Furthermore, if William of Tyre is implying that the artillery teams were working in relays, as he states the fighting men were, there can be little doubt that these were traction- rather than counterweight-powered machines, the former being far more physically taxing and thus crews would more likely benefit from encouragement. So once more William of

---

quibus erat defensionis spes maxima, edificiorum munimina. William of Tyre, *Chronicon* 15.1, ed. Huygens, 2:674, trans. Babcock and Krey, 2:94-95.

77 Niketas Choniates, *Historia*, trans. Magoulias, p. 18. John Kinnamos, born about five years after the siege, does not mention artillery at Shayzar, John Kinnamos, *Deeds* 1.8, trans. Brand, p. 24.

78 William of Tyre, *Chronicon* 15.1, ed. Huygens, 2:674, trans. Babcock and Krey, 2:95. Whether a common practice or literary convention, accounts of Byzantine commanders moving among their troops offering encouragement appear fairly regularly. For example, Leo Tornikios at the siege of Constantinople in 1047, Michael Psellus, *Chronographia*, trans. Sewter, p. 211.

Tyre's assessment of the degree of damage appears to conflict with the manner in which it was supposed to have been inflicted.

Paul Chevedden has looked to other sources to argue that counterweight trebuchets were used against Shayzar. His theory is based largely on Ibn al-'Adim's labelling of the Byzantine engines as *majaniq 'izam* (great mangonels) and Usama ibn Munqidh's designation of these engines as *majaniq ha'ilah* (huge/frightful mangonels).[79] However, Ibn al-'Adim's account was written in the mid-thirteenth century and is less detailed than those of Usama and William of Tyre. Although Usama was a native of Shayzar and was certainly the closest source to these events, he was absent during the siege in 1138 and appears to have relied on the testimony of others. Regardless, in neither case is the use of such a descriptive adjective conclusive evidence of a certain type of technology.

Usama claims that the Byzantine artillery threw stones weighing 20 to 25 *artal* to a distance greater than that of a bowshot.[80] If this loosely correlates to a weight of 50 kg (using the 2.14 kg *ratl* of Shayzar), and a range of close to 200 m, allowing for no exaggeration, this would be beyond the range of a traction trebuchet. While this range is open to question, so too is the mass of the projectiles. Although Usama was born in Shayzar, he held notable positions in both Damascus and Cairo during the mid-twelfth century. This allows for the possibility that he was using the 1.85 kg Syrian *ratl*, which would appear to cater to his intended audience, or even the 0.41 kg Egyptian *ratl*, closer to Hugh Kennedy's suggestion that the *ratl* was about half a kilogram.[81] If a 10 kg projectile was employed (using the Egyptian *ratl*), this could easily have been cast to a more reasonable range of 100-150 m by a traction trebuchet. While a significant counterweight trebuchet would have been able to throw the heavier mass the greater distance, there is only limited evidence to suggest the presence of one apart from the fairly general references to structural damage. Although Usama asserts that one shot destroyed an entire house and William of Tyre similarly states that large stones soared over the walls and caused destruction among the buildings within the enceinte, there is little descriptive, extrapolative or archaeological evidence to corroborate these colourful descriptions of destruction.

In retelling stories told to him by one of his father's mamluks, Usama provides a grisly personal perspective. As the defenders regrouped at one point

79 Chevedden, "Invention," pp. 92-93.

80 Usama ibn Munqidh, *Kitab al-I'tibar*, trans. Cobb, p. 125. The range of a traction trebuchet relative to that of an archer is discussed above in relation to the siege of Jaffa in 1123.

81 Kennedy, "Shayzar," p. 12.

during the siege, the man next to the storyteller turned to relieve himself and while doing so was struck on the head by a projectile cast by one of the Greek trebuchets. The force of the stone was sufficient to stick the man's head to the wall upon which he was leaning at the time. On a different occasion, a man was struck in the leg by a projectile and while others rushed to find a 'bonesetter' to tend to the broken leg another stone hit the wounded man on the head, crushing his skull.[82] Although gruesome, these casualties do not convey a dramatic sense of power; there is nothing to suggest that the projectiles responsible for these injuries were any closer to 50 kg than to 5 kg or that they were thrown with a velocity closer to 50 m/s than 30 m/s. By comparison, Ammianus describes almost identical injuries inflicted by classical torsion-powered scorpions about eight centuries earlier.[83] Furthermore, the impact of these personal injuries appears to overshadow any notion of colossal destruction. So while there is no direct evidence to rule out the use of small counterweight trebuchets at this siege, neither is there particularly strong evidence to support their presence.

William of Tyre's notion that this was a "vigorous but ineffective siege of the city" is slightly misleading as the assailants eventually managed to overwhelm the town defences and break into Shayzar.[84] Both Usama and Niketas Choniates claim that the Byzantine artillery was responsible for opening a breach in the town wall but only the former claims that this was stormed. It may have been that what played out was quite similar to the final stages of the siege of al-Sawr in 1134, only in this instance it was the town defences that were damaged rather than those of the citadel. John Comnenus was eventually inclined to take a payoff from the Banu Munqidh rather than continue to besiege the citadel, in part motivated to do so by the half-hearted support of his Frankish allies, Raymond of Antioch and Joscelin II of Edessa.[85]

After a siege of twenty-three days, according to Ibn al-Qalanisi, the Graeco-Frankish army withdrew back to Antioch.[86] Ibn al-'Adim claims that the Byzantines burnt their siege engines before departing but that they left behind some large trebuchets, which Zanki subsequently had taken to the citadel of

82 Usama ibn Munqidh, *Kitab al-I'tibar*, trans. Cobb, p. 126.

83 Ammianus, *Rerum gestarum* 19.2.7, ed. Eyssenhardt, p. 144, trans. Yonge, p. 188.

84 William of Tyre, *Chronicon* 15.1, ed. Huygens, 2:675, trans. Babcock and Krey, 2:96.

85 William of Tyre, *Chronicon* 15.2, ed. Huygens, 2:675-76, trans. Babcock and Krey, 2:96-97. William of Tyre is generally hostile to both Frankish characters; but in their defence, Raymond was essentially being compelled to fight for a new fief that would exclude Antioch if Shayzar were to fall.

86 Ibn al-Qalanisi, *Dhail*, trans. Gibb, p. 252.

Aleppo.[87] If indeed Zanki salvaged certain Byzantine engines, these may have been among those that he employed at Baalbek when he besieged the city from August to October 1139, although it is at least as likely that these engines were brought from elsewhere or constructed on site.

While the increasing frequency with which artillery is mentioned is noteworthy, what is perhaps most significant about the siege of Shayzar is that for the first time Ibn al-Qalanisi quantifies the number of engines that were used – fourteen.[88] This figure is repeated by both Ibn al-Athir and Ibn al-'Adim, and supported by the *Anonymous Syriac Chronicle*, which provides an emotive description of the siege, stating that a thousand stones were thrown against the citadel each day.[89] It is tempting to suggest that Ibn al-Qalanisi, at this point in his sixties and still living in Damascus, had begun to take a greater interest in siege technology as Zanki's army posed an increasing threat to Damascus.

Zanki had likely set his sights on Damascus soon after the death of Tughtakin in 1128 and through the 1130s he extended Aleppan control southward at the expense of Damascene influence. The defection of the Damascene governor of Banyas in 1137, followed by his capture of Baalbek in 1139, provided Zanki with bases of operation about 60 km to the north and south of Damascus. This allowed Zanki to camp against the city, imposing a loose blockade, through the winter and spring of 1139-40.

Banyas: 1140

In late spring 1140, the army of Jerusalem moved to support Damascus against the aggression of Zanki. In return for their assistance, the Damascenes had promised to deliver Banyas to the Franks. Taking a significant contingent of his army, Mu'in al-Din Unar, regent of Damascus, accompanied the Franks to Banyas and together they invested the town. The Damascene forces took up positions to the east of the town, while the Frankish forces were arrayed to the west.

The artillery employed by the besiegers is described in similar terms to that used previously at Montferrand (1137) and Shayzar (1138). "From the hurling engines called *petrariae* they threw huge stones of great weight, which shook the walls and demolished buildings within the city itself," wrote William of Tyre.[90] He adds that there was no safety to be found within the town, due to

---

87 Ibn al-'Adim, *Zubdat al-Halab*, RHC Or 3, p. 678. For a brief discussion of the siege and the sources, see Kennedy, "Shayzar," pp. 11-16.

88 Ibn al-Qalanisi, *Dhail*, trans. Gibb, pp. 254-55.

89 Ibn al-Athir, *al-Kamil*, trans. Richards, 1:349; Ibn al-'Adim, *Zubdat al-Halab*, RHC Or 3, p. 681; *Anonymous Syriac Chronicle*, trans. Tritton, pp. 273-74.

90 ... *et machinis iaculatoriis, quas petrarias vocat, immissis magnae quantitatis molaribus, moenia concutiunt* ... William of Tyre, *Chronicon* 15.9, ed. Huygens, 2:686, trans. Babcock

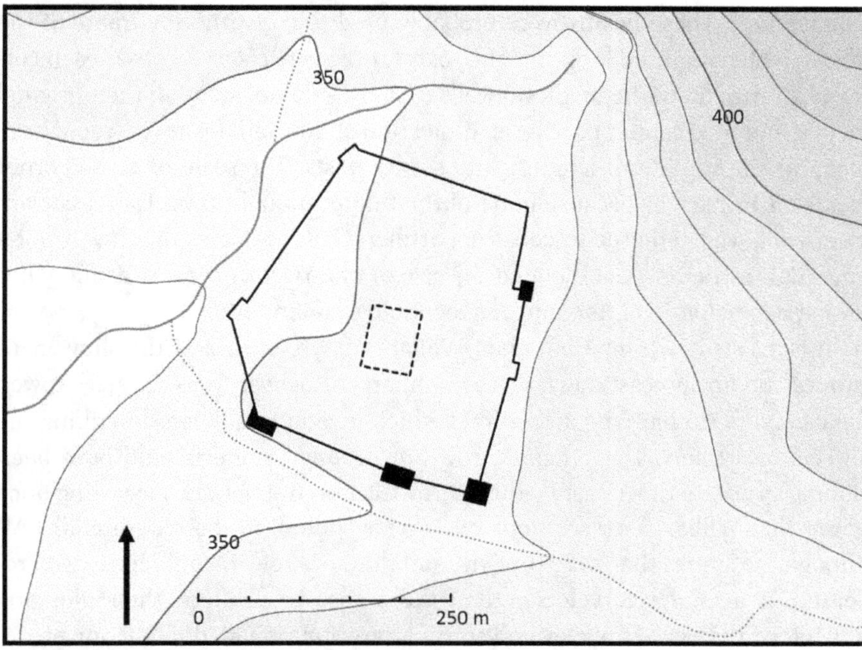

PLAN 8 *Banyas: town defences, including Ayyubid and Mamluk additions*

the showers of arrows and other projectiles, and even those manning the ramparts scarcely dared to look out from behind the crenellated parapet. Here, as elsewhere, William may have exaggerated the degree of material destruction in order to impress upon his readers the psychological impact of such a barrage. Notwithstanding what was evidently an impressive artillery bombardment, confirmed by Ibn al-Qalanisi,[91] both chroniclers testify that it achieved little in advancing the siege. With their initial efforts producing few tangible results, the besiegers turned to a staple in the Frankish siege arsenal: the siege tower.

Although the topography around Banyas was suitable for the use of a siege tower, there was apparently a shortage of suitable timber. This was a fairly lush and fertile area, benefiting from a number of springs and the runoff from the heights to the north; however, the local timber was apparently inadequate as the wood that was used to build the siege tower was sent for and brought from

---

and Krey, 2:108. This is William of Tyre's first use of the term *petraria* since his account of the First Crusade.

91 Ibn al-Qalanisi, *Dhail*, trans. Gibb, pp. 259-61.

Damascus.[92] These beams were probably used to construct the frame of the tower, which required long straight posts for the vertical supports at each corner. Unfortunately, the medieval walls of Banyas do not stand to their original height and thus cannot provide an indication of how tall the tower would have been or the minimum length of its corner posts. The scale of these corner beams is important because none of the timber brought from Damascus appears to have been used to construct artillery. This indicates that it was not a material restriction that dictated the size of mid-twelfth-century artillery but rather a perception of its value or a technological limitation.[93]

It is also noteworthy that when William of Tyre speaks of the showers of stones and arrows cast down on the ramparts of Banyas from the siege tower, he uses very similar language to that which he employs when describing the effects of artillery. These hand-thrown or dropped stones would have been similar in size to those cast by the nearby traction trebuchets, suggesting once more that William's descriptions of damage should not be read literally. Although the stones that were thrown from the siege tower would have been released at a far lower velocity than those cast by artillery, their plunging trajectory, thanks to the elevated starting position provided by the tower, denied the defenders the protection of their battlements. This gave the besiegers a decisive advantage and ultimately compelled the garrison to surrender.[94]

Edessa: 1144

Zanki more than compensated for the loss of Banyas with his acquisition of Frankish Edessa in 1144.[95] Seizing an opportune moment, he implemented

92 William of Tyre, *Chronicon* 15.9, ed. Huygens, 2:686-87, trans. Babcock and Krey, 2:108-9. Although William states that these beams were kept for just such a purpose, when considering the rarity with which Muslim forces employed timber-framed siege towers during the twelfth century, this seems unlikely. These, among other 'war materials', were probably part of a general stockpile of materials, used to patch up defences and perhaps raise counter-towers during a siege.

93 Unlike counterweight trebuchets, which could be built as large as the building materials and construction methods permitted, traction trebuchets had mechanically determined maximum dimensions, see Chapter 2.

94 William of Tyre, *Chronicon* 15.10-11, ed. Huygens, 2:687-91, trans. Babcock and Krey, 2:109-12. Ibn al-Qalanisi does not mention the construction of a siege tower at any point in his account of the approximately twenty-three day siege. He instead attributes the town's eventual surrender to the effects of the blockade.

95 Edessa had been besieged by a number of figures in the century leading up to this: Alp Arslan (using a siege tower and mining) in 1070/71; Turkish forces (using artillery) in 1095/96, and then by Ridwan of Aleppo and Taghisiyan of Antioch after they broke the Turkish siege; Jokermish of Mosul in 1105/6; Kilij Arslan in 1106; Mawdud of Mosul in 1110

what can be seen as a relatively standard siege tactic employed by Muslim forces in northern Syria: when a frontal assault was impractical, he turned to mining, supporting his sappers with light artillery. Although such a generalisation risks oversimplifying the broad variety of available siege tactics and particular preferences of various personalities, mining was frequently employed by Muslim commanders when frontal assaults failed. Whether it was used, or simply noticed, more often, artillery is found with increasing frequency providing cover for sapping efforts through the twelfth century, often identified alongside archers and slingers.

Gregory the Priest, followed by Bar Hebraeus, states that Zanki erected seven stone-throwers against the defenders manning the parapet of the city, but elaborates no more as he clearly implies that sapping was the focus of Zanki's siege plan.[96] According to the *Anonymous Syriac Chronicle*, each of the leading figures in Zanki's force erected a trebuchet. These engines, as well as the miners, were responsible for inflicting damage on the city's walls; however, it was the latter who opened the breach that led to the city's fall.[97] The descriptions left by Ibn al-Qalanisi and William of Tyre, as well as the more extensive versed account of Nerses Shnorhali, reveal little additional information about the engines themselves or how they were used, but all three confirm that mining contributed most to the city's fall.[98] The breach that led to the storming of Edessa appears to have been made near the north gate. Seven towers in this area required repairs after the siege, though the extent of this damage and whether it was inflicted by Zanki's sappers or the artillery supporting them is unclear.[99]

---

and 1112; al-Bursuqi in 1115; and briefly by Turkish forces in 1138, Matthew of Edessa, *Patmut'iwn*, trans. Dostourian, pp. 131-32, 162-63, 197, 199, 209-11, 215-16; Ibn al-Athir, *al-Kamil*, trans. Richards, 1:106, 166; Bar Hebraeus, *Makhtebhanuth Zabhne*, trans. Budge, 1:265; *Anonymous Syriac Chronicle*, trans. Tritton, pp. 82-86.

96 Gregory the Priest, in Matthew of Edessa, *Patmut'iwn*, trans. Dostourian, pp. 243-44; Bar Hebraeus, *Makhtebhanuth Zabhne*, trans. Budge, 1:268.

97 *Anonymous Syriac Chronicle*, trans. Tritton, pp. 282-83.

98 Ibn al-Qalanisi, *Dhail*, trans. Gibb, pp. 266-68; William of Tyre, *Chronicon* 16.4-5, ed. Huygens, 2:717-21, trans. Babcock and Krey, 2:141-43; Nerses, *Voghb Yedesyo* ll. 835-1,146, RHC Ar 1, pp. 246-62. See also Ibn al-Athir, *al-Kamil*, trans. Richards, 1:372-73; Ibn al-'Adim, *Zubdat al-Halab*, RHC Or 3, pp. 685-86.

99 *Anonymous Syriac Chronicle*, trans. Tritton, pp. 289-90.

The Second Crusade and the Ebb of Frankish Dominance

Zanki's successful siege of Edessa precipitated the Second Crusade. Despite the involvement of the leading barons of Outremer and many of the most prestigious lords of France and Germany, there are no indications that artillery, or any other type of siege engine, was used during the climactic push against Damascus in 1148.[100] However, the preceding siege of Lisbon by certain contingents on their way to the Holy Land is one of the most informative sieges to any study of twelfth-century artillery.[101]

Lisbon: 1147

Fifty years after the siege of Nicaea, and a continent away, a fleet of Anglo-Norman, German and Flemish contingents sailed into Lisbon in June 1147 on its way to the Levant. Despite the geographic separation, the ensuing episode clearly demonstrates the similarities between the siege tactics concurrently employed in Europe and the Near East. Importantly, the most detailed surviving eyewitness account of the ensuing siege, the anonymous *De Expugnatione Lyxbonensi*, provides revealing descriptions of the crusaders' artillery.

Disembarking against Muslim Lisbon, the crusaders surrounded the city: the Anglo-Normans took up a position to the west of the town while the Germans, with the Flemings on their left, faced the eastern side of the city.[102] Fifteen days after the siege had begun, the Franks began to construct siege engines. The Anglo-Normans built a siege tower and the contingents from Cologne and Flanders built a penthouse, ram and siege tower, as well as five *fundae Baleares* to support them. These 'Balearic slings' were mechanical stone-throwers, directed against the walls and towers of Lisbon. The ensuing attack met with complete failure on all fronts and the Anglo-Norman siege tower, which became stuck in the sand, was an easy target for three defending trebuchets, which played upon it for four days and nights until it was finally burnt in a sally.[103] The inability of the defending artillery to destroy the immobilised tower suggests that it was traction-powered; in this regard, there are clear similarities with the siege of Tyre in 1124.

---

100 See William of Tyre, *Chronicon* 17.1-2, 5-7, ed. Huygens, 2:760-61, 65-69, trans. Babcock and Krey, 2:184-86, 190-95; Ibn al-Qalanisi, *Dhail*, trans. Gibb, pp. 283-87; Ibn al-Athir, *al-Kamil*, trans. Richards, 2:21-22.

101 For Eurocentric interpretations of the importance of this account, see Köhler, *Die Entwickelung des Kriegswesens*, 1:164; Gillmor, "Introduction of the Traction Trebuchet," p. 2.

102 *De Expugnatione Lyxbonensi*, ed. and trans. David, pp. 124-27.

103 *De Expugnatione Lyxbonensi*, ed. and trans. David, pp. 134-37. For a similar use of the term, see Ralph of Caen, *Gesta Tancredi* 125, RHC Oc 3, p. 692.

Around September 1147, the Anglo-Normans employed two *fundae Baleares* of their own to cover mining efforts against an exposed section of the western wall:

> one on the river bank which was operated by seamen, the other in front of the Porta do Ferro, which was operated by the knights and their table companions. All these men having been divided into groups of one hundred, on a given signal the first hundred retired and another took their places, so that within the space of ten hours five thousand stones were hurled.[104]

This passage has formed the basis of many interpretations of mid-twelfth century artillery; but, unfortunately, some scholars have accepted it at face value and failed to analyse it more thoroughly.

These two sentences reveal a number of interesting details. First, that one engine was worked entirely by seamen supports a link between artillery and nautical traditions.[105] This link is logical given the levered basis of trebuchet mechanics and the critical use of cordage. This was not a limiting factor, however, as the knights and their retainers seem to have had no difficulties operating the other engine.

Second, the machines were again employed in an antipersonnel capacity, protecting mining efforts from sorties launched by the garrison and defenders along the parapet. Scholars who have dwelt on the number of men involved in the operation of these machines seem to have lost sight of this, preferring instead to see them as something larger and more threatening to the city's fortifications.[106] This leads to the third and most misinterpreted aspect of this episode: that the teams operating the machines were each composed of one hundred men, does not mean that these were traction trebuchets powered by one hundred men all pulling at the same time – the essential mechanics of this technology rule out this possibility. The distance that each man is able to pull limits the length of the short arm of the beam, which in turn limits the remaining mechanical variables and the space available for the men to pull from. Thus

104 *Insuper due funde Balearice a nostris eriguntur, una supra ripam fluminis a nautis trahebatur, altera contra portam ferream a militibus et eorum convictualibus. Hii omnes per centenos divisi, audito signo exeuntibus primis centenis, alii centeni subintrassent, ut inter decem horarum spatia v milia lapidum iactarentur. De Expugnatione Lyxbonensi*, ed. and trans. David, pp. 142-43.

105 For a broader introduction to this relationship, see Rogers, *Latin Siege Warfare*, pp. 201-7. Cf. William of Tyre, *Chronicon* 8.9, ed. Huygens, 1:399, trans. Babcock and Krey, 1:357.

106 For example, Chevedden, "King James I," pp. 324-25.

although adding a second puller might double the potential power of a single man, adding ninety-nine men will not produce one hundred times as much power.[107] Accordingly, there is no reason to believe that any more than approximately a dozen individuals provided the traction power for each of these engines at any one time. This is logical considering there were other tasks associated with the operation of such machines, including protecting the engine and gathering ammunition – a significant task considering the potential shooting rate of a traction trebuchet.

The fourth point of interest is the number of stones that were thrown: 5,000 shots over the course of 10 hours, or one shot every 7.2 seconds. If this rate is anywhere near accurate the number of men gathering ammunition, let alone trying to shape any of it, would need to be many times that of the men working the two engines. It is unlikely, however, that the anonymous author counted the number of stones that were thrown. His estimation is thus either an impression of the engines' shooting rate or an expression of the total number of stones thrown, two sides of the same coin. Accordingly, the author's figures should be interpreted as a reflection of his qualitative impressions rather than a quantitative reality.

The physical exertion required to achieve a shooting rate impressive enough to inspire such figures probably limited the relay periods to a matter of minutes rather than hours. If the author indeed meant that all one hundred men of each relay team took part in providing pulling power, though not all at once, and that the total number of shots was divided between the two engines, it stands to reason that for each engine to support a shooting rate of around four shots per minute, a rate within the realm of possibility, each team was further subdivided. For instance, sub-crews of ten might have taken six-minute shifts, allowing everyone to take a turn pulling once every hour. At any given time, the other ninety men could rest, gather ammunition and guard their engine. Alternatively, roles may have been more defined, leaving only a portion of each group to operate its engine, resting between shifts, while the others supported them. Regardless, short shifts would have been necessary to maintain a rapid shooting rate, although longer shifts could have been maintained if the vigour of the pullers was tempered slightly. In any case, the emphasis placed on crew size and shooting rate appears to confirm that these were not counterweight trebuchets.[108]

---

107 See Chapter 2. See also Fulton, "Myth of the Hybrid-Trebuchet."

108 That they were is suggested in Bradbury, *The Medieval Siege*, p. 260. Nicolle has grasped this and suggests that either these machines were quite light or that this whole episode

The anonymous author makes it clear that the Frankish artillery was employed in a supporting role, and what it was supporting was not going well. With their mine more trouble than it was worth, the Anglo-Normans gave it up to build another siege tower, somewhat shorter than their earlier one. Steadfast in their approach, the men of Cologne and Flanders managed to undermine a section of the opposite wall of the city, but the defenders held fast. The Anglo-Norman siege tower, supported by traction trebuchets and capable of resisting the defenders' artillery, eventually compelled Lisbon's surrender after it was pushed to within a few metres of the city.[109]

At the contemporary Iberian sieges of Almeria (1147) and Tortosa (1148), Caffaro witnessed the similar use of *mangana* to support siege towers, which were in turn targeted by similar defensive engines.[110] At one point, Caffaro claims that stones weighing 200 pounds were thrown against one of the siege towers erected at Tortosa, initially damaging one of its corners. Later, he appears to contradict the notion that the defenders were making use of particularly powerful engines when he asserts that rope netting, which was subsequently hung on the sides of the tower, provided sufficient protection against these projectiles.

The siege of Lisbon reveals many similarities to the siege practices of the Franks in the Holy Land. The preference for siege towers shown by the Anglo-Normans seems to reflect that demonstrated by the Franks of Outremer during the first half of the twelfth century, while both made use of light artillery to support these towers and sapping operations. It is the preference for mining shown by the contingents from Cologne that stands out. Although mining had been a tactic employed by Frankish forces during the First Crusade, and can be found at Frankish sieges in western Syria during the following decades, the rarity with which it is associated with forces of the kingdom of Jerusalem has led mining to be viewed more as a Muslim tactic.[111] Although artillery was not employed by the Frankish forces that went on to besiege Damascus in 1148, it was used at Ascalon a few years after the Second Crusade collapsed.

---

may be little more than the author displaying his knowledge of the efficiency of Balearic slingers in the classical Roman army, Nicolle, *Medieval Siege Weapons*, p. 15.

109 *De Expugnatione Lyxbonensi*, ed. and trans. David, pp. 142-65. For a secondary account of the siege, see Rogers, *Latin Siege Warfare*, pp. 183-88.

110 Caffaro, *Ystoria*, ed. Belgrano, pp. 80-88, trans. Hall and Phillips, pp. 128-35.

111 The topography of coastal Palestine and availability of Italian naval support influenced the apparent Frankish preference for siege towers during the first two decades of the twelfth century. Following the conquest of the coast, mining appears more frequently as a siege tactic than does the construction of siege towers.

Ascalon: 1153

Numerous sieges took place between 1148 and the Frankish siege of Ascalon in 1153; however, the sources reveal little about the nature of any artillery that was used.[112] William of Tyre references, very generally, the use of 'throwing-engines' by both Baldwin III and his mother's supporters during the former's siege of the Tower of David.[113] During Nur al-Din's pushes against Damascus between 1150 and 1154, neither he nor the city's defenders appear to have made use of artillery,[114] even though it would seem such engines were being stored inside the city by this point in time.[115]

Ascalon was an ancient strongpoint on the southern coast of Palestine. Its medieval fortifications were founded on Bronze Age defences and contained elements of late Roman, Umayyad and Fatimid construction.[116] Like most twelfth-century fortifications along the coast of Palestine, the defences of Ascalon were built using the local *kurkar*, a soft sandstone composed of fossilised sand washed into the Mediterranean from the Nile. The mortar was a compound of lime, sand, potsherds and ash, which formed a bonding agent slightly harder than the stone itself and more resistant to the natural elements. These defences were manned by a sizable professional garrison during the first half of the twelfth century, regularly supplied and rotated out by the Egyptian fleet.[117]

Baldwin III of Jerusalem led his army against Ascalon in January 1153, while Gerald of Sidon attempted to impose a naval blockade.[118] Additional ships were procured and stripped for their timber with the arrival of a wave of pil-

112 These include a number of sieges conducted by Nur al-Din and the Seljuks of Rum as they mopped up what remained of the county of Edessa, which was largely under Byzantine control at this point, Ibn al-Azraq, *Tarikh*, trans. Hillenbrand, pp. 132-34; William of Tyre, *Chronicon* 17.16-17, ed. Huygens, 2:781-85, trans. Babcock and Krey, 2:208-12; Gregory the Priest, in Matthew of Edessa *Patmut'iwn*, trans. Dostourian, p. 260; Ibn al-Qalanisi, *Dhail*, trans. Gibb, pp. 309-12. For Nur al-Din's second push into this region a decade later, see Gregory the Priest, in Matthew of Edessa, *Patmut'iwn*, trans. Dostourian, pp. 269-70, 276-77; Ibn al-Qalanisi, *Dhail*, trans. Gibb, pp. 356-57.

113 William of Tyre, *Chronicon* 17.14, ed. Huygens, 2:778-80, trans. Babcock and Krey, 2:206-7.

114 Ibn al-Qalanisi, *Dhail*, trans. Gibb, pp. 297-300, 302-8, 318-20; Ibn al-Athir, *al-Kamil*, trans. Richards, 2:71.

115 Fulton, "Development of Prefabricated Artillery," pp. 56-57. See also Ibn al-Qalanisi, *Dhail*, trans. Gibb, pp. 310, 312-13.

116 Pringle, "Crusader Castles and Fortifications," pp. 357-72; Benvenisti, *The Crusaders*, pp. 123-28; Pringle, "Richard I," p. 135.

117 Pringle, "Town Defences," pp. 84-85; William of Tyre, *Chronicon* 17.22, ed. Huygens, 2:790-92, trans. Babcock and Krey, 2:219-20. William of Tyre repeatedly mentions the city's reliance on a regular schedule of supply shipments from Egypt in books 13 and 17.

118 William of Tyre, *Chronicon* 17.21, ed. Huygens, 2:789-90, trans. Babcock and Krey, 2:217-18.

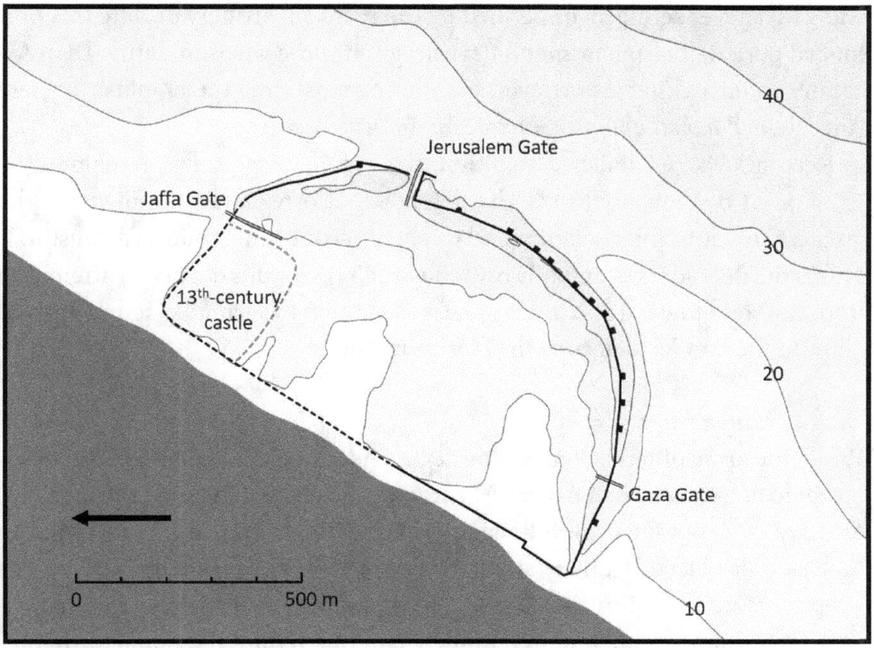

PLAN 9 *Ascalon: surviving defences*

grims around Easter. The masts of these vessels were used to form the frame of a siege tower and the remaining wood was used to make penthouses and artillery. William of Tyre provides another of his conflicting descriptions when he states that the artillery was arrayed in ideal positions to batter down the walls but then clearly depicts these engines in a supporting role, providing covering fire for the siege tower and those preparing the ground ahead of it.[119] The later *Anonymous Syriac Chronicle* claims that the Franks employed a stone-throwing machine on top of the siege tower and emphasises the threat that this posed to those walking around inside the city.[120]

The Frankish attack wore on into the summer and the barrage seems to have been most intense around the east-facing Jerusalem Gate. Despite William of Tyre's assertion that "Volleys of mighty rocks hurled from the casting machines threatened to weaken the walls and towers and to overthrow from their very foundations the houses within the city",[121] it seems that the integrity of the

119 William of Tyre, *Chronicon* 17.23-24, ed. Huygens, 2:792-94, trans. Babcock and Krey, 2:220-23.

120 *Anonymous Syriac Chronicle*, trans. Tritton, p. 301.

121 ... *tormentis nichilominus iaculatoriis turres ac menia debilitare et infra urbem non sine strage multa funditus dissolvere, immissis magnis molaribus, domicilia*. William of Tyre, *Chronicon* 17.27, ed. Huygens, 2:797, trans. 2:225-26.

walls was never seriously imperilled by the Frankish artillery despite this prolonged period of bombardment. Revealingly, it is the garrison, rather than Ascalon's defences, that is portrayed as suffering most from the Frankish artillery (mislabelled *balistae*) in the *Auctarium Aquicinense*.[122]

Keeping with established tradition, the Franks' siege tower remained the focal point of their offensive. The defenders also regarded the tower as the greatest threat to their security and directed most of their efforts against it. It seems the defenders eventually breached their own walls during an attempt to burn the siege tower. This catastrophe forced them to surrender soon after and allowed the Franks to occupy the city on 12 August.[123]

Banyas: 1157

While the army of Jerusalem was besieging Ascalon, Nur al-Din had briefly attempted to take Frankish Banyas.[124] Although he failed, he gained the far greater prize of Damascus the following year.[125] In 1157, Nur al-Din returned to Banyas and placed it under siege. Ahead of this, he travelled to Damascus where, according to Ibn al-Qalanisi, "He gave orders to dispatch what mangonels and weapons were required to the victorious *'askar*", then rallied what remaining forces he could from the city and made for Banyas.[126] In surprising similarity, William of Tyre states, "He summoned his cavalry, had his engines of war moved to the place, and suddenly appeared before the city."[127] Thus, both Frankish and Muslim sources confirm that artillery was being stockpiled in

122 *Auctarium Aquicinense*, MGH SS 6, p. 396. See also *Auctarium Affligemense*, MGH SS 6, pp. 401-2.

123 William of Tyre, *Chronicon* 17.27, ed. Huygens, 2:797-99, trans. Babcock and Krey, 2:225-27.

124 The town, about 300 m square, was surrounded by a wall prior to the arrival of the Franks. Many of the more impressive elements of the town's fortifications, interpreted as Frankish in the nineteenth century, appear more likely to have been built by the Ayyubids or Mamluks. Even if the large quadrangular towers and some stretches of the curtain postdate Nur al-Din's siege, the footprint of the town was the same. The western side was protected by the river while a seasonal stream ran through a small valley to the south. The town is most approachable from the north, where springs emerge from the rocks below a great cliff, and from the east, along which front the medieval ditch, complete with a counterscarp, is readily discernable. See Conder and Kitchener, *Survey of Western Palestine*, 1:110-11; Pringle, *Secular Buildings*, no. 42, p. 30; Pringle, "Town Defences," p. 92.

125 William of Tyre, *Chronicon* 17.26, ed. Huygens, 2:796, trans. Babcock and Krey, 2:225; Ibn al-Qalanisi, *Dhail*, trans. Gibb, pp. 314-16.

126 Ibn al-Qalanisi, *Dhail*, trans. Gibb, pp. 333-34. Abu Shama places his departure from Damascus on 12 May 1157, Abu Shama, *Kitab al-Raudatain*, RHC Or 4, p. 86.

127 William of Tyre, *Chronicon* 18.12, ed. Huygens, 2:827, trans. Babcock and Krey, 2:257.

Damascus at the start of the second half of the twelfth century. To warrant storage, these engines must have been viewed as sufficiently valuable to incur the costs associated with their storage, transportation and maintenance relative to those involved with building new engines *in situ*.[128]

Having surrounded the town and set up his artillery, Nur al-Din began the siege on 18 May 1157. The sources emphasise that a barrage of stones and arrows was unleashed on the defenders but again the artillery seems to have inflicted greater psychological damage than physical. The Muslim army gained entrance to the town either after a mine was successfully lit, causing a section of wall to collapse, or a careless sally by the garrison allowed their assailants to overcome them before they were able to withdraw back into the town and secure the gate behind them.[129] The garrison withdrew to the citadel but were soon relieved by a force led by Baldwin III in mid-June.

Nur al-Din was able to defeat this relief force soon after its arrival, compelling the king to seek refuge in Safed and allowing the Muslim army to return to Banyas.[130] According to William of Tyre, Nur al-Din once more encircled the town "and set up his numerous engines in strategic positions. The mighty blows of the stone missiles shook the towers and weakened the walls. At the same time, showers of darts and arrows fell like hail and prevented all resistance on the part of those within."[131] However, it is questionable whether Nur al-Din would have been able to bring any significant artillery to bear against Banyas in this second effort. It seems unlikely that these would have been the same trebuchets that had been employed in the first phase of the siege as both the surprise appearance of Baldwin and the mobility that Nur al-Din's forces demonstrated in defeating the army of Jerusalem suggest that he would not have been moving with a siege train. Had he not burnt his engines prior to

128 See Fulton, "Development of Prefabricated Artillery," 51-72. This appears to differ from the timber that was used to construct the siege towers when Banyas was besieged in 1140, see above.

129 William of Tyre, *Chronicon* 18.12, ed. Huygens, 2:827-28, trans. Babcock and Krey, 2:257-58; Ibn al-Qalanisi, *Dhail*, trans. Gibb, pp. 333-35; Abu Shama, *Kitab al-Raudatain*, RHC Or 4, pp. 87-88. William of Tyre offers the latter scenario but claims that mines were used to slight the town's defences after it fell, leaving the possibility that work on these mines might have begun before the town was taken.

130 William of Tyre, *Chronicon* 18.13-14, ed. Huygens, 2:828-32, trans. Babcock and Krey, 2:258-62.

131 ... *machinis frequentibus et ordine congruo dispositis turres concutit, menia debilitat et crebra sagittariorum opera et telorum instar grandinis immissione eos, qui intus se receperant, vices prohibet exercere resistentis.* William of Tyre, *Chronicon* 18.15, ed. Huygens, 2:832, trans. Babcock and Krey, 2:262-63.

leaving Banyas, they would almost certainly have been appropriated or destroyed by the garrison following Nur al-Din's initial departure. Despite his defeat, Baldwin was still able to break this second siege with the arrival of aid from Antioch and Tripoli; the weakened garrison having opted to defend only the citadel this time.[132]

Shayzar and Harim: 1157

Nur al-Din's move on Banyas took place during a devastating series of earthquakes that were felt throughout Syria in 1156-57.[133] In October 1157, the Franks took advantage of this destruction to move first against Shayzar and then Harim. In addition to the damage caused by the tremors, the Franks were emboldened by the arrival of Thierry of Flanders, on his third of four crusades to the Holy Land, and a bout of illness that afflicted Nur al-Din.

Artillery was used to support frontal assaults at the siege of Shayzar, which proved unsuccessful, and to help sustain a blockade and cover the mining efforts that led to the capture of Harim.[134] These engines may again have been most effective as psychological weapons, as at neither siege are they portrayed as being particularly powerful. Nur al-Din appears to have employed artillery with a similar psychological intent when he retook Harim in 1164,[135] while his use of stone-throwers at Banyas later the same year is another clear example of their effective use to support sapping efforts.[136]

---

132 William of Tyre, *Chronicon* 18.15, ed. Huygens, 2:832-33, trans. Babcock and Krey, 2:262-64.

133 Meanwhile, forces of Sultan Mahmud besieged Baghdad through the first half of 1157, making use of artillery, Nishapuri, *Jami' al-Tawarikh*, trans. Luther, p. 130.

134 For Shayzar, Ibn al-Athir, *al-Kamil*, trans. Richards, 2:89; William of Tyre, *Chronicon* 18.18, ed. Huygens, 2:836, trans. Babcock and Krey, 2:267. For evidence of Nur al-Din's reconstruction efforts at Shayzar in the aftermath of these earthquakes, see Tonghini and Montevecchi, "Shayzar: Recent Archaeological Investigations," pp. 140-50. For Harim, William of Tyre, *Chronicon* 18.19, ed. Huygens, 2:838-40, trans. Babcock and Krey, 2:269-71; Ibn al-Qalanisi, *Dhail*, trans. Gibb, p. 344; Abu Shama, *Kitab al-Raudatain*, RHC Or 4, p. 96.

135 William of Tyre, *Chronicon* 19.9, ed. Huygens, 2:874-75, trans. Babcock and Krey, 2:306-8; Ibn al-Athir, *al-Kamil*, trans. Richards, 2:146-48; *Anonymous Syriac Chronicle*, trans. Tritton, pp. 303-4.

136 William of Tyre, *Chronicon* 19.10, ed. Huygens, 2:877, trans. Babcock and Krey, 2:309-10; Ibn al-Athir, *al-Kamil*, trans. Richards, 2:148-49; Ibn al-Furat, *Tarikh*, trans. Lyons and Lyons, 2:47.

Egypt: A New Frontier

Alexandria: 1167

In the 1160s, King Amalric of Jerusalem's foreign policy was largely concerned with Egypt. During a campaign in 1167, in what had become a fight with Nur al-Din's representative, Shirkuh, for control of the weakened Fatimid realm, Amalric besieged Alexandria. Shirkuh led the initial defence of the city until he left for Upper Egypt, leaving his nephew, Saladin, in command.

The Franks initially imposed a blockade from a distance, content to pillage the surrounding area. However, a more active approach was adopted following Shirkuh's departure.[137] It is only during this second stage of the siege that any kind of machinery is noted. William of Tyre describes the situation in fairly familiar terms:

> The besieging host assembled before the city now collected an immense number of masts, summoned craftsmen and carpenters, and caused them to erect a tower of great height from whose top the entire city could be surveyed. Machines called *petrariae* which hurled forth enormous stones of great weight were also placed in strategic positions around the walls. From these, almost incessantly, were hurled immense stones of great weight which shattered the walls and terrified the people almost beyond endurance.[138]

The statement is essentially a compilation of stock phrases, from the ability to survey the town from the tower to the throwing of large stones that terrorised those within and wrought damage on the town's defences. It is worth reiterating that, with the possible exception of the Graeco-Frankish siege of Shayzar in 1138, none of these supposed breaches or weakened walls was subsequently stormed by assailing Frankish forces. While this indicates that the impact of these projectiles was somewhat less dramatic than William's language would suggest, his reference to the constant shooting of stones is another tenuous indication that these were traction trebuchets. In this light, William's repeated

137 William of Tyre, *Chronicon* 19.26, ed. Huygens, 2:901-2, trans. Babcock and Krey, 2:334-35; Abu Shama, *Kitab al-Raudatain*, RHC Or 4, pp. 132-33. For the wider campaign see Ibn al-Athir, *al-Kamil*, trans. Richards, 2:163-64. See also Omran, "King Amalric," pp. 191-96.

138 *Hic demum sumptis navium malis inmense quantitatis, vocati artifices, lignorum cesores, castellum erigunt mire altitudinis, unde totam despicere erat civitatem; machine quoque, quas vulgo petrarias vocant, unde missi molares graves et magni muros cedebant, congruis stationibus locate intolerabilem civibus, horis pene omnibus, inferebant terrorem.* William of Tyre, *Chronicon* 19.28, ed. Huygens, 2:903-4, trans. Babcock and Krey, 2:337.

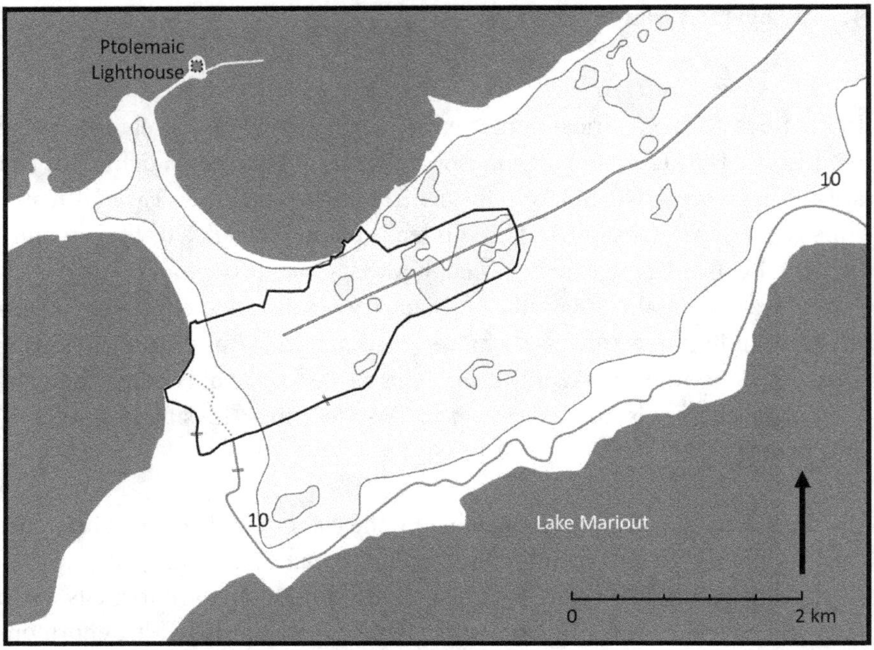

PLAN 10 *Alexandria: trace of discernable defences c. 1800*

claims that projectiles 'smashed walls', may more accurately be read as 'smashed *against* walls' and the 'great' size of the stones should perhaps be seen in relation to the scale of a stone that a slinger would typically cast or one that might be thrown by hand from a distance.

The Franks built their artillery using wood from the surrounding orchards and fruit trees.[139] Even if the beams and mains struts had been fashioned from multiple spars lashed together, the short and far from straight nature of these trees would have limited the size and strength of any traction trebuchets, and essentially rules out the possibility that counterweight engines were employed. As at Caesarea (1101), Banyas (1140) and Ascalon (1153), the timber that would have been suitable for a counterweight trebuchet was instead earmarked for the construction of a siege tower.

The inability of the Franks to inflict more damage here is noteworthy when considering the security that they enjoyed during this siege – the local population of Alexandria was as hostile towards its Syrian 'defenders' as it was to the assailing Franks. This discouraged the garrison from launching sallies against their besiegers, leaving the Franks free to conduct their siege in relative securi-

139 William of Tyre, *Chronicon* 19.28, ed. Huygens, 2:903-4, trans. Babcock and Krey, 2:337.

ty.[140] Despite the poor selection of building materials, which would have been similarly dismal in most areas south of the Lebanon, the siege appears to reveal the fairly meagre power of mid-twelfth century artillery, highlighting the limits of contemporary technology.

Damietta: 1169

In 1169, the Franks launched their first campaign against Damietta, doing so in conjunction with a Byzantine force dispatched by Emperor Manuel Comnenus. Damietta was an important port city, positioned on one of the two main branches of the Nile Delta about 5 km inland from the Mediterranean.[141] The city controlled maritime traffic between the sea and the rest of Egypt to the south through a tower that was set in the middle of the river. This was connected to the town by a chain that extended across the main shipping lane. They had no way of knowing in 1169, but this would not be the last time a Frankish army would come up against this tower.

Upon their arrival in the region the Franks paused, inadvertently allowing support and provisions to pour into Damietta.[142] When siege efforts began, "It soon became apparent that without the assistance of machines (*machinae*) and engines of war (*tormenta bellica*) Damietta could not be taken, although on the arrival of the Christians it had seemed scarcely able to sustain the first attack."[143] Having opted not to make an initial general assault, the attacking forces had committed themselves to a set-piece siege.

According to William of Tyre, the Graeco-Frankish army relied heavily on siege engines.[144] The showpiece of these was a supposedly seven-storey Frankish siege tower. In addition to this, penthouses and trebuchets were also constructed. The artillery is described in familiar terms, damaging the battlements

140 William of Tyre, *Chronicon* 19.29-32, ed. Huygens, 2:905-9, trans. Babcock and Krey, 2:339-43; Ibn al-Athir, *al-Kamil*, trans. Richards, 2:164-65.

141 Then, as now, the other main branch entered the Mediterranean at Rosetta.

142 Ibn al-Athir and Baha' al-Din credit this support to Nur al-Din and Saladin respectively, Ibn al-Athir, *al-Kamil*, trans. Richards, 2:183-84; Baha' al-Din, *al-Nawadir*, trans. Richards, p. 46. According to John Kinnamos, the Frankish delay occurred before they arrived in Egypt, providing the Byzantine marine force the opportunity to seize Tinnis. The ultimate failure to take Damietta was blamed on what was seen to be Amalric's deliberately late arrival, John Kinnamos, *Deeds* 6.9, trans. Brand, pp. 208-9.

143 William of Tyre, *Chronicon* 20.15, ed. Huygens, 2:929, trans. Babcock and Krey, 2:364.

144 William of Tyre was a notable figure in the kingdom of Jerusalem by this time; however, he was out of the realm on diplomatic missions during much of 1169, William of Tyre, *Chronicon* 20.17.

and buildings inside the city.[145] In addition to the engines built by the Franks, the Byzantine fleet also brought siege machines. Baha' al-Din appears to have noticed this when observing that the Christians brought their siege equipment with them.[146] Niketas Choniates downplays the significance of the Frankish siege tower, emphasising instead the frontal assaults made by the Byzantines using ladders. He clearly states that artillery was employed with some effect by at least some Byzantine elements of the army, although he fails to confirm that the engines had been prepared ahead of time.[147] All three sources allude to the garrison's use of artillery without explicitly describing it, ambiguously referring to these engines with terms akin to 'machines of war'.[148]

The siege was eventually abandoned as supplies wore thin and the Frankish siege tower failed in its assault against the most approachable, but most heavily defended, front of the city. Facing failure, the besiegers withdrew. Only five years later another Christian army would return in force to Egypt; however, this invasion was not orchestrated in Acre or Constantinople, but in Palermo.

Alexandria: 1174

At the end of July 1174, Norman forces from Sicily attacked Alexandria. Possibly intending to coordinate with a Shiite coup against Saladin, which had miscarried earlier in the year, the force landed near the Ptolemaic lighthouse north of the city.[149] A letter written by Saladin, which appears to form the basis of most Muslim accounts of the siege, claims that six ships of the Sicilian fleet, distinguished from the galleys and troop-carriers, bore siege engines and men who built artillery.[150] Having made their landing, the Sicilians pushed the Muslims back to the walls of Alexandria while their fleet occupied the harbour. By the following day, they had finished setting up their engines and advanced them towards the city walls. No less than three siege towers, each carrying a ram, and three trebuchets are said to have been employed by the attackers. On days four and five of the siege, 31 July and 1 August, the garrison launched successful sallies, first burning the siege towers and then defeating the besieging force itself.

145 William of Tyre, *Chronicon* 20.15, ed. Huygens, 2:929-30, trans. Babcock and Krey, 2:364.

146 William of Tyre, *Chronicon* 20.13, ed. Huygens, 2:927, trans. Babcock and Krey, 2:361; Baha' al-Din, *al-Nawadir*, trans. Richards, p. 46.

147 Niketas Choniates, *Historia*, trans. Magoulias, pp. 92-95.

148 Cf. Abu Shama, *Kitab al-Raudatain*, RHC Or 4, pp. 151-53; Ibn al-Athir, *al-Kamil*, trans. Richards, 2:183-84.

149 Although damaged by earthquakes over the years, this was still a prominent landmark at this point in time.

150 Abu Shama, *Kitab al-Raudatain*, RHC Or 4, pp. 165-66. See also al-Maqrizi, *al-Suluk*, trans. Broadhurst, p. 49.

The attackers who survived withdrew, abandoning their remaining machinery.[151]

The most detailed accounts of the siege come from Muslim sources and the way they deal with the Frankish artillery is tantalising and enigmatic. That only three trebuchets are mentioned, next to an equal complement of ram-equipped siege towers, and that some sources describe them as 'big', may suggest that there was something exceptional about these engines. This idea becomes more appealing when considering reports that they threw black stones brought from Sicily. It is unlikely that stone projectiles would have been imported for traction trebuchets, given their limited power and high shooting rate. At a reasonable rate of four shots per minute, a single engine would go through an astonishing 240 projectiles during an hour of continual use. Thus even if the Sicilians recognised that finding suitable stones to use as projectiles might be more difficult in the Nile Delta than other regions, it is unlikely that a sufficient number of stones could have been transported to Alexandria to sustain even one traction trebuchet over the course of multiple days. Instead, the prominent notice given to these three prefabricated machines, on equal footing with the siege towers, and the transportation of special ammunition, suggests that these may have been counterweight trebuchets. The slow shooting rate and high impact velocity of such engines would certainly warrant special projectiles and their use may even have overshadowed the presence of lighter traction engines that escaped notice.[152]

The stones thrown by these three engines are described as being large and hitting the walls of Alexandria with impressive violence, a portrayal with long-standing precedent. Rather than their reported size or impact, it is the colour of these stones that is significant. The black, presumably volcanic, rock brought from Sicily would have been harder not only than any stones shaped locally, but also the masonry of Alexandria's defences.[153] This gave the Sicilians the

---

151 Abu Shama, *Kitab al-Raudatain*, RHC Or 4, pp. 164-67; Ibn al-Athir, *al-Kamil*, trans. Richards, 2:229-30; al-Maqrizi, *al-Suluk*, trans. Broadhurst, pp. 48-50. See also Baha' al-Din, *al-Nawadir*, trans. Richards, p. 50; William of Tyre, *Chronicon* 21.3, ed. Huygens, 2:963, trans. Babcock and Krey, 2:399-400.

152 The only explanation that I can think of for why traction trebuchets might throw rocks brought from Sicily is if these were ballast stones. If this was the case, it would imply the Sicilians were extremely confident in their ability to take the city, perhaps believing that they could compensate for the lost weight of the stones with booty on their return home. This seems an unlikely possibility when considering that the Norman force was so small that it was defeated by the garrison before Saladin arrived to relieve it.

153 Randall Rogers has proposed that these black stones were supposed to fragment upon impact; however, this seems unlikely given the geology involved, Rogers, *Latin Siege War-*

same advantage that Richard I of England would enjoy seventeen years later, when he too made the effort to gather stone projectiles before departing from Sicily to join the siege of Acre.[154]

Although there is reasonable circumstantial evidence to propose the possibility that counterweight trebuchets were used by the Sicilians in 1174, the logistics of the siege expose this to question. According to the time scheme repeated by the various Muslim sources and confirmed by William of Tyre,[155] the Sicilians landed and then unloaded, assembled and deployed their siege engines in less than thirty-six hours. Although this would have restricted the potential size of the assailing machinery, evidence from the late 1180s onward suggests that small prefabricated counterweight trebuchets could have been set up in such a narrow window of time. Having arrived on 28 July and begun the siege the following day, 30 July was the only full day that the Sicilians were able to bombard the city uninhibited. A number of engines were burnt during a sortie by the garrison on 31 July and the besiegers were routed the following day ahead of Saladin's arrival on 2 August. This timetable would have prevented any medieval artillery, regardless of its type or power source, from threatening contemporary fortifications, let alone allowing a multi-storey siege tower to do so. It took the Franks attacking the cities of Palestine fifty to seventy-five years earlier the better part of a month to construct their siege towers. At Jerusalem, it took Godfrey of Bouillon all of three days to reassemble his tower from components just light enough to be moved in the field, rather than any so small that they could be stowed on board a ship.

The Normans' use of what may have been relatively unique engines in 1174 is significant when considering their siege of Thessalonica in 1185. Eustathius of Thessalonica claims that the Sicilians employed an array of artillery that included large and small engines of a traditional type, which were used in conjunction with mining against one side of the city, as well as some of a new variety.[156] Although the latter proved cumbersome and ineffective on account of their size, it is likely that these, and possibly even some of the large 'traditional' engines, were counterweight trebuchets.

Despite certain embellishments, the accounts of the siege of Alexandria provide plausible evidence that counterweight trebuchets might have been

---

fare, p. 121.

154 See Chapter 5.

155 William of Tyre claims that the Sicilians' siege lasted five or six days, William of Tyre, *Chronicon* 21.3, ed. Huygens, 2:963, trans. Babcock and Krey, 2:399-400.

156 Eustathius of Thessalonica, *Capture of Thessalonica*, ed. and trans. Melville Jones, pp. 72-105.

employed by a Latin force in the theatre of the crusades before the Third Crusade. This is not to say that this was the first time that such engines may have been used in or around the Latin East, only that this appears to be the first recorded instance that is supported by a reasonable amount of circumstantial evidence. Likewise, this does not imply that these were monstrous machines comparable to those constructed a century later, only that they were strong enough that the value of their greater power, when coupled with imported ammunition, outweighed their cost, complexity, and slower shooting rate. In any case, this is a rare example of Latin forces clearly employing prefabricated artillery in the Levant during the twelfth century. A more typical example of Frankish artillery tactics in the closing decades of the twelfth century is found at the siege of Harim.

Harim: 1177

Philip of Flanders, son of Thierry, arrived in Palestine in August 1177 on the first of what would be two trips to the Holy Land. That autumn, Philip campaigned to the north alongside Raymond III of Tripoli and Bohemond III of Antioch. After a brief move against Hama, the Frankish force invested Harim, then in revolt against al-Salih of Aleppo. Upon arrival, the Franks built huts to show their determination and artillery to support frontal assaults made using ladders – the topography rendered the use of a siege tower impractical.[157] The siege was not pressed particularly hard and after a lengthy stay, perhaps as long as four months, Bohemond was satisfied with a payoff and Philip was content just to leave. With their departure, al-Salih moved in to finish off the beleaguered garrison, which quickly agreed to terms.[158]

Despite the stronghold's inaccessibility, William of Tyre notes that the Frankish artillery that was arranged around the castle could still harass it from all sides, as appears to have been the case during Nur al-Din's siege of the castle in 1164.[159] The castle sits on top of a tell, roughly 40 m high, and it would appear that access was unobstructed to the base of its sides, which rise at an angle of

157 William of Tyre, *Chronicon* 21.18 (19), ed. Huygens, 2:986-87, trans. Babcock and Krey, 2:425-26; Ibn al-Athir, *al-Kamil*, trans. Richards, 2:255-56. For the defences of Harim, see Gelichi, "The Citadel of Harim," pp. 184-200; Gelichi, "Die Burg Harim," pp. 211-20.

158 William of Tyre, *Chronicon* 21.18 (19), 24 (25), ed. Huygens, 2:986-87, 994-96, trans. Babcock and Krey, 2:425-26, 434-35; Ibn al-Athir, *al-Kamil*, trans. Richards, 2:254-56; Abu Shama, *Kitab al-Raudatain*, RHC Or 4, pp. 191-93.

159 William of Tyre, *Chronicon* 21.24 (25), ed. Huygens, 2:995, trans. Babcock and Krey, 2:434-35. For Nur al-Din's siege of Harim in 1164, see William of Tyre, *Chronicon* 19.9, ed. Huygens, 2:874-75, trans. Babcock and Krey, 2:306-8; Ibn al-Athir, *al-Kamil*, trans. Richards, 2:146-48; *Anonymous Syriac Chronicle*, trans. Tritton, pp. 303-4.

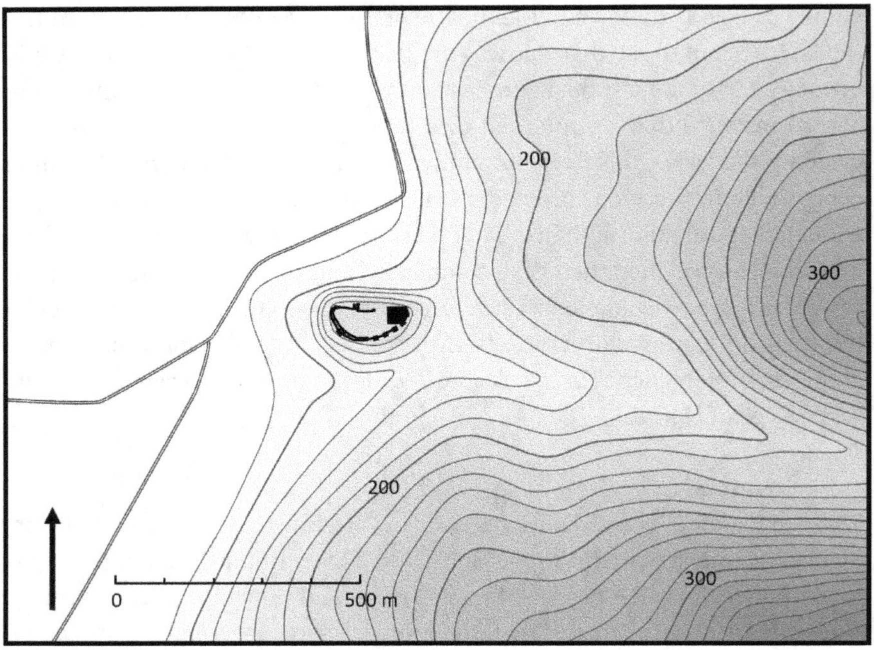

PLAN 11 *Harim*

about 45°. If William's account is accurate and some shots cleared the castle's battlements, these projectiles must have been released at a minimum velocity of 34 m/s (assuming the curtain wall was approximately 10 m tall). If the Frankish engines were capable of releasing projectiles at 40 m/s, they could achieve the same result at a distance of 40 m from the base of the tell (80 m from the vertical plane of the wall) using a release angle of 55°. If it was the walls rather than the battlements that were assailed, an engine in this farther position could employ a release angle of 45°, increasing the horizontal velocity of its projectiles by more than 5 m/s (Fig. 9).

Although an apparently small difference, projectiles released at 45° would be travelling 23% faster when they reached the plane of the wall than those released at 55°. When considering that kinetic energy is half the product of mass and velocity squared [$Ek = \frac{1}{2} m \cdot V^2$], the faster shot carries 52% more energy as it reaches the wall. While this is an important consideration when dealing with large projectiles, the issue is rather trivial if they are fairly small (Fig. 10).

Regardless of a projectile's size, its real velocity would be at its lowest when it reached the highest point along its flight path; however, if it struck the vertical exterior face of a wall at this point, the perpendicular angle of impact would allow for the most efficient transfer of energy. It might be argued that a more

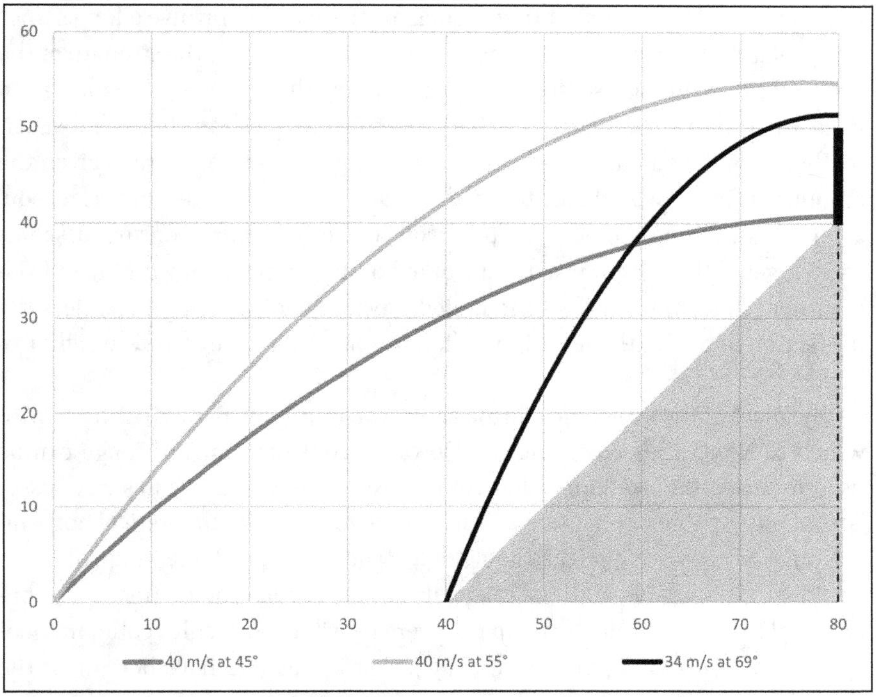

FIGURE 9 *Hypothetical trajectories at Harim*

Release angle (α)	45°	55°	69°
Release velocity (V)	40.0 m/s	40.0 m/s	34.0 m/s
Horizontal velocity (Vh)	28.3 m/s	22.9 m/s	12.2 m/s
Max. horizontal range (Rh)	163.3 m	153.4 m	78.9 m
Max. vertical range (Rv)	40.8 m	54.8 m	51.4 m

FIGURE 10 *Energy of projectiles travelling 28.3 m/s and 22.9 m/s*

Projectile Mass	Kinetic Energy at 28.3 m/s (45°)	Kinetic Energy at 22.9 m/s (55°)
1 kg	400 J	262 J
5 kg	2,002 J	1,311 J
10 kg	4,004 J	2,622 J
50 kg	20,022 J	13,110 J

acute or obtuse horizontal angle of impact might be ideal in order to target the corner of a polygonal structure, such as a tower or even a merlon, given the lack of material behind the point of impact to absorb the energy transferred by the projectile. However, shooting at such an angle would involve a longer shot and probably a sharper vertical angle of impact. Likewise, if the strength of the targeted structure was so great and the mass of the projectile so small as to make this approach appear attractive, the engine in question could almost certainly be employed more productively shooting at a different target, given the amount of time it would take to inflict any significant damage. For this reason, as engines became increasingly powerful in the thirteenth century, it seems they targeted the tops of walls and worked down, rather than working in from a corner. This approach had the added advantage of denying defenders the protection of the battlements immediately opposite a large counterweight trebuchet.

The mass of the projectiles thrown at Harim is unknown, thus the force with which they were tossed is unclear. However, considerations of range can be used to assess the likelihood that counterweight-powered engines were used. An engine capable of releasing a projectile at 34 m/s has a theoretical horizontal range of 118 m (at a release angle of 45°) and a comfortable range of 102 m (+/-15°) – comparable with the range of modern reconstructed traction trebuchets.[160] If a release velocity of 40 m/s were possible, projectiles could in theory be thrown up to 163 m (141 m +/-15°). This appears to have been near the limit for most traction trebuchets. Any stones tossed at or near this speed, however, would have been quite small, as it is a challenge to throw projectiles much heavier than 10 kg at velocities much greater than 30 m/s. By comparison, a robust counterweight trebuchet of similar dimensions should be able to throw stones twice as large at around the same velocity, albeit at a slower rate. Without the restrictions imposed by the human element providing the power, a counterweight trebuchet can make use of much more energy (provided by the suspended mass of the counterweight) relative to the mass of its projectiles.

Although there appears to be little evidence to suggest that counterweight trebuchets were used at Harim in 1164 or 1177, suggesting it was the defenders on the parapet rather than the walls below them that were targeted, neither is there enough evidence to rule out their presence conclusively. There is more persuasive evidence that Saladin employed counterweight engines in the late 1180s, but it is pertinent to look at his earlier use of artillery before moving on to these sieges.

160 See Appendix 4.

The Rise of Saladin

Saladin rose to prominence through the scramble for Egypt in the 1160s. When his uncle Shirkuh died in 1169, he succeeded him as vizier of Egypt, ruling on behalf of Nur al-Din. Through the early 1170s, Saladin's power in Egypt and the surrounding region increased, but so too did signs that he was interested in ruling autonomously. Open confrontation with his nominal overlord seemed increasingly likely until Nur al-Din died in 1174, allowing Saladin to take possession of Damascus, and its arsenal, before the year was out.

Jacob's Ford: 1179

In 1175 and 1176, Saladin used artillery against a number of Muslim strongholds in Syria and 'Imad al-Din implies that Saladin intended to use it against the newly built Frankish castle at Jacob's Ford in 1179.[161] Although this siege has been the subject of some discussion, little attention has been given to this reference to artillery.[162] 'Imad al-Din states that wood and vine poles were gathered from around Safed (13 km to the southwest) to build protective screens for the Muslim artillery, a sentiment echoed by Ibn al-Athir.[163] This was apparently a routine procedure as, according to a Frankish source, Saladin had similar defences built for his artillery when he later besieged Jerusalem in 1187.[164] However, artillery is not mentioned following the start of the siege and not a single identifiable projectile has been uncovered during recent excavations at the castle.[165] The apparent absence of artillery seems to speak to the speed of

161 For an overview of the occasions on which Saladin employed artillery, see Appendix 1.

162 For discussions of Jacob's Ford, see Ellenblum, et al., "Crusader Castle," pp. 258-74; Barber, "Frontier Warfare," pp. 9-22. For descriptions of the castle, see Boas, *Crusader Archaeology*, pp. 118-20; Pringle, *Secular Buildings*, no. 174, p. 85; Benvenisti, *The Crusaders*, pp. 304-5; Conder and Kitchener, *Survey of Western Palestine*, 1:250-51; Guérin, *Description: Galilée*, 2:341. Unfortunately, no comprehensive record of the archaeological activities, which commenced in 1992, has yet been published. For contemporary descriptions see Abu Shama, *Kitab al-Raudatain*, RHC Or 4, pp. 197, 203-6; William of Tyre, *Chronicon* 21.25 (26), ed. Huygens, 2:997, trans. Babcock and Krey, 2:436-37; Ibn al-Athir, *al-Kamil*, trans. Richards, 2:264-65. Contrary to the assertions of William of Tyre (21.27, 30), the castle was never completed.

163 Abu Shama, *Kitab al-Raudatain*, RHC Or 4, p. 203; Ibn al-Athir, *al-Kamil*, trans. Richards, 2:265.

164 *De expugatione terrae sanctae*, ed. Stevenson, pp. 243-44, trans. Brundage, p. 160.

165 Pers. cor. with excavators Adrian Boas and Kate Raphael, May 2013. For a survey of the numerous arrowheads that have been found, see Raphael, "A Thousand Arrowheads," pp. 252-61. For general results from the excavations, see Ellenblum, et al., "Crusader Castle," pp. 303-6.

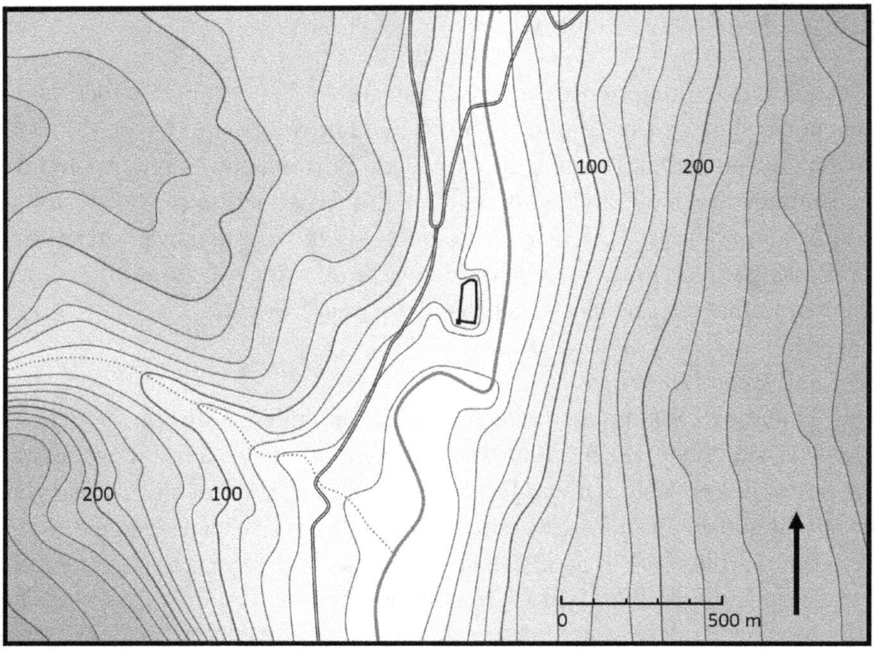

PLAN 12 *Jacob's Ford*

this siege: despite an initial mining setback, the castle fell to a frontal assault on 30 August, less than a week after the siege had begun.[166] While there have certainly been previous examples of artillery being erected in shorter periods of time than this, the rapid progress that Saladin's forces were making against the castle probably discouraged him from devoting further attention to these engines.

When Saladin crossed the Jordan during an earlier campaign in June, he left his baggage at Banyas and it stands to reason that any prefabricated artillery that he might have been moving with the army would have been left there as well. Only after he had lured Baldwin IV and the Frankish army into battle, defeating them on 15 June, did he move against Jacob's Ford. Accordingly, the early efforts to gather materials to establish a fortified artillery position in late August probably began before any engines arrived. Prefabricated artillery was

166 For Saladin's campaign against Jacob's Ford and the siege of the castle, see Abu Shama, *Kitab al-Raudatain*, RHC Or , pp. 195-205; Ibn al-Athir, *al-Kamil*, trans. Richards, 2:264-66; al-Maqrizi, *al-Suluk*, trans. Broadhurst, pp. 59-61; William of Tyre, *Chronicon* 21.26.28 (27-29), ed. Huygens, 2:998-1,002, trans. Babcock and Krey, 2:440-43. See also Benvenisti, *The Crusaders*, pp. 304-5; Boas, *Crusader Archaeology*, pp. 118-19; Ellenblum, *Crusader Castles*, pp. 270-73.

being stockpiled in Damascus before Nur al-Din established himself there in 1154 and it is possible that the engines that 'Imad al-Din implies Saladin intended to use at Jacob's Ford were drawn from this arsenal.[167] It is perhaps significant that the materials gathered from around Safed are the only ones mentioned in relation to artillery, suggesting the Muslim army had no intention of building any new trebuchets on site.

The ways in which artillery was employed by the end of the 1170s varied. While there is evidence that the Sicilian force that attacked Alexandria in 1174 did so with the aid of imported counterweight trebuchets, the Frankish forces that besieged Harim in 1177 probably did so with locally constructed engines, which may have been little more advanced than those used during the First Crusade. At Jacob's Ford, it would appear that Saladin had initially intended to use prefabricated artillery, but with siege efforts proceeding at a good pace, he opted neither to make use of these nor to build lighter engines from locally sourced materials. Despite what appears to be the growing importance of artillery, Saladin made no use of such weapons when he attacked Beirut a few years later.

Beirut: 1182

William of Tyre explicitly states that Saladin's forces did not bring artillery with them ahead of the assault launched against Beirut in 1182, nor does he mention its subsequent construction. Instead, archers were relied upon to cover ongoing sapping efforts and frontal assaults were undertaken in relays for three days until a Frankish fleet arrived to break the siege.[168] Unlike the scenario at Jacob's Ford in 1179, the topography between Beirut and a potential Muslim staging point, such as Banyas, would have posed significant challenges. Accordingly, Saladin may have opted to travel light and maximise the mobility of his forces. Alternatively, his strategy may have been similar to that which he had employed in 1179, choosing to keep his baggage on the far side of the Jordan until he had defeated the army of Jerusalem. As a third possibility, Saladin may have been sufficiently impressed by his miners and assault forces at Jacob's Ford to forgo the troubles of moving artillery with the army.

Whatever Saladin's motives were for not bringing artillery to Beirut, his decision to forgo constructing even light engines upon his arrival is interesting. William of Tyre's remark that the Franks built siege engines and ladders with wood harvested from the pine groves around the city when they besieged it in

---

167 Fulton, "Development of Prefabricated Artillery," pp. 58-59.

168 William of Tyre, *Chronicon* 22.18-19 (17-18), ed. Huygens, 2:1032-36, trans. Babcock and Krey, 2:475-79. Cf. Baha' al-Din, *al-Nawadir*, trans. Richards, p. 57; Abu Shama, *Kitab al-Raudatain*, RHC Or 4, p. 223; Ibn al-Athir, *al-Kamil*, trans. Richards, 2:283.

1110, a detail not found in Fulcher of Chartres' earlier account, suggests that appropriate sources of timber were to be found around Beirut at the time of Saladin's siege.[169] Two years later, in 1184, a confirmation of the bishop's right to take beams from a nearby pine forest leaves no doubt that suitable timber for the construction of traction trebuchets, if not large counterweight engines, was readily available.[170]

The fortifications of Beirut would have posed a much more significant obstacle than those of the incomplete castle at Jacob's Ford. Saladin probably recognised that he had little chance of taking the city before relief arrived; accordingly, his use of miners and waves of frontal attacks were a relatively inexpensive means of probing for an advantage, costing him little if they failed but bringing a huge reward if by chance the town fell rapidly or the Frankish relief force was delayed. Revealingly, the Muslim sources treat this attack as a far less significant episode than does William of Tyre. Despite the absence of stone-throwers at Jacob's Ford and Beirut, Saladin's competence with artillery is confirmed by his campaigns east of the Euphrates soon after.

Mosul and Amida: 1182-83

Mosul was the most significant city on the upper Tigris. Its ruler, 'Izz al-Din Mas'ud, a nephew of Nur al-Din, had arranged for artillery to be prepared for the city's defence before Saladin commenced his siege in November 1182. According to Ibn al-Athir, Saladin's nephew, Taqi al-Din, advocated they erect a large stone-thrower, to which Saladin responded, "No trebuchet can be set up against a city like this. If we erect it, they will seize it." Rejecting this, Taqi al-Din set one up anyway, only to have it targeted by nine defensive stone-throwers before it was seized in a sally.[171] The account reveals little of the power of Saladin's artillery, apart from the possibility that Taqi al-Din's engine may have been slightly more significant than the norm, though this is not stated.

Although absent at the time, Ibn al-Athir was a resident of Mosul and would have been familiar with the provisions for its defence. His account confirms the widespread use of defensive artillery but also touches on the difficulties of

169 William of Tyre, *Chronicon* 11.13, ed. Huygens, 1:515-16, trans. Babcock and Krey, 1:484-86; Fulcher of Chartres, *Historia Hierosolymitana* 2.42.1-2, ed. Hagenmeyer, pp. 534-36, trans. Ryan, pp. 196-97.

170 Pringle, *Churches*, no. 42, 1:115. My thanks to Denys Pringle for bringing this to my attention.

171 Ibn al-Athir, *al-Kamil*, trans. Richards, 2:285-87. Cf. Baha' al-Din, *al-Nawadir*, trans. Richards, pp. 57-58. Mosuli forces remained inside the city during most of this siege, rather than engaging the enemy in the field as typically occurred at the sieges of Aleppo and Damascus in the twelfth century.

disabling offensive artillery with similar engines. Unfortunately, Baha' al-Din, who left the city to gather support from Baghdad only days before Saladin arrived, included very few details concerning this siege when he later composed his history of events. After moving on from Mosul, Saladin erected artillery against Amida in early 1183, where it was once more used to support sapping efforts.[172]

Up to this point, one geographic region has been deliberately neglected in this investigation of twelfth-century artillery and its usage. Because Saladin devoted so much attention to Transjordan in the years leading up to 1187, it seems appropriate to deal with the sieges of Kerak and Montreal outside the general chronology.

Transjordan

At the start of the twelfth century, Baldwin I had led a reconnaissance force into Transjordan ahead of his coronation.[173] Starting in 1115, the Franks began to construct fortifications along the eastern ridge of the Wadi 'Araba and by the 1170s, the mighty castles of Montreal and Kerak anchored Frankish interests in this region.[174] What had once been a frontier area between the independent

172 Baha' al-Din, *al-Nawadir*, trans. Richards, p. 58; Ibn al-Athir, *al-Kamil*, trans. Richards, 2:287; al-Maqrizi, *al-Suluk*, trans. Broadhurst, p. 71. Baha' al-Din claims that the siege lasted eight days while Ibn al-Athir places its length at around fourteen days. The impressive remains of the basalt walls are some of the best preserved medieval defences in the region. See also Chevedden, "Invention," pp. 93-94.

173 Fulcher of Chartres, *Historia Hierosolymitana* 2.4, ed. Hagenmeyer, pp. 370-75, trans. Ryan, pp. 143-47; *Historia Nicaena* 64, RHC Oc 5, pp. 177-78. Cf. Albert of Aachen, *Historia Ierosolimitana* 7.41-43, ed. and trans. Edgington, pp. 546-51. Baldwin I followed up this initial reconnaissance in force with another in 1107, intercepting a Damascene force around Wadi Musa (Petra). In 1112/13, he returned to the region to prey on the caravan traffic between Cairo and Damascus, Ibn al-Qalanisi, *Dhail*, trans. Gibb, pp. 81-82, 130-31; Ibn al-Athir, *al-Kamil*, trans. Richards, 1:146, 164; Albert of Aachen, *Historia Ierosolimitana* 10.29-31, 12.8, ed. and trans. Edgington, pp. 744-47, 834-35. For the lordship and its lords, see Richard, *The Latin Kingdom of Jerusalem*, pp. 88, 95; Tibble, *Monarchy and Lordships*, pp. 29-36, 83-84. Cf. Riley-Smith, *The Crusades*, pp. 96-99.

174 In the extreme south, a castle was built as an outpost on the so called Ile de Graye (Pharaoh's Island, Jazirat Fara'un) at the north end of the Gulf of Aqaba, probably in the 1160s. In the Petra valley, al-Habis and Li Vaux Moise (Wadi Musa) were built by the Franks, overlooking the Nabataean-Roman basin and a few kilometres beyond the siq respectively. Montreal, situated between Petra and the Dead Sea, was the first major castle built in the area. From about 1142, Montreal was overshadowed by the equally impressive castle

powers of Jerusalem, Cairo and Damascus was transformed into a vital corridor between the two power-centres of Saladin's empire following his unification of Egypt and Syria in 1174. Under the aggressive leadership of Reynald of Châtillon from 1176/77, the region became a critical battleground between Egypt and Syria and led Kerak to become the most besieged castle of the kingdom of Jerusalem in the late twelfth century, having enjoyed relative peace until 1170.

Defences of Kerak

The fortifications of Kerak are not very sophisticated. Its towers do not protrude far from the line of the adjoining curtain walls and its Frankish masonry, composed of hard volcanic stone, appears poor compared to the use of finely cut limestone blocks by later Muslim masons in sections of the castle that were repaired or rebuilt. Kerak's strength is its position and solidity. High on a ridge, the castle is secured along its east and west flanks by a steep gradient. To the north, the castle is separated from the town by a fosse 20-30 m wide, which was reportedly sixty cubits (32.4 m) deep.[175] Behind the fosse is a wall 135 m long and up to 5 m thick, which rises from the rock-cut scarp of the ditch. It is secured at each end by a slight projection akin to a tower, each with a postern

---

of Kerak, just east of the southern end of the salt lake and much closer to Jerusalem. For the Ile de Graye and Aqaba, see De Meulemeester and Pringle, "Al-'Aqaba Castle," pp. 97-102; Pringle, "The Castle of Ayla," pp. 333-53; Pringle, *Secular Buildings*, no. P5, p. 113; Kennedy, *Crusader Castles*, p. 30; Deschamps, *Les Châteaux des Croisés*, 2:44; Rey, *Architecture militaire*, p. 276. For al-Habis see Pringle, *Secular Buildings*, no. 97, p. 49; Marino, "The Crusader Settlement in Petra," pp. 4-5; Kennedy, *Crusader Castles*, pp. 25, 29-30. For Li Vaux Moise see Pringle, *Secular Buildings*, no. 230, pp. 105-6; Kennedy, *Crusader Castles*, pp. 25-27; Vannini and Tonghini, "Medieval Petra," pp. 371-84. For Montreal, see Faucherre, " La forteresse de Shawbak," pp. 43-66; Pringle, *Churches*, 2:304-14; Pringle, *Secular Buildings*, no. 157, pp. 75-76; Brown, "Summary Report," pp. 225-46; Deschamps, *Les Châteaux des Croisés*, 2:43-45. Cf. Fulcher of Chartres, *Historia Hierosolymitana* 2.55, ed. Hagenmeyer, pp. 592-93, trans. Ryan, p. 215; Albert of Aachen, *Historia Ierosolimitana* 12.22, ed. and trans. Edgington, pp. 858-59; *Historia Nicaena* 76, RHC Oc 5, pp. 181-82; William of Tyre, *Chronicon* 11.29, ed. Huygens, 1:542, trans. Babcock and Krey, 1:513; Ibn al-Furat, *Tarikh*, trans. Lyons and Lyons, 2:65. For Kerak, see Deschamps, *Les Châteaux des Croisés*, 2:35-98; Müller-Wiener, *Castles of the Crusaders*, pp. 47-48; Pringle, *Churches*, 1:286-94; Pringle, *Secular Buildings*, no. 124, pp. 59-60; Brown, "Excavations", pp. 287-304; Kennedy, *Crusader Castles*, pp. 45-50. Cf. William of Tyre, *Chronicon* 15.21, 22.28, ed. Huygens, 2:703-4, 1056, trans. Babcock and Krey, 2:127, 499; Ibn al-Furat, *Tarikh*, trans. Lyons and Lyons, 2:50-51.

175 Ibn al-Athir, *al-Kamil*, trans. Richards, 2:300; Ibn al-Furat, *Tarikh*, trans. Lyons and Lyons, 2:51. Abu'l-Fida' claims that it was fifty cubits deep, Abu'l-Fida', *al-Mukhtasar*, trans. Holt, pp. 47-48. The fosse has not yet been excavated so its true depth remains unknown.

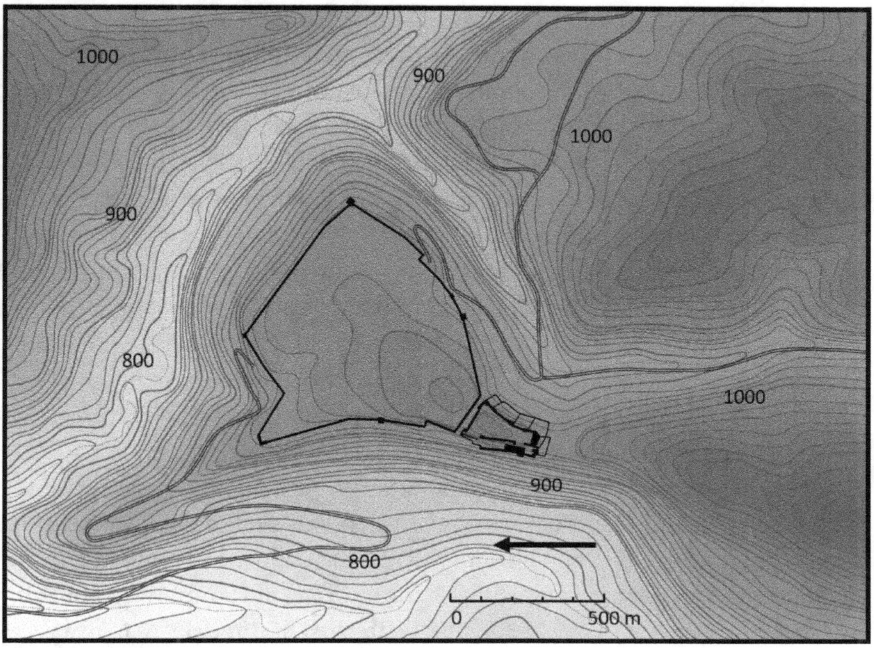

PLAN 13 *Kerak*

along its recessed inner flank. To the south, the castle is separated from the remainder of the ridge by a dip in the topography, which was further developed by the Franks. Most of the original Frankish work at this end of the castle has been obscured by the addition of the large Mamluk donjon. Beyond the southern walls, there was an external open cistern, both a source of water and a defensive obstacle. Along the western side of the castle, there was probably some kind of lower bailey or large barbican during the Frankish period, although this too has been obscured by subsequent Muslim rebuilding.

Although simple in design, Kerak was keenly planned. The castle is only approachable from the town, across the northern fosse. The present wall facing this front, presumably dating to the mid-twelfth century, is thick and pierced by at least two levels of embrasures below the parapet. The castle was thus almost impossible to assail directly provided the northern ditch remained unobstructed. Beyond the castle, it is hard to discern any remains of the town wall that would have stood during the Frankish period, having vanished in the wake of Mamluk rebuilding and modern urban development. However, the sources, and simple defensive necessity, suggest that the town was fortified to at least some extent during the twelfth century.

Early Sieges: 1170-73

Kerak was besieged no less than five times between 1170 and 1188.[176] Only months after the Frankish-Byzantine attack on Damietta in 1169, Nur al-Din led a screening manoeuvre into Transjordan in early 1170, shielding the movement of a large contingent of supplies, merchants and members of Saladin's family heading to join him in newly conquered Egypt. Kerak appears to have been attacked at this point and 'Imad al-Din claims that two stone-throwers bombarded the castle for four days before the siege was lifted with the approach of Frankish forces.[177] In 1171, Nur al-Din attempted to coordinate a joint assault on Montreal with Saladin but mutual mistrust led to its failure,[178] while in 1173 Saladin led the first assault on Kerak from the direction of Egypt.[179]

In spring 1182, Saladin led a force into Transjordan with the primary objective of shielding the movement of a large body of Egyptian troops, which he intended to use later that year in Mesopotamia. Although no siege appears to have taken place, al-Maqrizi notes that artillery was sent to the Egyptian force just before it set out.[180] The following year, Kerak endured Saladin's full might. Famously, this siege corresponded with the wedding of Humphrey III of Toron and Isabella of Jerusalem.[181] This would be the first of three sieges in five years.

Kerak: 1183

Saladin left Damascus on 22 October 1183, ordering his brother, al-'Adil, to meet him at Kerak. The latter was to bring Egyptian forces to aid siege efforts while simultaneously ferrying his family and treasury out of Egypt to his new post in

---

176 These took place in 1170, 1173, 1183, 1184 and 1187-88.

177 Abu Shama, *Kitab al-Raudatain*, RHC Or 4, pp. 153-54. See also Ibn al-Athir, *al-Kamil*, trans. Richards, 2:184. Cf. Baha' al-Din, *al-Nawadir*, trans. Richards, p. 46; Ibn al-Furat, *Tarikh*, trans. Lyons and Lyons, 2:51.

178 Ibn al-Athir, *al-Kamil*, trans. Richards, 2:198-200; al-Maqrizi, *al-Suluk*, trans. Broadhurst, pp. 38, 41-43.

179 Baha' al-Din, *al-Nawadir*, trans. Richards, p. 48; Ibn al-Athir, *al-Kamil*, trans. Richards, 2:213-14; Abu Shama, *Kitab al-Raudatain*, RHC Or 4, pp. 156-57; Ibn al-Furat, *Tarikh*, trans. Lyons and Lyons, 2:51; al-Maqrizi, *al-Suluk*, trans. Broadhurst, p. 43; William of Tyre, *Chronicon* 20.26-28, ed. Huygens, 2:948-52, trans. Babcock and Krey, 2:386-90. William of Tyre's discussions of events in Transjordan are rarely clear and his descriptions of Kerak are riddled with errors, suggesting that he never visited the castle. He claims that Muslim invasions of Transjordan took place in the eighth, ninth and tenth years of Amalric's reign (from 1170 to 1174), but they appear to be skewed within the chronology. Cf. the dating found in Richard, *The Latin Kingdom of Jerusalem*, p. 51.

180 Al-Maqrizi, *al-Suluk*, trans. Broadhurst, p. 67. See also Hamilton, *The Leper King*, p. 172.

181 Hamilton has suggested that Saladin deliberately planned the siege to take place while the castle was filled with wedding guests, Hamilton, "The Elephant of Christ," p. 105.

Aleppo.[182] The Egyptian forces arrived on 22 November, after the siege had already begun.

Upon its arrival, Saladin's main force established a blockade and began to set up artillery. These engines may have been erected initially against the town, which Reynald of Châtillon attempted to defend, but they were subsequently set up to assail the north face of the castle once the Muslims had forced their way into the town. In their panic to reach the castle, the fleeing refugees from the town tore down the bridge over the northern fosse. Although this firmly halted the besiegers, it also prevented the Franks from launching any sorties and provided Saladin with the security to erect his artillery as close to the fosse as he liked, having only to contend with the missiles and projectiles thrown from the castle.

William of Tyre states that Saladin brought siege equipment with him from Damascus,[183] which probably consisted of prefabricated artillery. Ibn al-Athir claims that Saladin erected seven stone-throwers and William of Tyre gives the similar figure of eight, specifying that six were set up within the town and two outside at a place known as 'Obelet'.[184] Deschamps suggested that this secondary position was on the spur to the south and that the engines placed here targeted the castle's southern defences, now dominated by the Mamluk donjon.[185] Although this appears to be the only sensible secondary position, at a distance of about 115 m, it is questionable whether contemporary artillery could have had any kind of effect from here.[186]

The bombardment from the town was clearly intense, allowing the besiegers to slaughter the Frankish livestock that had been gathered in the northern fosse. Although the physical effects of this artillery barrage seem to have been minimal, this was undoubtedly a psychologically exhausting affair and anyone who exposed himself along the parapet would have done so at great risk. At no

182 Baha' al-Din, *al-Nawadir*, trans. Richards, p. 62; Abu Shama, *Kitab al-Raudatain*, RHC Or 4, p. 248; Ibn al-Athir, *al-Kamil*, trans. Richards, 2:297; al-Maqrizi, *al-Suluk*, trans. Broadhurst, pp. 71-72.

183 William of Tyre, *Chronicon* 22.29 (28), ed. Huygens, 2:1,055-57, trans. Babcock and Krey, 2:498-501.

184 Ibn al-Athir, *al-Kamil*, trans. Richards, 2:297-98; William of Tyre, *Chronicon* 22.31 (30), ed. Huygens, 2:1,059, trans. Babcock and Krey, 2:503.

185 Deschamps, *Les Châteaux des Croisés*, 2:64.

186 The Ernoul account states that Saladin spared the tower in which the married couple were staying from the barrage of his *perrieres* and *mangonniaus*, *Ernoul* 9, ed. Mas Latrie, pp. 103-4. The context would suggest that this tower was in range of the attacking artillery; however, the author provides no details of where Saladin's engines were placed or which towers were in range, drawing into question the reliability of this anecdote.

point, however, did these engines endanger the integrity of the stronghold's fortifications. Either critically or apologetically, Ibn al-Athir claims that Saladin did not expect the main Frankish army to allow him to besiege Kerak for very long so had intentionally not brought a sufficient number of engines for a long siege. This observation is peculiar considering that this is the largest number of trebuchets that Ibn al-Athir places at a single siege since the Fatimid siege of Jerusalem in 1099. Saladin's lack of progress after a month of uninterrupted bombardment is revealing. Although he may have been employing prefabricated artillery, even these were evidently fairly light. Whether they were traction trebuchets or small counterweight engines, they appear to have done little damage to the castle's defences.

Inside the castle, the defenders attempted to build their own stone-thrower. This turned out to be a vain effort when it was quickly put out of action by the skilled crews working the Muslim artillery.[187] Those within found themselves with little more to do than trust in the static strength of Kerak's fortifications and endure the Muslim artillery barrage, hoping the masonry over their heads was strong enough to withstand the impact of any projectiles that sailed over the castle's outer walls.

The siege was lifted in early December 1183 when Baldwin IV arrived at the head of a relief force. With Egypt stripped of its defenders, brought out by al-'Adil to assist with the siege, Saladin sent his nephew, Taqi al-Din 'Umar, there while he departed with the remainder of his forces in the direction of Damascus, giving Kerak a short period of respite.[188]

Kerak: 1184

Saladin returned to Kerak around the start of August 1184 and the Arabic sources clearly convey the impression that he once more made extensive use of artillery.[189] William of Tyre omits any mention of this episode but Baldwin IV reported that the Muslim army employed fourteen *petrariae* in a letter written

187 Saladin's forces apparently faced none of the issues targeting defensive engines encountered by the Franks at Tyre in 1124.

188 William of Tyre, *Chronicon* 22.29 (28), 31 (30), ed. Huygens, 2:1,055-57, 1,059-60, trans. Babcock and Krey, 2:498-501, 503-4; *Ernoul* 8, 9, ed. Mas Latrie, pp. 80-82, 102-6; Baha' al-Din, *al-Nawadir*, trans. Richards, p. 62; Ibn al-Athir, *al-Kamil*, trans. Richards, 2:297-98; Abu Shama, *Kitab al-Raudatain*, RHC Or 4, p. 248; Ibn al-Furat, *Tarikh*, trans. Lyons and Lyons, 2:52; al-Maqrizi, *al-Suluk*, trans. Broadhurst, pp. 71-72.

189 Abu Shama, *Kitab al-Raudatain*, RHC Or 4, pp. 249-56; Baha' al-Din, *al-Nawadir*, trans. Richards, pp. 64-65; Ibn al-Athir, *al-Kamil*, trans. Richards, 2:299-300; Ibn al-Furat, *Tarikh*, trans. Lyons and Lyons, 2:52; al-Maqrizi, *al-Suluk*, trans. Broadhurst, p. 74. Cf. Ibn Jubayr, *Rihla*, trans. Broadhurst, pp. 300-1, 313-14.

soon after the siege.[190] Having failed the year before, there is little reason to doubt that Saladin brought his best and most powerful machines on this occasion.

'Imad al-Din suggests that the defenders again made an effort to defend the town, against which Saladin initially directed nine trebuchets. He claims that the engines destroyed part of a wall, leaving only the fosse of the castle to overcome.[191] Due to the source's poetic writing style, it is hard to tell if he is referring to a section of the town wall or the north wall of the castle. Ibn al-Athir and Ibn Shaddad, who probably made use of 'Imad al-Din's account, evidently preferred the former interpretation.[192] This damage must have been inflicted before 17 August, at which point efforts began to fill the castle's northern fosse. However, Baha' al-Din implies that Saladin did not erect his artillery until 23 August, when the Mesopotamian contingents of his army arrived. While it might be suggested that Baha' al-Din was simply referring to the point at which these engines were directed against the castle, rather than the town, a preserved section of 'Imad al-Din's lost *Kitab al-Barq al-Shami*, appears to imply that the various elements of the army had already gathered before the town was attacked, a sentiment repeated by Ibn al-Athir, who probably followed this earlier record. To further complicate the matter, al-Maqrizi, who was writing much later but had access to many earlier sources, asserts that artillery was not used until 3 September.[193] It is thus unclear how long these engines were in use before Frankish relief arrived and the siege was lifted around 4 September.

There are presently no indications that the northern wall of the castle was significantly damaged during this siege, or that of the previous year, while the Frankish town walls and gates have since disappeared, preventing an assessment of any repair work conducted along this outer trace. Although the weakness of these outer defences might have been what encouraged their subsequent rebuilding when the castle fell under Muslim rule, it still seems unlikely that late-twelfth-century artillery was the main factor leading to the capture of the town, given the implicitly short period of time during which it was pressed.

---

190 Letter of Baldwin IV to a Frankish delegation in Europe, in Ralph of Diceto, *Ymagines historiarum*, ed. Stubbs, 2:27-28.

191 Abu Shama, *Kitab al-Raudatain*, RHC Or 4, p. 254.

192 Ibn al-Athir, *al-Kamil*, trans. Richards, 2:300; Ibn Shaddad, in Ibn al-Furat, *Tarikh*, trans. Lyons and Lyons, 2:52.

193 Abu Shama, *Kitab al-Raudatain*, RHC Or 4, pp. 254-55; Baha' al-Din, *al-Nawadir*, trans. Richards, pp. 64-65; Ibn al-Athir, *al-Kamil*, trans. Richards, 2:300; al-Maqrizi, *al-Suluk*, trans. Broadhurst, p. 74. See also Ibn al-Furat, *Tarikh*, trans. Lyons and Lyons, 2:52. Cf. Deschamps, *Les Châteaux des Croisés*, 2:66-68.

When the sources are read collectively, the main thrust of the siege accompanied Saladin's attempt fill the northern fosse of the castle. Covered ways were constructed to assist those bringing up material to choke the ditch and most sources present the Muslim artillery as helping to provide cover for these operations. It is in the freer language of the letters preserved by Abu Shama that the destructive effect of the attacking artillery is most embellished. Despite the prevalent imagery of destruction, at times these letters definitively refer to artillery targeting the castle's parapets and its Frankish defenders.[194] Similarly, 'Imad al-Din specifically mentions the damaging of battlements, the chipping of merlons, the collapsing of roofs, and the enlarging of loopholes. Although the mass of the northern wall displays no evidence of obvious impact signatures or significant repairs, this cannot be said of the parapet, or what is left of it. The current state of these battlements is not conclusive evidence that the parapet of the northern wall was targeted during the siege of 1184, as the obvious damage and subsequent repairs carried out here could date to a number of later episodes.[195]

In contrast to the siege of 1183, most of the sources draw attention to the active defence mounted by the defenders. The garrison, which had apparently constructed defensive artillery since the last siege, showered the Muslims attempting to fill the fosse with stones as well as arrows. The limited power of these engines is suggested by their inability to break through the improvised mud-brick coverings built by the besiegers to protect their path to the fosse.

Kerak and Counterweight Artillery

Given the frequency with which Kerak was besieged through the 1170s and 1180s, a number of historians have included some of these sieges in their assessments of the power of late-twelfth-century artillery. Ronnie Ellenblum has suggested that the brevity of the siege in 1170, and 'Imad al-Din's claim that only two engines were used, is reflective of a wider policy whereby Muslim forces were beginning to integrate prefabricated heavy artillery.[196] It is unclear how the brevity of an unsuccessful siege is supposed to support the presence of particularly powerful siege engines, especially when most sources do not mention the use of artillery during this attack. This theory also fails to address the sieges of 1183 and 1184, where many more engines were used with equally little physical effect. Furthermore, if Saladin abandoned his artillery when he lifted the sieges of 1183 and 1184, as is indicated by William of Tyre and Baldwin IV

194 Abu Shama, *Kitab al-Raudatain*, RHC Or 4 pp. 251-55.

195 Brown, "Excavations," p. 290.

196 Ellenblum, *Crusader Castles*, p. 258.

IMAGE 1 *Kerak: northern wall of the castle* (AUTHOR'S PHOTO)

respectively,[197] this suggests that it was less valuable to him than the mobility that he gained without it, although on neither occasion was his force pressed by the Frankish relief army.

What has hitherto been neglected by historians is a simple investigation of the architectural remains. The outer face of Kerak's northern wall reveals neither notable blemishes nor subsequent repairs below the level of the ramparts. Not a single impact signature is discernable even though the attacking artillery could have been within 30 m of the wall.[198] While the rough masonry could easily disguise the minor scarring inflicted by traction trebuchets, if heavier counterweight engines were employed with the intention of damaging the wall, there should be evidence of this. Rather than the wall, it was the fosse in front of it that was considered to be the more formidable obstacle – so long as

197 William of Tyre, *Chronicon* 22.31 (30), ed. Huygens, pp. 1059-60, trans. Babcock and Krey, 2:504; Letter of Baldwin IV to a Frankish delegation in Europe, in Ralph of Diceto, *Ymagines historiarum*, ed. Stubbs, 2:28.

198 The castle was subsequently besieged at least twice in the thirteenth century, in 1246 by Egyptian forces and 1286 by a Mamluk detachment, though there appear to be no indications that artillery was used on either occasion, Ibn al-'Amid, *al-Majmu*, trans. Eddé and Micheau, p. 80; Maqrizi, trans. Broadhurst, pp. 281-82, trans. Quatrèmer, 2.1:87-88; Abu'l-Fida', *al-Mukhtasar*, trans. Holt, p. 13.

the castle's foundations remained beyond the reach of the Muslim miners, the castle and its defenders were secure.

Whether traction- or counterweight-powered, prefabricated or built locally, the inability of Saladin's artillery to make an impression on Kerak's defences following so many sieges clearly reveals the limited power of the artillery brought against them. This suggests that contemporary commanders did not consider artillery to be a breaching weapon in the late twelfth century, nor would they have tried to use it as such. These engines were instead used in a traditional manner: to create a hostile environment for defenders and perhaps to steadily chip away at battlements and the protection they provided.

Kerak: 1187

The final siege of Frankish Kerak began in 1187. As part of the multi-directional offensive that ultimately culminated in the battle of Hattin, Saladin took a force to Bosra that spring to protect the returning pilgrim caravans. This force then moved south into Frankish Transjordan where it was joined by contingents from Egypt. A detachment settled down to invest Kerak while the remainder raided throughout the Frankish lordship.[199] Having travelled to Bosra without baggage, it seems unlikely that Saladin made use of artillery or even intended to press the siege to an aggressive conclusion. Although it is possible that the Egyptian contingent might have brought siege equipment, there is no evidence of this. Saladin appears to have appreciated the relative resilience of Kerak to acute siege tactics and changed his strategy, content to allow the garrison to starve itself to the point of capitulation if events to the north failed to bring about a greater advantage there.[200]

Most contemporary references to twelfth-century artillery are far from detailed. From a Frankish perspective, this mirrors similar trends in Europe, where ambiguous terms such as *machinae* were often preferred and, even in instances where more specific terminology was used, accounts include few details of how these engines were employed.[201] However, the rather mundane or unremarkable manner in which these weapons were most often mentioned

199 Baha' al-Din, *al-Nawadir*, trans. Richards, pp. 71-72; Abu Shama, *Kitab al-Raudatain*, RHC Or 4, pp. 261-62; Ibn al-Athir, *al-Kamil*, trans. Richards, 2:318-19; al-Maqrizi, *al-Suluk*, trans. Broadhurst, p. 82; *Eracles* 23.54, RHC Oc 2, p. 81.

200 Since his devastating defeat at Montgisard in 1177, Saladin had been seeking a decisive battle with the kingdom of Jerusalem, but failed to lure the numerically inferior Frankish army into an ideal position despite repeated incursions into Galilee and Transjordan.

201 For an overview of contemporary Anglo-Norman artillery and the vocabulary used to identify it, see Fulton, "Anglo-Norman Artillery," pp. 1-31.

clearly indicates that these were familiar machines, a point plainly made by the number of occasions in which certain sources, such as Albert of Aachen, Matthew of Edessa and William of Tyre, claim that artillery was used.

From a Muslim perspective, references to artillery in Ibn al-Qalanisi's history, the only significant surviving contemporary source, increase from about 1138. This probably reflects, to some degree, the author's increased exposure to such engines, perhaps associated with the mounting pressure that Zanki was applying towards Damascus. A similar pattern can be found in Ibn al-Athir's record of events, which although composed in the early thirteenth century, probably made use of sources now lost. He mentions the use of artillery at only nine sieges in the Near and Middle East between 1100 and 1159 but on no less than ten occasions between 1160 and 1177. This dramatic increase, which corresponds with the height of Nur al-Din's power and Saladin's early career, is probably due in part to the mention of such engines in 'Imad al-Din's lost *Kitab al-Barq al-Shami*. Ibn al-Athir then mentions artillery at sixteen sieges during the 1180s, accompanying Saladin's rise to pre-eminence and the attention this received in the various histories that he used.

Although the number of Muslim sources increases with time, references to artillery appear to increase disproportionately. Accordingly, this was not just a case of there being a broader pool of sources, each with the potential to mention these engines, but rather individual sources independently noticing or highlighting the use of artillery with increasing frequency. While this might suggest that artillery was coming into broader use, when reconciled with the Frankish sources, which confirm the widespread use of these engines by all sides throughout the period, it instead speaks to the growing importance of these machines or a widening perception of this. By examining the events surrounding the Third Crusade, it is possible to gain an appreciation of how much these engines had developed since the First Crusade and a better sense of how they were being employed at the end of the twelfth century.

CHAPTER 5

Third Crusade

The period surrounding the Third Crusade has been documented from a fairly astonishing number of perspectives. The significance of the events and the personalities involved contributed to the compilation and preservation of numerous eye-witness accounts, providing detailed records of Saladin's campaigns of conquest in 1187-88 and the crusade these precipitated. The prestige and resources of the figures involved provide the opportunity to assess what may have been some of the most powerful mechanical stone-throwers of the period and examine their role in broader siege operations.

'Imad al-Din al-Isfahani, who had achieved a certain level of significance in Damascus under Nur al-Din, later joined Saladin's service and acted as his secretary. This position provided him with considerable access to information and the histories that he wrote following Saladin's death form the basis of most of the subsequent Arabic narratives relating to this period. While his *al-Fath al-Qussi fi'l-Fath al-Qudsi* has survived, only sections of his *Kitab al-Barq al-Shami* remain, preserved in part in Abu Shama's composite history and the works of later historians who similarly drew from it. Baha' al-Din ibn Shaddad was a contemporary of 'Imad al-Din and another respected scholar. He too found himself attached to Saladin when he was summoned to the sultan's service in early 1188. Whereas 'Imad al-Din's account was composed in deliberately ambiguous verse, often allowing passages to carry two quite different meanings, Baha' al-Din used clear prose. Although the influence of 'Imad al-Din's writings are quite apparent at times, especially relating to events before the Third Crusade, Baha' al-Din contributes considerable original information. The third critical Muslim account is that provided by Ibn al-Athir, at this point a contemporary but the youngest of the three authors. Although this section of his work is based heavily on the writings of the former historians, he appears to add his own perspective at times and may even have accompanied Saladin on campaign in 1188.[1]

1 For discussions of 'Imad al-Din, Baha' al-Din and Ibn al-Athir, their works and sources, see Richards, "A Consideration of Two Sources," pp. 46-65; Baha' al-Din, *al-Nawadir*, trans. Richards, pp. 1-7; Ibn al-Athir, *al-Kamil*, trans. Richards, 1:1-5; Richter-Bernburg, "'Imad al-Din al-Isfahani," pp. 29-51; Micheau, "Ibn al-Athir," pp. 52-83. See also Little, "Historiography," pp. 418-21.

 | DOI 10.1163/9789004376922_007

From a Frankish perspective, William of Tyre's account ends in 1184 but is continued in the Ernoul and subsequent *Eracles* continuations.[2] The author of the *De expugnatione terrae sanctae*, who appears to have been among those besieged in Jerusalem in 1187, provides another eyewitness account.[3] To these can be added the narratives provided by participants of the Third Crusade. An original versed account was penned by Ambroise and this was rendered into prose by the author of the *Itinerarium peregrinorum et gesta regis Ricardi*. Both men appear to have travelled to the Holy Land with elements of Richard I's army. Roger of Howden provides a third crusader account of the siege of Acre. Although Roger arrived at the siege with Richard I, he left the Holy Land following the city's capture in the company of Philip II. Back in England, Ralph of Diceto did not take part in the crusade but was nevertheless well informed of its progress, as is revealed in his chronicle.[4]

The number of accounts and their relative detail allow for a close examination of the artillery used during this period. Most sieges were observed and later described by multiple eye-witnesses, often providing perspectives from both sides. While most sources mention the continued use of fairly light antipersonnel engines, notions of increasingly powerful machines become apparent. Besieging forces were probably experimenting with counterweight trebuchets by at least the late twelfth century; however, the real impact of these weapons was yet to be appreciated. By the Third Crusade, these engines had almost certainly developed into practical and effective weapons, but were still far from the behemoths that would be built in later decades.

The Hattin Campaign

While Saladin was passively besieging Kerak, his son, al-Afdal, coordinated a raid into Galilee. The success of this party, which defeated a force of Franks near the Springs of Cresson, and the arrival of his Egyptian forces compelled Saladin to break the siege. Having invested Kerak for less than a month, Saladin

2 For a discussion of the Ernoul account and its relation the *Eracles* continuation, see Edbury, "*Ernoul, Eracles* and the Fifth Crusade."

3 For a discussion of the *Libellus de expugnatione de terrae sanctae per Saladinum*, an analysis of those sections dealing with the events of 1187, a brief examination of its author, and its relation to the Ernoul and *Eracles* continuations of William of Tyre's account, see Kane, "Wolf's Hair," pp. 95-112. My thanks to James Kane for sharing some thoughts on the *Libellus* as he, John Pryor and Keagan Brewer prepare a new edition and translation of the text.

4 For a discussion of these crusader accounts, see Ambroise, *Estoire*, trans. Ailes, pp. 1-23; *Itinerarium*, trans. Nicholson, pp. 1-15.

headed north and amassed his forces east of the Sea of Galilee in late May 1187, setting the stage for what would be the most influential campaign in the region since the First Crusade. Given this significance, it can be easy to lose sight of the immediate context. When Saladin crossed the Jordan in 1187, his campaign objectives may have been relatively short-sighted and far more similar to those that he had had when he invaded Galilee in 1179 and 1183 than is often considered.[5] But on 4 July 1187, just days after crossing the Jordan, Saladin was given the opportunity he had been looking for and completely defeated the army of Jerusalem near the Horns of Hattin.[6]

If Saladin's forces were travelling with artillery prior to the battle of Hattin, these engines were probably left east of the Jordan while the army attempted to draw the Franks into an engagement, as had been done in 1179. To do this, he sent a detachment to attack Tiberias. 'Imad al-Din is the only Muslim source who indicates that artillery might have been used at Tiberias; however, his choice of words seems to be influenced by the versed style of his writing.[7] The term he uses, *hajjarun*, can be understood as 'stoneworkers', which could identify sappers as well as those who shaped artillery ammunition.[8] It seems the Muslims relied on mining and a frontal assault rather than artillery, capturing the town, though not its citadel, in a single day.[9] Tiberias was probably taken on 2 July, a day before the first skirmishing began and two days ahead of the main battle.

Following the decisive Muslim victory, the citadel of Tiberias was quickly surrendered, leaving Saladin free to set off across Galilee. He arrived at Acre on

---

5 For the 1179 campaign, William of Tyre, *Chronicon* 21.27-29 (28-30), ed. Huygens, 2:1,000-4, trans. Babcock and Krey, 2:440-45; Abu Shama, *Kitab al-Raudatain*, RHC Or 4, pp. 198-206; Ibn al-Athir, *al-Kamil*, trans. Richards, 2:264-65; al-Maqrizi, *al-Suluk*, trans. Broadhurst, pp. 59-60. For the 1183 campaign, see William of Tyre, *Chronicon* 22.26-28 (25-27), ed. Huygens, 2:1,048-54, trans. Babcock and Krey, 2:492-98; Abu Shama, *Kitab al-Raudatain*, RHC Or 4, pp. 242-48; Baha' al-Din, *al-Nawadir*, trans. Richards, pp. 60-61; Ibn al-Athir, *al-Kamil*, trans. Richards, 2:297; al-Maqrizi, *al-Suluk*, trans. Broadhurst, p. 72.

6 For studies of this battle see France, *Hattin*, pp. 74-104; Kedar, "The Battle of Hattin Revisited," pp. 190-207.

7 See Fulton, "Development of Prefabricated Artillery," pp. 58-59.

8 See Khamisy, "Some Notes on Ayyubid and Mamluk Military Terms," pp. 73-92. See also Amitai, "Foot Soldiers," p. 236. For the siege of Tiberias, see 'Imad al-Din, *al-Fath*, trans. Massé, p. 24. For the Arabic, see Abu Shama, *Kitab al-Raudatain*, RHC Or 4, p. 264. Cf. Baha' al-Din, *al-Nawadir*, trans. Richards, p. 73; Ibn al-Athir, *al-Kamil*, trans. Richards, 2:320-21; Abu'l-Fida', *al-Mukhtasar*, RHC Or 1, p. 58; *Eracles* 23.31, RHC Oc 2, pp. 47-48.

9 For the defences of Tiberias, see Stepansky, "Das kreuzfahrerzeitliche Tiberias," pp. 384-95; Stepansky, "Tiberias, the Courtyard of the Jews"; Stepansky, "Tiberias, Map Survey"; Stepansky, "The Crusader Castle of Tiberias," pp. 179-81; Pringle, "Town Defences," pp. 94-95.

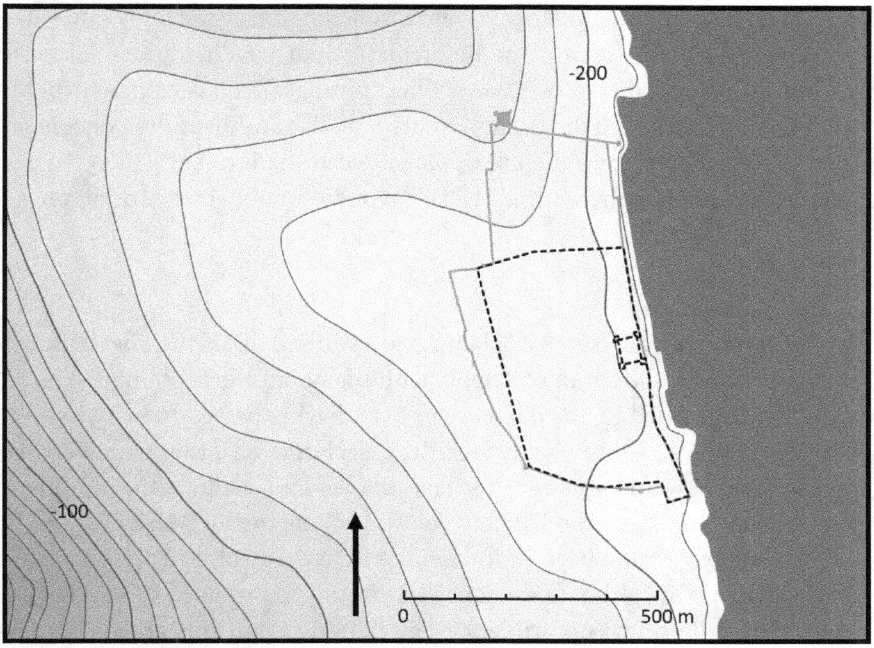

PLAN 14 *Tiberias: suggested Frankish defences and later Ottoman walls*

8 July and received the city's surrender the following day.[10] It is unclear if Saladin's artillery had reached him before he left for Acre, but Baha' al-Din and 'Imad al-Din note its use soon after at Toron.[11] It is possible that Taqi al-Din, or an otherwise unknown party that followed the main army, brought prefabricated engines to Saladin at Acre and that these engines were then taken to Toron, either by Taqi al-Din's advance party or by Saladin when he joined his nephew there. However, it is equally likely that new stone-throwers were built on site before the garrison of Toron surrendered on 26 July.[12] Artillery does not appear to have been employed during the subsequent push up the coast, dur-

10 'Imad al-Din, *al-Fath*, trans. Massé, pp. 31-33; Baha' al-Din, *al-Nawadir*, trans. Richards, p. 75; Ibn al-Athir, *al-Kamil*, trans. Richards, 2:324-26; Eracles 23.46, RHC Oc 2, pp. 69-70 (lower text). See also Abu'l-Fida', *al-Mukhtasar*, RHC Or 1, p. 58.

11 'Imad al-Din, *al-Fath*, trans. Massé, pp. 37-38; Baha' al-Din, *al-Nawadir*, trans. Richards, p. 76.

12 'Imad al-Din, *al-Fath*, trans. Massé, pp. 37-39; Baha' al-Din, *al-Nawadir*, trans. Richards, p. 76; Ibn al-Athir, *al-Kamil*, trans. Richards, 2:326-27. Baha' al-Din's account is the exception to the popular narrative, stating the castle fell to assault and its steadfast defenders were led into captivity. For a nineteenth-century description of the castle see Conder and Kitchener, *Survey of Western Palestine*, 1:133-35, for a more recent report, see Piana, "The Crusader Castle of Toron," pp. 177-91.

ing which Sidon, Beirut and Jubayl were taken, nor during the foray into Palestine that saw Ramla, Hebron and Bethlehem fall to Saladin's forces. Likewise, the Muslim sources do not mention artillery during al-'Adil's acquisition of Mirabel and Jaffa, although the contemporary Frankish author of the *De expugnatione terrae sanctae* notes the use of *machinae* at the former.[13] On 23 August, Saladin united his army with al-'Adil's Egyptian force before turning on Ascalon.

Ascalon: 1187

By the middle of August 1187, Saladin had captured all of the coastal towns north of Ascalon and south of Tripoli, with the notable exception of Tyre. He was likely impressed by the defences of Tyre, having passed the city twice in less than a month, and probably regarded Ascalon as an easier target. Possession of Ascalon would also open the coastal road and facilitate the movement of troops and supplies from Egypt to Palestine along the Via Maris.

The siege of Ascalon began with an offer to free Guy of Lusignan, a captive since the battle of Hattin, in exchange for the city's surrender. When this was rejected, Saladin set up his artillery.[14] The rather sudden appearance of these engines in most sources is peculiar. It is possible that the mention of artillery at this point reflects the greater significance of this siege, during which the defenders put up a stiff resistance at first. Alternatively, the engines employed here might have been more impressive than others used earlier, or that this was simply the first time that artillery had been used since the siege of Toron.

Despite the number of sources that mention the use of artillery, only a few portray it as exceptional. One of these is a letter found in the *History of the Patriarchs*, one of many sent out by Saladin following the siege announcing his latest victory. It states that Saladin ordered his trebuchets set up on 23 August, that they were ready on 25 August and only one day later they opened a breach in a certain large bastion.[15] The exaggerations contained in this letter are not

---

13 *De expugnatione terrae sanctae*, ed. Stevenson, p. 229; 'Imad al-Din, *al-Fath*, trans. Massé, p. 33; Ibn al-Athir, *al-Kamil*, trans. Richards, 2:326; al-Maqrizi, *al-Suluk*, trans. Broadhurst, p. 83.

14 Baha' al-Din, *al-Nawadir*, trans. Richards, pp. 76-77; 'Imad al-Din, *al-Fath*, trans. Massé, pp. 97, 100; Abu Shama, *Kitab al-Raudatain*, RHC Or 4, pp. 312-13; Ibn al-Athir, *al-Kamil*, trans. Richards, 2:329-30; al-Maqrizi, *al-Suluk*, trans. Broadhurst, p. 84; *De expugnatione terrae sanctae*, ed. Stevenson, pp. 236-37; *Itinerarium* 1.8, ed. Stubbs, pp. 19-20, trans. Nicholson, p. 37; *History of the Patriarchs*, trans. Khater and Khs-Burmester, 3.2:128-31. This is the first time that Ibn al-Athir clearly states that artillery was used by a component of Saladin's army since it crossed the Jordan months earlier.

15 *History of the Patriarchs*, trans. Khater and Khs-Burmester, 3.2:130.

unlike those found in similar propaganda intended for a distant audience, such as the letters dispatched after the siege of Kerak in 1184 and those that would be sent out a few weeks later after the capture of Jerusalem. The widespread notice of artillery by Muslim sources is probably related, to at least some degree, to the circulation of these letters.

Ascalon surrendered on 5 September after a siege of two weeks. No mention of artillery is made during the subsequent acquisition of many smaller towns and strongholds as contingents of the Muslim army spread out across Palestine. As resistance in the area was mopped up, Saladin prepared to invest Jerusalem.

Jerusalem: 1187

The defences of Jerusalem that Saladin encountered were similar to those faced by the Franks in 1099.[16] On 20 September 1187, Saladin encamped his army to the west of the city, much as Raymond of St Gilles had initially done eighty-eight years earlier. From this point, Saladin's archers began to discharge a shower of suppressing fire that would continue throughout the siege. After surveying the city's defences, Saladin moved the bulk of his army to the north of the city on 25 September, deciding to focus his assault east of the St Stephen's (Damascus) Gate, as the crusaders had in 1099. The open space and elevation of the ridges running northward from the city would have provided plenty of options when positioning his artillery.

Saladin began to set up his trebuchets on the night of 25 September and by the following morning they were operational.[17] Although artillery is included in all of the principal sources,[18] 'Imad al-Din's account is perhaps the most evocative:

16 Most Frankish fortification efforts at Jerusalem undertaken between 1099 and 1187 appear to have been repairs rather than more extensive rebuilding projects. For example, William of Tyre, *Chronicon* 21.24 (25), ed. Huygens, 2:996, trans. Babcock and Krey, 2:436. See also Boas, *Jerusalem*, p. 44; Pringle, "Town Defences," pp. 79-80.

17 Baha' al-Din, *al-Nawadir*, trans. Richards, p. 77; Abu Shama, *Kitab al-Raudatain*, RHC Or 4, pp. 326-27; *De expugnatione terrae sanctae*, ed. Stevenson, pp. 241-43, trans. Brundage, pp. 159-60; Ernoul, *Chronique* 18, ed. Mas Latrie, pp. 211-14; *Eracles* 23.55, RHC Oc 2, pp. 82-83.

18 'Imad al-Din, *al-Fath*, trans. Massé, pp. 44-46; Baha' al-Din, *al-Nawadir*, trans. Richards, pp. 77-78; Abu Shama, *Kitab al-Raudatain*, RHC Or 4, pp. 326-30; Ibn al-Athir, *al-Kamil*, trans. Richards, 2:330-33; *De expugnatione terrae sanctae*, ed. Stevenson, pp. 241-46, trans. Brundage, pp. 259-62; *Itinerarium* 1.9, ed. Stubbs, pp. 20-21, trans. Nicholson, pp. 38-39; Ernoul, *Chronique* 18, ed. Mas Latrie, pp. 211-14; *Eracles* 23.55-57, RHC Oc 2, pp. 83-85; Bar Hebraeus, *Makhtebhanuth Zabhne*, trans. Budge, 1:325.

> [Saladin] positioned the siege engines, and made calamities fall like water falls from the clouds ... against the whirling stones and missiles, [the defenders] had no shelter ... The courses of the walls and the crenellations of their battlements were destroyed and lowered by the stones from the siege engines. It seemed as though the trebuchets became crazy while they were operated; they were like the bravest of men. They seemed to be mountains drawn by ropes. It was like swarms of locusts set into action by men; mothers of misfortune and death, pregnant women who birthed calamities. There was no protection against their stones; no security near them, even for those who were on their guard. Only danger accompanied their projectiles; their passage created only bitterness for people. How many stars fell from high in the sky! How many rocks were thrust up from their positions! How many embers scattered their sparks. Nothing compared to the devastation caused by their projectile pockets [impact craters], to the marks of their wounds, to the results of their achievements, to the precision of their sudden blows, to the pulling of their ropes. Without ceasing, they tore away everything with their [slings?]; they battered with all of their framing; they pulled with enthusiasm on their ropes; they hit, demolished, overturned, and split. They drew from their reserve of stones, preparing the evil that they were about to do. They disarranged the assemblage of stones of the rampart with rocks thrown one after another. They destroyed the cohesion of the structures that they broke up, unfastened and separated the courses [of masonry] by reaching their foundations. They loosened keystones by taking them under the effect of their ropes and exhausted the cisterns by drinking at their hollows.
>
> In short, leaving the ramparts in a residual state that repulsed the defenders, the enemy found himself depressed and in disorder. The fosse was crossed; troops were thrown forward ...[19]

19 This author's translation from Massé's French translation of the Arabic, 'Imad al-Din, *al-Fath*, trans. Massé, pp. 45-46. Costello's translation of Gabrieli's Italian translation of the Arabic is slightly different: "He [Saladin] mounted catapults, and by this means milked the udders of slaughter ... The bases of the walls and the teeth of their battlements were battered and broken down by stones from the catapults' slings; they seem like madmen throwing stones at random, impregnable gallant knights, mountains crossed with ropes, living beings aided by others, mothers of disaster and death, pregnant with calamity. Their missiles were invincible, all precautions against them were useless. Their darts vibrated with menace, their flight nourished on the bile of perceptive men. How many boulders came down out of heaven upon them, how many blocks of sandstone plunged into the earth, how many blazing firebrands bespattered them! The damage caused by the

Although this appears to describe the use of traction trebuchets, it cannot be ruled out that these were instead counterweight engines. While the size of the machines and the level of destruction suggests these were engines of the latter type, the emphasis placed on the pulling of ropes and the number of stones that were thrown are indicative of the former.

Al-Tarsusi's treatise, in which he provides a description and illustration of a counterweight trebuchet, probably predates this siege.[20] Despite the likelihood that Saladin was familiar with counterweight trebuchet technology by this point, there is nothing exceptional in any of the Muslim or Frankish accounts to confirm the use of such engines. Accordingly, it is less important whether or not counterweight engines were used, but rather that there is no discernable difference in the power, role, or portrayal of artillery – even if early counterweight trebuchets were employed, their impact or influence is indiscernible in the contemporary records.

Letters were widely circulated following the capture of Jerusalem. Like those sent out after the sieges of Kerak (1184) and Ascalon (1187), these documents emphasise the destruction wrought by the besiegers' artillery, claiming they demolished the walls that stood in their way.[21] It is possible that 'Imad al-Din, Saladin's secretary, was responsible for writing some of these letters. Regardless of whether he penned them or not, the similarity between the points that are stressed suggests that 'Imad al-Din at least had access to these letters when composing his later versed history.

According to accounts from both sides, Saladin employed his artillery as a supporting weapon: it was used to restrict the garrison's ability to maintain an

---

catapults, the extraordinary extent of their devastation, the effects of their concentration, the whistling wind of their flight, the extent of their range were beyond compare. The attack from their catapults never ceased, the battery of their mangonels, the drawing of water with their ropes, the parading in their halters, the attacking and defending, prostrating and slashing open, shaking their buckets, becoming downcast at their misfortunes, dissolving the composure of strong men with the boulders that they shot one after another, smashing the huddles of buildings, breaking them down into ruins, demolishing their foundations, breaking their joints by hauling them within their ropes, exhausting the wells by drinking from them with their own cups, until they reduced the walls to a single line of bricks and drove their defenders away. The enemy's ordnance was smashed and broken, the moat crossed and the attack sustained ..." 'Imad al-Din, *al-Fath*, trans. in Gabrieli, *Arab Historians*, pp. 154-56.

20 See Appendix 2, **no. 34**.

21 For example, *History of the Patriarchs*, trans. Khater and Khs-Burmester, 3.2:133-34.

active defence.[22] This allowed miners to approach the city walls and work unmolested. Although more distant than some of his contemporaries, Ibn al-Athir clearly explains how archers advanced with the sappers as they crossed the fosse while artillery was used to bombard the parapet and keep it clear of defenders. The author of the *De expugnatione terrae sanctae* provides a very similar account of the Muslims' tactics. It is unclear how far the sappers had progressed when the city's defenders, fearing a breach, capitulated on 2 October.[23] With Jerusalem in hand, Saladin departed for Acre on 30 October and moved onwards to Tyre on 8 November. Tyre was now the Franks' last foothold on the coast south of Tripoli and a vital link with Europe.

Tyre: 1187

Saladin had already bypassed Tyre twice during the summer and it would appear he, like many of his contemporaries, was impressed by the city's defences.[24] He arrived on 12 November 1187 but was forced to wait until 25 November for his son, al-Zahir Ghazi, and artillery to arrive. Once his forces had assembled, Saladin established his camp on a hill near the famous causeway and began to attack the city.[25] The Ernoul account states that Saladin employed seventeen *perrieres* and *mangonniaus*, while the *Eracles* chronicle counts fourteen *perrieres* and *mangoneaus*. A letter from Brother Terricus of the Templars to Henry II of England notes that there were thirteen *perrariae*.[26] The Muslim accounts give no numbers; 'Imad al-Din states simply that there were both

22 *De expugnatione terrae sanctae*, ed. Stevenson, pp. 244-46, trans. Brundage, pp. 160-62; *Eracles* 23.55-57, RHC Oc 2, pp. 83-85; 'Imad al-Din, *al-Fath*, trans. Massé, pp. 45-46; Baha' al-Din, *al-Nawadir*, trans. Richards, p. 77; Ibn al-Athir, *al-Kamil*, trans. Richards, 2:331.

23 Benvenisti takes a literal interpretation and suggests that the sappers covered an incredible 30 m within two days, Benvenisti, *The Crusaders*, p. 44. Prawer places the mine at precisely the same point where the Franks broke into Jerusalem in 1099 and the Persians entered in 614, Prawer, "The Jerusalem the Crusaders Captured," p. 14.

24 For contemporary sentiments, see William of Tyre, *Chronicon* 22.30 (29), ed. Huygens, 2:1,057, trans. Babcock and Krey, 2:501; Ibn Jubayr, *Rihla*, trans. Broadhurst, pp. 319-20; Ibn al-Athir, *al-Kamil*, trans. Richards, 2:328; Wilbrand of Oldenburg, *Itinerarium terrae sanctae* 1.2, ed. Pringle, p. 117, trans. Pringle, p. 63. See also Ibn Battuta, *Tuhfat*, trans. Gibb, p. 58. For the medieval defences of Tyre, see Conder and Kitchener, *Survey of Western Palestine*, 1:72; Pringle, "Town Defences," pp. 85-88; Pringle, *Churches*, 4:178-80.

25 Baha' al-Din, *al-Nawadir*, trans. Richards, pp. 78-79; 'Imad al-Din, *al-Fath*, trans. Massé, pp. 63-65; *Eracles* 24.2, RHC Oc 2, pp. 104-5; Abu Shama, *Kitab al-Raudatain*, RHC Or 4, pp. 342-44; Ibn al-Athir, *al-Kamil*, trans. Richards, 2:335-38.

26 Ernoul, *Chronique* 20, ed. Mas Latrie, p. 237; *Eracles* 24.2, RHC Oc 2, p. 105; letter from Terricus of the Temple to Henry II of England, in Roger of Howden, *Chronica*, ed. Stubbs, 2:346-47, trans. Riley, 2:90-91.

large and small trebuchets and gives the impression that they were worked by teams in relay.[27]

Although some of the Frankish sources use two terms to describe Saladin's artillery, this on its own is not sufficient evidence to conclude that more than one type of engine was employed. Such a suggestion, however, gains traction when taking into consideration 'Imad al-Din's reference to machines of differing sizes. Regardless, whatever engines were used at Tyre were likely the same as had been used at Jerusalem only weeks earlier. Little more about these machines can be discerned as the sources instead focus mainly on the naval engagement in late December that ultimately compelled the sultan to end the siege and withdraw from Tyre.

What Saladin did with his artillery following his decision to lift the siege is unclear. According to many Frankish sources, Saladin burnt his engines before departing;[28] however, Baha' al-Din states that Saladin disassembled his artillery and withdrew it with the army, only burning what could not be taken away.[29] If Saladin was making use of prefabricated engines, having moved them from Ascalon to Jerusalem to Tyre, and especially if some of these were counterweight trebuchets, it would seem inexplicable why he would burn these as his force was not threatened or otherwise pressured by a hostile field army. If Saladin destroyed some of his artillery, it would most likely have been his smaller engines, perhaps those which had been built on site or others built earlier during the campaign that were not worth the trouble of storing over the winter.

Saladin spent a considerable portion of the winter of 1188, as well as that of 1189, in Acre, while most of his army was sent home for the season. Baha' al-Din Qaraqush al-Asadi, who had overseen the construction of Saladin's new citadel in Cairo, was summoned from Egypt to strengthen the fortifications of Acre. References to him bringing labourers and supplies with him from Egypt support the possibility that Saladin may have similarly brought artillery out of Egypt, to use in greater Syria, at some point over the preceding decade or even

27 'Imad al-Din, *al-Fath*, trans. Massé, pp. 64-65. Cf. Ibn al-Athir, *al-Kamil*, trans. Richards, 2:335-36.

28 *Eracles* 24.4, RHC Oc 2, p. 110; *Itinerarium* 1.10, ed. Stubbs, p. 25, trans. Nicholson, p. 41; William of Newburgh, *Historia rerum Anglicarum* 3.20, ed. Howlett, 1:264-65; Marino Sanudo, *Liber secretorum* 3.9.8, ed. Bongars, pp. 193-94, trans. Lock, pp. 307-8. Cf. Ernoul, *Chronique* 21, ed. Mas Latrie, p. 224.

29 Baha' al-Din, *al-Nawadir*, trans. Richards, p. 79. Cf. 'Imad al-Din, *al-Fath*, trans. Massé, pp. 79-80.

that al-ʿAdil had done so in 1187.[30] Regardless, Saladin's capture of the Palestinian coast allowed for the fairly free movement of traffic from Egypt along the Via Maris. Farther inland, the strongholds of eastern Galilee and Transjordan remained in Frankish hands, as did most of Lebanon and the entire Principality of Antioch.[31]

Eastern Galilee: 1188

In the second half of 1187, while Saladin was campaigning primarily against major urban centres, smaller contingents of his army were sent to oppose the castles of Belvoir, Safed, and Chastel Neuf, which overlook the upper Jordan Valley. Chastel Neuf surrendered in December 1187,[32] but the new Hospitaller stronghold of Belvoir and the hilltop castle of Safed, a Templar possession since 1167, which had been blockaded by at least August, defiantly held out. In December 1187 or January 1188, the garrison of Belvoir even launched a successful raid, capturing provisions and arms after falling on a small detachment of the Muslim army.[33] This led Saladin to move what forces he had retained over the winter against the castle in March 1188.[34] Despite the suitable topography around the castle, neither Saladin nor the earlier blockading force appears to

30 ʿImad al-Din, *al-Fath*, trans. Massé, pp. 108-9; Abu Shama, *Kitab al-Raudatain*, RHC Or 4, p. 393; Baha' al-Din, *al-Nawadir*, trans. Richards, p. 90; al-Maqrizi, *al-Suluk*, trans. Broadhurst, p. 87. ʿImad al-Din and Baha' al-Din, as well as al-Maqrizi, note that Qaraqush was overseeing the refortification of Acre by the time that Saladin left the city having spent time there during the winter of 1188-89. When considering ʿImad al-Din's account, it is possible that Qaraqush may have been summoned to undertake this project as early as the winter of 1187-88.

31 For the contemporary significance of these strongholds, see letter from the East to the Master of the Hospitallers of Italy, 1187, in Ansbert, *Historia*, FRA (1 Abt.) 5, p. 3, MGH SRG 5, pp. 3-4, trans. Loud, p. 35; letter of the Master of the Temple to Henry II of England, in Roger of Howden, *Chronica*, ed. Stubbs, 2:346-47, trans. Riley, 2:90-91.

32 ʿImad al-Din, *al-Fath*, trans. Massé, pp. 75-76; Baha' al-Din, *al-Nawadir*, trans. Richards, p. 79; Ibn al-Athir, *al-Kamil*, trans. Richards, 2:338; Abu Shama, *Kitab al-Raudatain*, RHC Or 4, p. 345. For discussions of Chastel Neuf, see Conder and Kitchener, *Survey of Western Palestine*, 1:123-24; Deschamps, *Les Châteaux des Croisés*, 2:130-31, pl. XXXIV A-C; Pringle, *Secular Buildings*, no. 164, pp. 79-80; Boas, *Crusader Archaeology*, p. 103. See also Ibn Jubayr, *Rihla*, trans. Broadhurst, p. 315.

33 Roger of Howden, *Chronica*, ed. Stubbs, 2:346-47, trans. Riley, 2:90; Baha' al-Din, *al-Nawadir*, trans. Richards, pp. 79-80; ʿImad al-Din, *al-Fath*, trans. Massé, pp. 81-82; Abu Shama, *Kitab al-Raudatain*, RHC Or 4, pp. 344-45; Ibn al-Athir, *al-Kamil*, trans. Richards, 2:339.

34 Baha' al-Din, *al-Nawadir*, trans. Richards, pp. 79-80; ʿImad al-Din, *al-Fath*, trans. Massé, pp. 103-4; Abu Shama, *Kitab al-Raudatain*, RHC Or 4, pp. 346-47; Ibn al-Athir, *al-Kamil*, trans. Richards, 2:344.

have employed artillery. Instead, the sultan sent his small force in waves against the castle, much as he had at Beirut in 1182 and would do again at Bourzey later in the year. When this proved ineffective, Saladin left a force to continue the blockade and made for Damascus, where he arrived on 5 May 1188.

For the time being, Belvoir and Safed remained Frankish outposts in Eastern Galilee. With blockades in place around these strongholds, Saladin was keen to press his advantage to the north and dedicated the campaign season of 1188 to Syria. As he had done in Palestine, his efforts were directed first against the coastal ports.

Indications of Range in Western Syria and Transjordan

Saladin left Damascus soon after arriving and mustered his army to the east of Crac des Chevaliers. Although not yet the great crusader castle that it would become, Saladin opted not to invest the Hospitaller stronghold. It was instead the local geography that encouraged Saladin to muster his forces there. A break between the Lebanon and Anti-Lebanon Mountains to the south and their continuation as the Syrian Coastal Mountains to the north, provided a natural causeway between the coast and the interior of western Syria, known as the Homs-Tripoli gap.

Al-Zahir Ghazi and Taqi al-Din were sent to oppose Antioch, tying up the principality's forces. Meanwhile, contingents of the main army raided as far west as Tripoli before it set out towards Tortosa on 1 July 1188. Saladin does not appear to have employed artillery during the early part of the campaign, during which the army moved up the coast from Tortosa to Latakia.[35] At Tortosa, frontal assaults were sufficient to take the town and one of its two tower-keeps. Neither offensive nor defensive artillery appears to have factored into the failed attempt to take the second tower, the Templars' keep, although neither did mining.[36] Continuing north, Maraclea was occupied unopposed. The town of Jabala was surrendered before Saladin had finished arranging his forces against it and the defenders of the citadel capitulated the following day after a single

35 For the inclusion of various smaller components that broke off and rejoined the main force, see Deschamps, *Les Châteaux des Croisés*, 3:127-31.

36 Baha' al-Din, *al-Nawadir*, trans. Richards, pp. 82-83; 'Imad al-Din, *al-Fath*, trans. Massé, pp. 124-25; Ibn al-Athir, *al-Kamil*, trans. Richards, 2:345; Abu'l-Fida', *al-Mukhtasar*, RHC Or 1, p. 59; letter of Hermenger to Leopold V of Austria, in Ansbert, *Historia*, FRA (1 Abt.) 5, p. 4, MGH SRG 5, pp. 4-5, trans. Loud, pp. 35-36.

attack.[37] 'Imad al-Din appears to be the only source to mention stones being thrown during this episode, but he does not specify if this was done mechanically. Mining led to the capture of Latakia, although the sappers responsible for this work seem not to have been supported by artillery.[38]

From a Frankish perspective, the *Itinerarium* understandably portrays Saladin's push up the coast as a series of attacks rather than sieges.[39] Although artillery is not mentioned as part of the more complex operations at Tortosa, or even Latakia, it suddenly becomes a critical element following the army's turn inland. Despite the apparent change of tactics, there was little difference between the length of these subsequent sieges and those that had taken place along the coast (Map 11).

Saone: 1188

Saone was the first castle that Saladin's army encountered after turning away from the coast. The stronghold was one of a number of Byzantine outposts in the Syrian hills that had been developed by the Franks.[40] The castle sits on a spur with wadis over 200 m wide to the north and south. The original Byzantine citadel is surrounded by a central enceinte and an additional bailey was built on either side of this. Dividing the castle from the remainder of the spur to the east is one of the most impressive medieval fosses, almost 40 m deep and about 20 m wide.[41] The outer eastern wall is pierced by thirty-two embrasures

---

37 Baha' al-Din, *al-Nawadir*, trans. Richards, p. 82; 'Imad al-Din, *al-Fath*, trans. Massé, pp. 125-28; Ibn al-Athir, *al-Kamil*, trans. Richards, 2:345; Abu'l-Fida', *al-Mukhtasar*, RHC Or 1, p. 59; Abu'l-Fida', *al-Mukhtasar*, RHC Or 1, p. 59.

38 Baha' al-Din, *al-Nawadir*, trans. Richards, pp. 82-83; 'Imad al-Din, *al-Fath*, trans. Massé, pp. 128-31; Ibn al-Athir, *al-Kamil*, trans. Richards, 2:346-47; Abu'l-Fida', *al-Mukhtasar*, RHC Or 1, p. 59.

39 *Itinerarium* 1.13, ed. Stubbs, pp. 26-27, trans. Nicholson, p. 43.

40 Matthew of Edessa, *Patmut'iwn*, trans. Dostourian, p. 32; Ibn 'Abd al-Zahir, *al-Rawd*, trans. al-Khowayter, 2: 708-9; A.W. Lawrence, "A Skeletal History," p. 218. See also Michaudel, "Étude historique," p. 6; Saadé, "Histoire du Château de Saladin," pp. 985-87; Deschamps, *Les Châteaux des Croisés*, 3:219-20. For the Frankish lordship of Saone, see Michaudel, "Étude historique," pp. 6-9; Saadé, "Histoire du Château de Saladin," pp. 987-91; Deschamps, *Les Châteaux des Croisés*, 3:227-28. Bourzey and Balatunus were among the other Byzantine strongpoints developed by the Franks in this region.

41 Baha' al-Din states that the fosse is 60 cubits deep (32.4 m according to the standard Egyptian measure). For studies and descriptions of Saone, see Mesqui, "Saône"; Michaudel, "Le château de Saône"; Saadé, "Histoire du Château de Saladin," pp. 980-1,016; Kennedy, *Crusader Castles*, pp. 84-95; Boase, "Military Architecture," pp. 145-49; Müller-Wiener, *Castles of the Crusaders*, pp. 44-45; Deschamps, *Les Châteaux des Croisés*, 3:217-47; T.E. Lawrence, *Crusader Castles*, ed. Pringle, pp. 42-49; Rey, *Étude sur les monuments*, pp. 105-13.

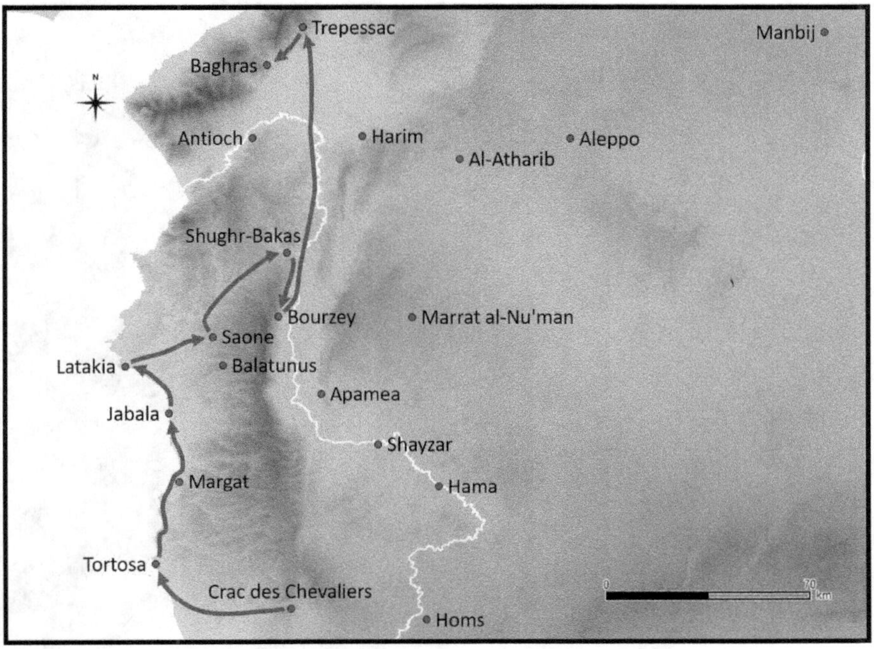

MAP 11 *Saladin's movement through Syria, 1188*

Origin	Destination	Siege
Crac (1 Jul)	Tortosa (3 Jul)	Tortosa: 3 Jul – 11 Jul
Tortosa (11 Jul)	Jabala (15 Jul)	Jabala: 15 Jul – 16 Jul
Jabala (20 Jul)	Latakia (21 Jul)	Latakia: 21 Jul – 22 Jul
Latakia (24 Jul)	Saone (26 Jul)	Saone: 27 Jul – 29 Jul
Saone (30 Jul)	Shughr-Bakas (2 Aug)	Shughr-Bakas: 2 Aug? – 12 Aug
Shughr-Bakas (>13 Aug)	Bourzey (20 Aug)	Bourzey: 21 Aug – 23 Aug
Bourzey (>23 Aug)	Iron Bridge (?)	*Pause at the Iron Bridge*
Iron Bridge (?)	Trapessac (2 Sep)	Trapessac: 2 Sep – 16 Sep
Trapessac (17 Sep)	Baghras (17 Sep)	Baghras: 17 Sep – 26 Sep

along the ground floor level alone, indicating that its designer(s) regarded a massed assault to be more threatening than the percussive force of any artillery, as these apertures would have detracted from the wall's strength.

The Muslim army arrived at Saone on 26 July. Baha' al-Din states that Saladin erected six stone-throwers early the following day, and although he does not give Saladin's location, 'Imad al-Din indicates that he was positioned on the spur facing the castle's eastern fosse. This fairly level plateau close to the

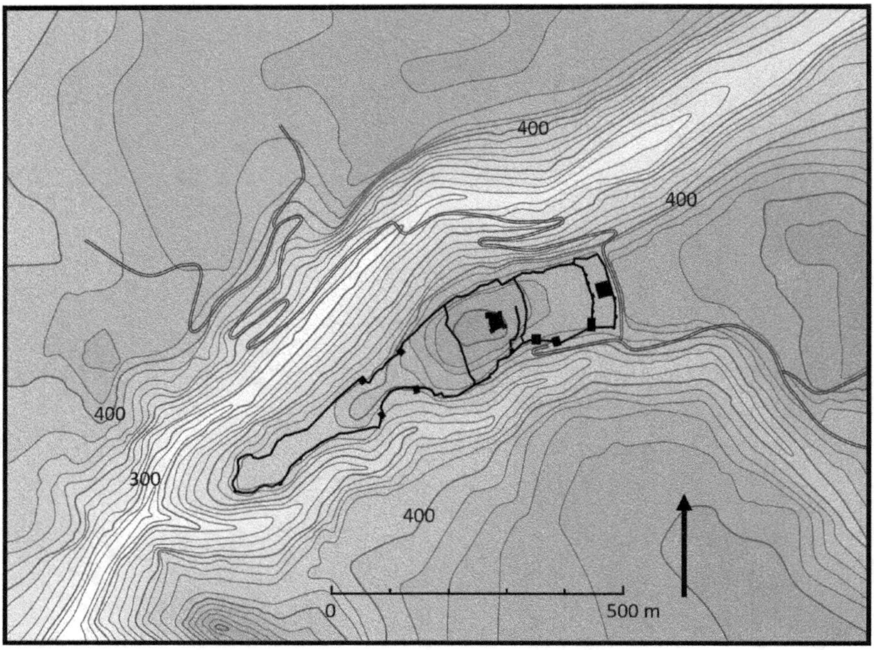

PLAN 15 *Saone*

castle would have been an ideal position to erect artillery. The sources unanimously place al-Zahir Ghazi elsewhere in a secondary position with another one or two trebuchets, ambiguously claiming that he was along the side of one of the wadis facing a certain salient of the castle. The sources claim that al-Zahir Ghazi and his Aleppan forces were able to damage a section of wall with their artillery and some appear to suggest that it was here that the Muslim army broke into the castle during the successful attack that was launched on 29 July.[42]

Interpreting the events of this siege, Gabriel Saadé argued that al-Zahir Ghazi's artillery was able to breach the north wall of the western bailey from across the northern wadi.[43] An alternative theory has been put forward by Benjamin Michaudel and Jean Mesqui, who suggest that the breach mentioned in the sources was created by Saladin's artillery at the north end of the eastern

42 Baha' al-Din, *al-Nawadir*, trans. Richards, pp. 84-85; 'Imad al-Din, *al-Fath*, trans. Massé, pp. 131-32; Abu Shama, *Kitab al-Raudatain*, RHC Or 4, pp. 365-67; Ibn al-Athir, *al-Kamil*, trans. Richards, 2:347-48.

43 Saadé, "Histoire du Château de Saladin," pp. 997-1,002.

IMAGE 2 *Saone: castle from the north* (STITCHED PHOTOS COURTESY OF DENYS PRINGLE)

wall.[44] Both theories, however, rely on misinterpretations of the power of artillery at this point in time.

Saadé believed that contemporary artillery not only had a range of over 200 m but that it could inflict significant damage from such a distance. Although there is circumstantial evidence that it may have been possible to throw small stones this far, there is little to suggest that this could have been done with sizable projectiles. Accordingly, it would appear impossible that one or two engines operating at this range could have opened a breach in less than forty-eight hours. Although counterweight trebuchets were extremely accurate, the prolonged flight time would magnify the variables that compromise precision.[45] Furthermore, the high angle of release necessary to throw a projectile this distance would lead to an equally sharp angle of impact, limiting the amount of energy transferred at the point of contact. With such considerations in mind, it is doubtful whether any trebuchet could ever have achieved such a task in so little time.

If the supposed artillery damage was instead inflicted at the north end of the easternmost wall, the issue is one of power alone. When placed in the broader historical context, there is no evidence that artillery powerful enough to breach defensive masonry over 2 m thick in only two days was ever employed during the twelfth century. Consider the prolonged periods of bombardment to which the northern defences of Kerak were subjected during 1183 and 1184 and the lack of any traces of this. If engines powerful enough to bring down a large section of wall were employed at Saone, it is inexplicable why they were not used with equal devastation at most other sieges. On this occasion, it seems the generally reliable Baha' al-Din borrowed heavily from the more exuberant 'Imad al-Din. He may have been inclined to incorporate the exaggeration, elevating the role of al-Zahir Ghazi, as he was under the patronage of al-Zahir by the time he was composing his account a number of years after these events.[46] Artillery probably played a much less significant role at this siege than has previously been thought, used to support the assault troops that ultimately stormed the castle rather than to create a breach through which they could enter.[47]

44 Michaudel, "Étude historique," pp. 15-16; Mesqui, "Saône," p. 14. See also Michaudel, "Le château de Saône," p. 3.

45 These variables include pre-release factors that influence a projectile's release trajectory as well as environmental elements that affect the projectile while in flight.

46 For these authors and the relation between their texts, see Richards, "A Consideration of Two Sources," pp. 46-65.

47 For a more thorough investigation of this siege and how artillery may have been employed, see Fulton, "A Ridge too Far," pp. 33-53.

Shughr-Bakas: 1188

Saladin took the main contingent of his army away from Saone on 30 July 1188 and on 2 August he encamped it on the banks of the Orontes near the stronghold of Shughr-Bakas. This castle, in the heights to the west of the important Jisr al-Shughr crossing, had two distinct wards, divided by a wide fosse that cut across the spur.[48] According to Baha' al-Din and 'Imad al-Din, Saladin took a detachment and assaulted Bakas until it was taken on 5 August. According to the former, it was acquired by force, whereas the latter claims that the garrison surrendered, while Ibn al-Athir states that the Franks abandoned this portion of the castle when the Muslims arrived.[49] Only Baha' al-Din mentions the use of artillery against the Bakas portion of the castle, but all three agree that it was used against al-Shughr.

Baha' al-Din claims that al-Shughr was bombarded from all sides, leaving the defenders no choice but to surrender. Although he does not explicitly state that Saladin's artillery was able to reach the castle, this would appear to be implied. 'Imad al-Din states that neither artillery nor archers could reach the castle; but despairing of taking it in a frontal assault, Saladin nevertheless resorted to setting up these engines. Ibn al-Athir appears to give the clearest description of events, stating that some shots were able to reach the castle but that they had no effect. When read together, the three accounts appear to agree that although a few stones thrown by the Muslims' artillery were able to reach al-Shughr, these were probably very light projectiles that had little impact. According to 'Imad al-Din and Ibn al-Athir, the situation was so hopeless that the Muslim forces were shocked when the garrison asked for terms on 9 August.[50]

If Bakas was the southern component of the castle and al-Shughr the more isolated northern part, as has been suggested by Van Berchem, Cahen and Deschamps,[51] this has interesting implications when considering Saladin's use of artillery. The higher southern enclosure might have been the more approachable if addressed from the southeast, but the northern component would have been easier to surround. If the Muslims' artillery was deployed in the valley around the castle and it struggled to reach al-Shughr, Bakas would

48 For Shughr-Bakas, see Deschamps, *Les Châteaux des Croisés*, 3:349-50; Boase, "Military Architecture," p. 162; Cahen, *Syrie du nord*, p. 159; Van Berchem and Fatio, *Voyages en Syrie* pp. 251-60.

49 Baha' al-Din, *al-Nawadir*, trans. Richards, p. 85; 'Imad al-Din, *al-Fath*, trans. Massé, p. 133; Ibn al-Athir, *al-Kamil*, trans. Richards, 2:348.

50 Baha' al-Din, *al-Nawadir*, trans. Richards, p. 85; 'Imad al-Din, *al-Fath*, trans. Massé, pp. 134-35; Ibn al-Athir, *al-Kamil*, trans. Richards, 2:348-49. See also Abu'l-Fida', *al-Mukhtasar*, RHC Or 1, p. 59.

51 Van Berchem and Fatio, *Voyages en Syrie*, 1:256 ; Cahen, *Syrie du nord*, p. 159 ; Deschamps, *Les Châteaux des Croisés*, 3:349-50

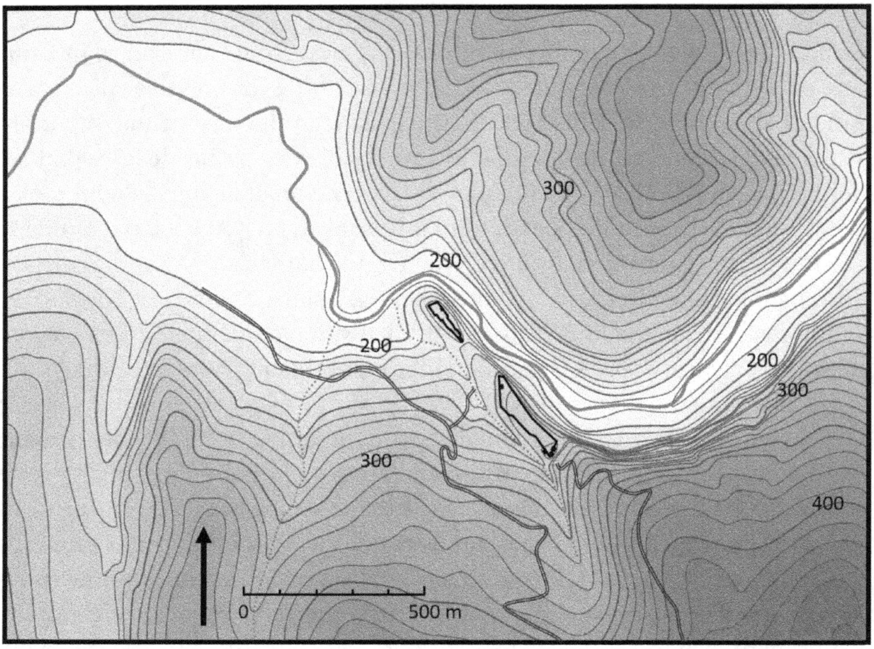

PLAN 16 *Shughr-Bakas*

almost certainly have been out of range from such positions. To the southeast, the spur rises steeply for about 300 m and does not appear to provide an obvious position where artillery might have been set up. Had such a position been found in 1188 and artillery been used to bombard Bakas from here, these engines would almost certainly have been brought within the southern enclosure following its capture. From the northern end of Bakas, they could easily have assailed al-Shughr, which does not fit with the accounts of the siege.

It seems more likely that Saladin kept his trebuchets in the valley below the castle, attacking the southern enclosure with a smaller, more mobile force without the close support of artillery. Due to the curve of the valley, which follows the north-eastern side of the castle before opening up to the west at the end of the spur, Baha' al-Din may have judged some of the easternmost engines to have shot at Bakas, as well as the slightly lower walls of al-Shughr. Although it is unlikely that these engines would have had any effect, if they were able to reach the castle at all, such an arrangement would support the assertion that al-Shughr was surrounded and attacked from all sides. Furthermore, it would appear to validate the reports of 'Imad al-Din and Ibn al-Athir regarding the difficulties the engines had in reaching the lofty walls of al-Shughr. Without a detailed survey of the landscape surrounding Shughr-Bakas, which would yield precise elevation and range variables, it is hard to gauge the evident limits of

Saladin's artillery as these engines attempted to shoot up at the castle. Fortunately, such information has been gathered at Bourzey.

Bourzey: 1188

Having occupied both parts of Shughr-Bakas, Saladin moved his army south to Bourzey. Sitting on a conical peak, 320 m above the floor of the Orontes valley to the east, Bourzey had been a notable Byzantine outpost before it was occupied by the Franks.[52] When Saladin arrived on 20 August, he again encamped the bulk of his army on the banks of the Orontes. The next day, after surveying the site, he moved a secondary assault force and his artillery up to a plateau on the saddle between the castle and the hills rising to the west.[53]

The plateau below the castle is about 420 m above sea level and comes as close as about 40 m horizontally from the plane of the castle's western wall. Above this position, the foundations of the south end of the castle's western face are about 460 m above sea level and the Byzantine citadel, in the northwest corner of the castle, crowns the highest point of the hill, about 500 m above sea level.[54]

Baha' al-Din states that men surrounded the castle and that artillery shot at it day and night. As at al-Shughr, he again implies that the besieging artillery had no trouble reaching the castle, although he does not mention any damage. This conflicts with 'Imad al-Din's account, which states that the besieging artillery was insufficiently powerful to assail the castle, as at al-Shughr, forcing Saladin to resort to frontal assaults two days later. Ibn al-Athir states that there was a position from which artillery could reach the castle, but that a defensive engine was able to put the Muslim trebuchets out of action on 22 August, events which he claims to have witnessed from one of the nearby hills overlooking the castle.[55]

---

52 For the castle and its early history, see Michaudel, "Burzaih," pp. 178-87; Mesqui, "Bourzeÿ," pp. 95-133; Mesqui and Michaudel, "Bourzey," pp. 4-27; Kennedy, *Crusader Castles*, pp. 81-83; Boase, "Military Architecture," p. 162; Deschamps, *Les Châteaux des Croisés*, 3:345-48. See also Anna Comnena, *Alexiad* 13.12, trans. Dawes, p. 354.

53 Baha' al-Din, *al-Nawadir*, trans. Richards, p. 86; 'Imad al-Din, *al-Fath*, trans. Massé, p. 136; Ibn al-Athir, *al-Kamil*, trans. Richards, 2:349-50; Abu'l-Fida', *al-Mukhtasar*, RHC Or 1, p. 59.

54 The Franco-Syrian expedition concluded that the elevation disparity between the point at which they suspected that Saladin planted his engines and the foundations of the walls which they assailed was about 60 m, Mesqui and Michaudel, "Bourzey," p. 11.

55 Baha' al-Din, *al-Nawadir*, trans. Richards, p. 86; 'Imad al-Din, *al-Fath*, trans. Massé, p. 136; Ibn al-Athir, *al-Kamil*, trans. Richards, 2:350. See also Abu'l-Fida', *al-Mukhtasar*, RHC Or 1, p. 59.

IMAGE 3 *Bourzey: castle from the southeast* (PHOTO COURTESY OF DENYS PRINGLE)

IMAGE 4 *Bourzey: castle from the saddle to the southwest* (PHOTO COURTESY OF DENYS PRINGLE)

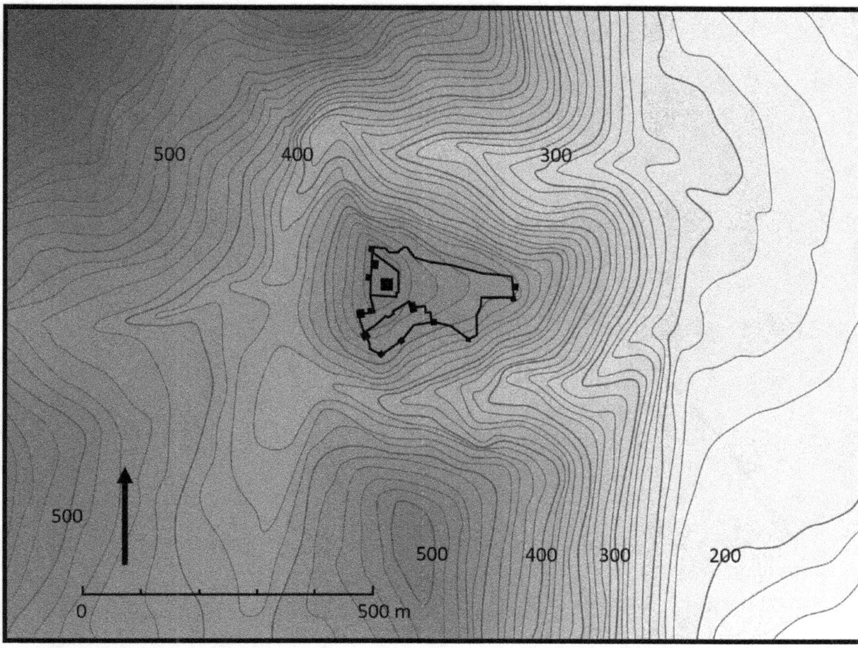

PLAN 17 *Bourzey*

Given the distances and elevations involved, it would appear possible for traction trebuchets to have reached the southern sections of Bourzey's western face; however, the citadel would have been harder to assail. From a horizontal distance of 40 m, a minimum release velocity of 34 m/s is necessary to reach the lowest battlements of the castle's western front (at an estimated elevation of 50 m). With a release velocity of 37 m/s, only slightly higher than the minimum required at Harim in 1177,[56] an engine could be backed up 20 m to a slightly more comfortable position and target the lowest parapet with a release angle of 58°, increasing horizontal velocity by 54%. To reach the battlements of the Byzantine citadel, a release velocity greater than 42 m/s is necessary, or almost 44 m/s from a horizontal distance of 60 m. Had Saladin's engines been capable of shooting stones at 44 m/s, projectiles of the same mass could have been thrown at the base of the lowest portion of wall at a horizontal velocity of almost 32 m/s (Fig. 11).[57]

56 See Chapter 4.

57 A slightly more generous data set, allowing the engines to have been set up slightly closer to the castle, can be found in Fulton, "A Ridge Too Far," pp. 46-48.

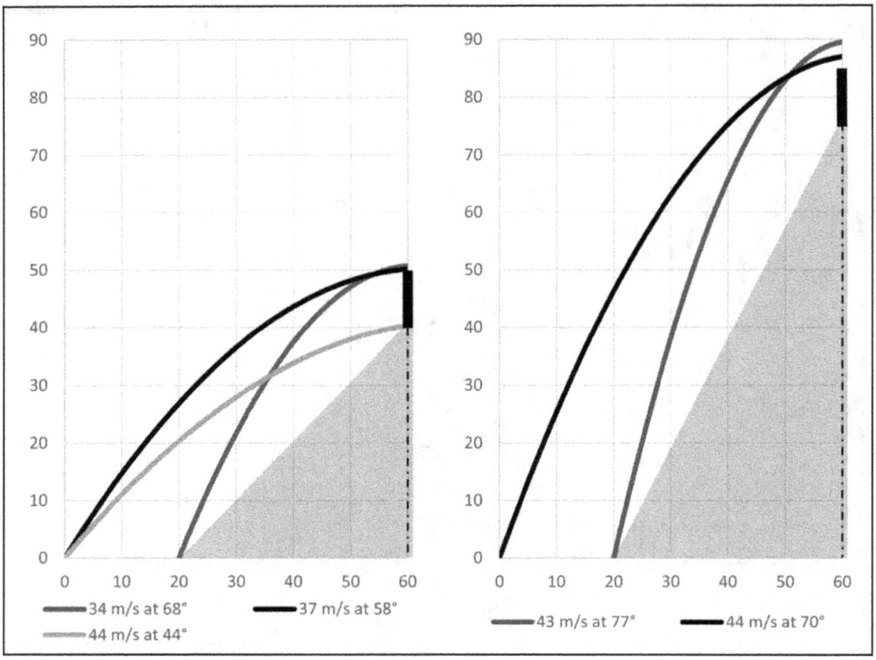

FIGURE 11 *Hypothetical trajectories at Bourzey*

Release angle	68°	58°	44°	77°	70°
Release velocity	34.0 m/s	37.0 m/s	44.0 m/s	43.0 m/s	44.0 m/s
Horizontal velocity	12.7 m/s	19.6 m/s	31.7 m/s	9.7 m/s	15.0 m/s
Max. horizontal range	81.9 m	125.6 m	197.4 m	82.7 m	127.0 m
Max. vertical range	50.7 m	50.2 m	47.7 m	89.6 m	87.2 m
Vertical range at 60 m	50.7 m	50.1 m	40.3 m	89.5 m	87.0 m

The size of the stones that Saladin's engines threw at Bourzey is unclear, as no projectiles have been identified at the site. The absence of this archaeological evidence prevents an accurate calculation of the energy these machines might have been able to channel and potential damage they may have been capable of inflicting. If Saladin's artillery struggled to reach the walls of Bourzey, it is almost certain that these were traction trebuchets; however, if the engines were not only able to reach the castle but inflict damage, which is not mentioned by any of the sources, these would necessarily have been counterweight trebuchets.

The lack of agreement between Baha' al-Din, 'Imad al-Din and Ibn al-Athir makes it impossible to say with certainty whether Saladin was using traction- or small counterweight-powered trebuchets. Given the absence of any suggestions of damage and 'Imad al-Din's uncharacteristic emphasis of their weakness, these would appear to have been quite light engines. This supports Ibn al-Athir's observations at al-Shughr that although the castle was theoretically within range, its elevation placed it beyond an *effective* range. All three sources agree that artillery alone would not compel the surrender of Bourzey's garrison. Although Saladin's artillery is presented as quite weak here, relative to its portrayal at Saone, it was integrated into the same tactical approach: the engines were set up soon after the army arrived, while forces continued to organise themselves, and then supported a general assault a few days later.[58]

On 23 August, Saladin divided the elements of his attack force into three divisions, arranging for them to assault the castle in rotation.[59] Around midday, after the Franks had repulsed the first wave with arrows and stones, the second wave, supported by elements of the third and first, scaled the western wall of the castle. While the Franks were forced back and eventually compelled to seek refuge in the citadel, contingents from the main army, still encamped next to the Orontes, had made their way up the steep eastern slope and managed to scale the outer eastern walls, which they found devoid of defenders.[60] From the citadel, the garrison surrendered itself into captivity, allowing Saladin to move on, taking his army north to the Iron Bridge and into the environs of Antioch.[61]

The Syrian Gates: 1188

In September, Saladin appears to have used artillery against both Trapessac and Baghras, the Templar castles that secured the Syrian Gates north of Antioch.[62] The former occupies a rocky rise similar to the tell at Harim and the latter, although secured by the steep topography to the west, north and east, is approachable from the southwest.[63] In a reversal, 'Imad al-Din, rather than

58 See Fulton, "A Ridge Too Far," p. 52.

59 For the use of such attacks in rotation, see the sieges of Montferrand (1137) and Beirut (1182) in Chapter 4 and Belvoir (1188) above.

60 Baha' al-Din, *al-Nawadir*, trans. Richards, p. 86; 'Imad al-Din, *al-Fath*, trans. Massé, pp. 136-39; Ibn al-Athir, *al-Kamil*, trans. Richards, 2:350-51. Cf. the account of Ibn Wasil, trans. in Mesqui and Michaudel, "Bourzey," p. 8. Ibn al-Athir describes the gradient beyond the outer eastern bailey as climbable but too steep to fight from.

61 Baha' al-Din, *al-Nawadir*, trans. Richards, p. 86; 'Imad al-Din, *al-Fath*, trans. Massé, p. 139; Ibn al-Athir, *al-Kamil*, trans. Richards, 2:351-52.

62 Baha' al-Din, *al-Nawadir*, trans. Richards, p. 87; 'Imad al-Din, *al-Fath*, trans. Massé, pp. 141-43; Ibn al-Athir, *al-Kamil*, trans. Richards, 2:352-53.

63 For the castle of Baghras, see A.W. Lawrence, "The Castle of Baghras," pp. 35-83. For Trapessac, see Edwards, *Fortifications of Armenian Cilicia*, p. 253.

Baha' al-Din, is the most generous with his praise of the artillery at these sieges. Baha' al-Din's account is exceptionally brief as he was not present during these events.[64] At Trapessac, Ibn al-Athir claims that the Muslims' trebuchets were able to inflict no damage while at Baghras they were rendered ineffective by the relative elevation of the castle. This latter observation is somewhat puzzling. Although the castles were afforded significant protection by their positions, both should have been assailable with contemporary artillery technology given the apparently greater distances reached elsewhere. While mining efforts forced the defenders of Trapessac to seek terms, it would appear the Templar garrison of Baghras agreed to negotiate quite quickly, securing favourable terms of surrender. Saladin then arranged a brief truce with Bohemond III before setting off for Damascus via Aleppo.[65]

The Syrian campaign of 1188 is an important window into Saladin's siege practices. None of the sieges were particularly grand and the accounts are far from the most detailed; however, having two, and possibly even three, eyewitness accounts of these events makes them extremely important.[66] The regularity with which Saladin employed his artillery confirms its apparent importance, despite the conflicting reports of its power. The speed with which the army seems to have moved suggests that these were prefabricated trebuchets, light enough to be set up overnight in most instances and conveyed through the hills of western Syria without too much difficulty.[67] It is interesting that the sources seem to have regarded elevation as a particularly challenging obstacle. Whereas artillery is presented as quite powerful at Saone, and earlier at Kerak, where engines were set up on the far side of a ditch, it is depicted as far less effective when struggling to reach the lofty defences of Shughr-Bakas, Bourzey and Baghras.

64 He accompanied one of the contingents that was dispatched to screen the main army from any attacks by the Franks by taking up a position in the plain north of Antioch.

65 Baha' al-Din, *al-Nawadir*, trans. Richards. p. 87; 'Imad al-Din, *al-Fath*, trans. Massé, pp. 144-45; Ibn al-Athir, *al-Kamil*, trans. Richards, 2:353-54.

66 Saadé has expressed this contentious belief, but Ibn al-Athir's contribution of information not found in the accounts of Baha' al-Din and 'Imad al-Din, such as the lady operating the defensive engine at Bourzey, and his use of the first person on occasion, make this a possibility. See Saadé, "Histoire du Château de Saladin," p. 996.

67 For the broader development and use of prefabricated artillery, see Fulton, "Development of Prefabricated Artillery," pp. 51-72.

The Absence of Artillery in Transjordan

Saladin left his brother, al-ʿAdil, to continue the siege of Kerak when he set out to join al-Afdal and the army amassing east of the Jordan in May 1187. Al-ʿAdil later joined Saladin in Palestine and responsibility for reducing the Frankish strongholds of Transjordan was delegated to Saʿad al-Din Kamshaba. The garrison of Kerak had refused to surrender when informed of the Frankish defeat at Hattin and by the autumn of 1187 Montreal was also subject to a close blockade. Neither direct assaults nor artillery appear to have been used against the two strongholds. Although a portion of the garrison of Kerak had left to fight at Hattin, the defenders were still in a strong enough position to secure their liberty, and that of Humphrey of Toron, when supplies ran out and they were forced to surrender sometime between 24 October and 23 November 1188.[68] To the south, the older, further isolated and less celebrated castle of Montreal continued to hold out. It was not until April or May 1189 that the castle's garrison surrendered the stronghold, having endured a year and a half of uninterrupted passive blockade.[69] So far from Frankish support, the determined resistance offered by the garrison is bewildering.

In his assessment of the extended blockade of Montreal, Hugh Kennedy has proposed that the castle's ability to hold out was due to its topographic isolation, which prevented the Muslims from bringing siege engines up to its walls.[70] Although the surrounding topography was one of the castle's greatest strengths, Muslim forces rarely used rams and siege towers in the Near East; mining, often supported by artillery, and frontal attacks were much more common tactics. The surrounding landscape provides only one position where artillery might be set up against the castle, which is now occupied by the visitors' centre (A on the plan). Similar to the ridge south of Kerak, the castle is less than 150 m away but at a slightly higher elevation, requiring projectiles be thrown with a release velocity of about 40 m/s. This would necessitate a counterweight trebuchet to batter the battlements or inflict any degree of damage, as 40 m/s was probably near the very edge of possibility for a traction trebuchet. Although heavier counterweight engines would soon be developed, it is important not to

68 Baha' al-Din, *al-Nawadir*, trans. Richards, p. 88; ʿImad al-Din, *al-Fath*, trans. Massé, pp. 104-7; Abu Shama, *Kitab al-Raudatain*, RHC Or 4, pp. 381-82; Ibn al-Furat, *Tarikh*, trans. Lyons and Lyons, 2:52; al-Maqrizi, *al-Suluk*, trans. Broadhurst, pp. 87, 88; *Eracles* 26.9, RHC Oc 2, p. 188. The smaller strongholds of the region also appear to have fallen around this point, Abu Shama, *Kitab al-Raudatain*, RHC Or 4, p. 382; Ibn al-Athir, *al-Kamil*, trans. Richards, 2:333, 354-55. See also Vannini and Tonghini, "Medieval Petra," pp. 371-84.

69 *Itinerarium* 1.15, ed. Stubbs, pp. 29-30, trans. Nicholson, pp. 45-46. See also, Baha' al-Din, *al-Nawadir*, trans. Richards, p. 91; *Eracles* 24.2, 26.9, RHC Oc 2, pp. 104-5 (lower text), 188.

70 Kennedy, *Crusader Castles*, p. 111.

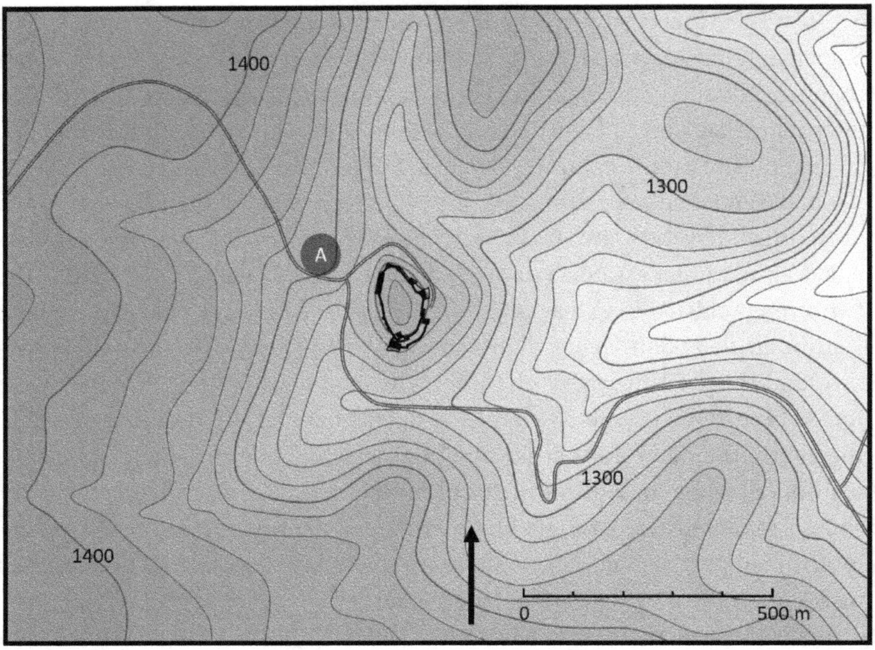

PLAN 18 *Montreal*

overestimate the power of these engines. It was not until the end of the thirteenth century that the northwest front of Montreal and southern defences of Kerak were significantly refortified.[71] The continued importance of these sites through the thirteenth century makes the decision not to strengthen these fronts earlier significant.

The Significance of Range

As was touched upon when dealing with the sieges of Harim (1177) and Bourzey (1188), evidence that projectiles were thrown at release velocities exceeding 40 m/s or to distances greater than 160 m, begins to indicate the use of counterweight engines. Even with an appreciation of range, the power of a trebuchet cannot be determined without a sense of the mass of its projectiles (Fig. 12).

With a known release velocity, an idea of scale can be determined if it is assumed that an engine was throwing light stones. Applying a simplified

71 For the development of the north-western defences of Montreal, see Faucherre, et al., "La forteresse de Shawbak," pp. 62-64. For the development of the southern defences of Kerak, see Deschamps, *Les Châteaux des Croisés*, 2:88-89. For the relation of these building efforts to artillery, see Chapter 8.

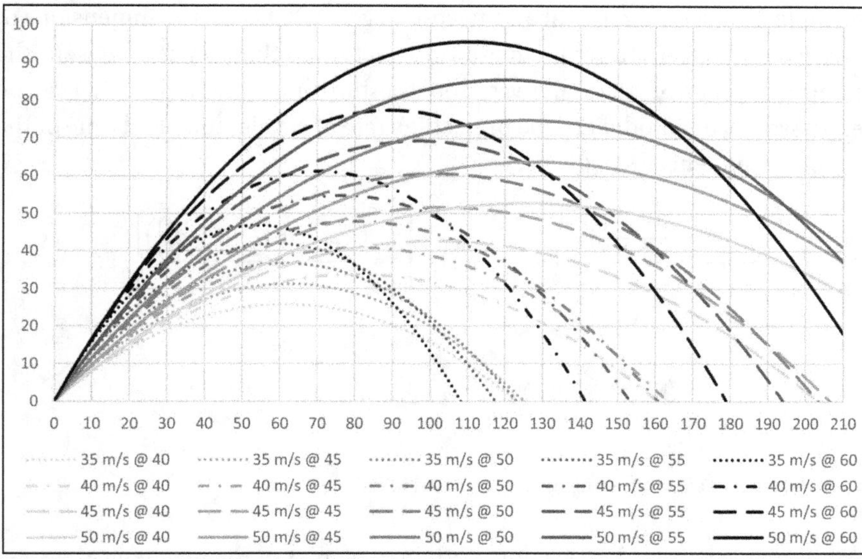

FIGURE 12 *Range by release velocity*

V	α	Vh	Rh max	Rv max	Rv at 50 m	Rv at 100 m	Rv at 150 m	Rv at 200 m
35 m/s	40°	26.8 m/s	123.1 m	25.8 m	24.9 m	15.8 m	-27.5 m	-104.8 m
	45°	24.8 m/s	125.0 m	31.2 m	30.0 m	19.2 m	-30.0 m	-120.0 m
	50°	22.5 m/s	123.1 m	36.7 m	35.4 m	22.4 m	-39.1 m	-148.9 m
	55°	20.1 m/s	117.5 m	41.9 m	41.0 m	21.2 m	-59.3 m	-200.7 m
	60°	17.5 m/s	180.2 m	46.9 m	46.9 m	13.2 m	-100.2 m	-293.6 m
40 m/s	40°	30.6 m/s	160.8 m	33.7 m	28.9 m	31.7 m	-8.4 m	-40.9 m
	45°	28.3 m/s	163.3 m	40.8 m	34.7 m	38.8 m	12.2 m	-45.0 m
	50°	25.7 m/s	160.8 m	47.9 m	41.1 m	45.0 m	11.0 m	-58.1 m
	55°	22.9 m/s	153.4 m	54.8 m	48.1 m	49.7 m	-4.8 m	-86.7 m
	60°	20.0 m/s	141.4 m	61.2 m	55.0 m	50.7 m	-15.8 m	-143.6 m
45 m/s	40°	34.5 m/s	203.5 m	42.7 m	31.6 m	42.7 m	33.1 m	2.9 m
	45°	31.8 m/s	206.6 m	51.7 m	37.9 m	51.6 m	41.1 m	6.4 m
	50°	28.9 m/s	203.5 m	60.6 m	45.0 m	60.6 m	47.0 m	4.1 m
	55°	25.8 m/s	194.2 m	69.3 m	53.0 m	69.3 m	48.7 m	-8.6 m
	60°	22.5 m/s	179.0 m	77.5 m	62.4 m	76.4 m	42.0 m	-40.8 m
50 m/s	40°	38.3 m/s	251.2 m	52.7 m	33.6 m	50.5 m	50.7 m	34.2 m
	45°	35.4 m/s	255.1 m	63.8 m	40.2 m	60.8 m	61.8 m	43.2 m
	50°	32.1 m/s	251.2 m	74.8 m	47.7 m	71.7 m	72.0 m	48.6 m
	55°	28.7 m/s	239.7 m	85.6 m	56.5 m	83.2 m	80.2 m	47.3 m
	60°	25.0 m/s	220.9 m	95.7 m	67.0 m	94.8 m	83.4 m	32.8 m

formula, based on the initial and release angles of a beam, the dimensions of the beam and the sling, and an arbitrary factor of efficiency to represent the retarding forces of friction, a *very* rough estimate of release velocity can be expressed as the product of the projectile's rate of acceleration and time of the shooting sequence [$V \approx Ap \cdot t$]:

$$V \approx \left[\frac{4\pi \cdot \mathrm{Ls} \left(\frac{180 - \psi}{360} \right)}{t^2} + \left(\mathrm{Acw} \cdot \frac{\mathrm{Lla}}{\mathrm{Lsa}} \right) \right] \cdot \left[\sqrt{4 \frac{\left(\mathrm{Lsa} \cdot \sin \left[\frac{1}{2} (\theta - \alpha) \right] \right)}{\mathrm{Acw}}} \right] \cdot \mathrm{Eff}$$

$$\text{Where } \mathrm{Acw} = \frac{[(\mathrm{Mcw}+\mathrm{Msa}) \cdot g] - \left[(\mathrm{Mp}+\mathrm{Mla}) \cdot g \cdot \frac{\mathrm{Lla}}{\mathrm{Lsa}} \right]}{\mathrm{Mcw}+\mathrm{Mp}+\mathrm{Mb}}$$

* Note, here, as elsewhere, angles are calculated in radians.

As this does not take into account certain dynamics of the mechanical system and presumes the beam has a uniform cross-section along its entire length, this is not a particularly accurate mathematical representation of a trebuchet's shooting sequence; however, it is suitable to provide a general sense of the capabilities of engines of varying sizes.[72] Using *Arrangements B* and *D*, if each is given a beam with a density of 700 kg/m3, the former with a cross-section 10 cm wide and 20 cm tall and the latter 15 cm wide and 24 cm tall, when loaded (θ) at 135° and shot (α) at 45°, the values found in Fig. 13 are produced.

Although imperfect, these rough figures reveal the mechanical scale required to produce certain release velocities.[73] Although maximising range would appear to have been important to the Muslims in 1188, this should not be taken as a given in every scenario. Whereas reducing the mass of an engine's projectiles will generally increase release velocity and maximise range, using heavier projectiles will typically increase efficiency, ensuring as much potential energy as possible is transferred to the projectile. Although less significant when throwing quite light stones, this will become an important consideration with the growth and increasing power of counterweight trebuchets through the thirteenth century.

72 See Chapter 2 for more exact calculations.

73 Although the beam of *Arrangement D* is only 2 m less than that of *Arrangement A*, the shorter sling of the former is a mathematical handicap. In reality, sling length is one of the easiest mechanical variables to alter. Although its optimal length, if shooting for distance, is determined by the proportions and masses of the other components, it can be easily adjusted to change the engine's range.

FIGURE 13 *General values for Arrangements B and D*

Arr.	Lb	Mcw	Mb	Mp	V (100%)	R (100 %)	V (70%)	R (70 %)
B	5 m	1,000 kg	70 kg	1 kg	48.1 m/s	236.1 m	33.7 m/s	115.7 m
				10 kg	46.8 m/s	223.4 m	32.8 m/s	109.5 m
				50 kg	40.8 m/s	169.5 m	28.5 m/s	83.0 m
D	10 m	5,000 kg	252 kg	1 kg	60.2 m/s	369.7 m	42.1 m/s	181.2 m
				10 kg	59.9 m/s	365.9 m	41.9 m/s	179.3 m
				50 kg	58.5 m/s	349.4 m	41.0 m/s	171.2 m

Saladin's Conquest of Northern Palestine

Safed: 1188

Around the start of November 1188, Saladin erected artillery against Safed.[74] Amidst heavy rains, Baha' al-Din states that Saladin was determined to have the work completed overnight and have five engines erected by morning. The task of setting up each machine was entrusted to a different component of his army and by morning only the *khanazirs* (the axles) were left to be fitted. This description suggests that the engines were sizable enough to require significant teams to assemble them but small enough so as to be set up in less than 18 hours, roughly the same amount of time as was required to erect artillery during the Syrian campaign a few months earlier. The Templar garrison sought terms on 30 November, their provisions exhausted from the extended blockade. From Safed, Saladin made for the Hospitaller castle of Belvoir.

Belvoir: Dec 1188

Belvoir, postdating 1168, was the most scientifically designed castle built by the Franks in Outremer during the twelfth century. Planned in two concentric squares, it was constructed with basalt blocks hewn from the surrounding fosse as well as lighter coloured limestone sourced elsewhere. The curtain and tower walls of the outer defences are 2.5-3.5 m thick and rise from a talus, cut from the bedrock in some areas and constructed of masonry in others. A tower stands at each corner and another was built in the middle of each wall. Many of the towers were provided with posterns hidden in their flanking walls,

74 Baha' al-Din, *al-Nawadir*, trans. Richards, pp. 31-32, 88-89; 'Imad al-Din, *al-Fath*, trans. Massé, pp. 149-50; Abu Shama, *Kitab al-Raudatain*, RHC Or 4, p. 384; Ibn al-Athir, *al-Kamil*, trans. Richards, 2:355. For descriptions and discussions of the castle, see Chapter 7.

IMAGE 5 *Belvoir: castle from the east* (PHOTO COURTESY OF MICHAEL EISENBERG)

allowing for rapid sallies from concealed positions, while a strong barbican dominates the main point of ingress.[75] These outer defences enclose a smaller quadrangular inner castle, which was laid out around a central courtyard with an entrance in both the eastern and western walls. A strong tower projects from each corner and a fifth was added in front of the western gate, forming a gate tower. The tower and curtain walls of the inner castle are about 3.5 m thick. This elaborate plan was necessary as the castle could be approached easily from the south, west and north.

Rather than developments in artillery technology, it was the contemporary political context that led to the incredibly sophisticated plan of the castle. Belvoir was constructed at a point when Frankish expansion had been checked and Nur al-Din, through Shirkuh, was making a bid to add Egypt to his realm, a dominion that already stretched from Damascus across northern Syria and parts of the Jazira. The designer(s) of Belvoir appreciated that the castle would

75 For the layout of the castle see Conder and Kitchener, *Survey of Western Palestine*, 2:117-19; Benvenisti, *The Crusaders*, pp. 298-99. For the castle's construction, see Benvenisti, *The Crusaders*, pp. 299-300; Pringle, *Churches*, no. 57, 1:123; France, *Western Warfare*, p. 92.

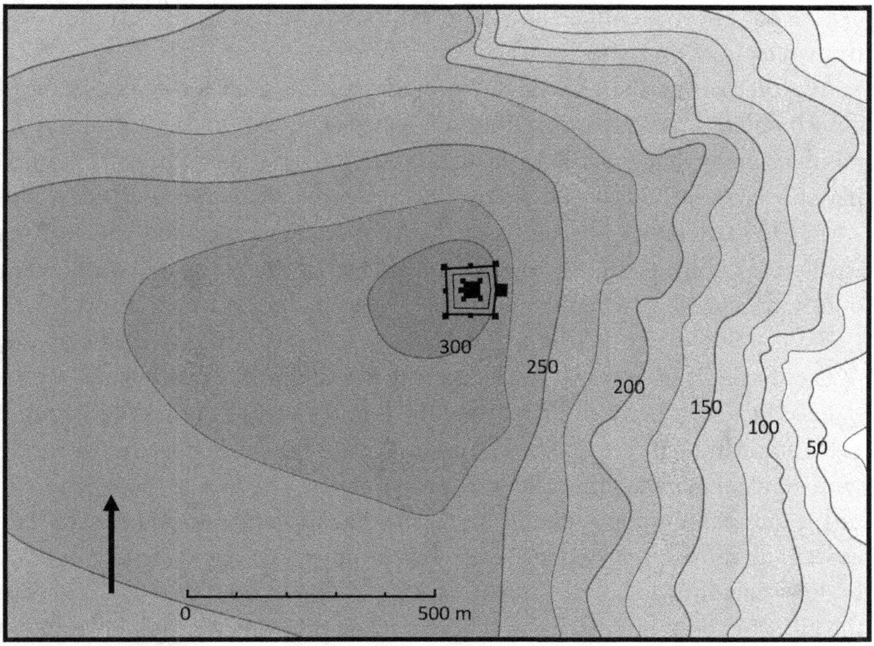

PLAN 19 *Belvoir*

need to be able to resist the manpower resources that could be assembled from such a vast empire. Twenty years later, the castle was put to the test.

From the start of December, Saladin arranged his forces and artillery around Belvoir's three approachable fronts. Indicative of his resolve, Saladin apparently encamped so close to the castle that his own tent was within bow-shot of the garrison.[76] Although this is almost certainly an exaggeration, it suggests that the Muslims' artillery was quite close to the castle, as these engines would have been ahead of their tents. Despite the clear use of artillery, it does not appear to have played a significant part in this siege; instead, it was frontal assaults and the undermining of the eastern barbican that led the garrison to sue for terms on 5 January 1189. Saladin probably used his artillery to support his frontline forces, which might explain why descriptions of artillery damage are largely absent in the various accounts. To date, only two possible artillery pro-

76 Baha' al-Din, *al-Nawadir*, trans. Richards, p. 89; 'Imad al-Din, *al-Fath*, trans. Massé, pp. 152-53; Ibn al-Athir, *al-Kamil*, trans. Richards, 2:356; Abu Shama, *Kitab al-Raudatain*, RHC Or 4, pp. 387-88.

jectiles have been identified and their size, 7.5 kg and 10.2 kg, suggest they were thrown by traction trebuchets.[77]

Like the north wall of Kerak, the surviving courses of masonry, consisting of large basalt blocks with protruding bosses, reveal no obvious indications of the bombardment that took place in 1188-89. Any impact signatures that might have been visible higher up the walls were obscured when the castle was slighted in the early thirteenth century.[78] Unfortunately, no detailed study of the castle was published during or after it was cleared in the 1960s, leaving many questions, including whether any projectiles were discovered, unanswered.

Frankish Palestine fell with Belvoir. When the campaign season of 1189 opened, Saladin moved his forces against Reynald of Sidon's mighty stronghold of Beaufort, perched high above the Litany River at the southern end of Lebanon. Although the castle was approachable from all sides but the east, with an ideal plateau for artillery to the south, the *Eracles* account appears to be the only contemporary source to suggest that artillery was used. Most sources were far more interested in the episode of intrigue orchestrated by Reynald to defer Saladin's assaults.[79] The siege was evidently more of a blockade and the garrison, even after its lord had been taken prisoner, was able to hold out until 22 April 1190, at which time it secured favourable terms of surrender. By this point, Saladin's attention had been drawn elsewhere: the Frankish siege of Acre had begun.

---

77 My thanks to Simon Dorso for bringing these stones to my attention. Both projectiles have a diameter of approximately 16-20 cm and were hewn from basalt. They probably date to this siege; however, it is possible that they relate instead to later Ayyubid activities. See Fulton, "Overlooked Ordnance."

78 Saladin appears to have ordered the castle's destruction in 1192, Abu Shama, *Kitab al-Raudatain*, RHC Or 5, pp. 87-88, cf. Ibn Shaddad's account, p. 91. See also Ambroise, *Estoire* ll. 6,826-60, ed. Ailes and Barber, pp. 110-11. 183, trans. Ailes, pp. 124-25; *Itinerarium* 4.23, ed. Stubbs, p. 280, trans. Nicholson, p. 261; Abu Shama, *Kitab al-Raudatain*, RHC Or 5, pp. 87-88. However, the damage was either very slight or the order was never executed as there is no record of the castle being rebuilt and it was occupied and defended by a Muslim garrison in the early thirteenth century. The destruction evident today was carried out under al-Mu'azzam 'Isa during the Fifth Crusade, see Chapter 6. One version of events holds that al-'Adil ordered the castle slighted in 1213 after capturing it from Damascene forces, Ibn al-Athir, *al-Kamil*, trans. Richards, 3:158; *History of the Patriarchs*, trans. Khater and Khs-Burmester, 3.2:195-97. Cf. Ibn al-'Amid, *al-Majmu al-Mubarak*, trans. Eddé and Micheau, p. 19; al-Maqrizi, *al-Suluk*, trans. Broadhurst, p. 155. See also Pringle, *Churches*, no. 57, 1:120-22; Boas, *Crusader Archaeology*, pp. 106-8.

79 *Eracles* 26.9, RHC Oc 2, pp. 187-88; Baha' al-Din, *al-Nawadir*, trans. Richards, pp. 90-91, 95-96, 97, 108; 'Imad al-Din, *al-Fath*, trans. Massé, pp. 159-62, 210; Abu Shama, *Kitab al-Raudatain*, RHC Or 4, pp. 397-400, 441; Ibn al-Athir, *al-Kamil*, trans. Richards, 2:360.

The First Great Siege of Acre

Saladin had initially delayed releasing Guy of Lusignan, a prisoner since the battle of Hattin and whose liberty had been a term of Ascalon's surrender. Once freed, and finding himself shut out of Tyre by Conrad of Montferrat, Guy and his supporters invested Acre from August 1189. With Acre's walls ahead of him, recently strengthened by Baha' al-Din Qaraqush al-Asadi, and Saladin to his rear, Guy had begun what would become the most protracted siege undertaken by Frankish forces in the Holy Land during the twelfth century.[80]

First Phase: 1189-90

Artillery does not appear to have been an immediate consideration of the Franks: the first notable assault was a frontal attack using ladders.[81] Although the motives for a rapid attack need no explanation, Saladin's arrival the day after this assault would have cut the Frankish force off from the larger trees beyond the plain of Acre, leaving them with the orchards around the city as their only source of timber. As the siege progressed, the attackers would have relied on wood brought by sea by Italian and crusading forces. Only one source appears to date the construction of artillery to the autumn of 1189,[82] most others emphasise the Franks' efforts to fortify and entrench their siege camp. From about December, both sides had begun to build or set up artillery, although the use of these engines is not mentioned until the spring of 1190.[83]

During the first winter of the siege and into the following spring, the Franks filled a section of Acre's ditch and began to build three siege towers; one was commissioned by Conrad of Montferrat and the Genoese, another by Guy of Lusignan, and the last by Landgrave Ludwig II of Thuringia. These, according to the *Itinerarium*, were designed to resist defensive artillery. Each contingent

80 *Eracles* 24.13-16, RHC Oc 2, pp. 123-27; *Itinerarium* 1.26-27, ed. Stubbs, pp. 61-63, trans. Nicholson, pp. 70-73; Baha' al-Din, *al-Nawadir*, trans. Richards, pp. 91, 96-98; 'Imad al-Din, *al-Fath*, trans. Massé, pp. 168-71; Ibn al-Athir, *al-Kamil*, trans. Richards, 2:364-65.

81 *Itinerarium* 1.27, ed. Stubbs, p. 62, trans. Nicholson, p. 71. For a complete breakdown of the Latin force's deployment, see Ralph of Diceto, *Ymagines historiarum*, ed. Stubbs, 2:79-80; Roger of Howden, *Chronica*, ed. Stubbs, 3:20-22, trans. Riley, 2:126-28.

82 Aymar the Monk, *Expeditione Ierosolimitana* ll. 81-104, ed. Falk and Placanica, pp. 8-10.

83 For these preparations, see *Itinerarium* 1.33, ed. Stubbs, pp. 77-79, trans. Nicholson, pp. 85-86; Ambroise, *Estoire* ll. 3,192-217, ed. Ailes and Barber, pp. 51-52, trans. Ailes, pp. 77-78; Aymar the Monk, *Expeditione Ierosolimitana* ll. 113-20, ed. Falk and Placanica, p. 12; Abu Shama, *Kitab al-Raudatain*, RHC Or 4, pp. 430-31, 441-42; 'Imad al-Din, *al-Fath*, trans. Massé, pp. 196-202; Ibn al-Athir, *al-Kamil*, trans. Richards, 2:370; Baha' al-Din, *al-Nawadir*, trans. Richards, p. 109.

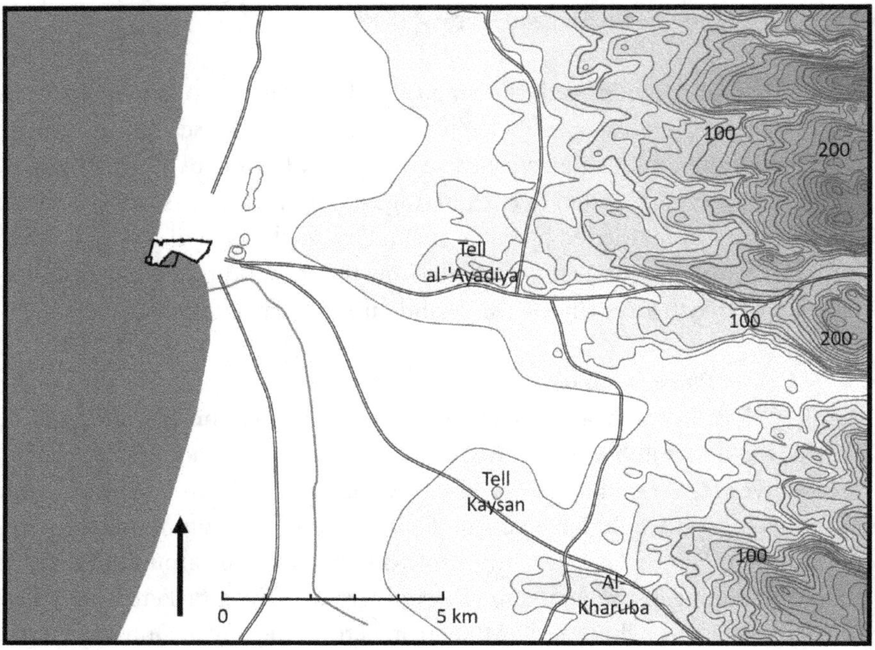

PLAN 20 *Acre region, 1189-91*

also had a number of *petrariae* to support its tower as it advanced and to batter the walls near it.[84] Ambroise calls these engines *pierieres* and *mangoniels*, and claims that the Muslim defenders made use of the same.[85] While it may appear conflicting that the same engines should be intended to provide cover for the advancing towers and batter the walls, they could have done both if they were used to target the parapet. Any damage done to the battlements would further expose defenders along the top of the wall.

Neither Ambroise nor the *Itinerarium* specifies where the Franks positioned their artillery. Baha' al-Din claims that each siege tower was capped with a trebuchet;[86] however, the *Itinerarium* states that this level was occupied, more sensibly, with archers and crossbowmen. Roger of Howden locates the besieging artillery behind the Franks' earthworks, implicitly quite far from the siege towers as they advanced towards the town walls.[87] If the attackers were employing traction trebuchets, these were probably brought forward to better assist the towers once the attack began. These trebuchets were most likely

84 *Itinerarium* 1.36, ed. Stubbs, pp. 84-85, trans. Nicholson, pp. 90-91.

85 Ambroise, *Estoire* ll. 3,196, 3,217, ed. Ailes and Barber, pp. 51-52, trans. Ailes, pp. 77-78.

86 Baha' al-Din, *al-Nawadir*, trans. Richards, p. 110.

87 Roger of Howden, *Chronica*, ed. Stubbs, 3:22, trans. Riley, 2:128.

lighter and less powerful than those used by Saladin in 1188: the best locally sourced timber, as well as any scavenged from the ships of Italians and crusaders that had joined the siege, would have been used to construct the three siege towers, leaving little choice lumber for artillery.[88]

On 5 May 1190, an attack was made and the Frankish siege towers were advanced to the walls of Acre. The towers initially fared well against the stones and containers of Greek fire that were thrown against them, but this success was short-lived and by nightfall they had been destroyed. According to the Arabic accounts, with the garrison's artillery failing to halt the advancing siege towers, Saladin permitted the son of a Damascene caldron-maker to approach Baha' al-Din Qaraqush, who commanded Acre's defence, and concoct a mixture with which he promised to burn the Frankish towers. The incendiaries that he produced were then thrown at the siege towers, consuming them in flames.[89] The Frankish sources give a far less dramatic account of this day's events, suggesting that the towers were burnt in a sally late in the day. Even if fanciful in their specifics, the Muslim accounts provide critical insight into how Acre's defensive engines were employed.

'Imad al-Din and Ibn al-Athir give the impression that the first series of shots were made to gauge the range and accuracy of the engine used to throw the special payload.[90] Once satisfied, the incendiary pots were cast against each tower in turn. Accuracy was critical to this exercise and adjustments would have been necessary whether the engine was traction- or counterweight-powered. Although it is tempting to judge the latter as the more likely, given the natural accuracy of such engines once properly ranged, there seems to be no conclusive evidence of this. The sources state unanimously that the Frankish siege towers had been pushed up to the edge of the fosse when they were burnt, so range would not have been a significant issue. If the towers were reasonably close together, they would likely have been within the range of a single engine located midway between them. Alternatively, it would have been pos-

88 The use of imported wood is emphasised by the Muslim sources, see Abu Shama, *Kitab al-Raudatain*, RHC Or 4, pp. 447-48; Ibn al-Athir, *al-Kamil*, trans. Richards, 2:372. See also 'Imad al-Din, *al-Fath*, trans. Massé, p. 214.

89 'Imad al-Din, *al-Fath*, trans. Massé, pp. 217-19; Baha' al-Din, *al-Nawadir*, trans. Richards, pp. 110-11; Ibn al-Athir, *al-Kamil*, trans. Richards, 2:373-74; *History of the Patriarchs*, trans. Khater and Khs-Burmester, 3.2:149-51. See also *Itinerarium* 1.36, ed. Stubbs, p. 85, trans. Nicholson, p. 91; Ambroise, *Estoire* ll. 3,390-427, ed. Ailes and Barber, pp. 54-55, trans. Ailes, p. 80; Ralph of Diceto, *Ymagines historiarum*, ed. Stubbs, 2:83; Roger of Wendover, *Flores historiarum*, ed. Hewlett, 3:23, trans. Giles, 2:92; Aymar the Monk, *Expeditione Ierosolimitana* ll. 201-12, ed. Falk and Placanica, p. 20.

90 'Imad al-Din, *al-Fath*, trans. Massé, p. 218; Ibn al-Athir, *al-Kamil*, trans. Richards, 2:373.

sible to move one or more traction trebuchets behind the curtain or even from tower to tower to occupy an ideal position to target each siege tower in turn, mitigating the inherent inaccuracy of such engines. The ability to move or elevate a traction trebuchet would also remove the complications that a small counterweight engine would face when trying to shoot at targets so close to the curtain wall.

As discussed in Chapter 3, it can be quite difficult for trebuchets to engage close targets on the far side of a wall. For a counterweight trebuchet to do so, it is necessary to increase the release angle, to produce a high trajectory shot, or to decrease release velocity, in order to maintain a more regular parabolic flight path. The former requires less labour but shortens the shooting sequence, making mechanical adjustments more sensitive, and extends a projectile's exposure to environmental influences. The latter requires that mechanical power be reduced by using a lighter counterweight. This is an inaccurate means of adjusting range that significantly increases the likelihood that an engine will misfire, as the projectile is accelerated slower and thus held in place by lower centrifugal and centripetal forces. Accordingly, there would have been a certain window of security for any siege tower once it was close enough, but not so close that its top levels became exposed, no longer obscured by the curtain wall. Although the tops of these siege towers would present vulnerable targets once they reached the wall, there was still the challenge of yawing, or even moving, a counterweight trebuchet in order to aim at the next tower approaching a different section of wall.

Second Phase: 1190

Later in the summer, the arrival of Henry II of Champagne, Theobald V of Blois, and a large party of French and English nobles, encouraged a second wave of siege engine construction. The arrival of these crusaders also compelled Saladin to withdraw to al-Kharuba on 1 August.[91] For the first time since the siege began, about a year earlier, the Franks initiated a significant artillery barrage.[92] From Baha' al-Din's account, it would appear that the barrage, maintained by teams working in relays, took a larger toll on Acre's defenders and their morale

91 *Itinerarium* 1.42-43, ed. Stubbs, pp. 92-94, trans. Nicholson, pp. 97-99; Ambroise, *Estoire* ll. 3,500-15, ed. Ailes and Barber, p. 56, trans. Ailes, pp. 81-82; Baha' al-Din, *al-Nawadir*, trans. Richards, p. 120; 'Imad al-Din, *al-Fath*, trans. Massé, pp. 243-44; Ibn al-Athir, *al-Kamil*, trans. Richards, 2:377.

92 Baha' al-Din, *al-Nawadir*, trans. Richards, p. 122; 'Imad al-Din, *al-Fath*, trans. Massé, pp. 245; Abu Shama, *Kitab al-Raudatain*, RHC Or 4, pp. 469-70; Ibn al-Athir, *al-Kamil*, trans. Richards, 2:378; Ambroise, *Estoire* ll. 3,524-26, ed. Ailes and Barber, p. 57, trans. Ailes, p. 82.

than its fortifications. This seems sensible when considering that almost all of the ammunition at hand would have been shaped from the same soft *kurkar* sandstone that made up Acre's defences.

According to the Muslim sources, the garrison launched an initial sortie in late August that successfully burnt two of the assailing stone-throwers. Henry's largest trebuchet appears to have escaped unscathed; still unfinished, it had not yet been brought forward. But it too fell victim to the garrison's pyrotechnics during another successful sally on the evening of 3 September, the flames then spread to a smaller trebuchet next to it.[93] In what appears to be a different version of the latter incident, Baha' al-Din relates that arrows with burning tips were used to light one of the Franks' trebuchets on fire and that the flames spread to another nearby engine.[94] Ibn al-Athir describes how Henry of Champagne, when attempting to build additional engines after the first sally, used a mound of earth to shield the two new trebuchets from Acre's archers. The mound was then advanced towards the walls until the engines, presumably those burnt in the second sally, were in a position to effectively shoot at the city and its defenders.[95]

If Chevedden's theory is correct, the use of the adjective 'great', which is found in certain Muslim accounts to refer to one or more of the Frankish trebuchets, suggests that at least some of the assailing engines were counterweight trebuchets.[96] Although there is no clear evidence to support or refute this, Baha' al-Din and Ibn al-Athir clearly state that Henry of Champagne's trebuchets were positioned within range of Acre's archers. The latter's emphasis on the Franks' efforts to construct screens to shelter the crews operating their artillery, rather than the engines themselves, speaks to the vulnerability of these individuals relative to the larger wooden engines next to them.

Baha' al-Din's claim that some trebuchets were completely set alight by incendiary arrows must be viewed critically as this would not have been an easy feat. The penetrating power of an arrow diminishes through the course of its flight, leaving combustibles attached behind the arrowhead farther from the material of the target the longer it is in the air. Likewise, arrows with heated heads, which would cool through flight, would embed themselves less deeply

93 Baha' al-Din, *al-Nawadir*, trans. Richards, pp. 122-23; 'Imad al-Din, *al-Fath*, trans. Massé, pp. 245-46; Ibn al-Athir, *al-Kamil*, trans. Richards, 2:378. Cf. Ambroise, *Estoire* ll. 3,656-94, ed. Ailes and Barber, p. 59, trans. Ailes, p. 83; *Itinerarium* 1.54, ed. Stubbs, p. 105, trans. Nicholson, p. 109; Aymar the Monk, *Expeditione Ierosolimitana* ll. 405-20, ed. Falk and Placanica, p. 40.

94 Baha' al-Din, *al-Nawadir*, trans. Richards, pp. 124-25.

95 Ibn al-Athir, *al-Kamil*, trans. Richards, 2:378.

96 Chevedden, "Invention," pp. 90-91. See Chapter 1 for a discussion of this theory.

and thus transfer their heat less effectively the farther they flew.[97] Furthermore, the essential mobility of a trebuchet's beam allows for any incendiaries that might become lodged near the end of the long arm, otherwise out of reach, to be dealt with fairly quickly. It is in part due to this resilience to arrows that encouraged the use of defensive artillery, which could throw much larger containers of combustibles as well as stones towards besieging engines, and led to artillery battles, such as that at 'Arqa in 1099 and the one that would later develop at Acre in 1191.[98]

The defensive artillery that appears during the events of May 1190 is described by Ambroise in one of his frequently quoted passages:

> There were within the city, as the history tells us in truth, many catapults that hurled [missiles] so well that you never saw the like. There was one which was so powerful that it did us much harm, continually breaking into pieces our catapults and cerceleia, for it hurled the stones so that they flew as if they had wings; it took two men to load the sling, according to the written word, and then the sling let fly, when the stone fell, it had to be looked for a full foot into the ground. This very catapult struck a man in the back and if the man had been a tree, or a column of marble, he would have been cut in half, but the good man did not feel it, for God did not permit it. One should have faith in such a Lord, for such a miracle inspires faith.[99]

Ambroise appears to suggest that the defenders of Acre possessed at least one trebuchet of comparable size to the largest employed by the Franks. Instead of praising the scale of this machine, likely concealed behind the town walls, he

97 The hardwood ideal for trebuchet components would have further resisted the penetration of arrowheads, although it is questionable whether such wood would have been available to the Franks besieging Acre in 1190.

98 For the siege of 'Arqa, see Chapter 3. For the 1191 stage of the siege of Acre, see below.

99 *Il aveit dedenz la citié, / Ço dit l'estorie en verité, / Mult perieres si bien jetantes / Que ainc ne vit l'en de tels tantes. / Une en i ot si jeteresse / Que trop esteit damajeresse, / Qui nus depeçoit toteveies / Noz perieres e noz cercleies, / Car el getoit les pieres teles, / Volenz com s'eles eussent eles, / Que dous gent coveneit a metre / En la funde, sulunc la letre; / E quant le piere descendeit / E la funde aval la rendeit / Que bien plein pié parfont en terre / Al chaeir la coveneit guerre. / Iceste meïsmes periere / Feri un home el dos deriere, / E si [li] home devenist arbre / Ou une columpne de marbre, / Si l'eüst el parmi colpee, / Tant i fud el dreit açopee; / E li prodom ne la senti, / Car Dampnedeus nel consenti; / En itel seignor doit l'om creire, / Que tel miracle fait a creire.* Ambroise, *Estoire* ll. 3,530-55, ed. Ailes and Barber, p. 57, trans. Ailes, p. 82.

emphasises its accuracy and the size of its projectiles. The reference to the man who was struck but spared, resembling the incident at Wark in 1174,[100] should not be taken as a literal record of events but rather a piece of fiction to stress the engine's power but the even greater power of God, in whose name the crusaders were fighting. In its similar rendition of these events, the *Itinerarium* also draws attention to the power of this engine but highlights its size and range, adding that the individual who was struck believed that he was out of range.[101] Both accounts state that the Muslim artillery, and this exceptionally large engine in particular, was targeting the Frankish artillery as well as the shelters protecting those working to fill the town's fosse and the men defending them. While counterweight engines would have been well suited to targeting the immobile artillery and shelters, traction trebuchets would have been more effective at harassing the personnel around them.

Following the destruction of their engines, the Franks attempted a naval assault against Acre's harbour defences later in September. According to the *History of the Patriarchs*, the Franks loaded a stone-thrower onto a large transport vessel and this was used, in conjunction with archers, to provide cover for the assault force.[102] Frankish and Muslim sources do not mention artillery during this episode and it seems more likely that a type of maritime siege tower, possibly containing large quantities of combustibles, was at the centre of this attack against the Tower of the Flies.[103]

Defensive engines appear to have been used against the rams built by Archbishop Theoderic of Besançon and Henry of Champagne. Frankish and Muslim sources state that artillery targeted these engines from the walls of the city on 14 October 1190.[104] If counterweight trebuchets were employed, they might have been used against the principal ram as it approached (only one ram is

100 Jordan Fantosme, *Chronique* ll. 1,240-55, ed. Howlett, pp. 307-8.

101 *Itinerarium* 1.47, ed. Stubbs, p. 98, trans. Nicholson, p. 103.

102 *History of the Patriarchs*, trans. Khater and Khs-Burmester, 3.2:151.

103 Ambroise, *Estoire* ll. 3,765-812, ed. Ailes and Barber, p. 61, trans. Ailes, pp. 84-85; *Itinerarium* 1.58, ed. Stubbs, pp. 109-11, trans. Nicholson, pp. 113-14; Baha' al-Din, *al-Nawadir*, trans. Richards, pp. 127-28.

104 Ambroise, *Estoire* ll. 3,813-90, ed. Ailes and Barber, pp. 61-63, trans. Ailes, pp. 85-86; *Itinerarium* 1.59, ed. Stubbs, pp. 111-13, trans. Nicholson, pp. 115-16; 'Imad al-Din, *al-Fath*, trans. Massé, pp. 255-57; Baha' al-Din, *al-Nawadir*, trans. Richards, pp. 130-31; Abu Shama, *Kitab al-Raudatain*, RHC Or 4, pp. 485-86; *History of the Patriarchs*, trans. Khater and Khs-Burmester, 3.2:151-52. Baha' al-Din misdates the assault to 4 October (3 Ramadan), rather than 14 October (13 Ramadan), so the ram and bore that he describes as being built by Frederick of Swabia, who arrived at Acre on the evening of 7 October, are likely equitable with the engines built by the archbishop and the count of Champagne.

highlighted in the attack); however, only lighter engines positioned on Acre's towers would have been able to effectively engage it as it neared the wall.[105] The limited power and scale of these engines is confirmed by their inability to significantly damage the ram. Once it reached the wall, heavier stones, as well as wood and incendiaries, were dropped from the parapet above while the lighter tower-mounted traction trebuchets provided antipersonnel support, creating a deadly environment for anyone bold enough to venture out of the ram. Disabled and burning, the archbishop's lumbering siege engine was eventually abandoned.

On the night of 4 Jan 1191, the first signs of significant damage appeared along Acre's walls when a section of the main curtain fell on the forewall ahead of it following a winter storm.[106] The Franks made an assault but were turned back and the defenders were able to patch the breach during the following days.[107] Traction trebuchets, in addition to archers, were used to protect the breach while it was being repaired. The inability of the Franks to exploit this gap in the wall, as well as its rapid repair, suggest that the damage did not extend far below the level of the parapet. Had the damage which led to this breach been caused by a large counterweight trebuchet in the lead up to the storm, it is inconceivable that the same engine would not have been able to repeatedly strike the same position, enlarging the breach or at least preventing its repair, let alone open a new one. Although the winter weather would have amplified any damage inflicted over the previous year, the breach may have been the result of a much older weakness in the structure, hidden within the masonry of the wall, which worsened with the rain and wind of each storm. While the siege of Acre continued into 1191, the principal leaders associated with the Third Crusade made their way to Palestine.

Third Phase: 1191

Before reaching the Holy Land, Richard I of England prepared his artillery, *perariae* and other engines according to Roger of Howden,[108] on Sicily during the

105 For the practice and limitations related to mounting trebuchets on top of towers, see Chapter 9.

106 If accurate, this would appear to contradict Jacoby's theory that Acre was surrounded by only a single line of walls until the early thirteenth century, Jacoby, "Montmusard," pp. 207-17. Cf. Fulcher of Chartres, *Historia Hierosolymitana* 2.22.1, ed. Hagenmeyer, pp. 456-57, trans. Ryan, p. 174; Pringle, "Town Defences," pp. 82-83; Pringle, *Churches*, 4:5-7.

107 Baha' al-Din, *al-Nawadir*, trans. Richards, p. 142; 'Imad al-Din, *al-Fath*, trans. Massé, pp. 277-78; Aymar the Monk, *Expeditione Ierosolimitana* ll. 557-68, ed. Falk and Placanica, pp. 54-56.

108 Roger of Howden, *Chronica*, ed. Stubbs, 3:72, 105, trans. Riley, 2:174, 200.

early months of 1191. Experience in France had taught Richard the benefits of speed and preparedness on campaign, while the landward blockade of the Christian force at Acre necessitated the importation of any materials that would be required during the siege. Richard also appears to have collected ammunition before departing Sicily: with the harder limestone beyond the coastal plain cut off by Saladin's forces, hard volcanic projectiles would have been far more effective against Acre's soft *kurkar* sandstone defences than projectiles hewn from the same type of stone.[109] From Tyre, Richard sent word ahead to Acre that his artillery was delayed due to southerly winds.[110] Baha' al-Din appears to confirm that Richard was transporting artillery with his army when he notes the presence of engines of war, among other supplies, on a ship sent ahead to Acre that was captured off the coast of Beirut.[111] Richard avenged this loss only weeks later when he intercepted a Muslim ship in the same area. Interestingly, among the thirty-five men that the king chose to spare were a number of experts at constructing siege machinery.[112] The only indication that Richard used his artillery before reaching Acre appears to be in Richard of Devizes's account of the siege of Buffavento on Cyprus; however, the sources more closely linked to these events do not corroborate this.[113]

Philip II of France arrived at Acre in mid-April 1191 as the siege was entering its third campaign season. Philip's siege strategy involved artillery from the outset. According to the *Itinerarium*, Philip erected *petrariae* and other engines opposite the Accursed Tower, which anchored the city's north east sa-

109 Ambroise, *Estoire* ll. 4,785-94, ed. Ailes and Barber, p. 77, trans. Ailes, pp. 98-99.

110 Ambroise, *Estoire* ll. 4,603-13, ed. Ailes and Barber, p. 75, trans. Ailes, pp. 95.

111 Baha' al-Din, *al-Nawadir*, trans. Richards, p. 147. Cf. 'Imad al-Din, *al-Fath*, trans. Massé, p. 292; Abu Shama, *Kitab al-Raudatain*, RHC Or 5, p. 8; Ibn al-Athir, *al-Kamil*, trans. Richards, 2:386-87.

112 Ambroise, *Estoire* ll. 2,260-72, ed. Ailes and Barber, p. 37, trans. Ailes, p. 64; *Itinerarium* 2.42, ed. Stubbs, p. 209, trans. Nicholson, p. 199. Cf. Roger of Howden, *Chronica*, ed. Stubbs, 3:112, trans. Riley, 2:206; Ralph of Diceto, *Ymagines historiarum*, ed. Stubbs, 2:93-94; Aymar the Monk, *Expeditione Ierosolimitana* ll. 757-80, ed. Falk and Placanica, pp, 74-76. Ibn al-Athir's timing of events agrees with that in the Frankish sources; however, Baha' al-Din and 'Imad al-Din place the ship's capture on 11 June, after Richard's arrival at Acre, Ibn al-Athir, *al-Kamil*, trans. Richards, 2:387; 'Imad al-Din, *al-Fath*, trans. Massé, p. 299; Baha' al-Din, *al-Nawadir*, trans. Richards, p. 151.

113 Richard of Devizes, *Chronicon* 61, ed. Howlett, p. 425, trans. in *Chronicles of the Crusades*, p. 39. See also Peter of Langtoft, *Chronicle*, ed. and trans. Wright, pp. 66-67. Cf. Ambroise, *Estoire* ll. 2.006-33, ed. Ailes and Barber, pp. 32-33, trans. Ailes, p. 60; *Itinerarium* 2.39, ed. Stubbs, p. 202, trans. Nicholson, p. 194; Roger of Howden, *Chronica*, ed. Stubbs, 3:111, trans. Riley, 2:205; Ralph of Diceto, *Ymagines historiarum*, ed. Stubbs, 2:92.

lient.[114] Roger of Howden also notes Philip's initial use of artillery but specifies that these engines had little impact prior to Richard's arrival.[115]

The Muslim sources claim that the Franks pressed Acre from 30 May with seven trebuchets. Saladin was able to interrupt the barrage when he advanced the core of his army from its wintering position and attacked the Frankish siege camp, but the bombardment resumed when his forces retired to rest.[116] While the Franks continued their attempts to fill parts of the ditch around Acre, the defenders divided themselves into four teams: one cut up the debris in the fosse, another threw it into the sea, a third stood guard and a fourth provided cover with artillery (traction trebuchets).

Richard landed at Acre on 8 June 1191, by which point, twenty-one months after the siege had begun, damage was becoming visible on one of the city's towers. Despite being struck down with a bout of illness, Richard saw to the deployment of *petrariae*, *mangunelli* and a *castellum*, which was erected in front of one of the city's gates.[117] Frankish artillery then played against the defences of Acre throughout June 1191. 'Imad al-Din states that most damage was sustained at the level of the parapet, though in some sections the walls were reduced to the height of a man.[118] Similarly, Baha' al-Din notes that the Franks placed increasing focus on their artillery and used the heaviest projectiles that they could, adding to the mental strain of the garrison.[119] Although 'Imad al-Din is a particularly difficult source to deal with, both he and Baha' al-Din appear to suggest that while lighter engines continued to suppress any defenders

114 *Itinerarium* 2.28, ed. Stubbs, p. 181, trans. Nicholson, pp. 177-78. See also Ambroise, *Estoire* ll. 1,338-42, ed. Ailes and Barber, p. 22, trans. Ailes, p. 50; Aymar the Monk, *Expeditione Ierosolimitana* ll. 729-40, ed. Falk and Placanica, pp. 70-72.

115 Roger of Howden, *Chronica*, ed. Stubbs, 3:113, trans. Riley, 2:207.

116 Baha' al-Din, *al-Nawadir*, trans. Richards, pp. 147,148-50; 'Imad al-Din, *al-Fath*, trans. Massé, pp. 293-94, 295-97; Abu Shama, *Kitab al-Raudatain*, RHC Or 5, p. 10; Ibn al-Athir, *al-Kamil*, trans. Richards, 2:387.

117 *Itinerarium* 3.4, ed. Stubbs, p. 214, trans. Nicholson, p. 204. Richard of Devizes states that the *castellum* was the same that Richard had disassembled before leaving Sicily, and it is likely the same armoured siege tower that the Muslim sources claim was burnt later in June and that which Ambroise and the *Itinerarium* mention among the events of late June and early July, Richard of Devizes, *Chronicon* 28, 64, ed. Howlett, pp. 403, 426-27, trans. in *Chronicles of the Crusades*, pp. 21, 41; Baha' al-Din, *al-Nawadir*, trans. Richards, pp. 151-52; 'Imad al-Din, *al-Fath*, trans. Massé, pp. 299-300; Ibn al-Athir, *al-Kamil*, trans. Richards, 2:387; Ambroise, *Estoire* ll. 4,775-80, ed. Ailes and Barber, p. 77, trans. Ailes, p. 98; *Itinerarium* 3.7, ed. Stubbs, p. 219, trans. Nicholson, p. 209.

118 'Imad al-Din, *al-Fath*, trans. Massé, pp. 306-8, 310-11, 314-15.

119 Baha' al-Din, *al-Nawadir*, trans. Richards, p. 156.

along Acre's ramparts, the Franks may have hoped that their heaviest engines would compromise the integrity of the walls below. The ability of the defenders to continually repair their damaged battlements was critical to their ability to maintain an effective defence; however, enough damage had apparently been inflicted by the start of July that the French forces felt confident enough to launch a frontal assault.

Acre was hosting a veritable convention for some of Europe's most powerful figures to demonstrate their prowess through the construction of siege engines. Philip of France had one trebuchet, named *Male Veisine/Mala Vieina*, which drew the attention of one of the garrison's best engines, dubbed *Male Cosine/Mala Cognata*, possibly the same impressive defensive engine noted during the autumn of 1190. The latter was able to knock the former out of action on a number of occasions but the French engine was repeatedly rebuilt and continued to shoot at the Accursed Tower. Although this is the only one of Philip's trebuchets that is mentioned, he probably had other lighter ones, which the predominantly Anglo crusader accounts fail to mention. Hugh of Burgundy, the Templars, and the Hospitallers each had a notable trebuchet and another was funded communally. Philip of Flanders, veteran of the siege of Harim in 1177, had a choice trebuchet along with a smaller one, both of which were appropriated by Richard following the count's death. Richard directed these two engines against a gate, presumably that which his *castellum* faced, and erected another two *perieres/petrariae*, likely those prepared on Sicily, and two *mangonels/mangunelli*.[120] Thus from mid-June, nine notable *petrariae* and at least two *mangunelli* were in use. This breakdown supports the Muslim sources' claims that seven engines were used by the Franks from the end of May 1191 – before Richard's arrival. Bar Hebraeus gives the same number but mistakenly dates this to after Richard's landing.[121]

For practical reasons, most of the trebuchets erected by the Franks through 1189-91 were probably traction-powered; however, this does not mean that the engines enumerated in the summer of 1191 were such. The wealth and resources of the European potentates and the natural competition between them for prestige make it likely that the trebuchets constructed in 1191 were particularly impressive for their time.

---

120 Ambroise, *Estoire* ll. 4,737-94, ed. Ailes and Barber, pp. 76-77, trans. Ailes, pp. 98-99; *Itinerarium* 3.7, ed. Stubbs, pp. 218-19, trans. Nicholson, pp. 208-9. See also Roger of Howden, *Chronica*, ed. Stubbs, 3:111, trans. Riley, 2:205; Aymar the Monk, *Expeditione Ierosolimitana* ll. 785-88, ed. Falk and Placanica, p. 76.

121 Bar Hebraeus, *Makhtebhanuth Zabhne*, trans. Budge, 1:334.

According to Ambroise, Richard was able to damage the top of the gate tower that he targeted with Philip of Flanders' engines. The same author states that Philip II's principal engine was able to inflict significant damage on the Accursed Tower and adjoining curtain, while the communal engine did further damage to a section of wall next to this tower, presumably on the other side of it.[122] Similarly, Roger of Howden states that the *perrariae* of the French king, the Templars and the Pisans opened a breach near the Accursed Tower and it was here that the French rushed the walls but were turned back in early July.[123] Ralph of Diceto is less specific, noting that walls were worn down by the assailing machines and *petrariae* but not where this damage was concentrated.[124] Despite the universal reports of destruction, which support the likelihood that counterweight trebuchets were used, the emphasis placed on the danger faced by members of the garrison when attempting to man their defences suggests a range of artillery was employed. The repeated references to shelters for attacking archers are a reminder that the parapet, although damaged in sections, remained largely defensible in July 1191.

The descriptions of Richard's engines provided by Ambroise and the *Itinerarium* are important to any discussion of terminology.[125] One of the projectiles loosed by one of Richard's *perieres/petrariae*, a river stone brought from Messina, is supposed to have killed twelve of Acre's defenders, causing it to be taken to Saladin for inspection.[126] The notion that either source could have seen the impact of this projectile is as unlikely as the possibility that a swimmer would have smuggled the stone out of the city during the peak of the siege. The responsible engine was probably relatively heavy: although the projectile supposedly killed a group of people, rather than destroying a notable section of the town's fortifications, the emphasis on the number of individuals suggests that it was the size of the stone that the authors were trying to highlight. Furthermore, the use of imported ammunition bears clear similarities to the Sicilians' use of such at Alexandria in 1174.[127] These special stones were probably used by the most powerful engines to transfer as much energy as possible to

---

122 Ambroise, *Estoire* ll. 4,739-46, 4,753-70, ed. Ailes and Barber, pp. 76-77, trans. Ailes, p. 98. Cf. *Itinerarium* 3.7, ed. Stubbs, pp. 218-19, trans. Nicholson, p. 209.

123 Roger of Howden, *Chronica*, ed. Stubbs, 3:116-17, trans. Riley, 2:210-11.

124 Ralph of Diceto, *Ymagines historiarum*, ed. Stubbs, 2:94.

125 Ambroise, *Estoire* ll. 4,761-94, ed. Ailes and Barber, pp. 76-77, trans. Ailes, pp. 98-99; *Itinerarium* 3.7, ed. Stubbs, pp. 219-20, trans. Nicholson, p. 209.

126 Cf. the description of a stone killing three of Tortona's defenders in 1155, Otto of Freising and Rahewin, *Gesta Friderici I* 2.21, ed. Waitz, p. 124, trans. Mierow, p. 135.

127 Abu Shama, *Kitab al-Raudatain*, RHC Or 4, pp. 165-66. For the siege of Alexandria in 1174, see Chapter 4.

their targets. Fieldstones would have been sufficient to feed lighter antipersonnel engines. This, along with the apparent damage caused by Philip of Flanders' larger engine, Philip II's engine and the communal trebuchet, suggest that the terms *perieres* and *petrariae* were used by these authors to describe the heaviest artillery, but not necessarily exclusively.

If the two *perieres*/*petrariae* that Richard erected at Acre were counterweight trebuchets, it is tempting to distinguish them from the two *mangonels*/*mangunelli*. It is possible that Ambroise simply used the term *mangonels* as a synonym for *perieres* in order to rhyme with *ignels* in the next line; however, later events suggest that these engines were distinct from Richard's four *perieres*. Unfortunately the only description of the *mangonels*/*mangunelli* is that one was particularly swift, and possibly capable of shooting projectiles across an impressive distance.[128] Swiftness is a characteristic attributable to a traction trebuchet but a remarkable range is not.

Ambroise's remark that *Quant sa piere voleit en Acre, Qu'ele aloit jusqu'en la maçacre*, has been interpreted as a statement praising the range of these *mangonels*, suggesting that they were able to shoot so far as the area of the city associated with a butcher (*macel*).[129] According to David Jacoby, Acre's butchery district was near the Church of St Demetrius, in the quarter of the Chain near the edge of the Venetian quarter.[130] This would place it at the north edge of Acre's inner harbour, more than 400 m from the city's northern wall and about 900 m from the eastern wall. Regardless of whether Richard was opposing a section of Acre's northern or eastern defences, it would have been impossible for his artillery to reach the harbour.[131] Accordingly, the remark might be a simple exaggeration; however, it may also be a reference to an area of more general slaughter associated with the ongoing siege, such as the stretch of parapet ahead of the English king.

The apparently comparable damage inflicted by the other enumerated *perieres*/*petrariae* suggests that these, like Richard's, were counterweight tre-

128 *Si fist fiare dous mangonels, / Dont li uns esteit si ignels, / Quant sa piere voleit en Acre, / Qu'ele aloit jusqu'en la maçacre*. Ambroise, *Estoire* ll. 4,781-84, ed. Ailes and Barber, p. 77, trans. Ailes, p. 98, trans. Hubert, pp. 202-3.

129 In 1229, when departing Acre, Frederick II was said to have boarded a ship in front of the *boucherie*. See Ambroise, *Estoire*, trans. Ailes, p. 98 n. 332, trans. Hubert, p. 203 n. 22. Cf. Philip of Novara, *Gestes des Chiprois* 138, ed. Raynaud, p. 50, trans. La Monte, p. 91.

130 Jacoby, "Three Notes," p. 90. My thanks to Adrian Boas for bringing this reference to my attention. See also Pringle, *Churches*, no. 402, 4:74-76, and map on pp. 16-17. For the trace of Acre's walls in the twelfth century, see Pringle, *Churches*, 4:4-6.

131 My thanks to John Hosler for sharing his thoughts regarding the deployment of Richard's forces in June-July 1191.

buchets. With Frankish numbers at their highest since the siege began, it is possible that they may have felt comfortable advancing their engines relatively close to the curtain in order to shoot particularly heavy projectiles. By using a lower trajectory and sacrificing release velocity for additional force and mechanical efficiency, a greater amount of energy would be transferred. The masonry would have been forced to absorb this added energy, manifesting in shockwaves that would break apart the bonds of the stone and mortar as they reverberated out from the point of impact. Despite the probable power of these engines, the term *perriere/petraria* does not appear to have been used exclusively to denote counterweight trebuchets, as it was also applied to the lighter engine used by the defenders against Philip's 'cat'[132] and those employed against the Frankish ram in the autumn of 1190.

The Franks' primary breaching weapon was their sappers, who worked to undermine Acre's defences from late June 1191, if not earlier.[133] It is an unexplored possibility that Philip and Richard may have used their heavy artillery to assist the sappers, working against the notoriously strong mortar of this region. Both kings attempted to sap the same towers that their trebuchets were targeting, perhaps using the percussive force of their artillery to spread small fractures through the mass of the walls being undermined, facilitating the collapse of these sections when the mines were lit. Richard may have done this later at Darum and, after his return to France, Philip employed a similar tactic during the final stages of his siege of Château Gaillard in 1204. Using a *petraria* called *Cabulus/Chadabula*, he knocked down a weakened section of the inner bailey that had been undermined but had not fallen completely.[134] By 12 July 1191, the defenders of Acre were no longer confident that they could withstand the Frankish assaults and surrendered before the town could be taken by force.

Legacy

"For if the ten-year war made Troy famous, and the Christian triumph made Antioch more illustrious, then Acre will certainly win eternal fame for the

132 Ambroise, *Estoire* ll. 4803-28, ed. Ailes and Barber, pp. 77-78, trans. Ailes, p. 99; *Itinerarium* 3.8, ed. Stubbs, pp. 220-21, trans. Nicholson, p. 210. Cf. the επιβαθρα, Vitruvius, *De architectura* 10.13.8, ed. Krohn, p. 253.

133 Ambroise, *Estoire* ll. 4,861-5,032, ed. Ailes and Barber, pp. 78-81, trans. Ailes, pp. 99-102; *Itinerarium* 3.9-14, ed. Stubbs, pp. 221-28, trans. Nicholson, pp. 211-16; Roger of Howden, *Chronica*, ed. Stubbs, 3:115, 116, 118-20, trans. Riley, 2:209-10, 212-14. Cf. Baha' al-Din, *al-Nawadir*, trans. Richards, pp. 156-58, 160; 'Imad al-Din, *al-Fath*, trans. Massé, pp. 306-13, 314-18.

134 William the Breton, *Gesta Philippi Augusti*, ed. Delaborde, p. 219.

whole globe assembled to fight for her."[135] Perhaps more than any other aspect of this extraordinary siege, it is the use of artillery during its final phase that has attracted the attention of modern historians. Although it is popularly held that counterweight trebuchets were employed, the evidence supporting this remains at best circumstantial. No new terminology is used by any of the eye-witness sources and similar descriptions of artillery damage can be found at a number of much earlier sieges. However, the wealth of the figures involved, the importation of prefabricated engines and ammunition, and what appears to be the use of both heavy and light varieties of artillery, make it very likely that both traction and counterweight trebuchets were used.

A commendable evaluation of the artillery used at Acre in 1191 has been put forward by Randall Rogers.[136] Considering the various descriptions of destruction in the sources, Rogers suggests that there is little firm evidence upon which to support any conclusions regarding the power source of these machines and instead emphasises the impact that they seem to have had on the city's fortifications. But this approach can be misleading as descriptions of artillery smashing fortified masonry can be found throughout the classical and medieval periods and even light engines would have left significant scarring on the soft *kurkar* walls of Acre after two years of bombardment.[137] Apart from 'Imad al-Din, the sources agree that mining was the greatest threat to Acre's fortifications.

Instead of relying on colourful and easily exaggerated adjectives and adverbs, the manner in which the engines at Acre were deployed should be stressed. But even when considering relative ranges, targets and the presence of imported ammunition, Rogers' conclusion holds firm as the lack of material evidence prevents any quantification of the power of these engines. Dispelling this caution, numerous historians have seized the chance to state unhesitatingly that counterweight trebuchets were employed, relying on no more evidence than a literal reading of the sources discussed above.[138]

Modern historians can be forgiven to some degree for their uncritical reading of the sources as some subsequent medieval authors added further embel-

---

135 *Itinerarium* 1.32, trans. Nicholson, p. 83.

136 Rogers, *Latin Siege Warfare*, pp. 227, 234-35.

137 For the effects of light artillery against this type of masonry, see the siege of Arsuf (1265) in Chapter 7.

138 Benvenisti and Nicolle do this without apparently considering any other alternative, while Chevedden, operating on the premise that such engines were employed much earlier in the twelfth century, sees the descriptions of destruction as ample evidence to confirm the use of counterweight trebuchets, Benvenisti, *The Crusaders*, pp. 284-85; Nicolle, "The Early Trebuchet," p. 272; Chevedden, *Citadel of Damascus*, p. 279.

lishment or injected descriptions better suited to their own context into their accounts. James of Vitry, writing only a few decades later, left a powerful, if brief, account of the siege which emphasises the kings' use of artillery and makes no mention of mining.[139] A similar description of what appear to be more powerful engines and larger projectiles than are described by the eyewitness sources can be found in Peter of Langtoft's later account, although this author makes it clear that mining was the impetus for the city's surrender.[140]

Despite the apparent power of certain engines at the siege of Acre, it is important to observe the consistent manner in which artillery was used throughout the twelfth century. As at Nicaea and Jerusalem in the last years of the eleventh century, the besieging artillery used at Acre appears to have targeted the city's parapets. This in itself should check any tendency to view these as breaching engines, even if the heavier ones were used deliberately to target the battlements and tops of walls rather than those defending them. Evidence to suggest that reciprocally heavy engines were commonly employed in defence is scarce; however, it would appear that garrisons were placing more thought into the use of defensive artillery. Although the tower-mounted engines employed at Acre did not disable the ram used in 1190, their use reflects more systematic defensive planning than appears to have been in place at the start of the century.

Later Sieges of the Crusade

Darum: 1192

After Acre had fallen, Richard I of England disassembled his artillery and prepared it for transport.[141] The king does not appear to have had the opportunity to use these engines until the following year, as most coastal defences and those of southern Palestine, with the notable exception of Jerusalem, had been

139 "King Philip of France battered the city walls, towers, and battlements incessantly, both by day and by night, with huge stones, breaking the enemy's machines, houses, and buildings within the city, and giving the besieged no rest. On the other side, the King of England frequently made perilous assaults on the besieged; wherefore, as the wall was giving way under the continual strokes of the great stones flung against it, the citizens, perceiving that they could not much longer resist, surrendered the city ..." James of Vitry, *Historia orientalis* 99, ed. Donnadieu, p. 454, trans. Stewart, p. 112.

140 Peter of Langtoft, *Chronicle*, ed. and trans. Wright, pp. 80-81.

141 Ambroise, *Estoire* ll. 5,372-76, ed. Ailes and Barber, p. 87, trans. Ailes, p. 106; *Itinerarium* 4.2, ed. Stubbs, p. 240, trans. Nicholson, p. 227. Cf. Roger of Howden, *Chronica*, ed. Stubbs, 3:122, trans. Riley, 2:215.

slighted on Saladin's orders ahead of the advance of the crusader army.[142] With the 1192 campaign season opening, Richard led a reconnaissance force from Ascalon to Gaza on Easter Tuesday and onward to Darum the following day.[143] The original small quadrangular castrum described by William of Tyre had grown since its initial foundation and the outer defences had probably been rebuilt since Saladin's attack in 1170. The stronghold is said to have boasted seventeen towers and a fosse by the time Richard viewed it in 1192.[144]

Richard appreciated that Saladin had been mustering his army since early April and grew impatient waiting for the French and Eastern Frankish forces to prepare themselves. Then at Ascalon, he decided to march what forces he had with him to Darum, which was still occupied by a Muslim garrison, and began besieging its outer defences on Tuesday 18 May 1192. Although Richard's force was insufficiently large to surround the stronghold, the king's artillery was deployed and an aggressive siege ensued. The engines had been transported by sea and alighted about 2 km from Darum. The prefabricated components of three machines were then reassembled in positions around the castle. These may have been the two *petrariae* that Richard had brought from Sicily and assembled at Acre and the notable trebuchet that the count of Flanders had built before his death. Richard maintained command of one of these engines while contingents from Normandy and Poitou each commanded another. As at Acre, Ambroise's version of events emphasises the fear that these engines impressed on the garrison while the *Itinerarium* stresses their destructive force.

The trebuchets are portrayed as being quite powerful yet at least Richard's engine was used to provide cover for the sappers who worked to undermine the largest of the mural towers. The defenders employed at least one defensive trebuchet; however, this machine, which had been set up on the large tower that Richard was attacking, was knocked out of action by the engine commanded by the king. One of the other attacking engines targeted a gate, which it managed to damage. Like the gate that Richard targeted with Philip of Flanders' engines at Acre, this might have been to prevent the garrison from

142 Ambroise, *Estoire* ll. 5,972-81, 6,826-60, ed. Ailes and Barber, pp. 96-97, 110-11, trans. Ailes, pp. 114, 124-25; *Itinerarium* 4.14, 23, 32, 35, ed. Stubbs, pp. 256, 297-98, 280, 303, trans. Nicholson, pp. 242, 261, 274-75, 278; Baha' al-Din, *al-Nawadir*, trans. Richards, pp. 177-80, 181, 183; 'Imad al-Din, *al-Fath*, trans. Massé, pp. 231, 345-48; Abu Shama, *Kitab al-Raudatain*, RHC Or 4, p. 462, RHC Or 5, pp. 40, 43-44, 50-51; Ibn al-Athir, *al-Kamil*, trans. Richards, 2:391, 392.

143 *Itinerarium* 5.7, ed. Stubbs, pp. 318-19, trans. Nicholson, pp. 289-90.

144 William of Tyre, *Chronicon* 20.19, ed. Huygens, 2:937, trans. Babcock and Krey, 2:372-73; Ambroise, *Estoire* ll. 9,223-29, trans. Ailes, p. 156; *Itinerarium* 5.39, ed. Stubbs, p. 353, trans. Nicholson, p. 317.

making a sally, as was done at the Bridge Gate during the siege of Antioch in 1097-98 and by the crusaders at Lisbon in 1147, or with hopes of damaging the defences that secured this natural aperture in the stronghold's fortifications. The position of the third engine is unclear.[145]

The garrison sought terms on the morning of the fourth day of the siege. Refusing to accept anything less than an unconditional surrender, Richard rejected the defenders' offer and renewed the barrage. The tower that his sappers were undermining was so weakened by this point that a sizable portion collapsed when his engine began to shoot at it again.[146] While Richard's forces stormed the breach, the garrison withdrew to a tower, surrendering from there before the end of the day.[147]

Although this episode reinforces many of the principles that can be discerned from the siege of Acre, there is no clear description of these engines and thus no definitive evidence of their mechanical type. Although the use of one engine to bring down a tower implies that it was quite powerful, the Frankish sources clearly state that it was only able to do so because the tower had been undermined. Revealingly, 'Imad al-Din and Baha' al-Din focus on Richard's use of miners and omit any mention of artillery, though they confirm the brevity of the siege.[148]

Jaffa: 1192

Later in the year, while Richard moved north towards Beirut, another coastal city to which he could bring his artillery by sea, Saladin moved on Jaffa. The Muslims arrived on 27 or 28 July 1192 and the siege began the following day.[149] The Frankish sources claim that the first three days of the siege were consumed with fighting in the suburbs around Jaffa until, on the fourth day, Saladin brought artillery into play. This, according to Ambroise, consisted of four strong and light *perieres* and two *mangunels*, or, according to the *Itinerarium*,

145 Ambroise, *Estoire* ll. 9,127-237, ed. Ailes and Barber, pp. 148-49, trans. Ailes, pp. 155-56; *Itinerarium* 5.39, ed. Stubbs, pp. 352-54, trans. Nicholson, pp. 316-17.

146 This bears similarities to the final stage of Philip II's siege of Château Gaillard in 1204, William the Breton, *Gesta Philippi Augusti*, ed. Delaborde, p. 219.

147 Ambroise, *Estoire* ll. 9,238-343, ed. Ailes and Barber, pp. 249-51, trans. Ailes, pp. 157-58; *Itinerarium* 5.39, ed. Stubbs, pp. 354-56, trans. Nicholson, pp. 318-19. Cf. Roger of Howden, *Chronica*, ed. Stubbs, 3:180, trans. Riley, 2:266.

148 'Imad al-Din, *al-Fath*, trans. Massé, pp. 378-79; Baha' al-Din, *al-Nawadir*, trans. Richards, p. 201; Abu Shama, *Kitab al-Raudatain*, RHC Or 5, p. 54.

149 Baha' al-Din, *al-Nawadir*, trans. Richards, p. 217. The Frankish sources place these events one day earlier than the Muslim accounts, Ambroise, *Estoire* ll. 10,778-82, ed. Ailes and Barber, p. 174, trans. Ailes, p. 175; *Itinerarium* 6.13, ed. Stubbs, pp. 400-1, trans. Nicholson, p. 349.

four notable and effective *petrariae* and two fast-firing *mangunelli*. The adjectives used to describe these engines suggest that there was a technical difference between the two types. While the *mangunelli* would appear to have been traction trebuchets, references by these sources to tower-mounted *petrariae* and the supposed difficulties encountered by Jaffa's defenders when attempting to operate additional *petrariae* should discourage any inclination to impose terminological guidelines too strictly.[150] Mentions of powerful 'fast-firing trebuchets' at Toulouse in the second decade of the thirteenth century complicate matters further.[151] But it is perhaps the number of engines noted by the sources that most exposes these Frankish sources to question: the same compliment of engines was assigned to Richard at Acre. Furthermore, the measure of the breach reportedly created near the eastern gate, two perches (about 10 m), is the same as the extent of the damage reportedly inflicted by the communal engine at Acre a year earlier.[152]

Baha' al-Din, a participant of this siege, provides the best account from a Muslim perspective. According to him, Saladin intended to employ artillery from the start, having his forces erect his trebuchets near Jaffa's eastern gate. He claims that this section of fortifications had been damaged during an earlier siege but had since been partly repaired.[153] Sappers were able to excavate an area under the weakened fortifications on the first day but the defenders successfully collapsed this mine before it could bring down a section of wall. Two of Saladin's trebuchets came into action on the morning of the second day of the siege and a third was ready later in the day. Although the sappers had begun to undermine a longer section of the town's defences, the Muslim army became disheartened as they came to appreciate that their artillery could have little effect before Richard arrived with relief. Even Saladin became worried after the extended mines brought down a part of the curtain but the improvised defences of the garrison proved sufficient to repel the attacks of his forces. That night, he resolved to have five trebuchets erected, either two more to bring his total to five or five more to bring his total to eight. The engines were ready by the morning of the fifth day of the siege (31 July), by which point stones had been scavenged from riverbeds and distant places for them to

150 Although counterweight trebuchets were mechanically more complex than the traction variety, they were much easier to operate accurately.

151 *Chanson de la croisade Albigeoise* 192, ed. Meyer, 1:285, trans. Shirley, p. 141.

152 Ambroise, *Estoire* ll. 10,783-19, ed. Ailes and Barber, pp. 174-75, trans. Ailes, pp. 175-76; *Itinerarium* 6.13, ed. Stubbs, p. 401, trans. Nicholson, pp. 349-50.

153 Baha' al-Din, *al-Nawadir*, trans. Richards, p. 217. Cf. 'Imad al-Din, *al-Fath*, trans. Massé, p. 231. The earlier damage was more likely inflicted when the town was slighted in 1190.

throw. Although Baha' al-Din claims that the engines were to be used to target the weakened curtain wall, the trebuchets are portrayed as supporting the attacks that were made that day and the miners were clearly responsible for bringing down the section of wall that allowed the Muslims to storm the town.[154] Having taken the town, Saladin does not appear to have used his artillery against the citadel, still without battlements since its slighting in 1190.

The Muslim and Frankish accounts loosely align, from the fighting beyond the walls and the eventual breach, to the notable number of Muslim trebuchets. Although the attackers' artillery is celebrated, it was the work of the miners that compromised the besieged defences, as had been the case at both Acre and Darum. This contributes to the difficulties facing an assessment of Saladin's artillery. The gathering of distant stones for ammunition can be interpreted in two ways: either that Saladin needed harder stones, such as limestone from the Judean-Samarian hills to the east, for counterweight trebuchets, which targeted the walls of Jaffa; or, that he was running out of field stones to feed his rapidly shooting traction trebuchets. A literal reading of the sources would suggest the former; however, Baha' al-Din's explicit statement that no stones were found around the town, and the use of these engines in conjunction with the frontal assaults supports the latter.

The period from 1187 to 1192 was an interesting and tumultuous one, documented by sources that were close to important figures and events. Their treatment of artillery over the five years between the battle of Hattin and the siege of Jaffa is enlightening. Although few mechanical descriptions are provided, the detailed week-to-week and day-to-day accounts reveal how these engines were used, although not the specifics of their designs. The apparent use of prefabricated artillery by Saladin and Richard I conveys a sense of value while Saladin's campaign through Syria appears to reveal certain limitations of these engines. The increasing use of what appear to be heavier engines in conjunction with mining, supplementing rather than replacing lighter antipersonnel engines, is revealed to be a popular tactic when the topography allowed.

At the end of the twelfth century, the heaviest artillery was still incapable of breaching masonry defences; however, there is evidence that it was being used to target fortifications, even if this was the thinner masonry of battlements. Defensively, light traction trebuchets were still popularly employed. Although most defensive machines appear to have been no more powerful than those

---

154 Baha' al-Din, *al-Nawadir*, trans. Richards, pp. 217-20. See also 'Imad al-Din, *al-Fath*, trans. Massé, pp. 384-85; Abu Shama, *Kitab al-Raudatain*, RHC Or 5, pp. 67-68.

used a century earlier, there are indications that heavier engines may also have been given a defensive function when available. The continued development of the heaviest engines would soon provide them with a definitively new role, entirely distinct from the antipersonnel function of lighter traction engines.

CHAPTER 6

Ayyubid Period

On average, the Franks were engaged in fewer sieges each year during the thirteenth century than they had been during the twelfth. The Third Crusade had allowed them to recover a small coastal strip of Palestine but little territory beyond this. While significantly weakened, the three remaining Frankish principalities endured, preserved more by the preoccupation of their Muslim neighbours than their own strength. Less than six months after Richard I had left the Holy Land, Saladin died and a sixty-year power struggle between the potentates of Cairo, Damascus and Aleppo ensued, with the Franks an occasional fourth party. The Franks also suffered from internal divisions; fractures evident in the late twelfth century developed, leading to a number of intra-Frankish sieges. Although artillery was probably used at many of the sieges undertaken by Muslim and Frankish forces in the first half of the thirteenth century, it often received little attention.

Events in the late twelfth century had a profound impact on the nature of siege warfare in the thirteenth century. In the decades leading up to the Third Crusade, the progressive consolidation of power by Nur al-Din and then Saladin allowed the Muslims to field increasingly large armies, forcing the Franks to rethink how they designed their fortifications. At the same time, the military orders, with their growing networks of resources that stretched across Europe, became increasingly responsible for defending the principalities' inland castles. Despite their wealth, their willingness to participate in most major campaigns and the regularity with which they raided neighbouring Muslim powers, the orders do not appear to have been powerful enough or had the desire to besiege any of the strongholds of their smaller neighbours.

Saladin's death in 1193 dealt Ayyubid authority a blow. His empire quickly fractured into a number of smaller principalities, which came to acknowledge the superiority, if only nominally, of the ruler of Egypt. Although this led to a number of sieges, these were conducted in the shadow of Saladin's legacy; accordingly, most accounts are briefer and less detailed than might be expected. The most emphasised campaigns of the first half of the thirteenth century were those involving significant European elements, namely the Fifth and Seventh Crusades, although certain Egyptian campaigns into Syria also drew considerable attention as the rifts between the Ayyubid powers widened and became more entrenched.

 | DOI 10.1163/9789004376922_008

The sources for this period present a challenge, as there are fewer staple contemporary narratives to rely on than there are for the half-centuries before and after. Ibn al-Athir's account ends abruptly in 1231, while Abu Shama's history becomes far less detailed following Saladin's death, at which point many of the sources that he was working from ended. Ibn Wasil provided a far more detailed contemporary account of events, which continues up to the end of the Ayyubid period; however, only parts of this have been translated from the Arabic. Similarly only fragments of the history written by Sibt Ibn al-Jawzi, who was based in Damascus during the first half of the thirteenth century, are available in translation.[1] Accordingly, it becomes necessary at times to look to the numerous histories composed under the Mamluks that relied heavily on these earlier authors.[2] The Frankish sources are slightly less interrupted, although often less detailed. The continuations of William of Tyre's history continue to provide critical narratives.[3] To these can be added Philip of Novara's contribution to the *Gestes des Chiprois*, particularly as it relates to the conflict between Frederick II and the baronial party of Jerusalem on Cyprus, and the brief entries that make up the *Annales de Terre Sainte*.[4] Often the most valuable Frankish sources are those provided by crusaders, including Oliver of Paderborn, James of Vitry and John of Joinville. Each brings a different perspective to events in the Holy Land and, perhaps more importantly, each man was well acquainted with contemporary European siege technologies.

The Fifth Crusade and New Latin Terminology

The first period of Frankish expansion in the aftermath of the Third Crusade coincided with a German crusade in 1197-98. There seem to be no indications that these forces used artillery when Sidon, Jubayl and Beirut were acquired

1 For a discussion of Abu Shama, Ibn Wasil, and Sibt Ibn al-Jawzi, their works and sources, see Little, "Historiography," pp. 418-2; Hirschler, *Medieval Arabic Historiography*, esp. pp. 10-14; Hirschler, "Ibn Wasil," pp. 136-60; Mallett, "Sibt Ibn al-Jawzi," pp. 84-108. See also Humphreys, *From Saladin to the Mongols*, pp. 393-99. My thanks to Rabei Khamisy for looking through the untranslated Arabic sources with me and helping me examine the references to artillery.

2 For these works, see Chapter 7.

3 For a discussion of these works, see *Rothelin/Eracles*, trans. Shirley, pp. 1-5. See also Edbury, "*Ernoul, Eracles* and the Fifth Crusade."

4 For discussions of these works, see Edbury, "Making Sense of the *Annales de Terre Sainte*," pp. 403-13; Templar of Tyre, *Gestes des Chiprios*, ed. and Trans. Minervini, pp. 1-20.

nor at the unsuccessful siege of Toron.[5] Similarly, there are few references to suggest that artillery was employed by al-ʿAdil when he took Jaffa in 1197, besieged Mardin in 1198 and 1203, attacked Crac des Chevaliers and Tripoli in 1207, or at the five Ayyubid sieges of Damascus between 1194 and 1201.[6] Despite the silence regarding artillery during this period, it was certainly employed during the Fifth Crusade.

Mount Tabor: 1217

Crusading armies under Andrew II of Hungary and Leopold VI of Austria arrived in the Levant in 1217. These forces, with Hugh I of Cyprus, John of Brienne, king-regent of Jerusalem, and sizable contingents of the military orders, launched three raids from Acre into Galilee that year. The second of these was directed towards Mount Tabor, which had recently been fortified by al-ʿAdil.[7] The exact details of this episode are not clear but it appears to have been more of an attack than a concerted siege: the garrison made a pre-emptive assault against their assailants and then held off a subsequent Frankish frontal attack. Sibt ibn al-Jawzi claims that the defenders threw Greek fire to hold off the Franks, but this was probably thrown by hand.[8] Although the *History of the*

5 James of Vitry, *Historia orientalis*, ed. Donnadieu, pp. 452-54, 464, trans. Stewart, pp. 111, 117-18; Arnold of Lubek, *Chronica Slavorum*, ed. Pertz, pp. 200-1, 205-7, 209-10; Wilbrand of Oldenburg, *Itinerarium terrae sanctae* 1.5, ed. Pringle, p. 118, trans. Pringle, p. 65; letter of the Duke of Lorraine to the Archbishop of Cologne (1197), ed. Pertz, *Annales Colonienses Maximi*, p. 805, trans. Munro, in *Translations and Reprints*, pp. 22-23; *Eracles* 27.6-7, 9, RHC Oc 2, pp. 224-26, 227; Roger of Howden, *Chronica*, ed. Stubbs, 4:28-29, trans. Riley, 2:406-7; Abu Shama, *Kitab al-Raudatain*, RHC Or 4, p. 462, RHC Or 5, pp. 117-18; al-Maqrizi, *al-Suluk*, trans. Broadhurst, pp. 102, 123; Philip of Novara, *Gestes des Chiprois* 127, ed. Raynaud, pp. 41-42, pp. 678-79, trans. La Monte, pp. 78-79; Deschamps, *Les Châteaux*, 2:225 n. 1; Müller-Wiener, *Castles of the Crusaders*, p. 70.

6 *Eracles* 27.3, RHC Oc 2, pp. 219-20; James of Vitry, *Historia orientalis*, ed. Donnadieu, pp. 464-66, trans. Stewart, p. 118; Roger of Howden, *Chronica*, ed. Stubbs, 4:25-26, trans. Riley, 2:404-5; Arnold of Lubek, *Chronica Slavorum*, ed. Pertz, p. 204; Abu Shama, *Kitab al-Raudatain*, RHC Or 5, pp. 118-19, 152; Ibn al-Athir, *al-Kamil*, trans. Richards, 3:67; al-Maqrizi, *al-Suluk*, trans. Broadhurst, pp. 118-19, 123-24, 144.

7 Ibn al-Athir, *al-Kamil*, trans. Richards, 3:79, 158; Abu Shama, *Kitab al-Raudatain*, RHC Or 5, pp. 156-57; al-Maqrizi, *al-Suluk*, trans. Broadhurst, pp. 153, 155-56; James of Vitry, *Historia orientalis*, ed. Donnadieu, p. 466, trans. Stewart p. 119 . Cf. Raphael, *Muslim Fortresses*, p. 17. For discussions of the castle, see Conder and Kitchener, *Survey of Western Palestine*, 1:390; Creswell, "Fortification in Islam," p. 123; Benvenisti, *The Crusaders*, pp. 358-62; Raphael, *Muslim Fortresses*, pp. 20-24.

8 Oliver of Paderborn, *Historia Damiatina* 2-4, ed. Hoogeweg, pp. 163-68, trans. Gavigan, pp. 53-56; James of Vitry, *Lettres* no. 3, ed. Huygens, pp. 98-99; *Eracles* 31.10-12, RHC Oc 2,

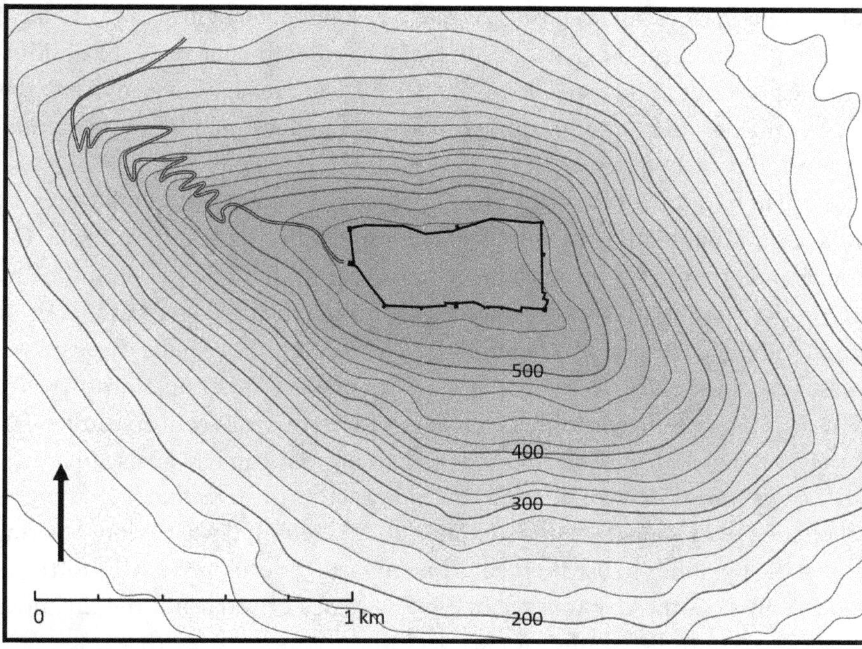

PLAN 21 *Mount Tabor*

Patriarchs mentions that the Franks prepared siege engines at Acre before setting out, Ibn al-Athir is the only source who mentions artillery, claiming it was used by the attackers.[9] The brief nature of the siege and challenges posed by the topography leave it doubtful that artillery was used; however, it was certainly employed against Damietta soon after.

Damietta: 1218-19

Tactically, Damietta offered many of the same options to a besieging force as did most cities along the Palestinian littoral: although its western front was protected by a branch of the Nile and Lake Manzala secured it to the east, it was easily approached from the south. These similarities, and the significance

pp. 321-25; Ernoul, *Chronique* 35, ed. Mas Latrie, pp. 411-12; *Cronica S Petri Erfordensis moderna*, ed. Holder-Egger, pp. 385-86; *Cronica Reinhardsbrunnensis*, ed. Holder-Egger, pp. 590-91; Ibn al-Athir, *al-Kamil*, trans. Richards, 3:174-75; Abu Shama, *Kitab al-Raudatain*, RHC Or 5, pp. 160-63; Abu'l-Fida', *al-Mukhtasar*, RHC Or 1, p. 88; al-Maqrizi, *al-Suluk*, trans. Broadhurst, pp. 164-65; *History of the Patriarchs*, trans. Khater and Khs-Burmester, 3.2:212-13. See also Marshall, *Warfare in the Latin East*, pp. 226, 228; Raphael, *Muslim Fortresses*, p. 18. Cf. Ibn al-'Amid, *al-Majmu*, trans. Eddé and Micheau, p. 18.

9 *History of the Patriarchs*, trans. Khater and Khs-Burmester, 4.1:39-40; Ibn al-Athir, *al-Kamil*, trans. Richards, 3:175.

of the 1218-19 siege, allowed Christopher Marshall to judge this siege as one of two that best reveal the dynamics of major thirteenth-century Frankish-Muslim sieges.[10] Not quite three decades after the last Frankish siege of Acre, this siege provides another opportunity to assess the development of contemporary artillery technology.

In March 1218, the crusaders who had remained in the Levant following the attack on Mount Tabor were joined by additional crusaders from Frisia, Cologne and other parts of Germany. With support from the local baronage and the military orders, this force set out by sea from Acre for Egypt. Advance units captured the port area of Damietta on 29 May and established a camp on the west side of the Nile before King John of Brienne and other notables arrived a few days later.[11] Since Amalric's campaigns in Egypt a half-century earlier, Damietta's defences had been strengthened. In 1181, Saladin had overseen repairs to the two towers that commanded the chain across the river and had strengthened the town's walls. Fifteen years later, al-'Adil reportedly considered tearing down the pyramids to use their cut stone to further develop the city's fortifications.[12] By 1218, the city was secured by two lines of walls and the infamous tower that controlled traffic on the river.[13]

The Franks landed on the west side of the mouth of the Damietta branch of the Nile and made camp upstream across from the city. The crusaders erected artillery on their bank of the river to provide cover for marine assaults against the Tower of the Chain, which was finally taken on 25 August 1218. A pause then ensued as the Franks expected the arrival of additional forces. Al-Kamil, who had positioned himself opposite the Franks on the east bank, likewise awaited the arrival of his father, al-'Adil. Although al-'Adil died on his way

---

10 Marshall, *Warfare in the Latin East*, pp. 250-53.

11 Oliver of Paderborn, *Historia Damiatina* 10, ed. Hoogeweg, pp. 175-78, trans. Gavigan, pp. 61-62. James of Vitry, *Lettres* nos. 3, 4, ed. Huygens, pp. 100, pp. 103-4. Cf. Ibn al-Athir, *al-Kamil*, trans. Richards, 3:176; Ibn al-'Amid, *al-Majmu*, trans. Eddé and Micheau, pp. 30-31; Abu Shama, *Kitab al-Raudatain*, RHC Or 5, p. 153; al-Maqrizi, *al-Suluk*, trans. Broadhurst, p. 166; *History of the Patriarchs*, trans. Khater and Khs-Burmester, 3.2:216.

12 Al-Maqrizi, *al-Suluk*, trans. Broadhurst, pp. 122, 144-46. These fears were not entirely unfounded as the Nile delta had been raided by Frankish forces earlier in the thirteenth century, Abu Shama, *Kitab al-Raudatain*, RHC Or 5, p. 153; al-Maqrizi, *al-Suluk*, trans. Broadhurst, p. 146; *History of the Patriarchs*, trans. Khater and Khs-Burmester, 3.2:193-95.

13 For contemporary accounts of this tower, see William of Tyre, *Chronicon* 20.15, ed. Huygens, 2:929, trans. Babcock and Krey, 2:363; Ibn al-Athir, *al-Kamil*, trans. Richards, 3:176; al-Maqrizi, *al-Suluk*, trans. Broadhurst, p. 166. See also Powell, *Anatomy of a Crusade*, p. 141.

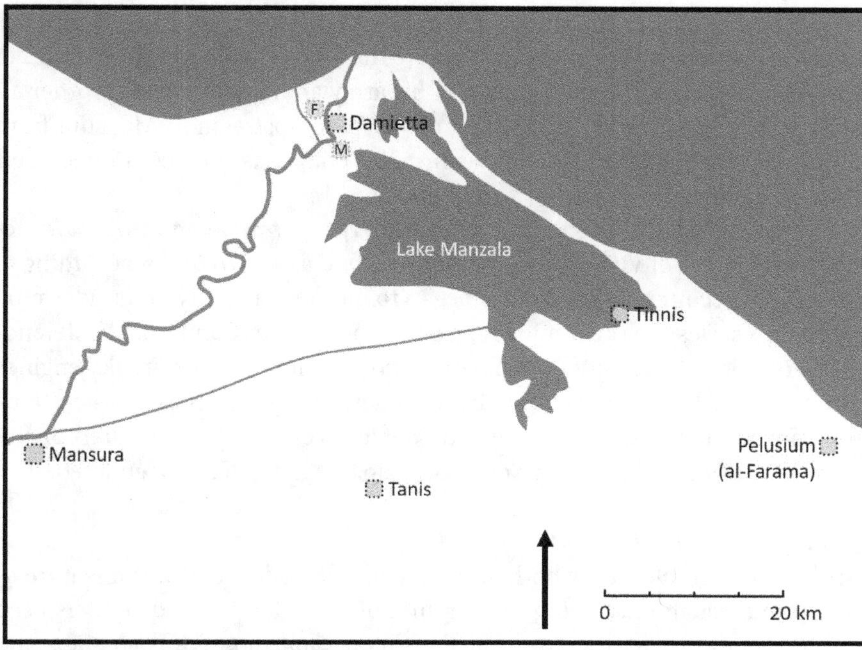

PLAN 22 *Damietta region (present shorelines)*

to support his son, al-Kamil would receive assistance from his brother, al-Mu'azzam 'Isa.[14]

To prevent the Franks from crossing the river, the Muslims had fortified the right bank south of Damietta. An earthen rampart was raised and ships and stakes were sunk in the river to further impede any attempt to cross or sail up the Nile. A number of sources also claim that these defences were supported by artillery.[15] The Franks avoided these obstacles by developing the narrow al-Azraq channel, but they found the right bank undefended when they crossed

14 Oliver of Paderborn, *Historia Damiatina* 11-14, 16, 18, 21, ed. Hoogeweg, pp. 179-91, 194-96 trans. Gavigan, pp. 63-72; James of Vitry, *Lettres* nos. 3, 4, 5, ed. Huygens, pp. 100, 105-8, 109, 114-16; Roger of Wendover, *Flores historiarum*, ed. Hewlett 2:229-34, trans. Giles 2:406-10; *Cronica S Petri Erfordensis moderna*, ed. Holder-Egger, pp. 386-87; *Cronica Reinhardsbrunnensis*, ed. Holder-Egger, p. 592; Ibn al-Athir, *al-Kamil*, trans. Richards, 3:176-77; Abu Shama, *Kitab al-Raudatain*, RHC Or 5, pp. 167-68; al-Maqrizi, *al-Suluk*, trans. Broadhurst, pp. 166-67; Ibn al-'Amid, *al-Majmu*, trans. Eddé and Micheau, p. 26; *History of the Patriarchs*, trans. Khater and Khs-Burmester, 3.2:215-19.

15 Oliver of Paderborn, *Historia Damiatina* 22, ed. Hoogeweg, p. 196, trans. Gavigan, pp. 73-74; James of Vitry, *Lettres* no. 5, ed. Huygens, p. 114; Roger of Wendover, *Flores historiarum*, ed. Hewlett, 2:238-39, trans. Giles, 2:415. For the river defences, see also Ibn al-Athir, *al-Kamil*, trans. Richards, 3:176-77; al-Maqrizi, *al-Suluk*, trans. Broadhurst, p. 172.

on 5 February 1219. Al-Kamil had fled in fear of a plot against him, leading his forces to abandon their camp hours before the Franks made the crossing.[16] Despite this, al-Kamil retained control of his army and, joined by his brother al-Mu'azzam of Damascus, established a new base upstream at Mansura from which he repeatedly sent forces against the Franks as they continued their siege of Damietta.[17]

On the right bank, the Franks built artillery – *petrariae* and *trabucula*, according to James of Vitry, and *perrieres*, *mangoniaus* and *trebuces*, according to the Ernoul account[18] – and efforts began to blockade the city and undermine its defences. Despite the diminishing provisions within Damietta, the defenders were able to launch a successful sally and burn many of the Franks' engines at one point.[19] Even without their machines, the Franks pressed the attack with fire and ladders, gaining the outer defences of the city at the start of November. A few days later, on 5 November 1219, the defenders surrendered.[20]

Artillery at Damietta

Artillery was employed by both attackers and defenders and was a feature of the siege throughout. During the summer of 1218, Damietta's defenders used trebuchets to protect the Tower of the Chain, damaging Frankish ships and ladders as they attacked it.[21] Oliver of Paderborn specifies that at least six

---

16 James of Vitry, *Lettres* no. 5, ed. Huygens, p. 118; Ernoul, *Chronique* 36, ed. Mas Latrie, pp. 418-20; *History of the Patriarchs*, trans. Khater and Khs-Burmester, 4.1:54-55; Ibn al-Athir, *al-Kamil*, trans. Richards, 3:177; al-Maqrizi, *al-Suluk*, trans. Broadhurst, pp. 173-74; Ibn al-'Amid, *al-Majmu*, trans. Eddé and Micheau, p. 33.

17 Oliver of Paderborn, *Historia Damiatina* 23, 25, 27, 29-31, ed. Hoogeweg, pp. 201-2, 205-7, 211-13, 213-24, trans. Gavigan, pp. 76, 78, 179-80, 81-84-86; James of Vitry, *Lettres* no. 5, ed. Huygens, pp. 118-19; al-Maqrizi, *al-Suluk*, trans. Broadhurst, pp. 174-75.

18 James of Vitry, *Lettres* no. 5, ed. Huygens, p. 119; Ernoul, *Chronique* 36, ed. Mas Latrie, p. 424. *Petrariarum* and *trebuculorum*) according to Roger of Wendover, *Flores historiarum*, ed. Hewlett, 2:243, trans. Giles 2:418.

19 Oliver of Paderborn, *Historia Damiatina* 28, ed. Hoogeweg, p. 211 trans. Gavigan, p. 80; *Cronica Reinhardsbrunnensis*, ed. Holder-Egger, p. 593.

20 James of Vitry, *Lettres* no. 6, ed. Huygens, pp. 125-26; Oliver of Paderborn, *Historia Damiatina* 32, ed. Hoogeweg, pp. 224-25, trans. Gavigan, pp. 86-87; Abu Shama, *Kitab al-Raudatain*, RHC Or 5, pp. 176-77; Ibn al-'Amid, *al-Majmu*, trans. Eddé and Micheau, p. 31. Cf. Roger of Wendover, *Flores historiarum*, ed. Hewlett, 2:248, trans. Giles 2:423; Marino Sanudo, *Liber secretorum* 3.11.8, ed. Bongars, p. 208, trans. Lock, p. 331; Abu'l-Fida', *al-Mukhtasar*, RHC Or 1, p. 91; al-Maqrizi, *al-Suluk*, trans. Broadhurst, p. 179. For complete accounts of this siege, see Oliver of Paderborn, *Historia Damiatina* 10-32, ed. Hoogeweg, pp. 175-226; James of Vitry, *Lettres* nos. 3-6, ed. Huygens, pp. 100, 103-10, 114-32; Ernoul, *Chronique* 36, ed. Mas Latrie, pp. 415-26; *Eracles* 31.14-32.14, RHC Oc 2, pp. 326-46.

21 Oliver of Paderborn, *Historia Damiatina* 11, ed. Hoogeweg, p. 181, trans. Gavigan, p. 64; James of Vitry, *Lettres* no. 4, ed. Huygens, pp. 105-6.

machine, which were mounted on towers, were used in this way but that the most powerful one broke after just a few shots.[22] These do not appear to have been particularly strong engines as netting hung in front of the ship-borne siege towers, which carried their own *lapidum iactatores*, was sufficient to resist their projectiles.[23] The defenders appear to have employed a broader range of artillery than just these machines, as the Franks found "four *trabucculi* with *petrariae* and many *mangonelli*" when they entered the city after the siege.[24] Oliver of Paderborn states that eight *mangonelli*, presumably the same as those discovered afterwards, were directed against the crusaders' camp in 1219, but he does not quantify the defenders' *petrariae*.

Oliver's selective use of terminology suggests that he categorised the defenders' artillery. The new term, *trabucculi*,[25] was most likely used to refer to counterweight trebuchets. These would have been used to target the Frankish siege engines and other stationary targets, while the *petrariae* and *mangonelli*, probably traction trebuchets, would have been more useful against ships and personnel. It is hard to gauge the strength of the defenders' strongest engines as their artillery is rarely emphasised in the surviving sources. Although stones are noted inflicting casualties and minor damage, it is often unclear whether these were thrown mechanically or by hand. This lack of emphasis seems to indicate that these defensive engines were either fairly ineffective or not particularly powerful.

Oliver of Paderborn and James of Vitry claim that the Franks made use of their own *trabuculi*. Taking the bombardment of the Tower of the Chain as an example, the influence of their accounts and their terminology is evident in subsequent works, although the traditional *mangoneaus* is found in the *Eracles* account in this instance:

Oliver of Paderborn	Non vero considerantes, turrim capi non posse petrariarum vel trabuculorum ictibus (hoc enim multis diebus fuit attemptatem), nec applicatione castri propter fluminis profunditatem, neque fame propter civitatis vicinitatem, neque suffossione propter circumfluentis aque importunitatem,[26]

22 Oliver of Paderborn, *Historia Damiatina* 13, ed. Hoogeweg, p. 183, trans. Gavigan, p. 66. See also Roger of Wendover, *Flores historiarum*, ed. Hewlett, 2:231-32, trans. Giles, 2:408-9.

23 This bears similarities to the siege of Spanish Tortosa in 1147, see Chapter 4.

24 Oliver of Paderborn, *Historia Damiatina* 39, ed. Hoogeweg, p. 239, trans. Gavigan, p. 96.

25 For the appearance of this term, see Chapter 1.

26 Oliver of Paderborn, *Historia Damiatina* 12, ed. Hoogeweg, p. 181.

James of Vitry	Mansimus autem in insula predicta IIII mensibus detenti in expugnatione cuiusdam turris mire fortitudinis, que nec petrariis nec instrumentis que trabucula dicuntur poterat superari nec a parte inferiori suffodi eo, quod in medio Nili fluminis inter insulam et civitatem sita erat,[27]
Roger of Wendover	Principes interea militiae Crucifixi de captione turris nimis solliciti viderunt eam capi non posse fame, propter civitatis vicinitatem; neque suffossione, propter aquae circumfluentis importunitatem; neque insultu petrariarum aut trebuculorum ictibus, quia, cum id multis esset diebus attentatum, parum vel nihil profecerunt.[28]
Eracles	Quant li Crestien si furent herbergé et atiré, si entendirent a faire assaillir la tor de la Cosbarie, si drecerent perrieres et mangoneaus et les firet geter a cele tor,[29]
Cronica Reinhardsbrunnensis	Cristiani vero considerantes turrim non posse capi petrariorum et tribuculorum ictibus neque fame propter vicinitatem civitatis,[30]

While the Frankish artillery was insufficiently powerful to threaten the defensibility of the Tower of the Chain, these same engines would appear to be the *machinae* responsible for damaging the bridge that connected the tower to the city.[31] Like Oliver of Paderborn's classification of the defending artillery, James of Vitry also seems to present the *trabuculi* as the heaviest engines and the *petrariae* as a lighter sort, which inflicted casualties among the Muslims when shot across the river.[32]

---

27 James of Vitry, *Lettres* no. 4, ed. Huygens, p. 105.

28 Roger of Wendover, *Flores historiarum*, ed. Hewlett, 2:229.

29 *Eracles* 31.14, RHC Oc 2, p. 327. Cf. Ernoul, *Chronique* 36, ed. Mas Latrie, p. 416.

30 *Cronica Reinhardsbrunnensis*, ed. Holder-Egger, p. 592.

31 Oliver of Paderborn, *Historia Damiatina* 12, ed. Hoogeweg, p. 182, trans. Gavigan, p. 65.

32 *… bellicis machinis, que petrarie dicuntur, lapides iaculando quosdam ex nostris interficientibus*, James of Vitry, *Lettres* no. 6, ed. Huygens, p. 130.

The heavier Frankish engines were apparently able to damage at least one of Damietta's twenty-eight outer towers in 1219. Oliver of Paderborn attributes this to the *trabuccus* of the duke of Austria while Roger of Wendover credits the *trebuculus* of the Templars.[33] The *Eracles* account states that John of Brienne erected one *trabuchet*, the Hospitallers another and a third was held communally, while many *perrieres* and *mangoneaus* were placed around the city; but among these, it was the Hospitallers' *trabuchet* that inflicted notable damage.[34] These accounts again suggest that the heavier *trabuculi*, distinct from the lighter *petrariae* and *mangonelli*, were newer counterweight engines. The latter were probably traditional traction trebuchets, although it is possible that some were smaller counterweight engines. Despite the specificity shown by some sources, this was not embraced by all. Unlike his contemporaries, the author of this part of the Ernoul account appears to have had little interest in artillery, mentioning none during the attacks against the Tower of the Chain and very generally grouping the Franks' use of *perrieres*, *mangoniaus* and *trebucés* during the 1219 phase of the siege.[35]

The significance and limitations of the attacking artillery are identified by the Frankish sources: although the heaviest trebuchets featured here appear to have been at least as powerful as any used in the region beforehand, none of the sources suggests that the damage inflicted by these engines led to the fall of the city.[36] For up to seven months these machines bombarded sections of Damietta's landward defences in 1219, yet no significant breaches were opened. Even the heaviest contemporary trebuchets probably targeted the battlements first, lowering their aim to the bulk of the wall below the parapet only after the wall-walk was no longer defensible in that section. Although these engines were now being used to damage and weaken towers and curtain walls, they were not yet powerful enough to compromise the structural integrity of these fortifications. Meanwhile, traction trebuchets continued to provide an important antipersonnel function. While the Frankish and Muslim sources generally agree in their assessment of the artillery used at Damietta, the *History of the Patriarchs* provides a series of strange details.

33 Oliver of Paderborn, *Historia Damiatina* 38, ed. Hoogeweg, pp. 237-38, trans. Gavigan, p. 94; Roger of Wendover, *Flores historiarum*, ed. Hewlett, 2:250, trans. Giles, 2:424-25.

34 *Eracles* 32.8, 14, RHC Oc 2, pp. 337-38, 345. Cf. *Chronica de Mailros*, ed. Stevenson, p. 136; Marino Sanudo, *Liber secretorum* 3.11.8, ed. Bongars, p. 208, trans. Lock, p. 331.

35 Ernoul, *Chronique* 36, ed. Mas Latrie, p. 421. For an examination of the Ernoul account and its relation with other accounts, see Edbury, "*Ernoul, Eracles* and the Fifth Crusade."

36 Roger of Wendover states this outright, Roger of Wendover, *Flores historiarum*, ed. Hewlett, 2:243, trans. Giles, 2:418. See also Marshall, *Warfare in the Latin East*, pp. 226-29.

The Coptic Account

According to the section of the continuation of Sawirus ibn al-Muqaffa's chronicle that deals with the life of John VI (Coptic pope of Alexandria, 1189-1216), the Franks erected a large trebuchet in their fortified camp on the west side of the river in 1218. This was destroyed by another stone-thrower erected by the defenders, leading the Franks to build an even larger trebuchet as well as four others. The largest of these targeted the city while the smaller ones shot at the Tower of the Chain. The author gives a sense of the scale of the smaller engines when stating that these proved incapable of reaching the tower. Despite their apparent weakness, the Franks built a few more during this phase of the siege.[37] None of this is particularly strange and certain parallels can be drawn with the eyewitness accounts; however, significant issues emerge when further details are considered.

In a later anecdote, the smaller trebuchets, which could not reach the Tower of the Chain, were used to throw the heads of dead Muslims into Damietta from across the river. Despite the apparent contradiction, these engines are still portrayed as fairly light compared to the Franks' larger trebuchet. This engine is described as having a lead box, weighing 2 Syrian *qintars* (370 kg), at its 'head' and throwing projectiles weighing 1 Syrian *qintar* (185 kg).[38] 600 men were also beneath the engine providing pulling power. These details are hard to reconcile.

If this large engine was a counterweight trebuchet, it is difficult to imagine that 600 people were required to pull its beam back against the weight of an implicitly enormous counterweight, let alone one less than 400 kg. If this was instead a traction trebuchet, it would be mechanically impractical to have 60 pullers, let alone 600, powering the engine.[39] While a liberal interpretation of the text might lead to the suggestion that this was a counterweight trebuchet with a power source comparable to the traction force of 600 pullers, it is hard to reconcile this with the stated mass of the counterweight, which is equivalent to the force of only about 6 pullers. Setting aside the impractically large pulling crew and assuming that this was a counterweight trebuchet, the disparity between the mass of the counterpoise and the mass of the projectiles is illogical. The mechanics of a counterweight trebuchet generally require the

37 *History of the Patriarchs*, trans. Khater and Khs-Burmester, 3.2:216-21.

38 *History of the Patriarchs*, trans. Khater and Khs-Burmester, 3.2:215-21.

39 Paul Chevedden has used this as part of a theory that this was a 'hybrid' engine, Chevedden, "The Hybrid Trebuchet," pp. 186-87. Cf. Fulton, "The Myth of the Hybrid Trebuchet."

counterweight to be at least ten times the mass of the intended projectile in order to shoot with any kind of success, while a disparity of around 100 times will typically allow for a fairly efficient shooting sequence and one of 1,000 times will yield a long-range shot. Thus even if the exceptional size of the projectiles is accepted, for which there appears to be no known precedent at this date, it remains hard to determine which elements of this description can be trusted as each seems to contradict the practicality of another. The description's abnormalities appear to suggest that the author had a vague appreciation of trebuchet technology, but only a superficial understanding of the mechanics. The author also seems to reveal a tendency to exaggerate, apparent in the suggested 600-member crew and elsewhere, such as his assertion that every nail used to construct the crusaders' naval siege tower, which was used against the Tower of the Chain, was 1.5-2 cubits long.[40]

In a subsequent portion of the *History of the Patriarchs*, which deals with the reign of Cyril III, who succeeded John VI after a nineteen-year interregnum, a clearer account of the siege is provided. Probably the work of a different author, the events of 1218 are recapped and the narrative of the siege is carried forward through 1219. According to this section, the Franks set up eight trebuchets against the Tower of the Chain. These are presented as providing a continuous bombardment alongside archers, pausing to allow for attacks, while some of their projectiles reportedly reached the centre of the city.[41] At least some of the Frankish engines are also described as gigantic, although the size of the archers' arrows is also emphasised. Intriguingly, it is stated that one of the Frankish stones was weighed and found to have a mass of 350 Egyptian *artal* (153 kg). The similarity of this figure to that provided in the earlier account might suggest that this is a fairly accurate measurement; alternatively, it might be the later author's conversion of the figure provided by his predecessor, giving the figure in Egyptian *artal* rather than as a Syrian *qintar*. Regardless, the later author provides a more detailed and tempered account of the siege, omitting many of the issues and problematic exaggerations found in his predecessor's version, while still impressing the significant scale of the engines employed during the siege. When describing the events of 1219, this later section of the *History* states that the Franks found artillery in the abandoned Muslim camp after crossing the river, apparently confirming that these were integrated into the Muslims' defence of the right bank. It is recorded that the Franks

40 *History of the Patriarchs*, trans. Khater and Khs-Burmester, 3.2:219.

41 *History of the Patriarchs*, trans. Khater and Khs-Burmester, 4.1:43-47.

subsequently erected artillery against the city but these engines are not mentioned thereafter.[42]

The Push to Mansura: 1221

After fortifying Damietta, the Franks marched south towards Mansura, where al-Kamil had taken up a position on the right bank of the Damietta branch just south of where the Ushmum canal (al-Bahr as-Saghir) diverts towards Tinnis to the northeast.[43] The Franks arrived at the fork in the river before the end of July 1221 and established a camp surrounded by earthworks. According to Oliver of Paderborn and al-Maqrizi, they also erected artillery.[44] Together with the Muslim army's use of artillery to defend the right bank of the river south of Damietta, these appear to be the first examples of artillery being used to defend fairly improvised field fortifications during the period of the crusades. Although plausible, such references should be regarded with caution, as the more parabolic trajectory and difficulties associated with yawing most trebuchets made them better suited to siege scenarios than more mobile forms of warfare.[45] However, Oliver of Paderborn's use of differing terms, Muslim *mangonelli* south of Damietta and Frankish *petrariae* at Mansura, and al-Maqrizi's notice of artillery in both instances appear to lend credence to this episode.[46] In both instances, the engines might have been used against improvised defences used by the other side.

The Frankish engines saw little action at Mansura. Reminiscent of their inaction through 1220, the Franks' did little to advance their progress towards Cairo or stem the influx of Syrian reinforcements joining al-Kamil. As their adversary's strength continued to grow, the Franks were eventually compelled to burn their camp on the night of 26 August and flee towards Damietta. The disorganised withdrawal allowed the Muslims to surround them, leading to

---

42 *History of the Patriarchs*, trans. Khater and Khs-Burmester, 4.1:53-67.

43 Oliver of Paderborn, *Historia Damiatina* 39, 40, ed. Hoogeweg, pp. 240-41, trans. Gavigan, pp. 96-97; James of Vitry, *Lettres* nos. 6, 7, ed. Huygens, pp. 128, 134-35, 139-40; al-Maqrizi, *al-Suluk*, trans. Broadhurst, p. 179; Pringle, "A Castle in the Sand," p. 190.

44 Oliver of Paderborn, *Historia Damiatina* 72, ed. Hoogeweg, p. 269, trans. Gavigan, p. 125; al-Maqrizi, *al-Suluk*, trans. Broadhurst, p. 185. For the movement of the Franks, see Oliver of Paderborn, *Historia Damiatina* 57-59, ed. Hoogeweg, pp. 259-61, trans. Gavigan, pp. 114-16; Ibn al-'Amid, *al-Majmu*, trans. Eddé and Micheau, pp. 31-32; al-Maqrizi, *al-Suluk*, trans. Broadhurst, pp. 179-80.

45 Cf. Marino Sanudo, *Liber secretorum* 2.4.22, ed. Bongars, p. 79, trans. Lock, pp. 134-35.

46 Oliver of Paderborn, *Historia Damiatina* 30, 72, ed. Hoogeweg, pp. 219, 269, trans. Gavigan, pp. 84, 125; al-Maqrizi, *al-Suluk*, trans. Broadhurst, pp. 185.

their surrender and the eventual return of Damietta in exchange for their lives.[47]

'Atlit and Caesarea: 1218-20

While considerable numbers of crusaders and eastern Franks were occupied with the siege of Damietta, forces from Damascus took the opportunity to attack Caesarea and the newly constructed Templar castle at 'Atlit.[48] According to James of Vitry and the *Eracles* account, al-Mu'azzam 'Isa besieged Caesarea and 'Atlit in 1218 before making for Egypt to support his brother.[49] While James of Vitry characterises the attack on Caesarea as more of a sack, the *Eracles* account states that three *perrieres* were used, implying an additional facet of complexity. Oliver of Paderborn, however, states that al-Mu'azzam 'Isa took Caesarea following his return from Egypt around March 1219, destroying it before going on to besiege 'Atlit.[50] The similarities suggest that the sources are describing the same set of events, although the correct year is unclear as al-Mu'azzam 'Isa briefly returned to Egypt in the autumn of 1219.[51]

47 Oliver of Paderborn, *Historia Damiatina* 71, 74-78, ed. Hoogeweg, pp. 267-69, 270-75, trans. Gavigan, pp. 123-24, 126-31; Ibn al-'Amid, *al-Majmu*, trans. Eddé and Micheau, pp. 33-34; Ibn al-Athir, *al-Kamil*, trans. Richards, 3:181; Abu Shama, *Kitab al-Raudatain*, RHC Or 5, p. 184; al-Maqrizi, *al-Suluk*, trans. Broadhurst, p. 181, 184-87.

48 For the fortification of these sites, see Oliver of Paderborn, *Historia Damiatina* 5-6, 26, ed. Hoogeweg, pp. 168-72, 207-11, trans. Gavigan, pp. 56-58, 78-79; James of Vitry, *Lettres* no. 3, ed. Huygens, p. 99; Roger of Wendover, *Flores historiarum*, ed. Hewlett, 2:261, trans. Giles, 2:434; *Cronica S Petri Erfordensis moderna*, ed. Holder-Egger, p. 386. For descriptions and discussions of 'Atlit castle, see Oliver of Paderborn, *Historia Damiatina* 6, ed. Hoogeweg, pp. 169-72, 290-91, trans. Gavigan, pp. 57-58, and letter no. 3 (1218), ed. Hoogeweg, pp. 290-91; Conder and Kitchener, *Survey of Western Palestine*, 1:293-96; Rey, *Étude sur les monuments*, pp. 97-99; Deschamps, *Les Châteaux*, 2:24-34; Johns, *Guide to 'Atlit*; Johns, "Excavations, (1932)," pp. 145-64; Johns, "Excavations: Faubourg," pp. 111-29; Boase, "Military Architecture," pp. 157-58; Benvenisti, *The Crusaders*, pp. 176-78; Pringle, "Town Defences," pp. 91-92; Kennedy, *Crusader Castles*, pp. 125-26; Boas, *Crusader Archaeology*, pp. 110-12. For the limited fortification of Caesarea see, Oliver of Paderborn, *Historia Damiatina* 5, ed. Hoogeweg, pp. 168-69, trans. Gavigan, p. 56. See also Wilbrand of Oldenburg, *Itinerarium terrae sanctae* 2.2, ed. Pringle, pp. 131-32, trans. Pringle, p. 86.

49 James of Vitry, *Lettres* no. 4, ed. Huygens, pp. 101-2; *Eracles* 32.5-6, RHC Oc 2, pp. 334-36 (lower text). Cf. Johns, *Guide to 'Atlit*, pp. 18, 20, and notes on pp. 1-2.

50 Oliver of Paderborn, *Historia Damiatina* 41, ed. Hoogeweg, pp. 244-45, trans. Gavigan, p. 99. See also Conder and Kitchener, *Survey of Western Palestine*, 1:299.

51 For a secondary account of al-Mu'azzam 'Isa's movements, see Humphries, *From Saladin to the Mongols*, pp. 162-66.

Al-Mu'azzam returned to 'Atlit in October/November 1220. On this occasion, according to Oliver of Paderborn, he fortified his position and erected artillery:

> He set up one *trabuculus*, three *petrariae* and four *mangonelli*, and harassed the fortification night and day by blows of the machines. However, he could not move one stone from its place in the new towers and the middle wall. But the *trabuculus* of the [castle], with a *petraria* and a *mangonellus* placed next to it, battered and broke the *trabuculus* and the *petraria* of the enemy.[52]

The mention of only one *trabuculum* on each side suggest that these were the most exceptional engines, probably counterweight trebuchets, but what differentiated the *petrariae* and *mangonelli* is once again more difficult to explain. The attacking engines appear to have had little impact on 'Atlit's thick but soft walls. Hugh Kennedy has concluded that the Ayyubid artillery was used against the castle's outer wall, unable to reach the inner line,[53] but the above account more accurately describes the castle's inner line of defences, specifically its two towers and connecting wall (*turribus novis ac muro medio*). This suggests that the castle's outer curtain may not have been built at this point or that Oliver of Paderborn, then in Egypt, was unaware of its completion.

The engines used at 'Atlit appear to have been no more powerful than those concurrently used at Damietta, although the accuracy of the defenders' engines is noteworthy. The castle's passive strength and some 300 crossbowmen were sufficient to hold back the besiegers until reinforcements began to arrive by sea.[54] Unable to overwhelm the castle with a frontal assault or weaken its defences with his artillery, al-Mu'azzam 'Isa cut his losses and withdrew before the end of November.[55]

---

52 *Coradinus obsidione fimata metuens excursum castensium fossatum fieri iussit inter castrum et sua tentoria, erigens trabuculum unum, petrarias tres, mangonellos quatuor, diebus ac noctibus ictibus machinarum munitionem infestans, sed de turribus novis ac muro medio nec unum lapidem de suo loco movere valuit. Trabuculus autem castri cum petraria iuxta se posita ac mangonello trabuculum hostium et petrariam concutiens fregit.* Oliver of Paderborn, *Historia Damiatina* 52-53, ed. Hoogeweg, pp. 254-55, trans. Gavigan, pp. 108-9. The vagueness in Oliver of Paderborn's account is excusable given that he was in Egypt during these events. Cf. Abu'l-Fida', *al-Mukhtasar*, RHC Or 1, p. 94; Richard of Morins, *Annales Prioratus de Dunstaplia*, ed. Luard, p. 63.

53 Kennedy, *Crusader Castles*, p. 127.

54 Oliver of Paderborn, *Historia Damiatina* 53, ed. Hoogeweg, p. 256, trans. Gavigan, p. 110.

55 Abu'l-Fida' claims that al-Mu'azzam 'Isa went on to besiege Caesarea after departing 'Atlit in 1220, adding to the confusion surrounding these campaigns in 1218-20, Abu'l-Fida',

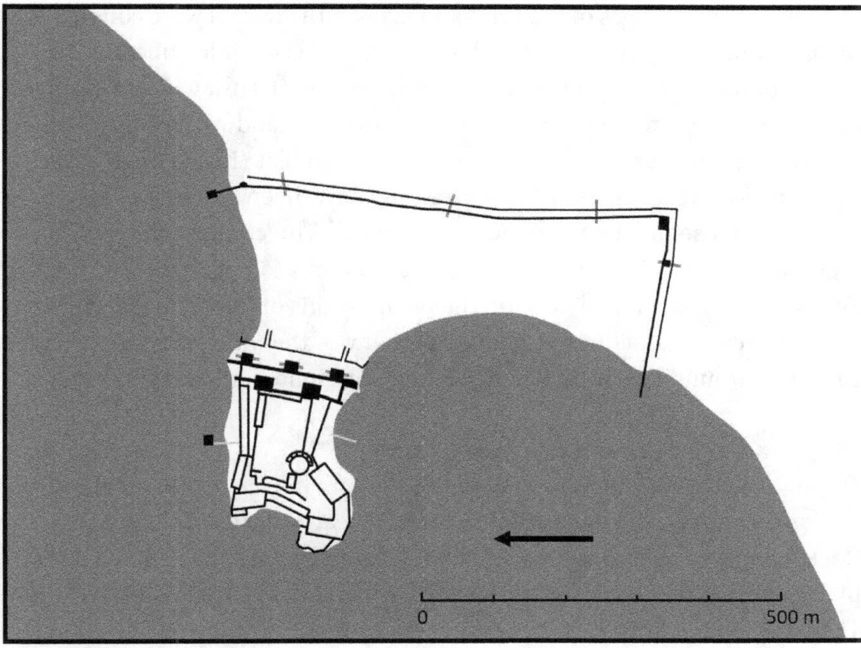

PLAN 23 *'Atlit*

The Sixth Crusade and War on Cyprus

During the Fifth Crusade, with the concentration of Frankish forces in Egypt, al-Mu'azzam 'Isa opted to slight many of the fortifications of interior Palestine, fearing this area and its strongholds might be recaptured by the Franks or returned to them through a negotiated peace deal. This campaign of destruction extended to the town walls of Jerusalem as well as the strongholds of Toron, Safed, Belvoir, and Mount Tabor, which had been built by al-'Adil only a decade earlier.[56] While this would have future consequences, the Muslims turned their attentions away from the Franks as the threat posed by the armies of the Fifth Crusade dissipated.

---

al-Mukhtasar, RHC Or 1, p. 94. Cf. Abu'l-Musaffar Sibt al-Djauzi, in Abu Shama, *Kitab al-Raudatain*, RHC Or 5, p. 178.

56 Oliver of Paderborn, *Historia Damiatina* 24, 42, 52, ed. Hoogeweg, pp. 203, 245, 254, trans. Gavigan, pp. 76-77, 100-1, 108; *Eracles* 32.10, RHC Oc 2, pp. 339-40; Ernoul, *Chronique* 36, ed. Mas Latrie, p. 417; Abu Shama, *Kitab al-Raudatain*, RHC Or 5, pp. 141, 165-66, 173-74; Abu'l-Fida', *al-Mukhtasar*, RHC Or 1, p. 91; al-Maqrizi, *al-Suluk*, trans. Broadhurst, p. 181.

Most recorded sieges of the 1220s and 1230s in the Levant were conducted by Muslim armies against the large urban centres of Syria. Unfortunately, detailed descriptions of artillery in sources that have been translated into European languages are scarce during this period. Notwithstanding, artillery was evidently significant at the Khwarizmian siege of Akhlat (Khilat) led by Jalal al-Din Mankubirti in 1229 and it appears to have been a notable weapon at al-Kamil's subsequent siege of Amida in 1232.[57] The continued Ayyubid conflict and arrival of Frederick II allowed the Franks to negotiate the return of Toron and Jerusalem and develop the fortifications of Jaffa and Sidon.[58] Frederick, however, was a divisive figure and many of the most informative Frankish sieges around 1230 were fought between his opponents and supporters.

Kantara, Beirut and Kyrenia: 1229-33

During the siege of the Cypriot castle of Kantara in 1229-30, the Ibelin besieging force employed a *grant trabuc* built by Anceau of Brie.[59] The topography of the site suggests that this engine would most likely have been erected on the plateau below the heavily fortified northern end of the castle's eastern front. This position is relatively close to the castle but significantly lower, limiting the amount of energy that a projectile could have transferred at the point of impact. Despite these difficulties, the engine itself was apparently so impressive that John of Beirut left the siege of Deidamour (St Hilarion) to view it.[60] More numerous engines were used soon afterwards at Beirut.

---

57 Bar Hebraeus, *Makhtebhanuth Zabhne*, trans. Budge, 1:394-95; Ibn al-'Amid, *al-Majmu*, trans. Eddé and Micheau, pp. 44-47; al-Maqrizi, *al-Suluk*, trans. Broadhurst, pp. 211, 213-14, 216-18. For the use of artillery at Akhlat, see also Khamisy and Fulton, "Manjaniq Qarabugha," p. 180.

58 Abu Shama, *Kitab al-Raudatain*, RHC Or 5, p. 186; Ibn al-'Amid, *al-Majmu*, trans. Eddé and Micheau, pp. 38-41; al-Maqrizi, *al-Suluk*, trans. Broadhurst, pp. 198-207. See also Bartlett, *Islam's War*, pp. 215-17. For the broader preoccupation of the Muslims, see also Abu Shama, *Kitab al-Raudatain*, RHC Or 5, pp. 190-91; Ibn al-'Amid, *al-Majmu*, trans. Eddé and Micheau, pp. 42-43; al-Maqrizi, *al-Suluk*, trans. Broadhurst, pp. 210-11. For the fortification of Jaffa, see Philip of Novara, *Gestes des Chiprois* 157, ed. Raynaud, p. 77; *Annales de Terre Sainte*, ed. Röhricht and Raynaud, A and B p. 14 (438), ed. Edbury, pp. 152-53; Roger of Wendover, *Flores historiarum*, ed. 2:365-69, trans. Giles, 2:522-24. See also Pringle, "Town Defences," p. 94. For the fortification of Sidon, see *Eracles* 32.25, RHC Oc 2, p. 365; Marino Sanudo, *Liber secretorum* 3.11.10, ed. Bongars, p. 211, trans. Lock, p. 335; al-Maqrizi, *al-Suluk*, trans. Broadhurst, pp. 204-5. See also Kalayan, "The Sea Castle of Sidon," pp. 81-82; Mesqui, "La fortification des Croisés," p. 11; Deschamps, *Les Châteaux*, 2:229-32.

59 Philip of Novara, *Gestes des Chiprois* 148, ed. Raynaud, p. 63, trans. La Monte, p. 105.

60 Philip of Novara, *Gestes des Chiprois* 148-51, ed. Raynaud, pp. 62-67, trans. La Monte, pp. 105-7. See also *Eracles* 33.11, RHC Oc 2, p. 377.

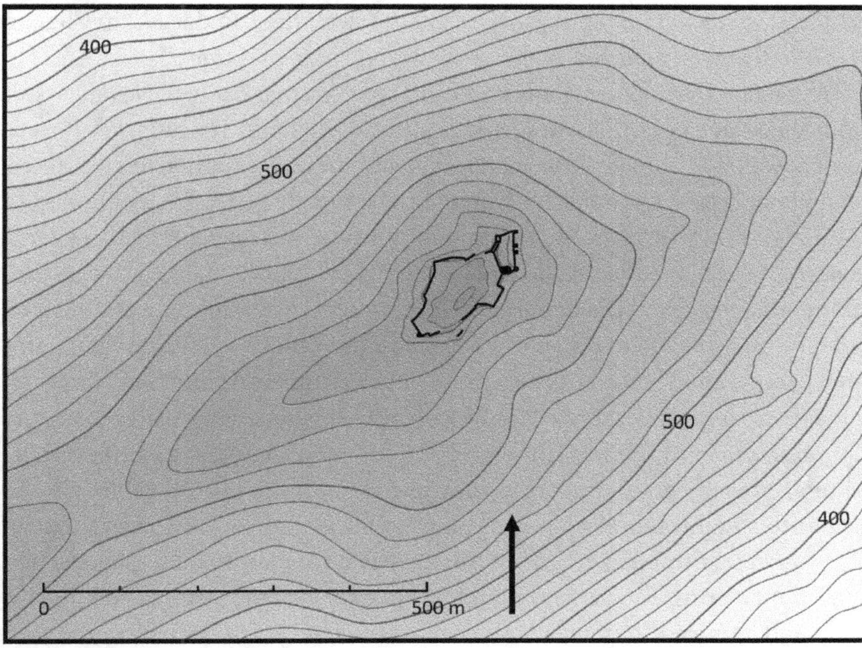

PLAN 24 *Kantara*

The town walls and citadel of Beirut had been rebuilt and refortified respectively by the time that Wilbrand of Oldenburg visited in 1211.[61] Wilbrand claims that the recently built towers had been reinforced with iron to resist the impact of heavy stones thrown by artillery, a measure that appears to have been tested two decades later. In 1231, imperial forces launched a sudden attack against Beirut, securing the town quite quickly, but the siege of the city's citadel was a protracted affair. During this siege, which saw the use of many different siege engines, the *Eracles* account claims that the besieging artillery included a large *trabuchet*, capable of throwing stones weighing a *quintat*, three small *trabuches* and six *tunbereaus*.[62] Philip of Novara also uses three terms (*perieres*, *mangueneaus* and *grans trabucs*) to identify the Ibelin artillery

61 Wilbrand of Oldenburg, *Itinerarium terrae sanctae* 1.5, ed. Pringle, pp. 118-19, trans. Pringle, p. 65.

62 *Eracles* 33.22, RHC Oc 2, p. 388. See also Philip of Novara, *Gestes des Chiprois* 159, 162, 164, ed. Raynaud, pp. 78-79, 83, 86, trans. La Monte, pp. 120-21, 129, 133; For the broader siege, see Philip of Novara, *Gestes des Chiprois* 158-70, ed. Raynaud, pp. 76-89, trans. La Monte, pp. 118-37; *Eracles* 33.19-29, RHC Oc 2, pp. 385-96; *Annales de Terre Sainte*, ed. Röhricht and Raynaud, A and B p. 15 (439).

at the siege of Kyrenia in 1232-33, while the defenders, under Philip Chenart, are said to have employed *trabus*, *perieres*, and *mangueneaus*.[63]

At each of these sieges, Kantara, Beirut and Kyrenia, Philip of Novara asserts that the attacking artillery inflicted significant damage. He claims that a section of wall was brought down at Kantara, but the rocky approaches were steep enough to repel a frontal assault and hard enough to resist the Ibelin sappers. Although Philip gives no description of a *trabuc*, or *trabucher*,[64] these terms again appear to denote the most powerful type of artillery. Like the *trabuculi* at Damietta in 1219 and the *petrariae* at Acre in 1191, the use of only one such engine by an army or contingent appears, in most cases, to indicate the relative power and significance of this engine relative to others. While these larger engines were almost certainly counterweight trebuchets, the smaller machines, found in greater numbers, were probably traction trebuchets, employed in a similar capacity and numbers as they had been in the twelfth century. References to small *trabuches* by some sources might indicate the use of smaller counterweight trebuchets, which others may have continued to identify with traditional terms. The *perieres* noted by Philip of Novara, distinct from the *trabucs* and *mangueneaus*, may have been counterweight trebuchets in the same way that the notable *petrariae* found at Acre in 1191 probably were.

Although counterweight trebuchets of differing sizes may have been built, the power of even the largest was still limited: Philip of Novara praises John of Caesarea's party of crossbowmen just as highly as the *grant trabuc* at the siege of Kantara.[65] Although the artillery used at Beirut is credited with inflicting notable damage, the besiegers' decision to abandon the siege when the defenders successfully countermined contextualises its power – unlike mining, it was not yet a practical breaching weapon. Similarly, when the defenders thwarted the use of siege towers at Kyrenia, a year-long passive blockade ensued.[66] In this light it is clear that although heavy artillery was growing stronger it was still not powerful enough to be the focus of siege efforts. Had it been more powerful at this point, it would be difficult to explain why it was not used after mining efforts failed at Beirut and the use of siege towers proved unsuccessful at Kyrenia.

63 Philip of Novara, *Gestes des Chiprois* 198, 199, ed. Raynaud, pp. 108, 109, trans. La Monte, pp. 157, 158.

64 As it is alternatively labelled, Philip of Novara, *Gestes des Chiprois* 151, ed. Raynaud, p. 66, trans. La Monte, p. 108.

65 Philip of Novara, *Gestes des Chiprois* 152, ed. Raynaud, pp. 68-69, trans. La Monte, pp. 109-10.

66 Philip of Novara, *Gestes des Chiprois* 198-201, ed. Raynaud, pp. 108-11, 116-17, trans. La Monte, pp. 157-61, 167-78.

Continued Infighting and New Arabic Terminology

Amid the ongoing Ayyubid power contest, new terms appear in Arabic sources that seem to classify the artillery used by Muslim armies. At the siege of Akhlat in 1229, al-Nasawi, a contemporary and quite possibly an eyewitness, notes the use of an engine called *qarabughra* (presumably an attempt to render the Turkish for 'black camel' into Arabic). This may originally have been the name of a particularly large or powerful engine; however, the term, which was quickly corrupted into the easier to say *qarabugha* (closer to the Turkish for 'black bull'), came to refer to a particular type of light trebuchet.[67] Writing in the fourteenth century, Ibn al-Dawadari notes the use of *shaytani* and *gharbi* trebuchets at the earlier Frankish siege of Damietta during the Fifth Crusade, though he gives no indications of what might have distinguished these engines. The latter term appears to be a corruption, as *maghribi* engines are mentioned at a number of sieges over the following decades, including al-Kamil's sieges of Harran and Edessa in 1236.[68]

Majaniq of the *shaytani* type were almost certainly light traction-powered trebuchets, often employed in significant numbers. *Maghribi* trebuchets were just as clearly the most powerful variety of their day. These are presented as exceptional engines, often mentioned in the singular when appearing in the late Ayyubid period. The first quantifiable notions of strength appear at the siege of Homs in 1248-49, and will be discussed below, before references to *maghribi* engines disappear twenty years later during the reign of Baybars.

In the meantime, the Ayyubid power struggle entered a new phase. Al-Kamil, who had successfully fought off the Fifth Crusade and established himself as the most powerful among his Ayyubid brothers and other relations, died in 1238. A contest for power ensued between his brother, al-Salih Isma'il, who claimed Damascus, and sons, al-'Adil II and al-Salih Najm Ayyub, who inherited Egypt and the Jazira respectively.

Jerusalem: 1239

In 1239, relations between al-Salih Ayyub and his brother al-'Adil II were tense. Meanwhile, the Franks, following the arrival of significant contingents of crusaders, suffered a crushing defeat at the hands of the Egyptians near Gaza.[69]

---

67 Khamisy and Fulton, "Manjaniq Qarabugha," pp. 179-201.

68 *History of the Patriarchs*, trans. Khater and Khs-Burmester, 4.1:147. For others, see Chevedden, "Citadel of Damascus," p. 280.

69 Ibn al-'Amid, *al-Majmu*, trans. Eddé and Micheau, pp. 60-62, 65-68; al-Maqrizi, *al-Suluk*, trans. Broadhurst, pp. 242-44, 245-49, 252-53. For the Frankish defeat, see Abu Shama,

Amid this environment of Frankish weakness and open hostility between Cairo and Damascus, al-Nasir Da'ud of Kerak took the opportunity to besiege Jerusalem towards the end of the year. Al-'Ayni states that al-Nasir directed artillery against the Tower of David and Ibn al-Furat, using Ibn Shaddad's account, appears to imply the same; however, al-Maqrizi gives the impression that a much more comprehensive attack took place, involving a three-week siege of the city.[70] Regardless of how the siege progressed, if the Tower of David was targeted by the Muslims' artillery, no impact signatures can be identified along the contemporary sections of rusticated masonry.[71]

Beaufort: 1240

When al-Salih Ayyub seized Egypt from his brother al-'Adil II in 1240,[72] their uncle, al-Salih Isma'il, then of Damascus, appealed to the Franks. Among the

Kitab al-Raudatain, RHC Or 5, p. 193; al-Maqrizi, *al-Suluk*, trans. Broadhurst, p. 251; *Rothelin* 27-29, 35, RHC Oc 2, pp. 541-48, 555, trans. Shirley, pp. 47-51, 57-58; *Eracles* 33.44-45, RHC Oc 2, pp. 413-15, trans. Shirley, pp. 123-24; Matthew Paris, *Chronica maiora*, ed. Luard, p. 25, trans. Giles, 1:272-73; Marino Sanudo, *Liber secretorum* 3.11.15, ed. Bongars, p. 215, trans. Lock, p. 241.

70 Al-'Ayni, *'Iqd al-Juman*, RHC Or 2a, p. 196; Ibn al-Furat, *Tarikh*, trans. Lyons and Lyons, 2:62; Al-Maqrizi, *al-Suluk*, trans. Broadhurst, p. 251. See also *Rothelin* 21, RHC Oc 2, pp. 529-30, trans. Shirley, p. 40. The *Rothelin* account states that the Franks had made improvements to the St Stephen's (Damascus) Gate before the siege, indicating that the town walls were not entirely indefensible since their destruction, but mentions only the citadel in the context of the siege proper, *Rothelin* 1, RHC Oc 2, p. 489, trans. Shirley, p. 13. For the Frankish fortification efforts during the 1230s, see also Pringle, "Town Defences," p. 80; Benvenisti, *The Crusaders*, p. 51.

71 For descriptions of the Tower of David, see Fulcher of Chartres, *Historia Hierosolymitana* 1.26.4, ed. Hagenmeyer, pp. 284-85, trans. Ryan, p. 117; Wilbrand of Oldenburg, *Itinerarium terrae sanctae* 2.6, ed. Pringle, p. 133, trans. Pringle, p. 88; *Guidebook to Palestine*, trans. Bernard, p. 23. See also Boas, *Jerusalem*, pp. 66-75; Benvenisti, *The Crusaders*, p. 53. For archaeological investigations, see Johns, "The Citadel, Jerusalem," pp. 121-90; Solar, "Jerusalem, citadel moat," pp. 47-48. For a radical, if unlikely, theory of the citadel's development, see Ellenblum, "Frankish Castles," pp. 93-109. For Jerusalem's Ayyubid defences more broadly, see Hawari, *Ayyubid Jerusalem*, pp. 22-26; Boas, *Jerusalem*, pp. 44-49; Pringle, "Town Defences," pp. 79-80; Benvenisti, *The Crusaders*, pp. 50-51. For archaeological evidence of these early thirteenth-century defences, see Weksler-Bdolah, "The Fortification System," pp. 105-30; Reich and Shukron, "Excavation in the Mamillah Area," pp. 128-30, 145-50. See also Wilbrand of Oldenburg, *Itinerarium terrae sanctae* 2.6, ed. Pringle, p. 133, trans. Pringle, pp. 87-88; Abu Shama, *Kitab al-Raudatain*, RHC Or 5, pp. 82-83; *Rothelin* 9, RHC Oc 2, p. 505, trans. Shirley, p. 22.

72 Ibn al-'Amid, *al-Majmu*, trans. Eddé and Micheau, pp. 61-63; al-Maqrizi, *al-Suluk*, trans. Broadhurst, pp. 245, 254, 256-57.

territorial rights exchanged for an alliance, al-Salih Isma'il agreed to return Beaufort and Safed;[73] the former had been developed by the Ayyubids but the latter had remained in ruins since the time of the Fifth Crusade.[74] Beaufort's defenders were less than willing to surrender their impressive castle, compelling al-Salih Isma'il to besiege the stronghold on behalf of the Franks. According to the *Rothelin* account, artillery brought from Damascus threw large and small stones against the castle.[75] Although there is no indication of how powerful these engines were, the account suggests not only that machines of two distinct scales were employed, but that both were stockpiled in Damascus. While the Ayyubids used artillery against each other into the 1240s, so too did the Franks.

Tyre: 1243

Balian of Beirut took advantage of the departure of Richard Filangieri (Frederick II's marshal in the East) to attack Tyre in 1243.[76] Philip of Novara's account states that after gaining the city, by a combination of surprise and collusion, the Ibelin party laid siege to the citadel with engines and *perieres*.[77] If Philip used a consistent set of terms to refer to specific types of artillery, these would appear to have been lighter than the heavier *trabucs* that he notes elsewhere.

---

73 Abu Shama, *Kitab al-Raudatain*, RHC Or 5, p. 193; Ibn al-Furat, *Tarikh*, trans. Lyons and Lyons, 2:6; Ibn al-'Amid, *al-Majmu*, trans. Eddé and Micheau, pp. 68-69, 71, 73; al-Maqrizi, *al-Suluk*, trans. Broadhurst, pp. 261-63; *Eracles* 33.42-43, RHC Oc 2, pp. 417-18, trans. Shirley, pp. 125-26; *Rothelin* 32, RHC Oc 2, p. 552, trans. Shirley, pp. 55-56; *Annales de Terre Sainte*, ed. Röhricht and Raynaud, A and B p. 16 (440); Matthew Paris, *Chronica maiora*, ed. Luard, 4:64-65, 288-91, trans. Giles, 1:303, 482-84; Marino Sanudo, *Liber secretorum* 3.11.15, ed. Bongars, pp. 215-16, trans. Lock, pp. 341-42.

74 For the Ayyubid contributions to the fortification of Beaufort, see Corvisier, "Les campagnes de construction," pp. 243-66; Deschamps, *Les Châteaux*, 2:197-208. See also Chapter 7.

75 *Rothelin* 32, RHC Oc 2, pp. 552-53, trans. Shirley, pp. 55-56. Cf. Ibn al-Furat, *Tarikh*, trans. Lyons and Lyons, 2:8.

76 For the dating of this episode to 1243 rather than 1242, see Philip of Novara, *Gestes des Chiprois* 227, trans. La Monte, p. 184 n. 1. Cf. Folda, *Crusader Art*, pp. 168-69.

77 Philip of Novara, *Gestes des Chiprois* 227, ed. Raynaud, pp. 130-32, trans. La Monte, pp. 178-80. Engines are mentioned in the account of Marsiglio Georgio but are absent from most others, Marsiglio Georgio, part. trans. La Monte, in Philip of Novara, *The Wars of Frederick II*, p. 207. Cf. *Eracles* 33.52, 55, RHC Oc 2, pp. 422, 426-27, trans. Shirley, pp. 129, 131-32; Marino Sanudo, *Liber secretorum* 3.11.16, ed. Bongars, p. 216, trans. Lock, p. 343; Templar of Tyre, *Gestes des Chiprois* 242, ed. Minervini, p. 52, trans. Crawford, p. 17.

Unfortunately, the obscure medieval landscape of Tyre prevents any further analysis of how large these engines might have been.[78]

'Ajlun: 1243/44

Artillery appears to have been used when 'Ajlun was besieged on behalf of al-Salih Isma'il of Damascus. Ibn al-Furat gives no indication of how these engines were employed but, as at Jerusalem, there are no obvious impact signatures on the surviving sections of Ayyubid masonry. It may be that the bossed masonry has obscured any damage inflicted by particularly light engines or that the targeted areas were among those rebuilt after one of the subsequent earthquakes.[79]

The Consequences of Forbie

The Khwarizmians, a Turkic group displaced by the Mongols, had become a semi-autonomous force of mercenaries operating in the Jazira by the 1230s. It was there that al-Salih Najm Ayyub first dealt with them and it was he who later summoned them to Palestine in 1244, seeking their support to confront the alliance forged against him by his uncle. The Khwarizmians sacked Jerusalem soon after arriving, but do not appear to have employed artillery when they took the city.[80] A few months later, al-Salih Isma'il's coalition, which included the Franks, al-Nasir Da'ud of Kerak and al-Mansur Ibrahim of Homs, was soundly defeated at the battle of Forbie on 17 October 1244,[81] allowing al-

78 No difficulties, such as those faced by the operators of Jerusalem's northern defensive engines in 1099, are noted.

79 Ibn al-Furat, *Tarikh*, trans. Lyons and Lyons, 2:1. Cf. al-Maqrizi, *al-Suluk*, trans. Broadhurst, pp. 278. For studies of 'Ajlun see Johns, "Medieval 'Ajlun," pp. 21-33; Yovitchitch, "The Tower of Aybak," pp. 225-42; Yovitchitch, "Die Aiyubidische Burg 'Aglun," 118-25.

80 For this agreement, see Ibn al-Furat, *Tarikh*, trans. Lyons and Lyons, 2:1-2; al-Maqrizi, *al-Suluk*, trans. Broadhurst, pp. 271-72. For the sack of Jerusalem, see *Rothelin* 41, RHC Oc 2, pp. 562-66, trans. Shirley, pp. 63-66; *Eracles* 33.56, RHC Oc 2, pp. 427-28, trans. Shirley, p. 132; Ibn al-Furat, *Tarikh*, trans. Lyons and Lyons, 2:2-3; Ibn al-'Amid, *al-Majmu*, trans. Eddé and Micheau, pp. 75-76; al-Maqrizi, *al-Suluk*, trans. Broadhurst, p. 273; Matthew Paris, *Chronica maiora*, ed. Luard, 4:288-91, 300-11, 337-44, trans. Giles, 1:482-84, 491-500, 522-28; *Annales de Terre Sainte*, ed. Röhricht and Raynaud, A and B p. 17 (441). The city remained defenceless at the time of Ibn Battuta's visit, Ibn Battuta, *Tuhfat* 1.1, trans. Gibb, p. 56.

81 *Eracles* 33.57, RHC Oc 2, pp. 429-31, trans. Shirley, pp. 133-34; *Rothelin* 41, RHC Oc 2, pp. 564-66, trans. Shirley, pp. 64-66; Templar of Tyre, *Gestes des Chiprois* 252, ed. Raynaud, pp. 145-46, trans. Crawford, pp. 19-20; Abu Shama, *Kitab al-Raudatain*, RHC Or 5, pp. 193-94; Ibn al-Furat, *Tarikh*, trans. Lyons and Lyons, 2:4-7; Ibn al-'Amid, *al-Majmu*, trans. Eddé and Micheau, pp. 76-77; al-Maqrizi, *al-Suluk*, trans. Broadhurst, p. 274. See also John of

Salih Ayyub to move first on Damascus and then on Homs. Artillery appears to have been used against the former,[82] but the disgruntled Khwarizmians do not appear to have employed such engines when they besieged Damascus in turn not long after.[83]

Ascalon: 1247

By 1240, coinciding with peaceful relations with Damascus and the crusade of Theobald IV of Champagne (king of Navarre), work was underway to refortify Frankish Ascalon with a castle.[84] Construction was completed under the supervision of Richard of Cornwall, who saw to its initial security by concluding a truce with the Egyptians before departing for Europe in the spring of 1241.[85] After the battle of Forbie, the castle, which provided refuge for some of the Franks who had escaped the battle, appears to have been besieged, although a more concerted effort to take the castle did not materialise until 1247.[86] That year, an Egyptian force crossed into Palestine, targeting Tiberias and Ascalon.[87] Tiberias, which had been recently refortified by Odo of Montbeliard, fell to a

Joinville, *Vie de Saint Louis* 528-38, ed. and trans. Monfrin, pp. 260-67. For the battle of Forbie, see Berkovich, "The Battle of Forbie," pp. 9-44.

82 Al-Maqrizi, *al-Suluk*, trans. Broadhurst, pp. 275-77; Ibn al-'Amid, *al-Majmu*, trans. Eddé and Micheau, pp. 77-78; *Eracles* 33.59, RHC Oc 2, p. 432, trans. Shirley, pp. 134-35; Chevedden, *The Citadel of Damascus*, pp. 73-74.

83 Al-Maqrizi, *al-Suluk*, trans. Broadhurst, pp. 278-81; Ibn al-'Amid, *al-Majmu*, trans. Eddé and Micheau, pp. 79-80; Chevedden, *The Citadel of Damascus*, pp. 74-75.

84 *Rothelin* 22, 25, 33, RHC Oc 2, pp. 531-32, 537-38, 552, trans. Shirley, pp. 41-42, 45, 56. Cf. Ibn al-'Amid, *al-Majmu*, trans. Eddé and Micheau, pp. 71-72; al-Maqrizi, *al-Suluk*, trans. Broadhurst, pp. 263-64. See also *Eracles* 33.49, RHC Oc 2, pp. 419-20, trans. Shirley, pp. 126-27; *Rothelin* 34, RHC Oc 2, p. 554, trans. Shirley, p. 57; Matthew Paris, *Chronica maiora*, ed. Luard 4:79-80, trans. Giles, 1:315; Templar of Tyre, *Gestes des Chiprois* 254, ed. Minervini, p. 58, trans. Crawford, p. 20; Marino Sanudo, *Liber secretorum* 3.11.15, ed. Bongars, p. 216, trans. Lock, p. 342; al-Maqrizi, *al-Suluk*, trans. Broadhurst, pp. 263-64, 267.

85 Matthew Paris, *Chronica maiora*, ed. Luard, 4:43-47, 71, 89, 138-45, trans. Giles, 1:287-90, 308-9, 323, 362-68; *Rothelin* 36, RHC Oc 2, pp. 555-56, trans. Shirley, p. 59. For the development of Ascalon's fortifications, see Pringle, "Richard I," pp. 133-47.

86 For the early siege efforts, see *Rothelin* 41, RHC Oc 2, p. 565, trans. Shirley, p. 65; Ibn al-Furat, *Tarikh*, trans. Lyons and Lyons, 2:8; al-Maqrizi, *al-Suluk*, trans. Broadhurst, pp. 274-75. See also Matthew Paris, *Chronica maiora*, ed. Luard, 4:559-60, trans. Giles, 2:174-75; Pringle, "The Walls of Ascalon."

87 *Eracles* 33.59, RHC Oc 2, pp. 432-33, trans. Shirley, p. 135; Templar of Tyre, *Gestes des Chiprois* 258, ed. Minervini, p. 58, trans. Crawford, p. 21; *Annales de Terre Sainte*, ed. Röhricht and Raynaud, A and B p. 18 (442); Abu Shama, *Kitab al-Raudatain*, RHC Or 5, p. 194; Ibn al-Furat, *Tarikh*, trans. Lyons and Lyons, 2: 10-11; Ibn al-'Amid, *al-Majmu*, trans. Eddé and Micheau, p. 82; al-Maqrizi, *al-Suluk*, trans. Broadhurst, pp. 283-84.

frontal assault, but Ascalon posed a greater challenge and a fleet was summoned to complete a blockade. The defenders appealed to Henry I of Cyprus and a naval force was dispatched under Baldwin of Ibelin. A maritime standoff took place until a storm drove many of the Egyptian vessels onto some rocks. This proved to be a mixed blessing when, according to the *Eracles* account, the Muslims used the wreckage of their ships to construct penthouses and trebuchets.[88] The stronghold eventually fell after miners dug a tunnel into the castle.[89]

Although artillery is not emphasised in the surviving accounts of the siege, this appears to be the first instance in the region where masts are recorded as having been used to construct trebuchets.[90] Whereas ship masts had been used at the sieges of Caesarea (1101), Tyre (1111-12), Ascalon (1153) and Alexandria (1167) to construct siege towers,[91] their use to build artillery indicates the increased scale and significance of the engines built at Ascalon. Although Muslim armies rarely built such siege engines, their part in the erection of the tower at Banyas in 1140 and Baybars' later use of a sizable penthouse, which may have had multiple levels, at Caesarea in 1265[92] suggest the construction of siege towers remained a practical option throughout the period. Accordingly, it is revealing that artillery was considered to be the wisest use of this timber.

If masts were used to construct artillery, or even if this reference is more figurative than literal, it provides a rough sense of the scale of these machines. The notion that they were significantly larger than those used at the end of the twelfth century is supported by the physical evidence. Two spherical granite projectiles, hewn from antique column-drums of unknown exact provenance, have been found in the fosse near the Jaffa Gate. Both have a diameter slightly greater than 30 cm and their corresponding masses are enough to confirm that they were thrown by a counterweight trebuchet. A number of smaller stones, which may have been thrown by lighter traction trebuchets, have also been

88 *Eracles* 33.60-61, RHC Oc 2, pp. 433-35, trans. Shirley, pp. 135-36. See also *Annales de Terre Sainte*, ed. Röhricht and Raynaud, B p. 18 (442), ed. Edbury, p. 154.

89 Cf. the Roman siege of Maogamalcha, Ammianus, *Res gestae* 24.4.2-23, ed. Eyssenhardt, pp. 303-7, trans. Yonge, pp. 357-61.

90 *Eracles* 33.61, RHC Oc 2, p. 434, trans. Shirley, p. 136. James of Aragon states that sailors from Marseilles built a *trebuchet* from the masts and wood of a ship in 1229 during the conquest of Majorca, James of Aragon, *Llibre dels fets* 69, trans. Smith and Buffery, p. 93.

91 Fulcher of Chartres, *Historia Hierosolymitana* 2.9.2, ed. Hagenmeyer, pp. 401-2, trans. Ryan, p. 153; Ibn al-Qalanisi, *Dhail*, trans. Gibb, pp. 125-26; William of Tyre, *Chronicon* 17.24, 19.28, ed. Huygens, 2:794, 903-4, trans. Babcock and Krey, 2:222, 337.

92 Ibn 'Abd al-Zahir, *al-Rawd*, trans. Sadeque, 2:557; Ibn al-Furat, *Tarikh*, trans. Lyons and Lyons, 2:70-71; Thorau, *The Lion of Egypt*, pp. 160-61.

IMAGE 6 *Ascalon: granite projectile* (PHOTO COURTESY OF DENYS PRINGLE)

identified in the fosse that surrounded the thirteenth-century castle, strongly indicating that they too date to the siege of 1247.[93]

The effect of these projectiles is hard to determine as the castle was destroyed following its capture and razed further by Baybars in 1270.[94] Nevertheless, this material evidence confirms the use of counterweight trebuchets and indicates they were of a substantial size. This technology would appear to have been fairly common, given the muted tone of the sources, rendering the emphasis placed on artillery the following year at Homs all the more significant.

Homs: 1248-49

In 1248, al-Nasir Yusuf of Aleppo and al-Salih Isma'il moved against Homs. The city was surrendered before an Egyptian relief force arrived, compelling

93 My thanks to Denys Pringle and Hannah Buckingham for bringing these stones to my attention. Although the granite stones have not been weighed, rough measurements of their circumference and an estimated density of 2,500 kg/m^3 suggest that the mass of the smaller stone is approximately 43 kg and the larger stone about 60 kg.

94 Ibn 'Abd al-Zahir, *al-Rawd*, trans. al-Khowayter, 2:737, 741; Ibn al-Furat, *Tarikh*, trans. Lyons and Lyons, 2:140, 142; Mufaddal, *Kitab*, ed. and trans. Blochet, 1:526; Ibn Battuta, *Tuhfat* 1.1, trans. Gibb, p. 57; Stager and Esse, "Ashkelon," p. 72. For archaeological examinations of this castle, see Pringle, "Town Defences," pp. 84-85; Pringle, "Richard I," esp. Appendix 1, pp. 144-46. Cf. Benvenisti, *The Crusaders*, pp. 120-28.

al-Salih Najm Ayyub to dispatch a larger force to retake it. According to Abu'l-Fida' and al-Maqrizi, fourteen trebuchets were arrayed against Homs during this second siege. Like the engines brought against Beaufort four years earlier, these had been transported from Damascus, some in as many as fifty disassembled components. One of these engines, a *manjaniq maghribi*, was supposed to have been able to throw projectiles weighing 140 Syrian *artal* (259 kg or 289 kg according to the measures of Damascus and Hama respectively),[95] more than four times the weight of the larger stone found at Ascalon. This figure, drawn from Ibn Wasil, who was himself informed of these events by participants of the siege, is almost certainly an exaggeration; however, it serves to distinguish this large engine from the lighter ones. Hill and Chevedden are thus justified in believing that this was a counterweight trebuchet.[96]

Despite the increasing evidence, both descriptive and physical, of the growing scale of artillery, it remains difficult to model them as energy systems without more variables, such as where the engines were shooting from or how much damage they were able to inflict. But even without this, it is clear that the largest engines were still incapable of effectively breaching fortifications: the garrison of Homs held out through the winter until a peace was eventually brokered. The conclusion of this agreement appears to have been encouraged more by news that Louis IX had arrived in the eastern Mediterranean than the threat posed by the Egyptian army's artillery.[97]

Louis IX and the Seventh Crusade

In the spring of 1249, Louis IX of France left Cyprus and landed with a force of crusaders at the mouth of the Damietta branch of the Nile. Unlike the landings of the Fifth Crusade in 1218, the Sicilian force in 1184, and Frankish-Byzantine force in 1169, Louis' army encountered little opposition. After defeating an initial force that opposed their landing, the crusaders found Damietta and its

95 Abu'l-Fida', *al-Mukhtasar*, RHC Or 1, p. 125; Al-Maqrizi, *al-Suluk*, trans. Broadhurst, pp. 285-86. See also Ibn al-Furat, *Tarikh*, trans. Lyons and Lyons, 2:11; Ibn al-'Amid, *al-Majmu*, trans. Eddé and Micheau, p. 85.

96 Hill, "Trebuchets," pp. 104, 106; Chevedden, *The Citadel of Damascus*, p. 280. My Thanks to Rabei Khamisy for translating the appropriate section of Ibn Wasil for me.

97 Ibn al-'Amid, *al-Majmu*, trans. Eddé and Micheau, p. 83; Vincent of Beauvais, *Speculum historiale* 32.95, ed. Koburger, trans. in Jackson, *Seventh Crusade*, no. 72, pp. 121-22; John of Joinville, *Vie de Saint Louis* 144, ed. and trans. Monfrin, pp. 70-71.

provisions abandoned to them.[98] Al-Salih Najm Ayyub had returned to Egypt from Syria in expectation of the French king's arrival. Now sick, he established himself at Mansura, where his father, al-Kamil, had halted the southward advance of the Fifth Crusade.[99] In December, Louis, like his predecessors twenty-eight years earlier, marched south to Mansura and dug in on the north bank of the Ushmum canal opposite his adversary.[100]

The Franks erected two siege towers (*chas chastiaus*), used as elevated shooting platforms, to provide cover for workers constructing a causeway across the river.[101] According to John of Joinville, the seneschal of Champaign and a leading knight among the crusaders, sixteen Muslim *engins* targeted the men working on the causeway. In response, the Franks constructed eighteen *engins* under Joscelin of Cornaut, a master engineer.[102] It is unclear how the Frankish and Muslim machines differed, but the latter were clearly employed with greater effect. Joinville claims that the Frankish artillery was unable to inflict any notable damage, while the Muslim engines were not only capable of harrying work on the causeway but also supported forces who crossed the river and attacked the Frankish camp.[103]

---

98 John of Joinville, *Vie de Saint Louis* 146-64, ed. and trans. Monfrin, pp. 72-81; *Rothelin* 59-60, RHC Oc 2, pp. 589-93, trans. Shirley, pp. 85-88; *Eracles* 34.1, RHC Oc 2, pp. 436-37, trans. Shirley, p. 137; Templar of Tyre, *Gestes des Chiprois* 263, ed. Minervini, p. 60, trans. Crawford, pp. 21-22; *Annales de Terre Sainte*, ed. Röhricht and Raynaud, A and B pp. 18-19 (442-43); Vincent of Beauvais, *Speculum historiale* 32.97-98, ed. Koburger, trans. in Jackson, *Seventh Crusade*, no. 72, pp. 122-24; letter of Jean de Beaumont (1249), trans. in Jackson, *Seventh Crusade*, no. 58, pp. 85-86; letters of the Count of Artois, Guy (in service of the count of Melun), and William of Sonnac, in Matthew Paris, *Chronica maiora*, ed. Luard, 6:152-62, trans. Giles, 3:409-18; Abu Shama, *Kitab al-Raudatain*, RHC Or 5, p. 195; Ibn Wasil, *Mufarrij*, trans. Jackson, no. 73, pp. 130-31; Ibn al-Furat, *Tarikh*, trans. Lyons and Lyons, 2:15-16; al-Maqrizi, *al-Suluk*, trans. Broadhurst, pp. 288-90.

99 Ibn Wasil, *Mufarrij*, trans. Jackson, no. 73, pp. 130-31; Ibn al-Furat, *Tarikh*, trans. Lyons and Lyons, 2:12, 17; Ibn al-'Amid, *al-Majmu*, trans. Eddé and Micheau, pp. 83-84; al-Maqrizi, *al-Suluk*, trans. Broadhurst, pp. 287-88, 290-92.

100 John of Joinville, *Vie de Saint Louis* 179-86, 191, 197, ed. and trans. Monfrin, pp. 88-91, 94-95, 96-97; *Rothelin* 62, RHC Oc 2, pp. 594-99, trans. Shirley, pp. 89-92; *Eracles* 34.1, RHC Oc 2, p. 437, trans. Shirley, p. 137; Ibn Wasil, *Mufarrij*, trans. Jackson, no. 73, pp. 141-42; Ibn al-Furat, *Tarikh*, trans. Lyons and Lyons, 2:20; al-Maqrizi, *al-Suluk*, trans. Broadhurst, p. 301.

101 Cf. Amalric's attempt to bridge the Nile farther upstream, intending this bridge to be fortified with wooden towers provisioned with artillery, William of Tyre, *Chronicon* 19.22, 23, 29, ed. Huygens, 2:892-93, 895, 905, trans. Babcock and Krey, 2:325-26, 328, 339.

102 John of Joinville, *Vie de Saint Louis* 192-95, ed. and trans. Monfrin, pp. 94-97.

103 John of Joinville, *Vie de Saint Louis* 200-1, ed. and trans. Monfrin, pp. 98-99.

Although effective against Frankish personnel, the Muslims' *engins* were not strong enough to threaten the siege towers. Against these, the Muslims erected a *perriere*, to use Joinville's term. This was probably a counterweight trebuchet, while the *engins* used by both the Franks and Muslims were traction trebuchets, elsewhere differentiated from large bolt-shooting weapons (*arbalestre a tour*).[104] Despite the evidence that sizeable stones were thrown by Egyptian artillery at Ascalon and Homs, Joinville clearly communicates that the *perriere* threw incendiaries (containers of Greek fire) rather than stones at the Frankish towers.[105] Only the lighter *engins* appear to have hurled stones.[106]

The first shot discharged by the *perriere* passed between the two towers. The Franks who attempted to extinguish the subsequent fire were initially sheltered by wings extending from the bases of the towers; however, the Muslims adjusted the release angle of their *engins*, shooting high into the air to subject the Franks to a shower of plunging stones.[107] The apparently rapid shooting rate of these engines, targeting soldiers and workers, is another indicator that these were traction trebuchets, whereas the *perriere* discharged only three shots during the first night that it came into service. The Muslims appear to have slowly adjusted the range of the *perriere*: after the first shot fell behind the Frankish towers, the next fell short.[108]

Intriguingly, Joinville claims that the Muslims' *perriere* was used only at night; it was thus exceptional when it was brought forward one day. Supported by a hail of stones thrown by the lighter *engins*, the *perriere* was advanced close to the river, the implicit suggestion being that this move contributed to the machine's ability to set both of the Frankish towers on fire that day. With the towers destroyed, Louis used timber from his barons' ships to build a new penthouse to continue work on the causeway. But after the men around the penthouse had been driven away by the sixteen *engins*, this too was set alight by the *perriere*.[109]

Joinville's clear distinction between the Muslim's antipersonnel stone-throwing *engins* and the single machine-targeting Greek fire-throwing *perriere*

104 John of Joinville, *Vie de Saint Louis* 203-4, 206, ed. and trans. Monfrin, pp. 100-1.

105 This is similar to the use of such by the defenders of Acre in May 1190, 'Imad al-Din, *al-Fath*, trans. Massé, pp. 217-19; Baha' al-Din, *al-Nawadir*, trans. Richards, pp. 110-11; Ibn al-Athir, *al-Kamil*, trans. Richards, 2:373-74.

106 John of Joinville, *Vie de Saint Louis* 210, ed. and trans. Monfrin, pp. 102-3.

107 John of Joinville, *Vie de Saint Louis* 205, ed. and trans. Monfrin, pp. 100-1.

108 John of Joinville, *Vie de Saint Louis* 208, ed. and trans. Monfrin, pp. 102-3.

109 John of Joinville, *Vie de Saint Louis* 209-13, ed. and trans. Monfrin, pp. 102-5.

must be evaluated alongside other sources. The *Rothelin* account states, rather non-specifically, that Louis employed *perrierez*, *mangongiax*, *trebuchez* and other *enginz* while the Muslims made use of *anginz*, which shot large and small stones as well as Greek fire and javelins.[110] In a letter composed by a Templar (or possibly a Hospitaller), the Muslims are noted as using many *machinae* and *fundibula* against the Franks as they attempted to cross the river.[111] Similarly, in a letter sent back to France in 1250, Louis IX mentions the use of *machinae* by both the Franks and Muslims. Although he does not describe these engines, he notes that the Muslims' artillery threw stones that damaged the Frankish towers but that it was Greek fire that ultimately destroyed them.[112] Florence of Worcester simply refers to Louis' use of *machinae* to facilitate the crusaders' efforts to cross the river.[113] Ibn Wasil, Ibn al-Furat, al-'Ayni and al-Maqrizi confirm that the Franks fortified their position and used artillery, without classifying it, but omit any mention of siege towers, penthouses or Muslim artillery.[114] Thus, while some sources appear to support certain aspects of Joinville's account, none fully corroborates it.

The crusaders abandoned their causeway following the destruction of the penthouse and eventually found a ford across the channel in early February 1250. Taking part of his force across the river, Louis occupied the Muslims' camp on the south bank, directly opposite the main Frankish force. Here, the advance force could be supported by the duke of Burgundy from the crusaders' original camp on the north side of the channel.[115] The Muslims had apparently abandoned their artillery, fourteen *engins* according to the *Rothelin* account. These were evidently of little value to the crusaders as they were broken up to build a palisade around the new camp while a bridge was built across the river

110 *Rothelin* 63, RHC Oc 2, p. 600, trans. Shirley, p. 93. Cf. Florence of Worcester, *Chronicon ex chronicis*, ed. Thorpe, 2:181.

111 Matthew Paris, *Chronica maiora*, ed. Luard, 6:191-97.

112 Letter of Louis IX to his subjects in France (1250), ed. Duchesne, in *Historiae Francorum Scriptores*, 5:428-32, trans. in Jackson, *Seventh Crusade*, no. 70, pp. 108-14. See also Vincent of Beauvais, *Speculum historiale* 32.99, ed. Koburger.

113 Florence of Worcester, *Chronicon ex chronicis*, ed. Thorpe, 2:181-82, trans. Forester, p. 325.

114 Ibn Wasil, *Mufarrij*, trans. Jackson, no. 73, pp. 141-42; Ibn al-Furat, *Tarikh*, trans. Lyons and Lyons, 2:20; al-'Ayni, *'Iqd al-Juman*, RHC Or 2a, p. 208; al-Maqrizi, *al-Suluk*, trans. Broadhurst, p. 301.

115 John of Joinville, *Vie de Saint Louis* 215-48, ed. and trans. Monfrin, pp. 104-23; *Rothelin* 64, RHC Oc 2, pp. 602-9, trans. Shirley, pp. 94-98; *Eracles* 34.1, RHC Oc 2, pp. 437-38, trans. Shirley, p. 138; Ibn al-'Amid, *al-Majmu*, trans. Eddé and Micheau, p. 86; al-Maqrizi, *al-Suluk*, trans. Broadhurst, pp. 302-3.

to connect the two Frankish positions.[116] Louis was initially able to defend the camps, which secured each end of the bridge, but momentum swung against him later in February when al-Mu'azzam Turanshah arrived to succeed his father, al-Salih Ayyub, who had died at Mansura three months earlier.[117] The Franks withdrew back across the Ushmum canal in early April and a disastrous effort to escape back to Damietta followed, in which most of the army was captured or killed.

In his account of the negotiations between the Franks and Muslims that followed Louis' initial withdrawal across the river, before the attempted retreat to Damietta, Joinville gives the impression that the king was attempting to move with his artillery. During one of these dialogues, he claims that one of Louis' conditions for exchanging Damietta for Jerusalem was that al-Mu'azzam Turanshah would mind his siege engines until he was able to collect them.[118] These talks broke down when Louis would not personally serve as a guarantor and the Franks' chaotic flight towards Damietta ensued.[119]

The unorganised rush towards Damietta resulted in the death or capture of most of the crusaders. The French king was among those apprehended and Damietta was included in the sizable ransom demanded for his release. In a twist of fate, al-Mu'azzam Turanshah was murdered before Damietta was surrendered in early May 1250; having spent most of his adult life in Hisn Kayfa, he

116 John of Joinville, *Vie de Saint Louis* 244, 254-61, 263-66 269-70, ed. and trans. Monfrin, pp. 120-21, 126-29, 130-33, 132-35; *Rothelin* 64, RHC Oc 2, pp. 607-8; trans. Shirley, p. 98; letter of Louis IX to his subjects in France (1250), ed. Duchesne, in *Historiae Francorum Scriptores*, 5:428-32, trans. in Jackson, *Seventh Crusade*, no. 70, pp. 108-14; Marino Sanudo, *Liber secretorum* 3.12.2, ed. Bongars, p. 219, trans. Lock, p. 347. Cf. al-Maqrizi, *al-Suluk*, trans. Broadhurst, p. 303.

117 For the death of al-Salih Najm Ayyub and arrival of al-Mu'azzam Turanshah in Egypt, see Ibn Wasil, *Mufarrij*, trans. Jackson, no. 73, pp. 133-43, 145-46; Abu Shama, *Kitab al-Raudatain*, RHC Or 5, pp. 195-96; Ibn al-'Amid, *al-Majmu*, trans. Eddé and Micheau, pp. 84, 86-87; Ibn al-Furat, *Tarikh*, trans. Lyons and Lyons, 2:23-26; Sibt Ibn al-Jawzi, trans. in Jackson, *Seventh Crusade*, no. 74a, p. 155; al-Maqrizi, *al-Suluk*, trans. Broadhurst, pp. 292-94, 306; John of Joinville, *Vie de Saint Louis* 196-97, 287-88, ed. and trans. Monfrin, pp. 96-97, 142-43. See also Thorau, *The Lion of Egypt*, pp. 32-36; Bartlett, *Islam's War*, pp. 227-30.

118 John of Joinville, *Vie de Saint Louis* 301-3, ed. and trans. Monfrin, pp. 148-51. Cf. Ibn al-Furat, *Tarikh*, trans. Lyons and Lyons, 2:28; al-Maqrizi, *al-Suluk*, trans. Broadhurst, p. 307.

119 John of Joinville, *Vie de Saint Louis* 304-25, ed. and trans. Monfrin, pp. 150-61; *Rothelin* 66, RHC Oc 2, pp. 612-16, trans. Shirley, pp. 100-3; *Eracles* 34.1, RHC Oc 2, p. 438, trans. Shirley, p. 138; Ibn al-'Amid, *al-Majmu*, trans. Eddé and Micheau, pp. 87-88; Ibn Wasil, *Mufarrij*, trans. Jackson, no. 73, pp. 147-8; Ibn al-Furat, *Tarikh*, trans. Lyons and Lyons, 2:28; al-Maqrizi, *al-Suluk*, trans. Broadhurst, pp. 307-8.

was never given the opportunity to employ artillery in the Levant.[120] Following Damietta's return to the Muslims, it was deemed too vulnerable to defend and its fortifications were torn down.[121]

The Aftermath of Defeat in Egypt

With the collapse of the Seventh Crusade, the Ayyubid princes quickly turned their attentions back to each other. This allowed Louis IX, interested in leaving a positive legacy in the Holy Land, the opportunity to finance and oversee a major wave of fortification efforts along the Palestinian coast. The town walls of Caesarea, Jaffa, Acre and Sidon were all developed, and there is evidence some of their citadels were also strengthened.[122] There can be little doubt that

---

120 For the murder of al-Mu'azzam Turanshah, see John of Joinville, *Vie de Saint Louis* 344-53, ed. and trans. Monfrin, pp. 168-75; Ibn Wasil, *Mufarrij*, trans. Jackson, no. 73, pp. 150-52; Abu Shama, *Kitab al-Raudatain*, RHC Or 5, pp. 198-200; Rashid al-Din, *Ta'rikh-i Ghazani*, trans. Boyle, p. 234; Ibn al-Furat, *Tarikh*, trans. Lyons and Lyons, 2:33-34; Ibn al-'Amid, *al-Majmu*, trans. Eddé and Micheau, p. 88; al-Maqrizi, *al-Suluk*, trans. Broadhurst, pp. 311-13. See also Bartlett, *Islam's War*, pp. 230-31. For the surrender of Damietta, see *Rothelin* 69, RHC Oc 2, p. 623, trans. Shirley, p. 108; Ibn Wasil, *Mufarrij*, trans. Jackson, no. 73, p. 154; Ibn al-Furat, *Tarikh*, trans. Lyons and Lyons, 2:35-36. For the subsequent role of the party of murderous mamluks in Palestine during the 1250s, see Ibn 'Abd al-Zahir, *al-Rawd*, trans. Sadeque, pp. 80-93; Ibn al-'Amid, *al-Majmu*, trans. Eddé and Micheau, pp. 89-97.

121 Matthew Paris, *Chronica maiora*, ed. Luard 5:258, trans. Giles, 2:458; Ibn al-Furat, *Tarikh*, trans. Lyons and Lyons, 2:38. The city was still indefensible in the early fourteenth century, Ibn Battuta, *Tuhfat* 1.1, trans. Gibb, pp. 46, 49; al-'Umari, *Masalik al-absar*, trans. Richards, p. 55.

122 For the refortification of Caesarea, see John of Joinville, *Vie de Saint Louis* 470, 493,ed. and trans. Monfrin, pp. 232-33, 252-45; letter of the patriarch of Jerusalem to Blanche of Castile, queen of France (1251), in *Annales de Burton*, ed. Luard, p. 296, trans. in Jackson, *Seventh Crusade*, no. 115, p. 209; *Eracles* 34.2, RHC Oc 2, p. 440, trans. Shirley, p. 139; *Annales de Terre Sainte*, ed. Röhricht and Raynaud, A and B p. 21 (445), ed. Edbury, p. 155. For modern studies of Caesarea's defences, see Faucher et al., *Caesarea, 2010*; Faucher et al., *Caesarea, 2009*; Faucher et al., *Caesarea, 2008*; Faucher et al., *Caesarea, 2007*; Faucher, Merlin and Mesqui, "Césarée (Israël)," pp. 1-4; Mesqui, "L'enceinte médiévale de Césarée," pp. 83-94; Pringle, "Town Defences," pp. 89-91; Benvenisti, *The Crusaders*, pp. 131-44; Frova, "Caesarea," pp. 150-51; Negev, "Caesarea," (1961), pp. 81-83; Negev, "Caesarea," (1960), pp. 264-65; Conder and Kitchener, *Survey of Western Palestine*, 2:13-28; Rey, *Étude sur les monuments*, pp. 221-25. For the refortification of Jaffa, see John of Joinville, *Vie de Saint Louis* 515-17, 561-62, ed. and trans. Monfrin, pp. 254-57, 278-79; *Eracles* 34.2, RHC Oc 2, p. 440, trans. Shirley, p. 139; *Annales de Terre Sainte*, ed. Röhricht and Raynaud, A and B p. 21 (445), ed. Edbury, p. 155; Ibn 'Abd al-Zahir, *al-Rawd*, trans. al-Khowayter, 2:641; Mufaddal, *Kitab*, ed. and trans. Blochet, 1:504-5; al-'Ayni, *'Iqd al-Juman*, RHC Or 2a, p. 227. For a more comprehensive history of Jaffa's defences, see Pringle, "Town Defences," p. 94. For the

artillery was taken into consideration when these fortifications were constructed.

Louis and his deputies would have been familiar with the capabilities of European artillery and had just experienced Ayyubid artillery at Mansura. An anecdote recounted by Joinville about the Count of Eu and his small model of a stone-thrower (*bible*), which shot pebbles into Joinville's tent and broke his glasses and pots, is an example of the close familiarity that the knightly class had with this technology.[123] These were the men who commissioned, if not designed on occasions, both siege machines and fortifications. Similarly, when engines of a certain type were not built at some sieges the reason cannot have been ignorance.[124]

The Mongols and the War of St Sabas

Mongols

When Louis IX left the Holy Land in late April 1254,[125] the East that he was leaving was in transition. The Mongols' westward migration was well underway and on 17 January 1258 forces under Hulagu, a grandson of Genghis Khan, crushed the army of Baghdad and entered the city less than a month later. Hulagu's push into Syria continued and in January 1260 his army besieged Aleppo,

refortification of Acre, see John of Joinville, *Vie de Saint Louis* 615-16, ed. and trans. Monfrin, pp. 304-5; *Eracles* 34.1, RHC Oc 2, p. 438, trans. Shirley, p. 138. For the refortification of Sidon, see John of Joinville, *Vie de Saint Louis* 582, ed. and trans. Monfrin, pp. 288-89. For modern studies of the fortifications of Sidon, see Mesqui, "La fortification des Croisés," pp. 11, 24-26; Pringle, "Town Defences," pp. 81-90; Boase, "Military Architecture," p. 161; Kalayan, "The Sea Castle of Sidon," pp. 81-83; Deschamps, *Les Châteaux*, 2:227-33; Pringle, "A Castle in the Sand," p. 188. For the Muslim attack during construction at Sidon, see John of Joinville, *Vie de Saint Louis* 551-53, ed. and trans. Monfrin pp. 272-75; *Eracles* 34.2, RHC Oc 2, pp. 440-41, trans. Shirley, p. 139; Marino Sanudo, *Liber secretorum* 3.12.4, ed. Bongars, p. 220, trans. Lock, pp. 348-49. Cf. the siege in 1249, Ibn al-Furat, *Tarikh*, trans. Lyons and Lyons, 2:18; al-Maqrizi, *al-Suluk*, trans. Broadhurst, p. 292.

123 John of Joinville, *Vie de Saint Louis* 583, ed. and trans. Monfrin, pp. 288-91. For variants of this term, see Chapter 1.

124 John of Joinville, *Vie de Saint Louis* 563-81, ed. and trans. Monfrin, pp. 278-89.

125 John of Joinville, *Vie de Saint Louis* 584-87, 603, 615-17, ed. and trans. Monfrin, pp. 290-93, 298-301, 304-7; *Eracles* 34.2, RHC Oc 2, p. 441, trans. Shirley, p. 441; *Annales de Terre Sainte*, ed. Edbury, pp. 155-56.

employing artillery and sappers against its defences. The town was captured on 25 January but the citadel held out for another month.[126]

The Mongols would have learned about trebuchet technology during the course of their campaigns in China, had they not known about it beforehand. According to Bar Hebraeus, a large body of Chinese men, "skilled in the working of engines of war and throwing naphtha", had been sent west with Hulagu in 1252[127] and al-Makin Ibn al-'Amid's first reference to the use of artillery, in proximity to the Latin East, is at the Mongol siege of Irbil in 1258-59.[128] It has been suggested that the Mongols learned about counterweight artillery from the Muslims. During the protracted siege of Fancheng (Xiangyang) from 1268 to 1273, Rashid al-Din notes that no 'Frankish trebuchets' had formerly been employed in this region along the Chinese frontier, but that an artillery-maker, who had come from Baalbek and Damascus with his three sons, constructed seven large trebuchets that were used against the strongly fortified town.[129] As will become apparent in Chapter 7, engines identified as *majaniq ifranji* were the most powerful used in the Levant during the early Mamluk period. Whether the Mongols had prior knowledge of the heaviest artillery used in western Syria or not, their allies Hethum I of Armenia, Bohemond VI of Antioch and al-Ashraf Musa of Homs, all present when Hulagu entered Aleppo in January 1260, certainly would have.[130] There are few indications that the Mongols employed trebuchets through the remainder of their conquest of western Syria: few strongholds resisted them and those that did, such as Subayba, often fell to a frontal assault or surrendered after a relatively brief siege.

126 For the Mongol migration in relation to Ayyubid matters, see Thorau, *The Lion of Egypt*, pp. 59-62, 66-67; Bartlett, *Islam's War*, pp. 233-35.

127 Bar Hebraeus, *Makhtebhanuth Zabhne*, trans. Budge, 1:419.

128 Ibn al-'Amid, *al-Majmu*, trans. Eddé and Micheau, pp. 103-4. Cf. Rashid al-Din, *Ta'rikh-i Ghazani*, trans. Boyle, p. 190. See also the account of a Mongol siege by papal emissary John of Plano Carpini, in Raphael, *Muslim Fortresses*, p. 54.

129 Rashid al-Din, *Ta'rikh-i Ghazani*, trans. Boyle, pp. 290-91. See also Hill, "Trebuchets," p. 104; Chevedden, "Black Camels," p. 236 n. 17; Chevedden, "King James I," p. 318; Needham and Yates, *Science and Civilization in China*, p. 221.

130 Hethum of Armenia had solicited the support of the Mongols in the 1250s and recognised their hegemony, his son-in-law-Bohemond VI initially rebuked the Mongols in 1244 but subsequently submitted to their rule, *Eracles* 34.2, RHC Oc 2, pp. 440, 442, trans. Shirley, pp. 139, 140; Vahram, *Chronicle*, trans. Neumann, pp. 48-49; Templar of Tyre, *Gestes des Chiprois* 302, ed. Minervini, pp. 80-82, trans. Crawford, p. 34; Matthew Paris, *Chronica maiora*, ed. Luard, 4:389-90, trans. Giles, 2:31; *Chronicle of Melrose*, ed. Stevenson, p. 158, trans. Stevenson, p. 188.

As the Mongols turned south from Aleppo, the leading citizens of Damascus fled the city, allowing it to be occupied without opposition.[131] It seems the city was only half-heartedly occupied as the Ayyubid commander of the citadel was still in a position to revolt when Hulagu was pulled back to Mongolia following the death of his brother Mongke, the Great Khan. Paul Chevedden has argued that artillery was responsible for compelling the citadel's garrison to surrender. Having spent the night of 25 April 1260 cutting wood to construct trebuchets and the following day gathering ammunition, the Mongols moved more than twenty engines into position to the west of Damascus on the night of 26 April. The bombardment that followed, according to Chevedden, was what led the garrison to seek terms after about twenty-four hours.[132] Although the technology would have been known and suitable materials to construct trebuchets were probably available along the banks of the Barada River, there is little to suggest that the implied degree of damage could have been inflicted in so little time. Besides the lack of any evidence to support the existence of artillery powerful enough to compel this type of reaction, the period of time in which the Mongols erected their engines was similar to that required by Saladin seventy years earlier, suggesting the Mongol engines may not have been much larger.[133] The fundamental issue with this theory, however, is the same facing the suggestions that al-Zahir Ghazi's artillery was able to breach the walls of Saone in a similar period of time in 1188: if engines this powerful existed, why were they not used with equal effect elsewhere?

The considerable rebuilding work evident along the western and northern sides of the citadel of Damascus is associated with a later destruction phase. These fronts, which form part of the city's town walls, were slighted by the Mongols later in 1260 and were besieged a number of times during the following centuries.[134] Contrary to Chevedden's theory that the citadel's large towers,

131 Ibn al-Furat, *Tarikh*, trans. Lyons and Lyons, 2:41-42; *Menkonis Chronicon*, ed. Weiland, pp. 547-49, trans. Barber and Bate, in *Letters from the East*, pp. 153-56; Ibn al-'Amid, *al-Majmu*, trans. Eddé and Micheau, pp. 114-16.

132 Chevedden, *The Citadel of Damascus*, pp. 78-79. This version of events appears to come from Abu Shama's *al-Dhayl 'ala 'l-rawdatayn*, the sequel to his celebratory history of Nur al-Din and Saladin that has not yet been translated into a European language. Cf. Ibn al-'Amid, *al-Majmu*, trans. Eddé and Micheau, p. 117; Ibn al-Furat, *Tarikh*, trans. Lyons and Lyons, 2:41; Amitai, *Mongols and Mamluks*, p. 32.

133 See, for example, the sieges of Saone (1188), Safed (1188) and Jaffa (1192) discussed in Chapter 5.

134 For Chevedden's discussion of this slighting, derived largely from Ibn Shaddad, see Chevedden, *The Citadel of Damascus*, pp. 87-89.

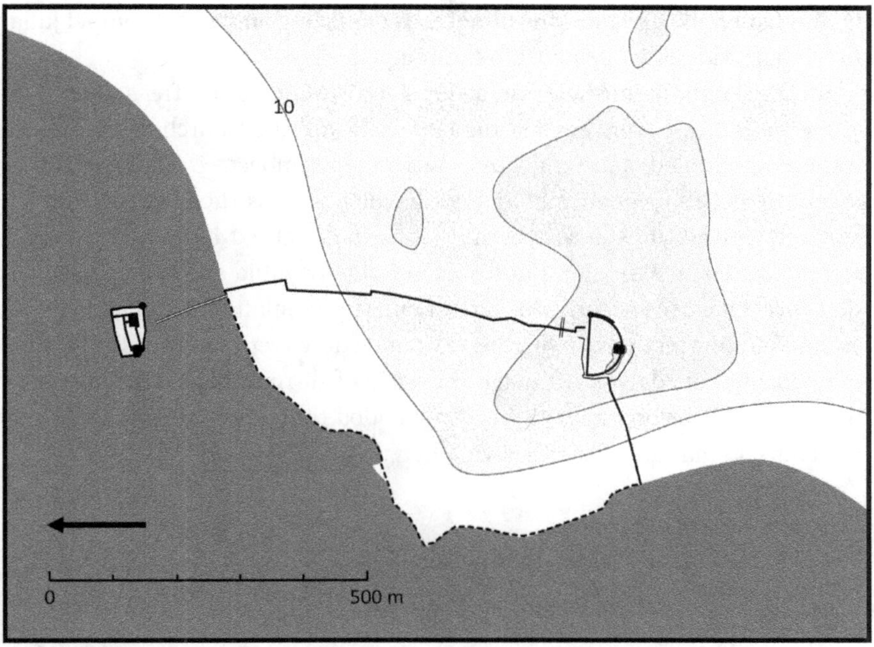

PLAN 25 *Sidon*

built by al-ʿAdil, were meant to be platforms for counterweight trebuchets, defensive artillery does not appear to have played a significant role in this siege.[135]

The Franks managed to evade the Mongols' gaze during the latter's campaign through Syria in 1259-60. Unlike Bohemond VI of Antioch, the Franks of Palestine rejected any notion of Mongol suzerainty.[136] Sidon was the only Frankish possession to be attacked, an act of reprisal after Julian of Sidon led a raid into the Biqaʿ valley from Beaufort. The Mongols do not appear to have employed artillery on this occasion: they broke into Sidon following a frontal assault but were unable to take either of the castles.[137] Although Sidon re-

135 For more on this, see Chapter 10.

136 *Rothelin* 80, RHC Oc 2, pp. 635-36, trans. Shirley, pp. 118-19; Ibn al-ʿAmid, *al-Majmu*, trans. Eddé and Micheau, p. 116. For interpretations of the Mongol invasion of Syria-Palestine in 1260, see Jackson, "Crisis in the Holy Land in 1260," pp. 481-513. For the brief Mongol administration of Syria see Amitai, "Mongol Provincial Administration," pp. 117-43.

137 *Eracles* 34.3, RHC Oc 2, pp. 444-45, trans. Shirley, pp. 141-42; *Annales de Terre Sainte*, ed. Röhricht and Raynaud, B p. 25 (449); Marino Sanudo, *Liber secretorum* 3.12.6, ed. Bongars, p. 221, trans. Lock, p. 350; *Menkonis Chronicon*, ed. Weiland, p. 549; Hayton, *La flor des estoires* 3.24, RHC Ar 2, p. 174; Abu Shama, *Kitab al-Raudatain*, RHC Or 5, p. 203.

mained in Frankish hands, the attack was damaging enough to compel Julian to sell both Sidon and Beaufort to the Templars.

In Acre, preparations were made for a similar encounter. Trees around the city were felled, towers beyond the city walls and the Church of St Nicholas were disassembled, and tombstones were even removed.[138] These measures, meant to deny the enemy materials with which to construct a siege camp and engines, proved unnecessary as no attack materialised before the Mongols were defeated at ʻAyn Jalut. But it was not only Ayyubid and Mongol artillery that threatened the security of Acre's Frankish population. Engines set up by various feuding factions within the city rained down terror at times as tensions periodically boiled over. Although the wave of disruption that accompanied the Mongol invasion of Syria effectively ended the period of Ayyubid rivalry, the Franks remained relatively divided.

The War of St Sabas

Around thirty days of open fighting took place between Pisan and Genoese forces in Acre in 1249. According to the *Eracles* account, "they shot at each other with twenty-two different engines, *perrieres*, *trebuches* and *mangouniaus*" while the *Annales de Terre Sainte* similarly states that twenty-two engines, large and small, were used to throw stones at each other.[139] The Templar of Tyre, who also notes the use of large and small engines, states that "there was one such engine which hurled a stone so large that it weighed 100 *rotas*, and each engine had its own name."[140] While the former accounts emphasise the number of engines, the latter appears to give some indication of the size of the largest trebuchets. Crawford suggests that the *rota* might be associated with the Cypriot *oka*, implying the mass of these projectiles was around 130 kg.[141] The *ratl* might also be considered; however, it seems more likely that the heavier Syrian measures would have been used, rather than the lighter Egyptian

138 *Annales de Terre Sainte*, ed. Röhricht and Raynaud, p. 449, ed. Edbury, p. 157; Marino Sanudo, *Liber secretorum* 3.12.6, ed. Bongars, p. 221, trans. Lock, p. 350. For the church and cemetery, see Pringle, *Churches*, no. 438, 4:151-55.

139 *Eracles* 34.1, RHC Oc 2, p. 437, trans. Shirley, p. 137; *Annales de Terre Sainte*, ed. Röhricht and Raynaud, B p. 19 (443). Cf. Marino Sanudo, *Liber secretorum* 3.12.2, ed. Bongars, p. 218, trans. Lock, p. 346.

140 ... *geterent les uns as autres de plusors manieres d'engins et grans et petis, et tel engin avet quy getet une piere si grant quy pezet c rotes, et avoient les engins chascun son nom.* Templar of Tyre, *Gestes des Chiprois* 270, ed. Minervini, p. 64, trans. Crawford, p. 23. The largest Genoese engines were named *Boverel* and *Vincheguerre*, and another was named *Peretin*. One of the Venetian engines was called *Marquemose*.

141 Templar of Tyre, *Gestes des Chiprois*, trans. Crawford, p. 24 n. 4.

measure. Although there is mounting evidence for the use of increasingly large counterweight trebuchets, there is no physical evidence to support the use of projectiles much heavier than about 60 kg. Furthermore, the use of such a clean figure suggests it was intended to communicate an impression of scale rather than an accurate measure of weight.

The conflict between the Genoese and Venetians entered a new phase from 1256, triggered by a dispute over possession of a building in Acre that belonged to the Orthodox monastery of St Sabas. According to the *Rothelin* account,

> And all that year at least sixty engines, every one of them throwing down onto the city of Acre, onto houses, towards towers and turrets, and they smashed and laid level with the ground every building they touched, for ten of these engines could deliver rocks weighing as much as 1,500 pounds weight of Champagne. This meant that nearly all the towers and strong houses of Acre were destroyed ...[142]

Like many others, this description appears to be emotive rather than literal. Although buildings within the city were certainly damaged during the fighting between the Italian communities, the size of the largest projectiles appears to be a clear exaggeration or a copyist's error. If the measure in question roughly corresponds with the modern Troy pound, these stones would have weighed around 660 kg. This is far too heavy to have been thrown by engines of this period. However, when considering the size of the stones found at Ascalon, it is not unreasonable that stones around 66 kg could have been used, allowing for the possibility that a copyist mistakenly, or for added effect, wrote *XV C* rather than *CXV*, which corresponds with a more reasonable 43 kg.

At the climax of this struggle in 1258, the Templar of Tyre notes that the Genoese gathered a fleet of forty-eight galleys and four *nefs*, each of the latter supported an *engin*, adding a naval element to the artillery battle taking place in Acre.[143] Although the Venetians were eventually able to push the Genoese

142 *Et tout cel an ot bien* LX *angins, qui tuit gitoient a val la cite d'Acre seur les messons et sor les torz et seur les tornelles, et abatoient et fondoient jusques en terre quan qu'il les consivoient, car il i avoit tiex* X *engyns, qui ruoient si grosses pierres et si pesanz que eles pesoient bien* XV C *livres aus poiz de Champainge. Dont il avint que prez que toutes les torz et les forz maissonz d'Acre furent ... Rothelin* 79, RHC Oc 2, p. 635, trans. Shirley, p. 117.

143 Templar of Tyre, *Gestes des Chiprois* 281, ed. Minervini, p. 68, trans. Crawford, p. 27. See also *Eracles* 34.3, RHC Oc 2, p. 443, trans. Shirley, p. 141; Marino Sanudo, *Liber secretorum* 3.12.5, ed. Bongars, pp. 220-21, trans. Lock, pp. 349-50. This was not the first occasion that war had broken out in Acre: Frederick II had besieged the Templar complex in 1229 and the Ibelins had besieged the Hospitaller complex for around six months in the early 1240s,

out of Acre, the rivalry and factional fighting would drag on and continue to divide the Franks,[144] even while another major power was rising in Egypt.

There appears to be evidence that heavy artillery was used with increasing regularity in the decades following the Third Crusade. New terminology was used by Frankish sources to identify engines at the siege of Damietta in 1218-19 and at the intra-Frankish sieges around 1230, while the *manjaniq maghribi* also appears with increasing frequency in Muslim accounts, used to identify only the heaviest trebuchets.[145] Although the most powerful engines were still incapable of breaching fortifications, even after months of bombardment on some occasions, there appears to have been a continual process of technical development and experimentation. This is suggested by the use of only one, or very few, of these larger engines and their perceived impressiveness, evident at sieges such as those of Acre (1189-91), Damietta (1218-19), Kantara (1229-30), Beirut (1231-32), and Homs (1248-49), as well as at Mansura (1250).

The familiarity that the Franks and Muslims had with their rivals' technology, clearly apparent during the Third Crusade, seems to have continued. The frequency with which al-ʿAdil and his descendants moved between the power-centres of Egypt and Syria would have contributed to this diffusion of knowledge: even if certain engines and experts remained tied to specific regions, high-level military figures would have been able to examine closely the engines in each arsenal. Even certain Franks were privy to these peacetime opportunities to investigate a neighbour's artillery. As part of their alliance with al-Salih Ismaʿil, the Franks were permitted to send officials to buy arms, siege engines and war materials from the inhabitants of Damascus. Less than a decade later, Louis IX's artilleryman, John the Armenian, was allowed to visit Damascus to buy horn and glue for crossbows.[146] Such peacetime interactions would have added to the more obvious exchanges of information during hostilities, as various powers arrayed their artillery technology against each other in plain sight.

It is still hard to discern the size of the largest mid-thirteenth-century engines: while masts of ships may have been used to build some, others were apparently erected in just twenty-four hours. Although the projectiles found at

---

Philip of Novara, *Gestes des Chiprois* 138, 222-23, ed. Raynaud, pp. 50, 126-27, trans. La Monte, pp. 91, 172-74. See also *Annales de Terre Sainte*, ed. Röhricht and Raynaud, A and B p. 17 (441).

144 For an extremely hostile opinion of this rivalry and its ultimate cost, see Ludolph of Suchem, *De itinere Terrae sanctae liber* 26, ed. Deycks, p. 42, trans. Stewart, pp. 54-55.

145 See Chapter 7 and Appendix 1.

146 Al-Maqrizi, *al-Suluk*, trans. Broadhurst, pp. 262-63; John of Joinville, *Vie de Saint Louis* 446, ed. and trans. Monfrin, pp. 218-19.

Ascalon indicate the use of substantial engines, many variables are left unknown, inhibiting a better sense of the scale of these machines or the amount of energy they were capable of harnessing. It is during the Mamluk period when many of the various indicative elements that have been examined so far, such as topography, archaeology, and terminology, come together at certain sieges to provide a much clearer impression of the scale and capabilities of these engines.

CHAPTER 7

Mamluk Period

Following the death of al-Salih Ayyub and the murder of his son al-Muʿazzam Turanshah in 1250, ʿIzz al-Din Aybak established Mamluk rule in Egypt. The ongoing intra-Muslim struggle for regional power continued, however, until Hulagu's invasion of western Syria and Palestine in 1260.[1] Qutuz, who had been the effective ruler of Mamluk Egypt since Aybak's murder in 1257, halted the Mongol expansion when he defeated the forces left behind by Hulagu at ʿAyn Jalut on 3 September 1260. Following this watershed event, Qutuz remained in power just long enough to extend Mamluk authority across most of western Syria before he too was murdered in October 1260. It was left to his successor, Baybars, to reveal the potential of Mamluk artillery.[2]

The ascendancy of Baybars marks the start of the final period of Frankish rule in the Levant, culminating in their expulsion from Acre in 1291. This period is documented in the continuations of William of Tyre's chronicle and the *Annales de Terre Sainte*, which remain important sources, while the so called Templar of Tyre provides the most valuable eye-witness account of events from a Frankish perspective. Numerous later historians, including Marino Sanudo, used these works as the basis for their own accounts.

Under Baybars, Egypt and western Syria were once more united under a strong and stable ruler. Like Saladin, Baybars was an attractive subject for biographers and more general historians could hardly ignore his campaigns and military successes. Ibn ʿAbd al-Zahir, Baybars' secretary, wrote a celebratory biography of Baybars and another dedicated to his ultimate successor, Qalawun. Another important eye-witness source was ʿIzz al-Din ibn Shaddad al-Halabi, whose works on the history and geography of the region can be characterised by their direct style, precision and reliability. Unfortunately, significant sections of his work related to Baybars have been lost. Following these men was a

1 For the last portion of these conflicts, see Ibn ʿAbd al-Zahir, *al-Rawd*, trans. Sadeque, pp. 80-93. See also Thorau, *The Lion of Egypt*, pp. 54-69.

2 Qutuz, then sultan, and Baybars, who would soon succeed him, were entertained in Acre ahead of the battle, inadvertently giving Baybars the opportunity to view some of Acre's defences from within, *Rothelin* 81, RHC Oc 2, pp. 637-38, trans. Shirley, pp. 118-19; *Eracles* 34.3, RHC Oc 2, p. 444; Ibn ʿAbd al-Zahir, *al-Rawd*, trans. Sadeque, pp. 193-95; Bar Hebraeus, *Makhtebhanuth Zabhne*, trans. Budge, 1:437-38. Baybars' visit to Acre is emphasised in Boase, *Kingdoms and Strongholds*, p. 204 and Bartlett, *Islam's War*, p. 239. For the murder of Qutuz, see Ibn ʿAbd al-Zahir, *al-Rawd*, trans. Sadeque, pp. 96-98.

 | DOI 10.1163/9789004376922_009

second generation of Mamluk sources who were often well informed and appear to have had access to earlier accounts as well as original documents and oral testimony. Shafi' ibn 'Ali, nephew of Ibn 'Abd al-Zahir, was another important member of the Mamluk secretariat, while Baybars al-Mansuri, one of Qalawun's mamluks, was an active figure in the military. Other key sources include: al-Jazari, al-Yunini, who provided a continuation to Sibt ibn al-Jawzi's account, and Abu'l-Fida', who took part in Mamluk campaigns from the siege of Margat in 1285 before rising to become ruler of Hama in the early fourteenth century. To these Syrian perspectives can be added the work of al-Nuwayri, who was based in Cairo, and that of Ibn al-Dawadari. Later historians, such as Ibn al-Furat, Mufaddal, al-Maqrizi and al-'Ayni are also helpful, at times preserving information from lost accounts and offering different interpretations of events.[3]

Baybars' Early Artillery

In the spring of 1263, Baybars moved his army into Palestine, having resolved to acquire Kerak from al-Mughith 'Umar, his last remaining Muslim rival between Cairo and Aleppo. As a sideshow to this, he briefly tested the strength of Acre's defences from the morning of 14 April until he withdrew from the area on 16 April.[4] Peter Thorau has suggested that Baybars brought siege engines with him on this campaign, but Ibn 'Abd al-Zahir makes no mention of any during the assaults on Acre or at any other point during the ninety-five-day campaign. The source states that craftsmen travelled with the army but that Baybars called only for ladders when they reached Transjordan.[5] It would appear, therefore, that Baybars had little intention of using artillery against either Acre or Kerak, instead opting to use assault troops and miners.

Baybars faced a growing threat from the Mongols in 1263-64, as Hulagu received support from both the Byzantines and his brother Kublai, who had succeeded Mongke. The emboldened Franks were confident enough to launch raids through Palestine in 1264. Baybars responded by ordering limited raids

3 For a discussion of these historians, their works, sources and relation to each other, see Little, *An Introduction to Mamluk Historiography*; Haarmann, "L'étude de la chronique mamelouke," pp. 195-203; Guo, "Mamluk Historiographic Studies," pp. 15-43; al-Yunini, *Dhayl*, ed. and trans. Guo, 1:6-96; Northrup, *From Slave to Sultan*, pp. 25-58; Little, "Historiography," pp. 421-43; Bauden, "Taqi al-Din Ahmad ibn 'Ali al-Maqrizi," pp. 161-200.

4 Ibn 'Abd al-Zahir, *al-Rawd*, trans. Sadeque, pp. 139-40, 164-71, 173-75; Ibn al-Furat, *Tarikh*, trans. Lyons and Lyons, 2:43-44, 56-59; *Eracles* 34.4, RHC Oc 2, pp. 446-47.

5 Ibn 'Abd al-Zahir, *al-Rawd*, trans. Sadeque, pp. 178-82; Thorau, *The Lion of Egypt*, p. 138.

around Caesarea and 'Atlit, weary of committing his forces with the far greater Mongol threat looming beyond the Euphrates.[6]

Baybars was in Egypt when he received word that Mongol forces had laid siege to al-Bira in early 1265. He dispatched an advance force and followed with the bulk of his army, reaching Gaza on 9 February. There, Baybars learned that the Mongols had erected seventeen trebuchets against al-Bira, fifteen of which focused on a certain tower. A letter arrived six days later informing the sultan that the Mongols had withdrawn upon the arrival of the troops that he had sent ahead.[7] Among the instructions that Baybars issued for the refortification of al-Bira, was that wood and stones be collected for the defensive artillery within the town, suggesting that such engines had been present, and presumably used, during the siege.[8] He had additional engines sent to al-Bira from Shayzar, which were in turn replaced by trebuchets from Damascus.[9] Although artillery appears to have been stockpiled in certain strongholds since the middle of the twelfth century, Baybars would repeatedly exploit this network of arsenals with far greater effect than any of his predecessors.

Caesarea: 1265

No longer obliged to relieve al-Bira, and with his army already in Palestine, Baybars took the opportunity to move against the Franks. On 25 February, under the pretence of hunting in the woods near Arsuf, Baybars reconnoitred the area around Caesarea. While he was away, timber to construct artillery and ammunition were brought to the Mamluk camp at al-'Auja. Upon his return, Baybars had 'Izz al-Din al-Afram, *amir jandar*, erect several trebuchets of the *maghribi* and *ifranji* types from the timber that had arrived. Four large engines, as well as some smaller ones, were ready by the end of the following day. The army then moved to an advanced position and at midnight it prepared for the short march to Caesarea, arriving before dawn on 27 February.[10]

---

6 Ibn 'Abd al-Zahir, *al-Rawd*, trans. Sadeque, pp. 186-91, 208-9, 216; Marino Sanudo, *Liber secretorum* 3.12.7, ed. Bongars, p. 222, trans. Lock, p. 351; Ibn al-Furat, *Tarikh*, trans. Lyons and Lyons, 2:67.

7 Ibn 'Abd al-Zahir, *al-Rawd*, trans. Sadeque, pp. 237-39, cf. trans. al-Khowayter, 2:547-49; Abu'l-Fida', *al-Mukhtasar*, RHC Or 1, pp. 473-74; Abu Shama, *Kitab al-Raudatain*, RHC Or 5, p. 205; Mufaddal, *Kitab*, ed. and trans. Blochet, 1:473-74, 503-4. See also Thorau, *The Lion of Egypt*, pp. 158-59.

8 Ibn 'Abd al-Zahir, *al-Rawd*, trans. al-Khowayter, 2:248-50, 252.

9 Ibn 'Abd al-Zahir, *al-Rawd*, trans. al-Khowayter, 2:253.

10 Ibn 'Abd al-Zahir, *al-Rawd*, trans. al-Khowayter, 2:554-55; Ibn al-Furat, *Tarikh*, trans. Lyons and Lyons, 2:69. See also Thorau, *The Lion of Egypt*, p. 160.

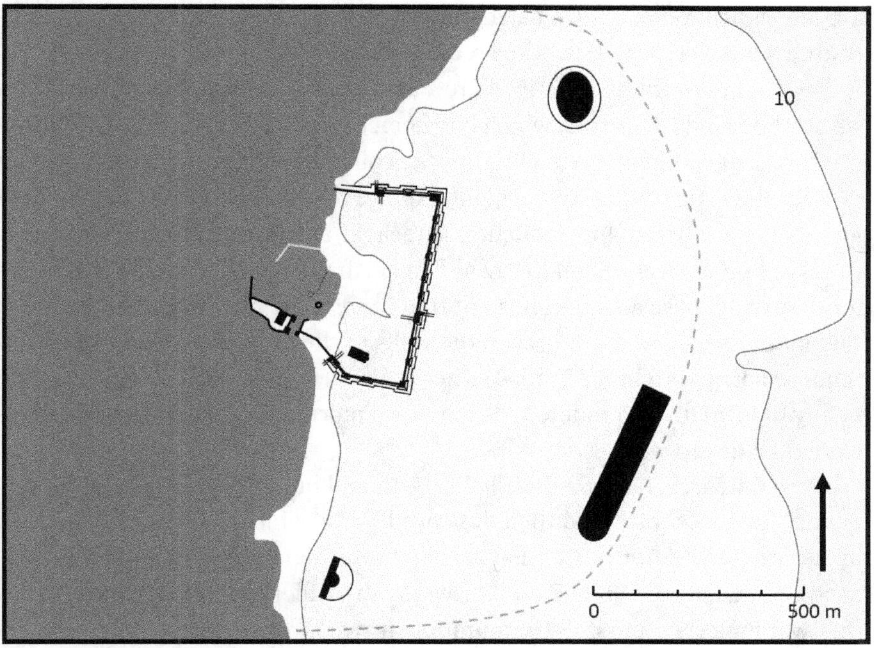

PLAN 26 *Caesarea: defences improved by Louis IX and earlier Roman wall*

Caesarea's twelfth-century defences, well within the trace of the earlier Byzantine wall, had been destroyed by Saladin in 1191 and the town appears to have remained largely defenceless until the refortification efforts of 1217-18. Louis IX later developed the town defences in 1251-52.[11] Despite these efforts, the Mamluks took the town in a frontal assault on the first day of the siege, compelling the Frankish defenders to withdraw to the citadel.[12] Al-'Ayni's assertion that mangonels were used against the town defences appears inaccurate; although the recently prepared engines would have accompanied the army, there would appear to have been little time to deploy and use them.[13] Furthermore, there

11 Pringle, "Town Defences," pp. 89-90; Benvenisti, *The Crusaders*, p. 138; Müller-Wiener, *Castles of the Crusaders*, p. 74; Negev, "Caesarea," (1961), pp. 81-83; Willbrand of Oldenburg 2.2, ed. Pringle, p. 131, trans. Pringle, p. 86. For the town defences and citadel, see Mesqui, "L'enceinte médiévale de Césarée"; Faucher, Merlin and Mesqui, "Césarée," pp. 1-3; Pringle, "Town Defences," pp. 90-91; Benvenisti, *The Crusaders*, pp. 141-44; Negev, "Caesarea," (1960), pp. 264-65. See also, Conder and Kitchener, *Survey of Western Palestine*, 2:13-16, 23-26, 28. Cf. Frova, "Caesarea," pp. 150-51.

12 Ibn 'Abd al-Zahir, *al-Rawd*, trans. al-Khowayter, 2:555-56; Ibn al-Furat, *Tarikh*, trans. Lyons and Lyons, 2:70.

13 Al-'Ayni, *'Iqd al-Juman*, RHC Or 2a, p. 219.

are few indications of impact signatures along the faces of the surviving fortifications. Artillery was, however, erected against the citadel.

Baybars' mastery of logistics is revealed during the second phase of the siege. Thousands of arrows, which had been brought from 'Ajlun, were distributed and arrangements were made for the collection of firewood and ammunition for the artillery. Baybars evidently found the transportation, deployment and then operation of his trebuchets agreeable, as Ibn 'Abd al-Zahir states that robes of honour were given to 'Izz al-Din al-Afram, who had been responsible for preparing these engines ahead of time, and to the men who then operated them. But artillery did not breach the walls of the citadel. Instead, it was surrendered on 5 March: the Franks either evacuated the citadel by sea the night before or formally surrendered the following morning. Caesarea's fortifications were then destroyed.[14]

From Caesarea, Baybars travelled to 'Atlit and burnt its gardens while a secondary force occupied and then destroyed Haifa, which had been abandoned by its garrison.[15] Shortly after Baybars had returned to Caesarea to oversee its destruction, Sayf al-Din al-Zayni arrived with artillery from Subayba, and further war materials, possibly ammunition, arrived from Damascus.[16] These engines were soon put to use against Arsuf.

Arsuf: 1265

John of Ibelin, lord of Beirut, had acquired Arsuf by marriage in the 1190s. The town lay in ruins from the time of the Third Crusade until at least 1211, and may

---

14 Ibn 'Abd al-Zahir, *al-Rawd*, trans. al-Khowayter, 2:556-57; Ibn al-Furat, *Tarikh*, trans. Lyons and Lyons, 2:70-71. See also al-'Ayni, *'Iqd al-Juman*, RHC Or 2a, pp. 219-20; Mufaddal, *Kitab*, ed. and trans. Blochet, 1:474; *Annales de Terre Sainte*, ed. Röhricht and Raynaud, A and B pp. 27-28 (451-52), ed. Edbury, p. 158; Templar of Tyre, *Gestes des Chiprois* 328, ed. Minervini, p. 96, trans. Crawford, p. 44; Marino Sanudo, *Liber secretorum* 3.12.7, ed. Bongars, p. 222, trans. Lock, p. 351; Thorau, *The Lion of Egypt*, pp. 160-61. The citadel was taken on 5 March (15 Jumada I) according to Ibn 'Abd al-Zahir and those following him (such as Ibn al-Furat), 7 March according to the Templar of Tyre, and 8 March (18 Jumada I) according to the Edbury edition of the *Annales de Terre Sainte*, al-Yunini and those following him (such as Mufaddal). The Franks departed by sea in the versions provided by Ibn 'Abd al-Zahir and al-Yunini while Ibn al-Furat claims that the garrison surrendered. The Frankish sources state that the garrison went by sea to Acre but do not specify if this occurred after a formal surrender or not. For testimony of the town's destruction a century and a half later, see John Poloner, *Descriptio Terrae Sanctae*, trans. Stewart, p. 29.

15 Ibn 'Abd al-Zahir, *al-Rawd*, trans. al-Khowayter, 2:561; Ibn al-Furat, *Tarikh*, trans. Lyons and Lyons, 2:72; al-'Ayni, *'Iqd al-Juman*, RHC Or 2a, p. 220.

16 Ibn 'Abd al-Zahir, *al-Rawd*, trans. al-Khowayter, 2:562; Ibn al-Furat, *Tarikh*, trans. Lyons and Lyons, 2:72.

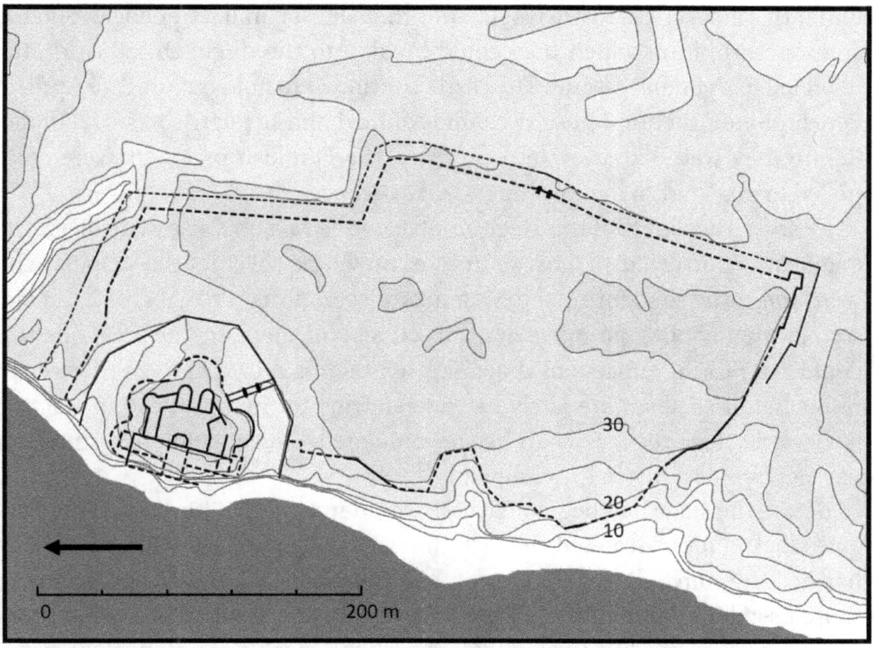

PLAN 27 *Arsuf*

not have been refortified until 1241, when it is recorded that John's son, John of Arsuf, strengthened the site. In 1253 John of Joinville remarked that the castle was strongly fortified. Balian of Ibelin, son of John of Arsuf, sold the site to the Hospitallers in 1261 and the order's subsequent work to develop the castle was used by Baybars as a pretext to reject Frankish overtures of peace in 1263.[17]

Baybars arrived at Arsuf on 15 or 21 March 1265.[18] The Muslims initially attempted to fill the fosse with timber; however, the Franks managed to tunnel

17 Wilbrand of Oldenburg, *Itinerarium terrae sanctae* 2.2, ed. Pringle, p. 132, trans. Pringle, p. 86; *Annales de Terre Sainte*, ed. Röhricht and Raynaud, A and B p. 16 (440), B p. 26 (450), ed. Edbury, pp. 153, 157; Philip of Novara, *Gestes des Chiprois* 220, ed. Gaston, p. 124; John of Joinville, *Vie de Saint Louis* 563, ed. and trans. Monfrin, p. 278; *Eracles* 34.4, RHC Oc 2, p. 446, trans. Shirley, p. 142; Templar of Tyre, *Gestes des Chiprois* 328, ed. Minervini, p. 96, trans. Crawford, p. 44; Marino Sanudo, *Liber secretorum* 3.12.6, ed. Bongars, p. 221, trans. Lock, p. 350; Ibn 'Abd al-Zahir, *al-Rawd*, trans. Sadeque, p. 168. See also Benvenisti, *The Crusaders*, p. 132; Deschamps, *Les Châteaux*, 2:18. See also the account of Ibn Shaddad, in Ibn al-Furat, *Tarikh*, trans. Lyons and Lyons, 2:73. For the site of Arsuf, see Tal, "Apollonia-Arsuf, 2006 and 2009," pp. 108-10; Roll, "Introduction," 41, 44-45; Conder and Kitchener, *Survey of Western Palestine*, 2:137-38.

18 The Frankish sources provide the former date while the latter is given by Ibn 'Abd al-Zahir and Ibn Shaddad, as well as their followers.

under this and set fire to the wood. They then dug a parallel trench beyond the counterscarp, from which they could break into the ditch. On 26 April, the town fell to a general assault. The castle continued to hold out until 30 April, at which point a section of wall was undermined and the garrison surrendered. Before they were led away into captivity, the Frankish prisoners were compelled to take part in demolishing Arsuf's defences.[19]

Nearly all accounts of this siege emphasise the Muslims' use of artillery. The engines are portrayed, in most sources, as providing cover for the sappers, who were the focal element of the Mamluks' siege efforts. Ibn 'Abd al-Zahir describes men hauling on ropes in his discussion of these engines.[20] While this could refer to the process of drawing back the main arm of a counterweight trebuchet, it seems more likely he was referring to the operation of traction trebuchets. Regardless, Baybars had accumulated a considerable assortment of engines by this point in the campaign.

Before the siege had begun, Baybars would have had at his disposal the engines that were constructed ahead of the siege of Caesarea as well as those that had arrived from Subayba.[21] Many of the former may have been no bigger than those used by Saladin in the 1180s, as they required a similar twenty-four hour period to prepare, while the latter were probably sizable counterweight trebuchets. Additional prefabricated engines had also arrived from Damascus, their disassembled components having been carried to Arsuf on men's shoulders. Another engine, which was built on site, is referred to as an engine of seven arrows, suggesting its throwing beam consisted of seven spars, perhaps owing to the shortage of tall and straight trees in the area. Command of this collection of trebuchets fell to 'Izz al-Din al-Afram, who was again celebrated for his conduct in this capacity.[22] The value associated with the prefabricated engines,

19 Ibn 'Abd al-Zahir, *al-Rawd*, trans. al-Khowayter, 2:562-64, 568-72; Ibn al-Furat, *Tarikh*, trans. Lyons and Lyons, 2:73-74, 76-78. See also Templar of Tyre, *Gestes des Chiprois* 328, ed. Minervini, p. 96, trans. Crawford, p. 44; *Eracles* 34.6, RHC Oc 2, p. 450; *Annales de Terre Sainte*, ed. Röhricht and Raynaud, A and B p. 28 (452), ed. Edbury, p. 158; Marino Sanudo, *Liber secretorum* 3.12.7, ed. Bongars, p. 222, trans. Lock, pp. 351-52; al-'Ayni, *'Iqd al-Juman*, RHC Or 2a, pp. 220-21; Amitai, "The Conquest of Arsuf," pp. 69-72; Roll, "Introduction," p. 17; Thorau, *The Lion of Egypt*, pp. 161-62; Benvenisti, *The Crusaders*, pp. 132-33. For the tunnels found under the castle, see Roll, "Introduction," p. 17 n. 36; Amitai, "The Conquest of Arsuf," pp. 70-71.

20 Ibn 'Abd al-Zahir, *al-Rawd*, trans. al-Khowayter, 2:564, 566; Ibn al-Furat, *Tarikh*, trans. Lyons and Lyons, 2:75.

21 Ibn 'Abd al-Zahir, *al-Rawd*, trans. al-Khowayter, 2:554-55, 562; Ibn al-Furat, *Tarikh*, trans. Lyons and Lyons, 2:69, 72.

22 Ibn 'Abd al-Zahir, *al-Rawd*, trans. al-Khowayter, 2:566; Ibn al-Furat, *Tarikh*, trans. Lyons and Lyons, 2:76.

suggests that these were counterweight trebuchets;[23] however, the impracticality of using such machines to support sapping efforts, along with the suggestion that small engines were built ahead of the siege of Caesarea, indicate that a number of traction trebuchets complimented, and probably outnumbered, the heavier engines. This is supported by the physical evidence.

By 2005, archaeological excavations within the castle had uncovered 2,747 stone artillery projectiles, 800 of which were weighed and measured. The average mass of the weighed sample is about 10 kg, with eleven stones weighing over 70 kg and almost fifty weighing less than 3 kg.[24] The impracticality of counterweight trebuchets throwing stones at the light end of this range and traction trebuchets throwing stones at the heavy end confirms the use of both types of engines. The vast majority of projectiles are between 2 and 20 kg, indicating that most were thrown by traction trebuchets.

A concentration of heavier stones, averaging about 25 kg (most 20-35 kg), was discovered just north of the castle's inner gate among some burnt cedar wood.[25] It seems a reasonably powerful engine was used to target a wooden shelter of some kind, which was positioned either between the castle's inner and outer walls or on top of one of them, its remains falling to the ground during or after the siege. The size of the stones suggests that the responsible engine was either counterweight-powered and shooting from a considerable distance or traction-powered and positioned quite close to the castle. In the former scenario, the trebuchet would have been located beyond the town walls, perhaps sited on the rise to the north of the castle or the plateau in front of the north end of the eastern wall. If the engine was instead a traction trebuchet, it would almost certainly have been positioned within the town to the east of the castle, in which case the stones would have been thrown after the Mamluks broke into the town on 26 April.

The surface finds suggest that at least as many projectiles await discovery in the fosse around the castle, which has not yet been excavated, as have been found inside the stronghold. The highest concentrations of identifiable projectiles on the surface of the fill are located in the northern sections of the ditch. Correspondingly, the outer northern walls of the castle reveal the highest concentrations of impact signatures. These indicate that most projectiles cast against the castle were thrown from the north and their number suggests that the bombardment of the castle began before the town was captured. Engines

23 Following the siege they were redistributed among certain strongholds, including Kerak and 'Ajlun, Ibn al-Furat, *Tarikh*, trans. Lyons and Lyons, 2:82.

24 Raphael and Tepper, "The Archaeological Evidence," pp. 85, 87, see also Chart 1A.

25 Raphael and Tepper, "The Archaeological Evidence," p. 87.

IMAGE 7 *Arsuf: projectiles in the castle fosse* (AUTHOR'S PHOTO)

were probably set up on the rise to the north of the castle, about 150 m from its outer walls, on the other side of what seems to have been a slight depression – a great ditch was cut through this area in the second half of the twentieth century, obscuring the natural topography. Following the Mamluks' capture of the town, it is possible that these engines might have been advanced closer to the town wall to a position about 50 m from the castle's northern outer wall.

A significant number of projectiles (1,112), 40% of those identified within the castle, along with 65% of the discovered arrowheads, were found around the inner gate, indicating that this area was targeted during the siege. The concentrations of impact signatures on the northern and eastern faces of the flanking towers reveal that most stones were thrown from the east or northeast. Accordingly, it is possible that these stones, like the larger projectiles used against the wooden shelter just north of the gate, were thrown from beyond the town walls or from inside the town after it had fallen. The latter appears more likely, as engines to the east would almost certainly have been used against the town defences until this line of fortifications was captured. Having gained entrance on 26 April, these engines were then brought within the town and set up in the area to the east and northeast of the castle as siege efforts focused on this last bastion of resistance. Impact signatures on sections of the castle's southern walls confirm that the Mamluks brought artillery into the town and that these machines were employed for a period of at least a few days before the castle surrendered. As engines bombarded the castle from the north and south, it seems the machines shooting at the inner gate were used either to inhibit the Franks' ability to move through the castle or to target something in this relatively open space between the castle's inner and outer walls.

The large number of projectiles recovered (and number still to be excavated from the surrounding fosse), their size and the apparent targeting of gateways and parapets, evident by the number of stones found inside the castle, appear to confirm that the majority of the Mamluk engines were traction trebuchets. Although limited, the power of these engines was still sufficient to warrant importing ammunition. The overwhelming majority of projectiles are limestone, which petrographic analyses suggest were most likely sourced from the base of the Samarian hills, about 15 km to the east.[26] In addition to these imported stones, much smaller numbers of locally sourced sandstone (*kurkar*) as well as a number of marble projectiles were also discovered.

Although most of the projectiles found inside the castle were probably thrown by the Mamluks, it is possible that some of these were intended for defensive use. It is hard to imagine that the garrison would have had much of its original stockpile left after five weeks of siege; however, they could have reused projectiles that were thrown into the castle and some stones may have been thrown back and forth a number of times. When ammunition ran short, the defenders appear to have begun harvesting projectiles from the marble columns in the castle. While it seems impossible to conclude with certainty how many of the marble projectiles were shaped by the defenders, the discovery of a number of columns inside the castle with rounded ends suggest that the defenders may have begun destroying parts of their own stronghold to produce ammunition before it fell. Like the limestone projectiles sourced by the Mamluks, these marble stones were much harder than the limited number carved from the soft local *kurkar*. This made them more effective against opposing engines, but also the soft masonry of the castle if thrown back by the besiegers. It is possible that the Mamluks were responsible for rounding the ends of the columns found inside the castle, taking advantage of the available hard marble for a future siege. This, however, appears unlikely when considering the haphazard way in which many were left incomplete, suggesting work stopped very abruptly, and the inherent difficulties associated with transporting these heavy stones around without a clear final destination – this was the final siege of the campaign.

26 Raphael and Tepper, "The Archaeological Evidence," pp. 87-88.

When Baybars left Arsuf on 11 May 1265, after the demolition of the castle was complete, Ibn al-Furat remarked that the back of Frankish power had been broken.[27] The campaign of 1265 had extended Mamluk influence along the Palestinian coast, leaving Jaffa, an ally at the time, the only port in Frankish hands south of 'Atlit. In so doing, Baybars had expanded and further secured the coastal corridor between Egypt and Syria. The following year, looking to build on these gains and extend his influence across eastern Galilee at the expense of the Franks, Baybars turned on Safed.

Safed: 1266

According to Ibn al-Furat, the castle of Safed was built by the Templars in 1101-2. While some modern historians have accepted this dating, believing it to have been commissioned by Tancred or Hugh of St Omer, others have favoured certain Frankish accounts, such as those of James of Vitry and Marino Sanudo, which credit the castle's construction to Fulk around 1140.[28] The castle was destroyed by Mu'azzam 'Isa but the Templars began to refortify it when it was returned to them in 1240, aided significantly by the efforts of Benedict (Benoit) of Alignan, archbishop of Marseille. Much is still unknown about Safed, including what it looked like before and after the Templars developed it in the twelfth century and the extent to which it was reshaped in the thirteenth century. The large cylindrical keep and much of the southwest gate complex reflect the castle's subsequent development by the Mamluks.[29]

Hulagu's death in 1265 provided Baybars with the opportunity to launch a second consecutive spring offensive against the Franks. In May 1266, he led his

27 Ibn al-Furat, *Tarikh*, trans. Lyons and Lyons, 2:82.

28 Ibn al-Furat, *Tarikh*, trans. Lyons and Lyons, 2:88; James of Vitry, *Historia orientalis* 49, ed. Donnadieu, p. 217, trans. Stewart, p. 25; Marino Sanudo, *Liber secretorum* 3.6.18, ed. Bongars, p. 166, trans. Lock, p. 264. For the earlier foundation date, see Kennedy, *Crusader Castles*, p. 40; Benvenisti, *The Crusaders*, pp. 202-3; Deschamps, *Les Châteaux*, 2:119, 125. For the latter, see Boas, *Crusader Archaeology*, p. 118; Boase, "Military Architecture," pp. 159-60. William of Tyre first mentions the castle as the place where Baldwin III took refuge in 1157, while there appears to be no administrative record of the castle earlier than when it was granted to the Templars by Amalric in 1168, William of Tyre, *Chronicon* 18.13-14, ed. Huygens, 2:828-32, trans. Babcock and Krey, 2:258-62; Pringle, *Churches*, 2:206.

29 *De constructione castri Saphet*, ed. Huygens, pp. 378-87, trans. Kennedy, 190-98; Ibn al-Furat, *Tarikh*, trans. Lyons and Lyons, 2:88-89; al-'Umari, *Masalik al-absar*, trans. Richards, p. 80. For studies of the castle, see Barbé, "Zefat"; Barbé and Damati, "Le château de Safed," pp. 77-93; Pringle, "Reconstructing the Castle of Safad," pp. 139-49; Benvenisti, *The Crusaders*, pp. 199-204; Huygens, "Un nouveau texte," pp. 360-77; Conder and Kitchener, *Survey of Western Palestine*, 1:248-50.

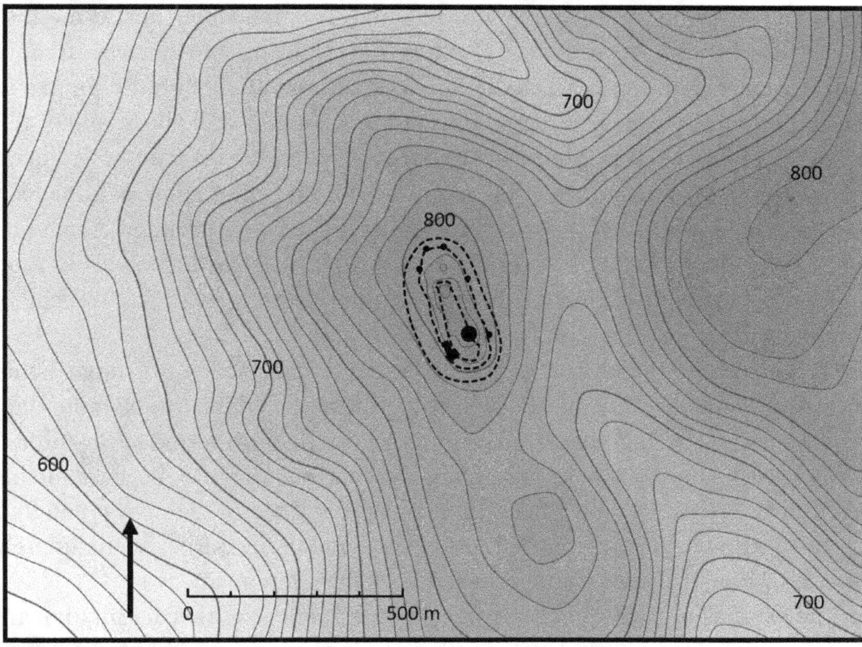

PLAN 28 *Safed*

forces out of Egypt and into Palestine. An advance force, which had been sent to Homs, raided around Tripoli while contingents of the main army raided the coast from Arsuf to Tripoli and inland towards Montfort.[30] At the same time, Syrian forces invaded eastern Galilee, blockading Safed and Beaufort. Coordinating his forces from a position outside the walls of Acre, Baybars sent for artillery from Damascus while new engines were constructed in his camp. With the return of his coastal raiders, Baybars marched on Safed, where he encamped his army on 13 June 1266.[31]

While Baybars waited for his artillery to arrive, receiving envoys in the meantime, his sappers began to work against the castle.[32] The engines from Damascus reached the Jordan on 26 June and ʿIzz al-Din al-Afram was sent to

30 Ibn ʿAbd al-Zahir, *al-Rawd*, trans. al-Khowayter, 2:582, 584-87; Ibn al-Furat, *Tarikh*, trans. Lyons and Lyons, 2:84-87; al-ʿAyni, *ʿIqd al-Juman*, RHC Or 2a, p. 221. See also Templar of Tyre, *Gestes des Chiprois* 346, ed. Minervini, p. 108, trans. Crawford, p. 50; *Eracles* 34.9, RHC Oc 2, p. 454; Marino Sanudo, *Liber secretorum* 3.12.8, ed. Bongars, p. 222, trans. Lock, p. 352; Mufaddal, *Kitab*, ed. and trans. Blochet, 1:490; Thorau, *The Lion of Egypt*, 167-68.

31 Ibn ʿAbd al-Zahir, *al-Rawd*, trans. al-Khowayter, 2:587-88; Ibn al-Furat, *Tarikh*, trans. Lyons and Lyons, 2:89; al-ʿAyni, *ʿIqd al-Juman*, RHC Or 2a, pp. 222-23.

32 Ibn ʿAbd al-Zahir, *al-Rawd*, trans. al-Khowayter, 2:588-91; Ibn al-Furat, *Tarikh*, trans. Lyons and Lyons, 2:89-91.

help convey them from Jacob's Ford to Safed. The first trebuchets came into action on 1 July but it was the less glorious work of the miners that was far more effective at compromising Safed's defences, ultimately forcing the Franks to countermine. On 19 July, a seam appeared in the outer wall above one of the mines. The Franks were compelled to burn and abandon their outer defences, allowing the Muslims to scale the outer wall. Mining efforts then commenced against the inner castle.[33] Shortly thereafter, the garrison surrendered on terms of safe conduct, but were executed after exiting the castle.[34] Upon gaining Safed, Baybars immediately used it as an armoury, storing within it the artillery that he had just used against the stronghold.[35]

The engines brought from Damascus seem to have been quite large, while those built outside Acre may have been significantly lighter. It took more than two weeks for the engines stored in Damascus to be transported to the Jordan, a distance of about 90 km by road. Once they reached the river, Baybars and his entourage are said to have taken part in dragging along the components of these machines that were carried by oxen the remaining 13 km (approximately 20 km by road) to Safed. Among the engines that were ready by 1 July was one of the *maghribi* type.[36] When considering that engines of this designation are noted at the siege of Arsuf the year before, it seems reasonable to postulate that this was a counterweight trebuchet that threw stones weighing upwards of 20 kg. Despite the likelihood that at least some of the Damascene engines were counterweight trebuchets, they are not described as inflicting any notable damage and Ibn 'Abd al-Zahir claims that their components were carried from Damascus to the Jordan on men's shoulders, suggesting these parts were fairly small.[37]

---

33 Ibn 'Abd al-Zahir, *al-Rawd*, trans. al-Khowayter, 2:591-95; Ibn al-Furat, *Tarikh*, trans. Lyons and Lyons, 2:91-93. Cf. al-'Ayni, *'Iqd al-Juman*, RHC Or 2a, p. 222. See also Mufaddal, *Kitab*, ed. and trans. Blochet, 1:490.

34 Ibn 'Abd al-Zahir, *al-Rawd*, trans. al-Khowayter, 2:595-99; Ibn al-Furat, *Tarikh*, trans. Lyons and Lyons, 2:94-96. Cf. Templar of Tyre, *Gestes des Chiprois* 346-47, ed. Minervini, pp. 108-10, trans. Crawford, pp. 50-51; al-'Ayni, *'Iqd al-Juman*, RHC Or 2a, pp. 222-23; Mufaddal, *Kitab*, ed. and trans. Blochet, 1:490-93. See also *Eracles* 34.9, RHC Oc 2, p. 454; *Annales de Terre Sainte*, ed. Röhricht and Raynaud, B p. 28 (452), ed. Edbury, p. 158; Marino Sanudo, *Liber secretorum* 3.12.8, ed. Bongars, pp. 222-23, trans. Lock, p. 352; Thorau, *The Lion of Egypt*, pp. 169-70.

35 Ibn 'Abd al-Zahir, *al-Rawd*, trans. al-Khowayter, 2:600; Ibn al-Furat, *Tarikh*, trans. Lyons and Lyons, 2:96 al-'Ayni, *'Iqd al-Juman*, RHC Or 2a, p. 223.

36 Ibn 'Abd al-Zahir, *al-Rawd*, trans. al-Khowayter, 2:591-92; Ibn al-Furat, *Tarikh*, trans. Lyons and Lyons, 2:91. See also Mufaddal, *Kitab*, ed. and trans. Blochet, 1:490.

37 Ibn 'Abd al-Zahir, *al-Rawd*, trans. al-Khowayter, 2:588, 591; Ibn al-Furat, *Tarikh*, trans. Lyons and Lyons, 2: 89, 91. Ibn 'Abd al-Zahir initially claims that the engines were carried

The artillery that Baybars was said to have commissioned outside Acre, before he turned inland, is not mentioned during the early part of the siege. It was constructed as the sultan paused, allowing elements of the army to raid and position themselves to block any attempts to relieve Safed once the siege began. Because these engines are not mentioned afterwards, it can probably be assumed that, if such engines had been built, they were light traction trebuchets, used to support the work of the miners. The sources' failure to notice these machines during the siege draws a critical distinction between them and what were probably far more impressive counterweight trebuchets brought from Damascus.

The siege of Safed marks a transition point in the use of certain Arabic terms. It is the last siege where sources note the use of a *maghribi* trebuchet, yet takes place after the first of the *ifranji* type are noted, having been mentioned ahead of the siege of Caesarea the previous year. Both terms were used to identify the most powerful engines of the day, but the lack of any overlap, except at Caesarea, suggests that a new technology quickly superseded another or that a newer term, having come into popular use, simply replaced an outdated or traditional one. If the former was the case, it is possible that the largest stones thrown against Safed were slightly smaller than those used at Arsuf, where the *ifranji* engines built in 1265, along with those of the *maghribi* type, were presumably used. If the latter was instead the reality, engines of a comparable scale probably threw stones similar in size to the larger ones found at Arsuf. Without clearly datable projectiles, it is difficult to determine how precise these classifying terms were employed.

It is hard to tell where the Mamluk artillery was positioned at Safed. Unlike at Arsuf, where Baybars is said to have walked between his artillery and siege works, he is described as riding to and from his artillery at Safed. This probably reflects the distance between these engines and other siege works, given the considerable circumference of this stronghold, rather than an exceptional distance between the trebuchets and the castle.[38] The only indication of range comes from Ibn 'Abd al-Zahir's claim that Baybars climbed up the first of the *maghribi* engines ready for use and ordered its crew to target a prominent house within the castle.[39] Although the position of the house is unknown, the

---

from Damascus on men's shoulders, he then states that as they neared Safed they could not be carried by camel, requiring men from all stations to help carry them to the castle. How the oxen that were involved in moving certain components relate is unclear.

38 For example, Ibn 'Abd al-Zahir, *al-Rawd*, trans. al-Khowayter, 2:593; Ibn al-Furat, *Tarikh*, trans. Lyons and Lyons, 2:92.

39 Ibn 'Abd al-Zahir, *al-Rawd*, trans. al-Khowayter, 2:592.

anecdote gives the impression that Baybars was relatively free to select his targets.

Judging from the topography, the plateau to the south of the castle, where the eighteenth-century Saraya was built, would have been the best place to position a besieging force, while the rise to the east, now crowned by the courthouse, would have offered a decent vantage point to observe the castle from another angle. There appear to have been only two positions where artillery could have been set up: to the south, on the fairly spacious saddle between the castle hill and the plateau, and on a much smaller level area just north of the castle. The latter position is within about 100 m of the castle's northern wall, making it an attractive spot to set up traction trebuchets, which could support sapping efforts at this end of the stronghold, but perhaps too restricted and close to the castle's defenders to risk erecting a valuable counterweight engine here. Future excavations will hopefully discover stratigraphically datable projectiles, shedding further light on which sections of the castle's fortifications were targeted and where Baybars positioned his various engines.

The defenders of Safed also appear to have employed artillery. According to the Templar of Tyre, this was used to throw back a gift that Baybars had sent to them at the start of the siege. The stone that killed a certain emir one night seems to have been similarly thrown by a trebuchet.[40] Little more is known about these machines.

Following its capture, Safed would become a critical Mamluk arsenal. However, Baybars does not appear to have taken any of the Frankish engines that he may have found inside the castle, nor any of those he had used to besiege it, on his subsequent campaign into Cilicia. Likewise, there is little to indicate that artillery was used against either Chastel Neuf or Toron, which were both taken shortly after the fall of Safed.[41]

In 1267, Baybars left Egypt and returned to Safed to oversee its development.[42] Although some sources emphasise that he focused on improving the castle ditch, it may have been at this point that construction began on the mas-

---

40 Templar of Tyre, *Gestes des Chiprois* 346, ed. Minervini, p. 108, trans. Crawford, p. 50; Ibn 'Abd al-Zahir, *al-Rawd*, trans. al-Khowayter, 2:592.

41 For Baybars' Cilician campaign, see Ibn 'Abd al-Zahir, *al-Rawd*, trans. al-Khowayter, 2:600, 608-13; Bar Hebraeus, *Makhtebhanuth Zabhne*, trans. Budge, 1:445-46. For the capture of Chastel Neuf and Toron, see Ibn al-Furat, *Tarikh*, trans. Lyons and Lyons, 2:97.

42 Abu Shama, *Kitab al-Raudatain*, RHC Or 5, p. 205; Ibn 'Abd al-Zahir, *al-Rawd*, trans. al-Khowayter, 2:622-23, 628; Ibn al-Furat, *Tarikh*, trans. Lyons and Lyons, 2:101; al-'Ayni, *'Iqd al-Juman*, RHC Or 2a, p. 224.

sive circular keep.[43] In response to a threat from the Mongols, Baybars briefly left Safed for Damascus, where he deposited certain trebuchets. It is unclear whether these engines had been brought from Egypt or were simply brought back from Safed, where they had been kept since the siege the previous summer. While refortification efforts at Safed continued, artillery does not appear to have been used during the assault that was made on Acre in May 1267, much as it had been absent in the similar attack of 1263.[44]

Beaufort: 1268

Threats of another Mongol invasion of Syria led Baybars to move his army from Egypt to Palestine in February 1268. With his army mobilised, he used it to further secure his control of the region and road network between northeast Palestine and Lebanon. Mamluk authority had been extended across eastern Galilee in 1266, but Beaufort still posed a threat to traffic moving to and from Damascus via Banyas and the Hula Valley, as well as those travelling north-south between the Jordan and Biqaʿ Valleys. Baybars wrote ahead to prepare his Syrian governors and ordered Damascene forces to establish a blockade around Beaufort. Coincidentally, the detachment sent from Damascus arrived just ahead of a contingent of the castle's defenders who were returning from the coast.[45]

Ibn ʿAbd al-Zahir calls Beaufort one of the strongest castles in the region and a threat to the territory of Subayba (Banyas).[46] Beaufort was a Templar possession by this point and a clear symbol of Frankish power, visible from Subayba, less than 20 km away across the Hula Valley. The surrounding topography, however, was hardly conducive to launching raids into the north end of the valley. The steep cliffs that rise from the right bank of the Litani, securing the eastern side of the castle, would have forced raiding parties to take a significant detour to the north or west in order to approach and then cross the river. Ibn ʿAbd al-Zahir's remark appears to reflect a pretence for action, rather than a genuine

---

43 See also Pringle, "Reconstructing the Castle of Safad," pp. 143-48; Pringle, *Churches*, 2:206; Pringle, *Secular Buildings*, no. 191, pp. 91-92; Conder and Kitchener, *Survey of Western Palestine*, 1:250; Ibn ʿAbd al-Zahir, *al-Rawd*, trans. al-Khowayter, 2:630.

44 *Annales de Terre Sainte*, ed. Röhricht and Raynaud, A and B, p. 29 (453), ed. Edbury, p. 159; *Eracles* 34.10, RHC Oc 2, p. 455; Ibn ʿAbd al-Zahir, *al-Rawd*, trans. al-Khowayter, 2:623-24; Ibn al-Furat, *Tarikh*, trans. Lyons and Lyons, 2:102-3.

45 Ibn ʿAbd al-Zahir, *al-Rawd*, trans. al-Khowayter, 2:636-37; Ibn al-Furat, *Tarikh*, trans. Lyons and Lyons, 2:105-6. For Haitum's petition for Mongol support, see also Bar Hebraeus, trans. Budge, *Makhtebhanuth Zabhne*, 1:447.

46 Ibn ʿAbd al-Zahir, *al-Rawd*, trans. al-Khowayter, 2:641; Ibn al-Furat, *Tarikh*, trans. Lyons and Lyons, 2:109.

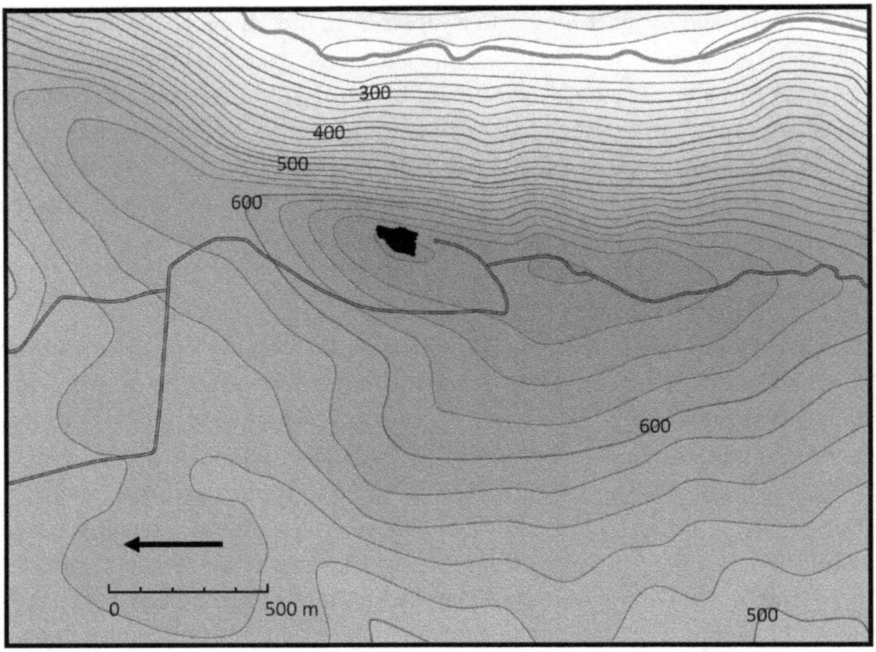

PLAN 29 *Beaufort*

evaluation of the threat posed by the castle. As the only Frankish stronghold within 45 km of Banyas, the capture of Beaufort would extend Mamluk influence as far west as the territory of Tyre and Sidon.

On his way north, Baybars took Jaffa on 7 March. Artillery did not factor into the siege of the town, which was taken in a rapid frontal assault, or the citadel, which surrendered later the same day. The town's defences had been rebuilt by Frederick II and Louis IX but were again destroyed by Baybars before he moved on.[47] The capture of Jaffa brought all of southern Palestine under Mamluk rule and left 'Atlit the southernmost Frankish stronghold on the coast.

While Baybars supervised the destruction of Jaffa, a contingent of the army was sent to Beaufort with orders to intensify the blockade and commence siege operations. This escalation was made possible in part by the arrival of artillery from Safed. Baybars followed with the main army, travelling via Safed to inspect the ongoing work, and reached Beaufort on 3 or 4 April. Two trebuchets

47 Ibn 'Abd al-Zahir, *al-Rawd*, trans. al-Khowayter, 2:638-39; Ibn al-Furat, *Tarikh*, trans. Lyons and Lyons, 2:108; Mufaddal, *Kitab*, ed. and trans. Blochet, 1:503-5 al-'Ayni, *'Iqd al-Juman*, RHC Or 2a, pp. 225-26; *Eracles* 34.11, RHC Oc 2, p. 456; *Annales de Terre Sainte*, ed. Röhricht and Raynaud, A and B p. 29 (453), ed. Edbury, p. 159. See also Thorau, *The Lion of Egypt*, pp. 187-8. Cf. events of 1263, Ibn 'Abd al-Zahir, *al-Rawd*, trans. Sadeque, pp. 237-38.

were ready for action the day after he arrived and a total of twenty-six engines were operational a few days later.[48] These were probably arranged around the western side of the castle. According to Ibn 'Abd al-Zahir, funds contributed to support the Holy War by emirs who could not take part in the campaign were used to help pay the crews working some of the engines.[49]

Beaufort sits on the edge of a plateau overlooking the Litani River. The site was under Damascene control from at least 1121 and appears to have been ceded to Fulk in 1139-40. Following Saladin's acquisition of the castle in 1190, it was developed during the reign of al-'Adil. Although James of Vitry claims that Beaufort, along with Safed and Banyas, was in ruins in 1220, it was evidently defensible when its garrison defied al-Salih Isma'il in 1240. In 1260, following the sack of Sidon, Julian of Sidon sold the castle to the Templars.[50]

The Templars built a second 'citadel' to the south of Beaufort, probably 60-170 m from the south wall, presumably to deny this area of level ground to any potential besiegers. This outwork did not have the desired effect: unable to defend it for more than a week after Baybars' arrival, the Templars evacuated it on the night of 10-11 April 1268, burning everything that they could not carry out with them. The Muslims occupied this secondary strongpoint the following morning and probably moved some of their trebuchets in later that day. Artillery is described as bombarding Beaufort from the outer citadel from 12 April, while the Franks shot back with at least one engine that was mounted on one of the castle's towers.[51]

Despite its proximity and relative safety within the Templar citadel, the Mamluks' artillery does not appear to have compromised the castle's fortifica-

48 Ibn 'Abd al-Zahir, *al-Rawd*, trans. al-Khowayter, 2:640-43; Ibn al-Furat, *Tarikh*, trans. Lyons and Lyons, 2:110-11. See also al-'Ayni, *'Iqd al-Juman*, RHC Or 2a, p. 227; Mufaddal, *Kitab*, ed. and trans. Blochet, 1:506.

49 Ibn 'Abd al-Zahir, *al-Rawd*, trans. al-Khowayter, 2:644; Ibn al-Furat, *Tarikh*, trans. Lyons and Lyons, 2:111-12. Cf. Ibn 'Abd al-Zahir, *al-Rawd*, trans. al-Khowayter, 2:635.

50 Ibn al-Athir, *al-Kamil*, trans. Richards, 1:227 and n. 25; Ibn al-Furat, *Tarikh*, trans. Lyons and Lyons, 2:109-110; Ibn 'Abd al-Zahir, *al-Rawd*, trans. al-Khowayter, 2:641; James of Vitry, *Lettres*, no. 6, p. 124; *Eracles* 34.3, RHC Oc 2, p. 445, trans. Shirley, p. 142; *Annales de Terre Sainte*, ed. Röhricht and Raynaud, B p. 25 (450), ed. Edbury, p. 157. For studies of the castle, see Conder and Kitchener, *Survey of Western Palestine*, 1:128-32; Rey, *Étude sur les monuments*, pp. 127-39; Deschamps, *Les Châteaux*, 2:197-208; Boase, "Military Architecture," p. 160; Pringle, *Secular Buildings*, no. 44, p. 31; Pringle, "Town Defences," p. 78; Kennedy, *Crusader Castles*, pp. 43, 128; Corvisier, "Les campagnes de construction," pp. 243-66; Yasmine, "Die Burg Beaufort," pp. 277-84.

51 Ibn 'Abd al-Zahir, *al-Rawd*, trans. al-Khowayter, 2:643-44; Ibn al-Furat, *Tarikh*, trans. Lyons and Lyons, 2:111.

tions. Instead, it seems to have contributed to what must have been a nightmarish environment inside Beaufort over the following days. The castle ultimately fell after dissent broke out among the defenders following Baybars' interception of a letter, which he substituted with one of his own before passing it along to the garrison. The defenders, as many as 480 fighting men, surrendered themselves into captivity on 15 April, while the women and children were escorted to Tyre.[52] The external citadel, which had proven to be a liability, was destroyed after the siege and a lower ward was subsequently developed on the east side of the castle.[53] After the castle had been occupied, Baybars took his army north, moving against the Mongols' Frankish ally, Bohemond VI of Antioch-Tripoli.

Antioch: 1268

When Baybars left Beaufort on 25 April 1268, he sent his heavy baggage, which probably included his artillery, to Damascus, while detachments of his army were sent out in various directions. Without the burden of a baggage train and the smaller detachments of raiders masking the movement of his main force, Baybars moved to Baalbek and from there was able to take the difficult road across Mount Lebanon and surprise the Franks of Tripoli by approaching from the south.[54] This allowed him to raid freely around the city; however, without his artillery, he declined to invest it or any of the nearby Frankish castles.[55] Instead, he continued on to Antioch, travelling via Homs, Apamea and Jisr al-Shughr.

Baybars arrived at Antioch on 15 May and baggage, from an unclear origin, arrived three days later. The city was stormed on 19 May and the citadel

52 Ibn 'Abd al-Zahir, *al-Rawd*, trans. al-Khowayter, 2:642-45; Ibn al-Furat, *Tarikh*, trans. Lyons and Lyons, 2:210-12; Mufaddal, *Kitab*, ed. and trans. Blochet, 1:506-7. Cf. *Annales de Terre Sainte*, ed. Röhricht and Raynaud, A and B p. 30 (453), ed. Edbury, p. 159; Templar of Tyre, *Gestes des Chiprois* 365, ed. Minervini, p. 124, trans. Crawford, p. 59; *Eracles* 34.11, RHC Oc 2, p. 456.

53 Ibn 'Abd al-Zahir, *al-Rawd*, trans. al-Khowayter, 2:645-46; Ibn al-Furat, *Tarikh*, trans. Lyons and Lyons, 2:112.

54 Ibn 'Abd al-Zahir, *al-Rawd*, trans. al-Khowayter, 2:646-48, 653; Ibn al-Furat, *Tarikh*, trans. Lyons and Lyons, 2:116. See also al-'Ayni, *'Iqd al-Juman*, RHC Or 2a, p. 227; Mufaddal, *Kitab*, ed. and trans. Blochet, 1:507; *Annales de Terre Sainte*, ed. Edbury, p. 160.

55 Ibn 'Abd al-Zahir, *al-Rawd*, trans. al-Khowayter, 2:654-56, 685-87; Ibn al-Furat, *Tarikh*, trans. Lyons and Lyons, 2:116-18. Cf. Mufaddal, *Kitab*, ed. and trans. Blochet, 1:507. Cf. Saladin's similar move against Beirut in 1182 (in Chapter 3) and Nur al-Din's acquisition of Munaytira, commanding the pass between Baalbek and Tripoli, in 1165-66, Ibn al-Athir, *al-Kamil*, trans. Richards, 2:161.

surrendered the following day.[56] Notwithstanding Bartlett's suggestion that the Mamluks broke into the city through a breach, the speed with which the Mamluks were able to enter the town, along with their numerical superiority and the extended circumference of Antioch's defences, strongly suggest that the city was stormed in a frontal attack the day after the baggage arrived.[57]

The remaining Frankish strongholds of the north fell shortly afterwards. Some were taken but most were abandoned, leaving Cursat, the patriarch's castle, the only Frankish stronghold north of Jisr al-Shughr by the end of 1268. With Frankish influence in the region all but removed, Baybars extended his authority farther north through a peace agreement with the Armenians. In return for freeing Hethum of Armenia's son and heir, Leo, who had been captured during his invasion of Cilicia in 1266, Baybars acquired a number of strongholds, further securing his northern flank.[58]

Although artillery does not appear to have been used at Antioch, it is highlighted in the letter that Baybars sent to Bohemond soon after the city's fall, bragging of his success. Among the other exploits that he recounts, Baybars claims that he cut all of the trees in the area save those that could be used for the beams of artillery.[59] Certain areas of Western Syria, unlike Palestine, would have supported tall coniferous trees. Although not typically as strong as certain hardwoods, these trees were often tall and straight, making them ideal for constructing large counterweight trebuchets. While Baybars does not appear to have used any of these trees at this point, his suggestion that he could is a reflection of his prowess: this region, effectively a farm for artillery timber, was now his. The association of these large trees with artillery, along with the

56 Ibn 'Abd al-Zahir, *al-Rawd*, trans. al-Khowayter, 2:656-58, 678; Ibn al-Furat, *Tarikh*, trans. Lyons and Lyons, 2:121-22, 125-26; Mufaddal, *Kitab*, ed. and trans. Blochet, 1:508. See also *Eracles* 34.11, RHC Oc 2, pp. 456-57; Templar of Tyre, *Gestes des Chiprois* 365, ed. Minervini, p. 124, trans. Crawford, p. 59; *Annales de Terre Sainte*, ed. Edbury, p. 159; Marino Sanudo, *Liber secretorum* 3.12.9, ed. Bongars, p. 223, trans. Lock, p. 353; Bar Hebraeus, *Makhtebhanuth Zabhne*, trans. Budge, 1:448. For the city's state of destruction in the fourteenth century, see Ibn Battuta, *Tuhfat* 1.1, trans. Gibb, p. 61.

57 Bartlett, *Islam's War*, p. 247.

58 Ibn 'Abd al-Zahir, *al-Rawd*, trans. al-Khowayter, 2:678-80, 682-85; Ibn al-Furat, *Tarikh*, trans. Lyons and Lyons, 2:126-27; Mufaddal, *Kitab*, ed. and trans. Blochet, 1:514; al-'Ayni, *'Iqd al-Juman*, RHC Or 2a, p. 228; *Eracles* 34.11, RHC Oc 2, p. 457; Templar of Tyre, *Gestes des Chiprois* 365, ed. Minervini, p. 124, trans. Crawford, p. 59; Vahram, *Chronicle*, trans. Neumann, pp. 50-53.

59 Ibn 'Abd al-Zahir, *al-Rawd*, trans. al-Khowayter, 2:659, trans. Gabrieli in *Arab Historians*, p. 310; Mufaddal, *Kitab*, ed. and trans. Blochet, 1:510; al-'Ayni, *'Iqd al-Juman*, RHC Or 2a, p. 230.

heaviest stones found at Arsuf, confirms the steadily increasing scale of the most powerful trebuchets. Through the 1270s, these heaviest engines appear to have been used in greater numbers.

Indications of Significant Power: 1271

The Ninth Crusade caused Baybars considerable worry. His fears of another invasion of Egypt were alleviated with the crusade's collapse in Tunisia and the death of Louis IX in August 1270, freeing Baybars to march his Egyptian army into Palestine in early 1271. After raiding around Tripoli and taking the Templar castle of Chastel Blanc, he prepared to invest the most formidable Frankish castles of the Homs-Tripoli corridor.[60]

Crac des Chevaliers: 1271

Crac des Chevaliers sits on a basalt spur at the south end of the Syrian Coastal Mountains. Originally constructed under Kurdish rule in the eleventh century, Crac had fallen to the Franks by the end of the first decade of the twelfth century and was acquired by the Hospitallers in the 1140s. The order may have begun to rebuild the castle from 1157; however, all that survives appears to postdate the earthquakes of 1170 and most of what is visible was built after the destructive tremors of 1202.[61] Crac was a frontier stronghold, as threatening to neighbouring Hama and Homs as it was threatened by them.[62] At the end of January 1270, Baybars had taunted the garrison, marching up to the castle walls

---

60 Ibn 'Abd al-Zahir, *al-Rawd*, trans. al-Khowayter, 2:742; Ibn al-Furat, *Tarikh*, trans. Lyons and Lyons, 2:143. For the castle of Chastel Blanc, see Rey, *Étude sur les monuments*, pp. 85-92; Deschamps, *Les Châteaux*, 3:249-58; Piana, "Die Templerburg Chastel Blanc," 293-301.

61 Ibn 'Abd al-Zahir, *al-Rawd*, trans. al-Khowayter, 2:746-47; Ibn al-Furat, *Tarikh*, trans. Lyons and Lyons, 2:144-45; Mufaddal, *Kitab*, ed. and trans. Blochet, 1:530; Rey, *Étude sur les monuments*, pp. 265-70; Boase, *Kingdoms and Strongholds*, p. 80; Müller-Wiener, *Castles of the Crusaders*, p. 60; Zimmer and Meyer, "Le Krak des Chevaliers," p. 360; Kennedy, *Crusader Castles*, pp. 145-47. For comprehensive studies of Crac des Chevaliers, see Deschamps, *Les Châteaux*, vol. 1; Biller, et al., *Der Crac des Chevaliers*; Zimmer, Meyer and Boscardin, *Krak des Chevaliers*; Mesqui, "Les programmes architecturaux"; Mesqui, "La troisième enceinte." See also Rey, *Étude sur les monuments*, pp. 39-67; Boase, "Military Architecture," pp. 153-56; Kennedy, *Crusader Castles*, pp. 145-63; Mesqui, "La fortification des Croisés," pp. 20-22.

62 See Major, "Al-Malik al-Mujahid," pp. 61-75. Cf. Ibn Jubayr, *Rihla*, trans. Broadhurs, pp. 265, 268.

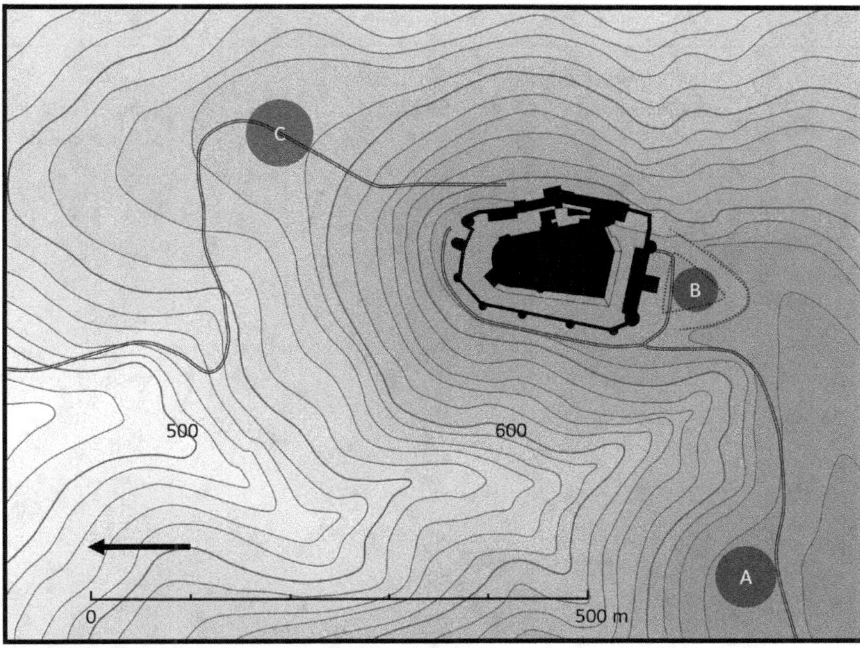

PLAN 30 *Crac des Chevaliers*

without his baggage.[63] When he returned the following year, his intentions were very different.

Baybars' siege of Crac des Chevaliers in 1271 can be divided into four phases, as the sources agree that certain obstacles were captured or overcome on 4/5 March, 21 March, and 29/30 March or 7/8 April.[64] However, the differing vocabulary used by the sources allows for different interpretations of events when reconciling the surviving accounts with the architectural evidence (Fig. 14).

63 Ibn 'Abd al-Zahir, *al-Rawd*, trans. al-Khowayter, 2:729-30; Ibn al-Furat, *Tarikh*, trans. Lyons and Lyons, 2:139, 141.

64 Ibn 'Abd al-Zahir, *al-Rawd*, trans. al-Khowayter, 2:743-45; Ibn al-Furat, *Tarikh*, trans. Lyons and Lyons, 2:145-46; Ibn Shaddad, from Deschamps, *Les Châteaux*, 1:132-34 and King, "The Taking of Le Krak," pp. 88-89; Mufaddal, *Kitab*, ed. and trans. Blochet, 1:527-29; Templar of Tyre, *Gestes des Chiprois* 376, ed. Minervini, pp. 136-38, trans. Crawford, p. 66; *Annales de Terre Sainte*, ed. Röhricht and Raynaud, B p. 31 (455), ed. Edbury, p. 160; Marino Sanudo, *Liber secretorum* 3.12.11, ed. Bongars, p. 224, trans. Lock, p. 355. Al-'Ayni gives an abbreviated version of Ibn 'Abd al-Zahir's timetable but dates events a month later. He also provides the account of Ibn Kathir, who confirms that the Franks surrendered ten days after the castle was breached, al-'Ayni, *'Iqd al-Juman*, RHC Or 2a, pp. 237-38.

FIGURE 14 *Timeline of the siege of Crac des Chevaliers, 1271*

	Ibn 'Abd al-Zahir **Ibn al-Furat**	**Ibn Shaddad** **Mufaddal**	***Annales de TS*** **Templar of Tyre** **Marino Sanudo**
18 February (6 Rajab)			Siege began
21 February (9 Rajab)	Baybars arrived/ castle first attacked		
3 March (19 Rajab)		Baybars arrived and artillery prepared	
4 March (20 Rajab)	*Rabad* taken		
5 March (21 Rajab)		First *bashura* fell	
14 March (30 Rajab)	A number of trebuchets had been prepared		
21 March (7 Sha'ban)	*Bashura* stormed	Second *bashura* fell	
29 March (15 Sha'ban)		Third *bashura* fell to mining Hospitallers massacred	
30 March (16 Sha'ban)	Breaches made in a tower and citadel taken by storm Artillery (possibly two engines) moved into the citadel to assail the keep		
7 April (24 Sha'ban)	Franks left after surrendering		
8 April (25 Sha'ban)			Castle fell

Paul Deschamps understood *bashura* to mean a defensive work ahead of a gate – a barbican in the European sense of the term. Accordingly, he concluded that the Mamluks attacked from the east, moving up the great ramp and past its three gates. The last of these gave the attackers access to the inner ward of the castle and it was here that Baybars intended to set up his artillery and bombard the three inner southern towers, to which the garrison had withdrawn.[65]

65 Deschamps, *Les Châteaux*, 1:132-34. See also Hillenbrand, *The Crusades*, p. 538, cf. p. 525.

D.J. Cathcart King challenged this theory and convincingly argued that the main thrust of the attack came from the south. For King, the triangular outwork to the south of the stronghold (B on the plan) and the castle's inner and outer walls were the three lines of defences identified by the sources. Accordingly, if the town beyond the castle, which Rey claims was walled,[66] fell on 4/5 March and the outwork was taken on 21 March, the outer wall of the castle would have been that which fell to mining on 29/30 March. The Franks thus continued to occupy a relatively strong position, still in control of the entire inner castle, when they capitulated just over a week later.[67] This theory explains why the outer southwest tower, which was undermined on 29/30 March, was rebuilt immediately after the siege and how the defenders were able to secure quite favourable terms of surrender.[68]

If the timing provided by the sources is correct,[69] Baybars appears to have initially erected his artillery to the south of the castle to play against the triangular outwork. Lighter engines, and possibly heavier ones, were probably then advanced onto or beside the outwork to support the sappers working against the outer enceinte from 21 March, although this is not mentioned in the sources. Symptomatic of his enthusiastic view of artillery, Kennedy has suggested that artillery was partly responsible for breaching the outer walls and that it was in part the efforts to move these engines inside the outer trace that compelled the defenders of the inner castle to sue for terms.[70] Besides the challenge of moving deconstructed components through the breach and then assembling them under fire from the defenders, artillery would have had little effect against the inner castle if it were erected between the two lines of walls, given the limited space and relative elevation of the inner enceinte. A more practical position from which to assail the inner castle would have been the outwork, south of the outer wall. This was such an ideal position that it may

66 Rey, *Étude sur les monuments*, p. 40. This town would appear to have contained a parish church, Delaville le Roulx, *Cartulaire*, nos. 3282, 3307, pp. 165-66, 183.

67 King, "The Taking of Le Krak," pp. 90-92. This theory has stood largely unchallenged, for a slight variant, see Mesqui, "Les programmes architecturaux," pp. 10-11.

68 For different theories of the development of Crac's outer defences, see Mesqui, "Le troisième enceinte"; Biller, et al., *Der Crac des Chevaliers*, pp. 185-253, 263-70; Zimmer, Meyer and Boscardin, *Krak des Chevaliers*.

69 Peter Thorau has observed that if attention is moved away from Ibn Shaddad and his followers towards Ibn 'Abd al-Zahir and Baybars al-Mansuri, who were present at the siege, the timing of events between the Muslim and Frankish accounts falls closely into line, Thorau, *The Lion of Egypt*, Appendix 5, pp. 265-66.

70 Kennedy, *Crusader Castles*, pp. 109, 150.

IMAGE 8 *Crac des Chevaliers: castle from the west side of the southern outwork* (PHOTO COURTESY OF DENYS PRINGLE)

have been here where, according to some sources, Baybars positioned himself at one point to shoot at the castle with his crossbow.[71]

The sources do not emphasise the use of artillery during this siege. The Muslim accounts agree generally that it was prepared at some point before 14 March and Ibn 'Abd al-Zahir adds that the Mamluks attempted to bring engines (two according to Ibn al-Furat's subsequent rendition) forward to use against the innermost part of the castle. The Frankish accounts of the siege are less detailed. One version of the *Annales de Terre Sainte* states simply that *enginz* were used in addition to mining.[72] If Ibn 'Abd al-Zahir's timing of events is correct and notable engines were ready by 14 March, eleven days after Ibn Shaddad claims that artillery preparations began, this was twenty-one days after the army had arrived and a week before the outwork was taken. This would

71 Ibn 'Abd al-Zahir, *al-Rawd*, trans. al-Khowayter, 2:744; Ibn al-Furat, *Tarikh*, trans. Lyons and Lyons, 2:145. Cf. the similar reference to Baybars shooting at the citadel of Caesarea from a church, presumably the Cathedral of St Peter, Ibn 'Abd al-Zahir, *al-Rawd*, trans. al-Khowayter, 2:557; Ibn al-Furat, *Tarikh*, trans. Lyons and Lyons, 2:71.

72 *Annales de Terre Sainte*, ed. Edbury, p. 160. For a study of this source and its various versions, see Edbury, "Making Sense of the *Annales de Terre Sainte*," pp. 403-13.

IMAGE 9 *Crac des Chevaliers: castle from the southwest* (PHOTO COURTESY OF DENYS PRINGLE)

IMAGE 10 *Image 10. Crac des Chevaliers: castle from the northwest* (PHOTO COURTESY OF DENYS PRINGLE)

have been sufficient time to both erect and employ machines of a significant size.

A number of artillery projectiles can be found inside the castle; however, a few stones, apparently unused ammunition, discovered to the southwest of the stronghold (A on the plan) more clearly date to the siege of 1271. Biller and Burger suggest that this was where Baybars placed his artillery, 250 m from both the triangular outwork and outer southwest tower.[73] They assume that thirteenth-century artillery was capable of ranges up to 500 m but cite no evidence to support this. Zimmer, Meyer and Boscardin place these stones 300-350 m from the castle and have postulated that Baybars' masons may have shaped projectiles here before they were brought forward to wherever the engines were positioned.[74] There is little corroborating evidence to support the notion that these projectiles, some weighing around 100 kg, could have been thrown at the castle from the position where they were found with any hopes of hitting it. To reach a target 250 m away, a projectile must be thrown upwards of 50 m/s, requiring 125,000 J be transferred to a stone weighing 100 kg.

The sources do not specify what Baybars' artillery was targeting, but the timing of events suggests that these engines were used first against the timber palisade or such other light defences that most likely surrounded the southern outwork. After the outwork had been taken, the Mamluks probably advanced their lighter engines to support the sappers who began working against the outer defences of the castle. Larger engines may also have been moved, although this would have been a more laborious task. Regardless, the outwork would have provided an ideal elevated platform for both traction and counterweight trebuchets. Accordingly, the castle may have been assailed from the very outwork that was originally constructed to deny that position to a besieger, as had happened at Beaufort three years earlier.

It is hard to imagine stones weighing 100 kg being thrown against the southwest tower at the same time as the sappers, who eventually brought down this tower, were working.[75] It is possible that the two strategies were alternated, using powerful engines to further weaken undermined sections of masonry, as may have occurred at Darum in 1192,[76] or that artillery was instead used to

73 Biller, et al., *Der Crac des Chevaliers*, pp. 285-89. See also Biller, et al., "The Crac des Chevaliers," p. 11 fig. 4.

74 Zimmer, Meyer and Boscardin, *Krak*, p. 285, see also p. 284 figs. 5.70-71. My thanks to Letizia Boscardin and Werner Meyer for confirming the size and position of these stones.

75 This tower appears to sit on the bedrock so it is unlikely that the sappers were working underground, this deprived the protection that naturally accompanied a subterranean tunnel.

76 See Chapter 5.

target the southern battlements, exposing the castle's defenders. Despite the likelihood that artillery was used against the outer enceinte in some way, there is neither textual nor physical evidence of such: the southern walls were refaced following the siege, removing any blemishes, and no projectiles have been identified in the southern ditch.

The only other direction from which artillery could have been employed against Crac was the north. There is a level area (C on the plan) some distance from the castle and at a significantly lower elevation, within the sprawl of the present town. The slope is less steep on this side than to the east and west and it is possible that a closer position has since been obscured by the development of the town. Regardless, there do not appear to be any impact signatures along the northern walls, suggesting that if any engines were used against this side of the castle they were light traction trebuchets. Although it is possible that damage inflicted by a counterweight trebuchet was concealed when the Mamluks developed the towers flanking the small northern gate, this damage would have been extremely limited given the extent of the rebuilding.

After the siege, Baybars appointed 'Izz al-Din al-Afram, along with 'Izz al-Din Aybak al-Salihi, to repair the castle.[77] Because Ibn 'Abd al-Zahir associates 'Izz al-Din al-Afram with Baybars' artillery more than any other emir, it is tempting to suggest that his selection to oversee the castle's repair may have been connected with the increasing power and effective use of such engines. However, no significant trebuchets appear to have been left at Crac at this point; instead, the artillery that had been brought to Crac was prepared for transport and, on 29 April, was moved with the army to 'Akkar.

'Akkar: 1271

On the opposite side of the Homs-Tripoli corridor, 'Akkar was tucked into the north end of Mount Lebanon, overlooking the Nahr al-Kabir valley. The site, fortified from the eleventh century, had been acquired by the Franks around the same time as Crac des Chevaliers. The castle was briefly taken by Nur al-Din's forces in the 1160s but thereafter remained under Frankish control.[78]

---

77 Ibn 'Abd al-Zahir, *al-Rawd*, trans. al-Khowayter, 2:745; Ibn al-Furat, *Tarikh*, trans. Lyons and Lyons, 2:145-46; Mufaddal, *Kitab*, ed. and trans. Blochet, 1:529. For 'Izz al-Din's earlier activities, see Ibn 'Abd al-Zahir, *al-Rawd*, trans. al-Khowayter, 2:554-55, 557, 566, 591; Ibn al-Furat, *Tarikh*, trans. Lyons and Lyons, 2:69, 71, 76, 91.

78 Ibn 'Abd al-Zahir, *al-Rawd*, trans. al-Khowayter, 2:751-52; Ibn al-Furat, *Tarikh*, trans. Lyons and Lyons, 2:147-48; Ibn al-Qalanisi, *Dhail*, trans. Gibb, p. 93. For the castle of 'Akkar, see Fournet and Voisin, "Le Château de Aakkar," pp. 149-63; Deschamps, *Les Châteaux*, 3:307-9; Müller-Wiener, *Castles of the Crusaders*, p. 50; Kennedy, *Crusader Castles*, pp. 20, 67-68.

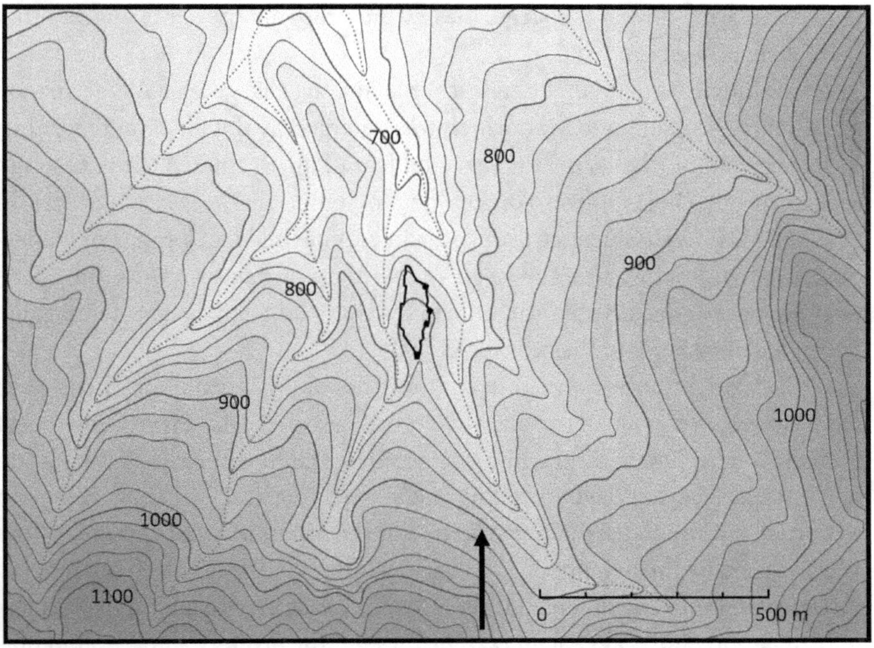

PLAN 31 *'Akkar*

Baybars is said to have ridden on top of the carts that transported his trebuchets as they crossed the rough terrain between Crac and 'Akkar. Upon arriving, the sultan determined where to deploy his artillery while masons worked to level the ground, creating roads for the carts and probably also developing suitable positions where the engines would be placed. Work began to erect the large trebuchets on 2 May, the same day that a stone from a defensive engine killed a certain emir. On 11 May, the defenders asked for terms, surrendering the following day in exchange for their freedom.[79]

This appears to be the first reference in these sources to artillery being transported by cart, rather than by camels, oxen or hand. Although it could be argued that this was the default means of moving these machines, the reference to 'large' engines, which were ready for use on 2 May, suggests that these were significantly bigger than smaller contemporary varieties, some of which may have come into action a day or two earlier. As no artillery appears to have

79 Ibn 'Abd al-Zahir, *al-Rawd*, trans. al-Khowayter, 2:748-49; Ibn al-Furat, *Tarikh*, trans. Lyons and Lyons, 2:148; al-'Ayni, *'Iqd al-Juman*, RHC Or 2a, p. 242. See also Abu'l-Fida', *al-Mukhtasar*, RHC Or 1, p. 153. The martyrdom of Rukn al-Din Mankuwars al-Dawadari in this way, while praying, bears resemblance to that of Badr al-Din al-Aydimuri at Safed in 1266, Ibn 'Abd al-Zahir, *al-Rawd*, trans. al-Khowayter, 2:592.

joined the army during the siege, these were probably the same engines that had been used against Crac.

According to Mufaddal, the trebuchet facing the east gate of ʻAkkar opened a breach in the wall on 10 May, which was enlarged by the following morning. In this version of events, the damage and fear of an assault led the garrison to seek terms.[80] The Frankish sources do not mention the Mamluks' use of artillery,[81] but its significance is once again emphasised in a letter sent by Baybars to Bohemond VI after the siege. Rather than focusing on the scale of the engines directly, he stressed the difficulties that had been overcome dragging these machines up into the hills to ʻAkkar.[82]

Baybars might have planned to continue his campaign, making further use of these engines against Tripoli, but he was distracted by Prince Edward of England's arrival in Palestine on 9 May, leading him to withdraw to Damascus. It is unclear whether Baybars left his engines at Crac and ʻAkkar or took them with him to Damascus; likewise, there is no mention of what became of the engines used by the defenders of ʻAkkar, or those that the Franks had probably incorporated into the defence of Crac. When Baybars returned to inspect the refortification efforts at Crac in 1272, he ordered the artillery stones found outside the castle to be brought inside. He then travelled to ʻAkkar, where he had the garrison shoot the engines that were then inside the castle, allowing him to examine their range.[83] While this reveals that artillery was stored at each in 1272, if trebuchets had been there all the while, it is unclear why these arrangements were not made when Baybars visited both sites in the autumn of 1271.[84] This ambiguity also prevents a clear appreciation of whether the engines used against Crac and ʻAkkar were those employed at Montfort in June 1271; after leaving ʻAkkar in May, Baybars made stops at Damascus and Safed, where trebuchets are known to have been stored, before investing the Teutonic castle.

---

80 Mufaddal, *Kitab*, ed. and trans. Blochet, 1:532.

81 Templar of Tyre, *Gestes des Chiprois* 376, ed. Minervini, pp. 136-38, trans. Crawford, p. 67; *Eracles* 34.14, RHC Oc 2, p. 460.

82 Ibn ʻAbd al-Zahir, *al-Rawd*, trans. al-Khowayter, 2:749-51.

83 Ibn ʻAbd al-Zahir, *al-Rawd*, trans. al-Khowayter, 2:778.

84 Ibn ʻAbd al-Zahir, *al-Rawd*, trans. al-Khowayter, 2:753-54; Ibn al-Furat, *Tarikh*, trans. Lyons and Lyons, 2:150; Mufaddal, *Kitab*, ed. and trans. Blochet, 1:532-36. For the arrival of Edward of England, see *Annales de Terre Sainte*, ed. Edbury, pp. 160-61; Templar of Tyre, *Gestes des Chiprois* 376, ed. Minervini, p. 138, trans. Crawford, p. 67. For Baybars' visit in late September 1271, see Ibn ʻAbd al-Zahir, *al-Rawd*, trans. al-Khowayter, 2:768-69.

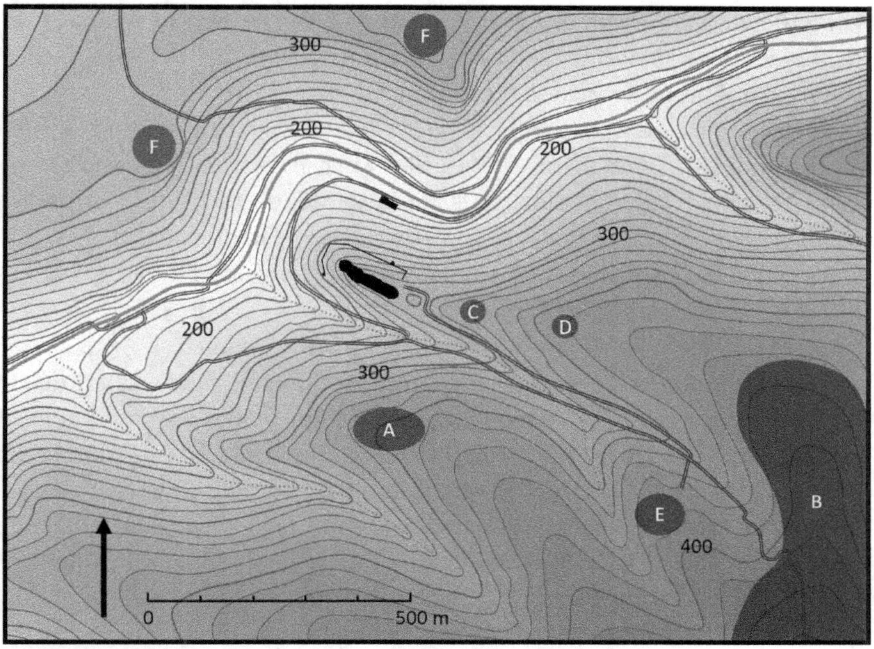

PLAN 32 *Montfort*

Montfort: 1271

In 1271, Montfort was the most impressive castle held by the Teutonic Knights. Construction had begun in about 1226, coinciding with the fortification efforts at Jaffa, Caesarea and Sidon during Frederick II's visit, and work on the castle appears to have continued through much of its history.[85] The stronghold sits on a spur, about 180 m above Wadi al-Qarn to the north and west, and one of its tributaries to the south.[86] The castle's eastern end is secured by a large D-shaped keep and a fosse, with a second ditch about 50 m farther east. Baybars

85 *Annales de Terre Sainte*, ed. Edbury, p. 152; letter of Frederick II to Henry III, 17 March 1229, in Roger of Wendover, *Flores historiarum*, ed. Hewlett, 4:189-93, trans. Giles, 2:522-24; *Eracles* 32.25, 33.7, RHC Oc 2, pp. 365, 372. See also Müller-Wiener, *Castles of the Crusaders*, p. 75; Benvenisti, *The Crusaders*, pp. 333-34; Pringle, "A Thirteenth-Century Hall," p. 52; Pringle, *Churches*, 2:40; Kennedy, *Crusader Castles*, p. 129; Mesqui, "La fortification des Croisés," p. 23; Boas and Khamisy, "Initial Thoughts," pp. 95-101.

86 For the castle, see Rey, *Étude sur les monuments*, pp. 143-51; Conder and Kitchener, *Survey of Western Palestine*, 1:186-88; Pringle, "A Thirteenth-Century Hall," pp. 54-55; Pringle, *Secular Buildings*, no. 156, pp. 73-75. Boas, *Crusader Archaeology*, p. 109; Mesqui, "La fortification des Croisés," pp. 10-11; Boas, "Interpretation of the Parts," pp. 102-19. See also Masterman, "A Visit to the Ruined Castles," p. 74.

had ordered Badr al-Din al-'Aydamuri and Badr al-Din Baysari to attack the castle in May 1266, during his campaign against Safed, and Ibn 'Abd al-Zahir claims that it posed a threat to Mamluk Safed thereafter.[87]

Baybars, with the force that he had with him at Damascus, moved to Safed in early June 1271. From there, where artillery may have been gathering beforehand, the sultan sent his trebuchets ahead to Montfort. When he arrived in person, Baybars received envoys while overseeing the erection of these engines, much as he had done at Safed in 1266. Like the siege of Crac, the sources provide different details that, collectively, appear to provide a fairly clear account of the siege. In this case, however, they are less detailed (Fig. 15).

The Muslims initially encountered difficulties erecting their artillery due to the rough topography, but eventually three large engines and a number of smaller ones were set up. The *rabad* (castle town) was taken on 11 June and the *bashura* (outer defences) fell the following day. Sapping then commenced against the inner defences and some damage was apparently inflicted before the garrison agreed to terms and departed the castle for Acre on 23 June, or possibly the preceding evening.[88] The night of the garrison's surrender, Baybars made a rapid move against Acre, appearing before the city at daybreak the following morning, but returned to Montfort after finding little to be gained. The Mamluks then spent twelve days deconstructing the castle, having spent fifteen days besieging it.[89] Although artillery is given relatively little attention by the sources, the physical evidence provides a rare quantitative indication of the power of contemporary trebuchet technology.

Excavation efforts led by Bashford Dean in 1926, discovered a number of spherical stones that were identified as artillery projectiles. Of the fifty-six stones, forty-one were reportedly found in the westernmost portion of the central range (Dean's chamber J) and fifteen more were uncovered close to the

87 Ibn 'Abd al-Zahir, *al-Rawd*, trans. al-Khowayter, 2:586-87, 756; Ibn al-Furat, *Tarikh*, trans. Lyons and Lyons, 2:86-87, 151; *Annales de Terre Sainte*, ed. Röhricht and Raynaud, B p. 28 (452).

88 Mufaddal appears to misdate this by ten days, in the same way that al-Yunini, or Ibn Shaddad before him, probably misdated Baybars' arrival at Crac des Chevaliers.

89 For accounts of the siege, see Ibn 'Abd al-Zahir, *al-Rawd*, trans. al-Khowayter, 2:755-57; Ibn al-Furat, *Tarikh*, trans. Lyons and Lyons, 2:151; Abu'l-Fida', *al-Mukhtasar*, RHC Or 1, p. 153; Mufaddal, *Kitab*, ed. and trans. Blochet, 1:539-43; al-'Ayni, *'Iqd al-Juman*, RHC Or 2a, p. 244, trans. Gabrieli in *Arab Historians*, p. 320; Templar of Tyre, *Gestes des Chiprois* 378, ed. Minervini, p. 138, trans. Crawford, p. 67; *Eracles* 34.14, RHC Oc 2, p. 460; edited and translated sections of the accounts of Baybars al-Mansuri, Abu'l-Fida', al-'Ayni and al-Maqrizi, as well as a broader discussion of the Arabic sources, can be found in Khamisy, "Montfort Castle," pp. 28-40.

FIGURE 15 *Timeline of the siege of Montfort, 1271*

	Ibn 'Abd al-Zahir Ibn al-Furat	Al-Yunini Ibn Shaddad?	Ibn al-Dawadari Mufaddal	Baybars al-Mansuri Abu'l-Fida'	Templar of Tyre
5 June (24 Shawwal)	Baybars left Damascus				
6 June (25 Shawwal)		Baybars left Damascus			
9 June (28 Shawwal)		Baybars arrived	Baybars arrived		
11 June (1 Dhu'l-Qa'da)	*Rabad* taken				
12 June (2 Dhu'l-Qa'da)	*Bashura* taken			Castle (*bashura*) attacked, garrison surrendered	Castle taken
23 June (13 Dhu'l-Qa'da)		Garrison surrendered			
3 July (23 Dhu'l-Qa'da)			Garrison surrendered		
4 July (24 Dhu'l-Qa'da)	Destruction complete				
5 July (25 Dhu'l-Qa'da)		Destruction complete			
6 July (26 Dhu'l-Qa'da)			Baybars departed		

keep. These varied from about 25 cm to 42 cm in diameter, averaging about 32 cm.[90] Two of these stones, which were brought back to New York and catalogued by the MET, weigh about 66 kg and have diameters of 35 cm and 37 cm.[91] More recently, a smaller projectile weighing only 32 kg was found *in situ* in 2012

90 Dean, "The Exploration of a Crusaders' Fortress," pp. 18, 23, 38-39; Boas, "Stone, Metal, Wood," pp. 197-99; Boas, "Archaeological Evidence," p. 49.

91 These stones are catalogued under accession numbers 28.99.41 and 28.99.42. They were part of an exhibit in 2012 and general reports were published at http://www.metmuseum.org/art/collection/search/651123; http://www.metmuseum.org/art/collection/search/26476. My thanks to Donald LaRocca at the MET for providing some additional information relating to these stones.

IMAGE 11
Montfort: projectiles discovered in 1926 (PHOTO COURTESY OF THE METROPOLITAN MUSEUM OF ART, ARCHIVES OF THE DEPARTMENT OF ARMS AND ARMOR. IMAGE © THE METROPOLITAN MUSEUM OF ART)

near the northwest corner of the central range.[92] A large projectile, comparable in size to the Dean stones, was also found in 2017. This most recent stone was damaged after it was thrown, perhaps upon impact, and has a flattened face, a not uncommon feature of projectiles this size. It was discovered resting on top of the collapse that had built up against the northern section of the outer wall, suggesting it had originated in the upper castle and rolled down at some point after the castle was slighted.[93]

The topography around Montfort limits the potential positions where attacking artillery could have been located. It might have been possible to set up

92 My thanks to Adrian Boas for allowing me to weigh the 2012 stone when I visited in the spring of 2013. I was fortunate enough to participate in the excavations of 2015, which were carried out in a structure that has been identified as a stable. Although a number of stones were set aside as possible artillery projectiles, none appear to me to be such. Despite my best efforts, I have not yet been able to locate any of the stones that have been identified as artillery projectiles in the valleys to the north and south of the castle. For brief mentions of these stones, see Boas, "Archaeological Evidence," p. 49; Khamisy, "Montfort Castle," p. 38.

93 This stone was found by the author while exploring the overgrown sections of the lower bailey during the 2017 excavation season led by Adrian Boas.

IMAGE 12 *Montfort: castle from Area C* (AUTHOR'S PHOTO)

smaller traction engines at the lip of the outer ditch (this area has been developed in recent decades to create a viewing platform, obscuring its natural form); however, a small plateau farther east along the spur (C on the plan), would appear to be the closest suitable place to position a significant counterweight trebuchet. This relatively flat area consists of two terraces about 150 m from the keep and around 13-18 m higher than its base.[94] Although there is sufficient space to position two engines here, one in front of the other, the risks associated with doing so make it more likely that only one large trebuchet would have been used from this position. The observation that the rough terrain initially prevented Baybars from positioning his artillery close enough to the castle may reflect the difficulties the Mamluks encountered finding this plateau. To reach this position, one must traverse the rugged spine of the spur or climb up to it from some point along the partially rock-cut quarry road, which ran from Tarphile (E on the plan) along the southern side of the spur,

94 Lewis has similarly identified this as a potential artillery position, although his measurements present it as much wider and farther from the castle, Lewis, "How Strong was Strong Mountain," pp. 285-86.

IMAGE 13 *Montfort: castle from the far side of the outer ditch* (AUTHOR'S PHOTO)

IMAGE 14 *Montfort: castle from Area A* (AUTHOR'S PHOTO)

through the outer ditch, to the gate on the north side of the keep.[95] It seems most likely that scouts would have identified the plateau while working their way down the top of the spur towards the castle; however, engine components were probably brought up from the quarry road, as the slope up to the plateau is conveniently climbable due south of this position.

There are a number of vantage points that overlook the castle from the north side of Wadi al-Qarn (F on the plan), but the only other potential artillery position within 300 m of the castle appears to be on the ridge across the southern wadi (A on the plan). It has previously been suggested that artillery was positioned here,[96] near where it is believed Baybars established his headquarters. Such suggestions, however, have been made without considering the capabilities of contemporary artillery technology.

At its very closest, the relatively flat top of the southern ridge, now largely occupied by an olive orchard, is 210 m from the keep, the closest part of the castle, and 54 m higher than its base.[97] In his preliminary study of the surrounding landscape, Lewis provides a similar distance of 230 m between the olive field and the castle and a difference in elevation of 60-80 m to the base of the keep.[98] From the edge of the ridge, a release velocity of 46 m/s is required to reach the concentration of projectiles found at the west end of the central range, approximately 260 m away and about 58 m lower. From a slightly more conservative position, 290 m from the western concentration of stones and 65 m above, a release velocity of 48 m/s is necessary. These are minimums and environmental factors such as drag, which vary based on weather conditions, and imperfect release angles would require slightly higher release velocities. To place this in perspective, a 70 kg stone travelling at 48 m/s carries 80,640 J. If, however, a significant engine were positioned on the plateau to the east of the castle, a release velocity of at least 42 m/s would be necessary to throw stones to the west end of the central range, about 200 m away and 20 m lower. At 42 m/s, a 70 kg stone carries 61,740 J, 23% less energy.

Release velocities exceeding 42 m/s may have been achieved by traction trebuchets before the end of the twelfth century and stones the size of those discovered at Montfort had been used earlier at Ascalon, Arsuf, Crac and probably elsewhere. However, this would appear to be the first instance in which it can be definitively concluded that stones of this mass were thrown at such

95 For this road, see Shotten-Hallel, Korngreen and Grossowicz, "The Stones of Montfort," pp. 139-41.

96 Khamisy, "Montfort Castle," p. 38; Lewis, "How Strong was Strong Mountain," p. 285.

97 My thanks to Rabei Khamisy for sharing his GPS survey data of this area.

98 Lewis, "How Strong was Strong Mountain," p. 284.

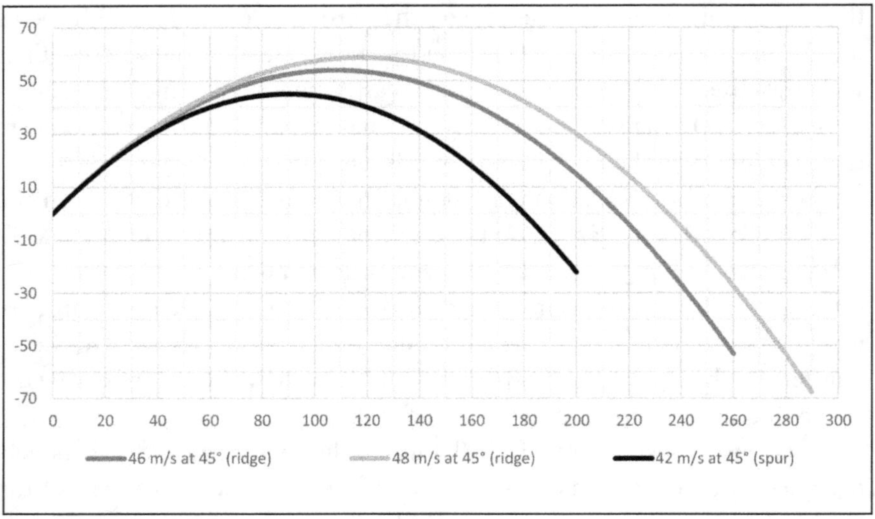

FIGURE 16 *Hypothetical trajectories at Montfort*

Engine position	**Southern ridge (A)**		**Spur (C)**
Release angle	45°	45°	45°
Release velocity	46 m/s	48 m/s	42 m/s
Horizontal range at impact	260 m	290 m	200 m
Δ elevation	-53 m	-68 m	-22 m

velocities. Mathematically, a trebuchet capable of throwing a 70 kg projectile at 42 m/s requires a counterweight of about 3,000 kg and a beam around 8 m long; however, mechanical trials suggest slightly larger components are typically necessary.[99]

From the plateau on the spur (C), the keep would have been the closest target. Built to defend the most exposed end of the castle, the keep was a significant obstacle. Its walls are around 7 m thick, although there is evidence of internal passages, and faced with hard limestone ashlars, which al-Yunini and Mufaddal note were linked with iron cramps welded with lead.[100] The base was

99 For hypothetical data sets and the capabilities of reconstructed engines, see Appendices 3 and 4.

100 Mufaddal, *Kitab*, ed. and trans. Blochet, 1:542; Khamisy, "Montfort Castle," p. 40. For archaeological evidence of the cramps used to join the stones, see Boas, "Stone, Metal, Wood," pp. 210-11.

further strengthened with a talus to discouraged mining. Unfortunately it is unclear if the keep was bombarded during the siege. The thorough dismantling of the tower following the garrison's surrender has obscured any impact signatures and buried any projectiles that may have fallen into the surrounding ditch. Regardless, projectiles weighing upwards of 60 kg (or even 100 kg) would have posed little threat to the masonry of such a strong tower. It is possible that the projectiles found in the central range had been aimed at the battlements crowning the keep, having flown over their intended target. However, it is at least as likely that these stones were thrown deliberately against this part of the castle with the objective of causing destruction and disruption within the stronghold, rather than attempting to bring down its outer walls.

There is little reason to believe that the Teutonic Knights were unaware of the capabilities of contemporary Mamluk artillery. As construction had continued at the castle through the 1260s,[101] it must have been believed that the south wall of the central range, which is 2 m thick along its entire length, was either out of range of any engines that might be erected on the ridge to the south or sufficiently strong to withstand a bombardment from this direction. The interior walls of the central range were strengthened over time but there appears to have been no attempt to thicken or otherwise strengthen the outer walls of the upper ward. If this interior work was inspired by the potential threat posed by artillery, which seems unlikely, the objective would have been to provide better protection against projectiles lobbed at the centre of the castle.

It has been suggested that a defensive trebuchet may have been erected by the garrison on the keep;[102] however, the projectiles discovered in 1926, including the fifteen found just west of the keep, are too large to have been thrown by anything other than a counterweight trebuchet. Furthermore, they are rather large for practical defensive use: there are few advantages to throwing a stone weighing 60 kg compared to throwing one a third as large with the same amount of force when targeting men, siege engines or improvised defences. The only reference to defensive artillery is found in al-'Umari's work, where it is stated that the defenders were unable to set up any trebuchets due to the besiegers' barrage. Accordingly, this may have been an assumption rather than a reality. The keep is the only practical position to place such an engine, given the surrounding topography. Although it may have been structurally strong enough to support a single eastward-facing counterweight trebuchet, the value of such an engine relative to the space it denied other means of defence is hardly justifiable. For these reasons, it seems very unlikely that the stones

101 Boas and Khamisy, "Initial Thoughts," pp. 95-101.

102 Khamisy, "Montfort Castle," pp. 32-33.

found near the keep were meant for defensive use. If they had been shaped by the Franks, they were more likely part of an offensive stockpile, although it remains difficult to explain their position.

There are few other insights into the use of artillery before Baybars' death in 1277. Edward of England and Hugh III of Cyprus do not appear to have used any siege equipment when they attacked Qaqun in 1271,[103] and there are few indications that Baybars employed engines during his successful siege of Cursat in 1275.[104] A decade later, however, trebuchets even larger than those used by Baybars appear to have been employed by his ultimate successor.

The Sultanate of Qalawun

Baybars was initially succeeded by his son, al-Sa'id, in 1177; however, it was the young sultan's father-in-law, Qalawun, who ultimately consolidated and added to Baybars' recent conquests when he took the thrown in 1279. Like Baybars, Qalawun had been brought up as a mamluk under the Ayyubids. He was a leading figure during Baybars' reign and mentioned during the campaign against Caesarea and Arsuf in 1265 and that of the following year against Safed. Although he does not appear in the records of the 1271 campaign, he played a critical role in the northern expeditions of 1266, 1272 and 1274/75. Given his position and relationship with Baybars, Qalawun would have been intimately familiar with contemporary siege technology and the logistical network that Baybars had established.[105]

Margat: 1285

During the early years of his reign, Qalawun was happy to continue peaceful relations with his neighbours, allowing him to deal with internal opponents. Aside from the Mongol invasion of 1281, which resulted in a Mamluk victory at the battle of Homs, there was relative peace. From 1285, Qalawun adopted a more confident and aggressive foreign policy. His actions against the Franks

103 Templar of Tyre, *Gestes des Chiprois* 381, ed. Minervini, p. 140, trans. Crawford, p. 68; *Eracles* 34.14, RHC Oc 2, p. 461; Ibn 'Abd al-Zahir, *al-Rawd*, trans. al-Khowayter, 2:770-71. For the structure, see Pringle, *Secular Buildings*, no. 168, pp. 83-84.

104 Ibn 'Abd al-Zahir, *al-Rawd*, trans. al-Khowayter, 2:831-33; Ibn al-Furat, *Tarikh*, trans. Lyons and Lyons, 2:165. For this castle, see Deschamps, *Les Châteaux*, 2:351-57; Boase, "Military Architecture," pp. 162-53; Kennedy, *Crusader Castles*, p. 84.

105 For the early career and reign of Qalawun, see Northrup, *From Slave to Sultan*, pp. 67-158.

began with the capture of Margat in the north, before steadily acquiring the remaining Frankish outposts along the Mediterranean coast to the south.

Margat had been acquired by the Hospitallers in 1186 and, like Crac des Chevaliers, was bypassed by Saladin as he moved his army up the Syrian coast in 1188.[106] The castle's double line of walls also impressed Wilbrand of Oldenburg when he passed by in the early thirteenth century.[107] Baybars had made two abortive attacks on Margat during the harsh winter rains of January 1270, apparently travelling without siege equipment on both occasions.[108] Another Mamluk attack was ordered by Qalawun (probably in late 1280) following the Mongol sack of Aleppo in October 1280; however, this was routed in a sally by the garrison.[109]

In 1285, Qalawun gathered his forces and arms in Egypt and enlisted a number of siege experts before setting out from Damascus to besiege the castle. The trebuchets stored in Damascus, as well as engines requisitioned from the surrounding strongholds, were carried on men's shoulders to his camp at 'Uyun al-Qasab before the army travelled by forced marches to Margat, arriving there on 17 April. Ibn 'Abd al-Zahir specifies that "three of the great 'Frankish' type, three 'Qarabughas' and four 'devils' [*shaytania*]" were deployed against the castle, while Abu'l-Fida' similarly claims that both great and small engines were erected. The *ifranji* trebuchets destroyed the defenders' artillery, allowing the Muslims to advance their trebuchets closer to the castle walls. This proved to be a reckless move as the Franks were able to repair their engines, which then destroyed a number of the Muslims' trebuchets and killed some of their

106 Caffaro states that the castle was built before the Franks arrived in the Levant. It fell to them in 1140 after a surprise attack, Caffaro, *De liberatione*, ed. Belgrano, pp. 115-16, trans. Hall and Phillips, p. 119. For the castle's ability to apply pressure to Saladin's army as it passed, see 'Imad al-Din, *al-Fath*, trans. Massé pp. 125-26; Ibn al-Athir, *al-Kamil*, trans. Richards, 2:345. Isaac Comnenus was imprisoned here in 1191, Roger of Howden, *Chronica*, ed. Stubbs, 3:116, trans. Riley, 2:210. See also Ralph of Diceto, *Ymagines historiarum*, ed. Stubbs, p. 92.

107 Wilbrand of Oldenburg, *Itinerarium terrae sanctae* 1.11, ed. Pringle, pp. 121-22, trans. Pringle, pp. 69-70. For the castle and its history, see Rey, *Étude sur les monuments*, pp. 19-38, 255-64; Deschamps, *Les Châteaux*, 3:259-85; Müller-Wiener, *Castles of the Crusaders*, p. 58; Kennedy, *Crusader Castles*, pp. 78-79, 164-79. Mesqui, "La fortification des Croisés," pp. 22-23.

108 Ibn 'Abd al-Zahir, *al-Rawd*, trans. al-Khowayter, 2:729; Ibn 'Abd al-Zahir, *Tashrif al-Malik al-Mansur*, trans. Gabrieli in *Arab Historians*, p. 334; Ibn al-Furat, *Tarikh*, trans. Lyons and Lyons, 2:139, 141. See also Thorau, *The Lion of Egypt*, pp. 200-1.

109 Abu'l-Fida', *al-Mukhtasar*, RHC Or 1, p. 158; Templar of Tyre, *Gestes des Chiprois* 406, ed. Minervini, p. 154, trans. Crawford, p. 77; *Annales de Terre Sainte*, ed. Röhricht and Raynaud, A p. 33 (457); Bar Hebraeus, *Makhtebhanuth Zabhne*, trans. Budge, 1:463.

IMAGE 15 *Margat: castle from the southeast* (PHOTO COURTESY OF DENYS PRINGLE)

operators.[110] Amidst the artillery duel, Muslim sappers worked against the castle's foundations.

After about three weeks, a mine was ignited under a tower at the southern apex of the castle, identified as the Tower of Hope by the Templar of Tyre. The mine compromised a section of wall but the attackers were still forced to scale the obstruction. The advantage was short-lived: the remaining part of the tower collapsed around sunset, somehow shoring up the initial breach. The Franks, fearing the progress of the Mamluk miners elsewhere, sought terms a day or two later and Qalawun seized the opportunity to take possession of the castle without damaging it any further (Fig. 17).[111] Among the sizable force that was installed to defend the castle thereafter, was a group of artillery experts, suggesting that at least some of the engines used against the castle, and any of the Franks' left in a decent state of repair, became a part of its arsenal.

110 Ibn 'Abd al-Zahir, *Tashrif al-Malik al-Mansur*, trans. Gabrieli in *Arab Historians*, pp. 334-35; Abu'l-Fida', *al-Mukhtasar*, trans. Holt, p. 12, RHC Or 1, p. 161.

111 Ibn 'Abd al-Zahir, *Tashrif al-Malik al-Mansur*, trans. Gabrieli in *Arab Historians*, pp. 335-38; Abu'l-Fida', *al-Mukhtasar*, trans. Holt, p. 12; Templar of Tyre, *Gestes des Chiprois* 429, ed. Minervini, p. 166, trans. Crawford, p. 85; Marino Sanudo, *Liber secretorum* 3.12.19, ed. Bongars, p. 229, trans. Lock, p. 364. 'Ibn 'Abd al-Zahir appears to date the mine to Wednesday 17 Rabi I but the castle's fall to Friday 18 Rabi I; the day of the week will be taken as correct, bringing this account in line with that of Abu'l-Fida'. See also Northrup, *From Slave to Sultan*, pp. 130-32.

FIGURE 17 *Timeline of the siege of Margat, 1285*

	Ibn 'Abd al-Zahir	Abu'l-Fida'	Templar of Tyre	Marino Sanudo
17 April (10 Safar)	Qalawun arrived		Siege began	
27 April (20 Safar)				Siege began
23 May (17 Rabi I)	Mine fired			
25 May (19 Rabi I)	Castle surrendered	Castle surrendered		
27 May (21 Rabi I)			Castle surrendered	Castle surrendered

The topography around the castle would have limited the locations where significant engines could have been set up. Unlike at Montfort, the engines deployed against Margat would have been at a lower elevation than their targets. Qalawun probably established his camp on the flat top of the rise to the south, about 300 m from the castle. There is little to suggest that the strongest counterweight trebuchets of the day were powerful enough to reach the castle from this position. To do so, stones would need to be thrown at velocities exceeding 54 m/s. The greatest engines of the Middle Ages may have been able to achieve this, throwing quite small stones, but this would appear to be about, if not beyond, the practical limit of this technology.[112]

The Mamluks probably set up their artillery in the saddle between the castle and rise to the south, about 30 m below its walls and within range of its defenders. The larger *ifranji* trebuchets were responsible for destroying the defenders' artillery, which might have been traction trebuchets or heavier counterweight engines stored in the castle that were pressed into a defensive role. This would have allowed the besiegers to bring up their *qarabugha* and *shaytani* traction trebuchets, which were used to suppress the defenders and support the work of the sappers. It is most likely that these were the engines damaged after the Frankish trebuchets had been repaired.

The use of traction trebuchets to support mining efforts had been employed since the First Crusade; however, it is possible that the Mamluks also attempt-

112 For the factors limiting the ultimate power of these engines, see Chapter 1. For results achieved with reconstructed engines, see Appendix 4.

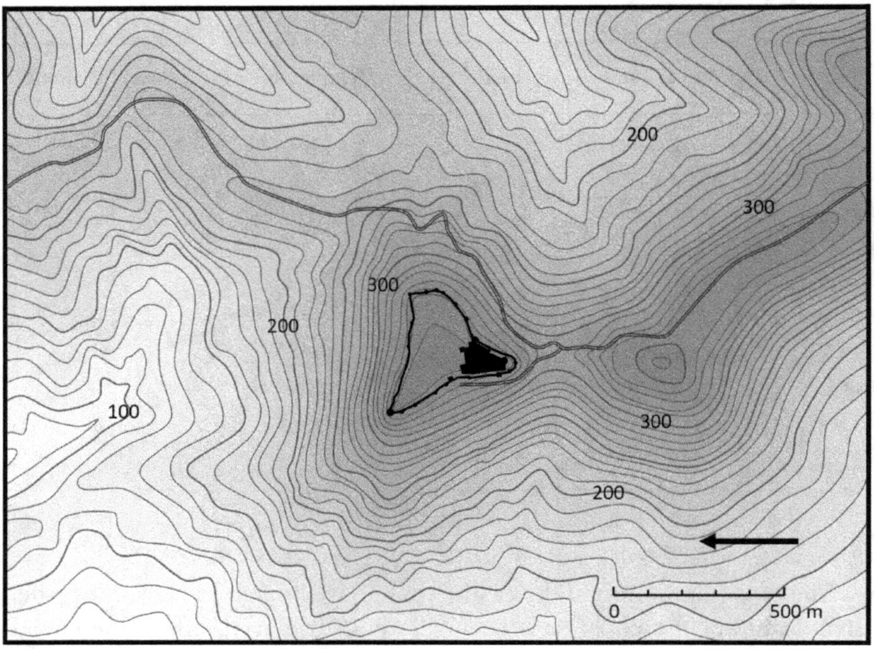

PLAN 33 *Margat*

ed to advance some of their heavier engines. Although there is little to indicate that these larger engines were brought forward, rather than just lighter ones, such a move would suggest that the besiegers were trying to use them to effect real structural damage, which may also have been attempted at Acre in 1191, Darum in 1192 and Crac des Chevaliers in 1271. By moving them closer, these engines could use lower release angles, increasing the horizontal velocity and destructive force of each shot. Their lower elevation, however, would have mitigated much of this advantage. Regardless of which engines were advanced, the emphasis placed on sapping suggest that even the most powerful engines, at least in the positions available, were still not considered effective breaching weapons.

After the siege, the outer wall around the castle's southern apex was rebuilt. Although this has obscured any damage that Qalawun's artillery might have inflicted here, where the Mamluks' siege efforts were focused, impact signatures inflicted by higher trajectory shots are discernible on the southern face of the donjon, which rises behind. Given the accuracy of counterweight trebuchets, which necessarily inflicted these marks, clusters of impact signatures would be expected; however, they are distributed erratically and fairly few in number, suggesting this was not a primary target. These may instead be the result of stones that were aimed at the battlements of the outer wall in front of the tower or reflect the use of inconsistent ammunition, which was lobbed into the castle for general effect rather than aimed at any specific target. These impact signatures provide a tangible indication of the power of late thirteenth-century artillery and help to contextualise the descriptions found in the sources. Although it would have required considerable power to leave such impressions in the hard basalt masonry, these are little more than cosmetic blemishes and posed no threat to the structural integrity of the tower.

The growing significance of artillery and equally impressive resources of the Mamluks are evident not only in Qalawun's ability to exploit the network of armouries that Baybars had developed, but also the notice given to the artillery experts that joined his army when setting out from Egypt. The use of large counterweight trebuchets was clearly becoming increasingly institutionalised. Following its capture, Margat was immediately incorporated into the Mamluks' network of arsenals, much as Safed and Crac had been in 1266 and 1271 respectively, and 'Ajlun, Subayba and Shayzar had been before 1265.[113]

Maraclea: 1285

After making arrangements at Margat, Qalawun led his forces to Maraclea. The stronghold was situated on an offshore island facing the settlement of the same name. The site had been a Frankish possession since the early twelfth century and appears to have been newly refortified in the decade leading up to 1285.[114] Ibn 'Abd al-Zahir claims that the stronghold had sides 25.5 cubits long and was two bow-shots from the mainland. Modern satellite imagery shows a square structure, with sides about 15 m long, just under the surface of the water on a shoal about 190 m off the present coast of Syria. Although this appears to match Ibn 'Abd al-Zahir's general description, it is harder to verify his claims that the walls were 7 cubits thick, that the tower was built on piles of stones held in place by ironwork, and that it was seven storeys tall. Only the last of these remarks warrants extra suspicion. Critically, he states that the roof was covered with sacking and hemp rope, as a defence against artillery, and that a smaller structure behind the main one supported three defensive trebuchets.[115]

Although Maraclea would appear to have been within range of contemporary artillery positioned along the shoreline, Qalawun recognised that he could not take the strongpoint without a naval force superior to that of the Franks, which he did not have. Rather than bombard the fortress needlessly, he

113 See Fulton, "Development of Prefabricated Artillery," pp. 65-70.

114 For this castle, see Deschamps, *Les Châteaux*, 3:323-26; Rey, *Étude sur les monuments*, pp. 161-62.

115 Ibn 'Abd al-Zahir, *Tashrif al-Malik al-Mansur*, trans. Gabrieli in *Arab Historians*, p. 339.

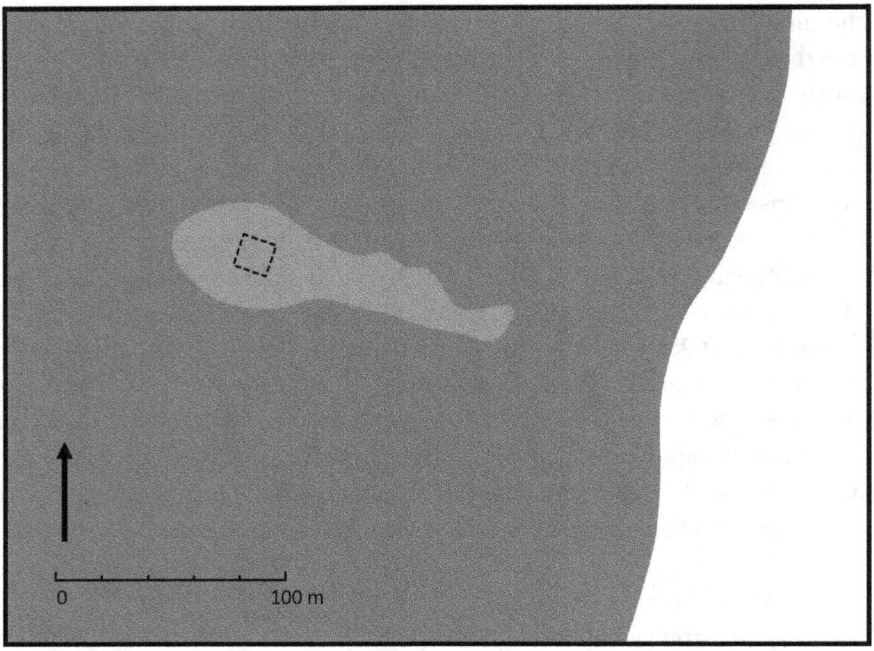

PLAN 34 *Maraclea*

persuaded Bohemond VII to destroy it, sparing the Frankish lands around Tripoli from raids in return.[116]

The castle's padded roof suggests that it was vulnerable to high-trajectory artillery shots; however, that this was also sufficient protection is equally telling. When placed alongside the earlier events at Montfort and Margat, a fairly clear impression of late-thirteenth-century artillery becomes apparent. At least the largest engines were capable of throwing projectiles weighing 70 kg up to 200 m, while moderately greater ranges would have been possible if lighter stones were used. The effect of these projectiles, however, was not significant by modern standards. Rather than seeking to breach fortified defences, against which they could inflict little more than superficial damage, they were probably used to target parapets and inhibit a garrison's ability to defend its stronghold, as was probably done at Crac. They could also be used more indiscriminately as psychological weapons, targeting structures behind the defensive perimeter to add to the chaos and stress faced by the defenders, weakening their ability to organise themselves and fight on effectively, as appears to have been done at Montfort.

116 Ibn 'Abd al-Zahir, *Tashrif al-Malik al-Mansur*, trans. Gabrieli in *Arab Historians*, pp. 340-41.

Saone: 1287

In 1287, Qalawun moved against Saone, which was held by Sunqur al-Ashqar, a dissident emir. Ibn ʿAbd al-Zahir mentions the use of a number of significant trebuchets at this siege, including several *ifranji* engines. One of these, named Biqashmar, is described as one of a number of prefabricated *Mansuri* trebuchets that had been constructed in Damascus. The proper adjective used to designate these engines suggests that they were commissioned by Qalawun (al-Mansur Qalawun al-Salihi).[117]

Saone's eastern wall, that most easily targeted by artillery, reveals evidence of multiple rebuilding phases. The likelihood that much of this damage was inflicted by earthquakes prevents any clear assessment of the potential destruction inflicted by Qalawun's artillery in 1287. Although these engines would have contributed to the plight of the defenders, they do not appear to have had a decisive impact on the siege and the castle was surrendered after about a month. From Saone, Qalawun moved to the coast and took Frankish Latakia, the defences of which had been damaged in a recent earthquake. The same tremors may also have weakened Saone, although identifying which sections of the castle were damaged is extremely difficult.

Tripoli: 1289

Tripoli had been a Frankish possession since 1109. The Franks had developed its fortifications over the years, but it is unclear what the trace of its defences looked like at the end of the thirteenth century. At the start of the century, Wilbrand of Oldenburg compared the city's layout to that of Tyre, complete with a citadel complex and two lines of town walls. Abu'l-Fida' confirms that the city was protected by its extension into the Mediterranean and that it was only along a narrow eastern front that the Mamluks could assail it.[118]

The Mamluk siege of Tripoli began at some point in March 1289. Before leaving Egypt, Qalawun had sent instructions to his Syrian deputies ordering them to prepare for the campaign. The artillery stored in various strongholds was also prepared and sent ahead of his army. After pausing at Damascus, Qalawun marched on Tripoli and by early April an assortment of trebuchets were

---

117 Khamisy and Fulton, "Manjaniq Qarabugha," p. 186. A translation of this section of Ibn ʿAbd al-Zahir's account has not yet been published. For the siege, see also Abu'l-Fida', *al-Mukhtasar*, trans. Holt, p. 13, RHC Or 1, p. 162; Templar of Tyre, *Gestes des Chiprois* 461-62, ed. Minervini, p. 186, trans. Crawford, pp. 94-95. For an analysis of the damage in this section of Saone's defences, see Fulton, "A Ridge too Far," esp. pp. 50-51.

118 Wilbrand of Oldenburg, *Itinerarium terrae sanctae* 1.8, ed. Pringle, p. 120, trans. Pringle, p. 67; Abu'l-Fida', *al-Mukhtasar*, trans. Holt, p. 14. See also Müller-Wiener, *Castles of the Crusaders*, p. 42.

deployed against the city. The engines appear to have focused their fire against the Bishop's Tower and that of the Hospitallers, while sappers attempted to undermine sections of the town's defences. The outer enceinte was taken first and then a single concerted attack carried the inner line on about 26 April 1289.[119]

Abu'l-Fida' and the Templar of Tyre claim that Qalawun employed both large and small trebuchets. Al-Maqrizi, probably working from Ibn al-Furat – who in turn made use of sources such as Ibn al-Dawadari and al-Yunini – claims that Qalawun employed nineteen trebuchets, associating 1,500 men with their operation (seventy-nine men to each engine if evenly distributed).[120] Six of the nineteen trebuchets are said to have been of the *ifranji* type and probably correspond with the four large engines identified in one version of the *Annales de Terre Sainte*. Certain Muslim accounts identify thirteen *qarabugha* trebuchets and the use of this type of engine is confirmed by the Templar of Tyre's mention of *carabohas*.[121] While the powerful *ifranji* engines were almost certainly counterweight trebuchets, the remainder were probably traction-powered. Thus, as appears to have been the case at Arsuf twenty-four years earlier, increasingly heavy artillery was still outnumbered by lighter, more adaptable engines that could effectively support sapping efforts. The Templar of Tyre relates that significant damage was sustained by both of the targeted towers before the town fell. Unfortunately, the city's complete destruction following the siege has rendered it impossible to assess the damage inflicted by these engines during the two months that they were in action.[122]

---

119 Abu'l-Fida', *al-Mukhtasar*, trans. Holt, pp. 14-15, RHC Or 1, p. 162; Templar of Tyre, *Gestes des Chiprois* 475-77, ed. Minervini, pp. 196-98, trans. Crawford, pp. 100-1; Marino Sanudo, *Liber secretorum* 3.12.20, ed. Bongars, p. 229, trans. Lock, p. 365; al-Maqrizi, *al-Suluk*, trans. Quatremère, 2:102-3; *Annales de Terre Sainte*, ed. Röhricht and Raynaud, A p. 36 (460). See also Khamisy and Fulton, "Manjaniq Qarabugha," pp. 183-84; Marshall, *Warfare in the Latin East*, p. 218. Marino Sanudo blends the earlier campaign against Saone and Latakia with that against Tripoli, claiming that Qalawun had engines prepared at Crac on his way to the latter.

120 If anywhere near accurate, it is unlikely that everyone counted in this total would have been directly involved with the working of these engines at any one time. If the counterweight engines were not provided with winches, large groups would have been needed to pull back the beam after each shot. See the comparable figure suggested at Lisbon in 1147 and associated analysis in Chapter 4.

121 For a study of this term and corresponding engine-type, see Khamisy and Fulton, "Manjaniq Qarabugha," pp. 179-201.

122 The city was quickly resettled farther inland around the castle founded by Raymond of St Gilles, see Templar of Tyre, *Gestes des Chiprois* 478, ed. Minervini, p. 198, trans. Crawford, p. 101; Ibn Battuta, *Tuhfat* 1.1, trans. Gibb, p. 60; al-'Umari, *Masalik al-absar*, trans. Richards, pp. 78-79; Müller-Wiener, *Castles of the Crusaders*, p. 42.

The Second Great Siege of Acre: 1291

A peace that had been established between the Franks of Acre and the Mamluks was violated in 1290.[123] This provided Qalawun with a pretext to move against Acre; however, he died in November while setting out from Cairo for Palestine. Before his death, Qalawun had arranged for the governor of Damascus to build siege equipment, including artillery, and for Shams al-Din al-A'sar to collect timber from Wadi Murabbin, a valley between 'Akkar and Baalbek where trees ideal for the construction of trebuchets grew – stout, straight and as tall as twenty-one cubits according to al-Nuwayri. Although the latter emir's efforts were complicated by an early snowfall, al-Yunini claims that he had been able to move a number of trees to the outskirts of Damascus by 23 December 1290.[124]

Qalawun was succeeded by his son, al-Ashraf Khalil, who lost little time carrying out his father's planned campaign. In March 1291, the Egyptian army set out for Palestine, bringing siege engines and technicians to assemble them. 'Izz al-Din al-Afram, who had received acclaim for handling Baybars' trebuchets more than two decades earlier,[125] was sent ahead to see to the artillery in Damascus. He was met there by the princes and emirs of Syria, along with their armies and artillery. There can be little doubt that the largest prefabricated engines were brought; one engine, a 'great' trebuchet collected by al-Muzaffar Mahmud II of Hama from Crac des Chevaliers, supposedly required a hundred cartloads to transport. The forces gathered at Damascus then moved to join those that had begun to assemble outside Acre.[126]

Artillery and the Siege

There are numerous accounts of the siege of Acre, which have been discussed and analysed by Donald Little and Andreas D'Souza.[127] According to al-Nuwayri, followed by Ibn al-Furat and al-Maqrizi, ninety-two trebuchets were set up against Acre, including some of the *ifranji*, *qarabugha* and *shaytani* types. Al-Jazari's account, however, states that there were seventy-two engines all

---

123 For this treaty and its violation, see Holt, "Qalawun's Treaty," pp. 802-12.

124 Little, "The Fall of 'Akka," pp. 167-68. See also Abu'l-Fida', *al-Mukhtasar*, trans. Holt, p. 15.

125 For a brief biography of 'Izz al-Din Aybak al-Afram, see Irwin, *Middle East*, p. 38.

126 Ibn Taghribirdi, trans. Gabrieli in *Arab Historians*, p. 347; al-Maqrizi, *al-Suluk*, trans. Quatremère, 2:121-24; Abu'l-Fida', *al-Mukhtasar*, trans. Holt, p. 16; Little, "The Fall of 'Akka," pp. 168-71; D'Souza, "The Conquest of 'Akka," p. 239. Abu'l-Fida' vividly describes the difficulties of moving the cart assigned to him in the harsh winter conditions.

127 Little, "The Fall of 'Akka," pp. 159-81; D'Souza, "The Conquest of 'Akka," pp. 234-50.

together, similarly mentioning the use of *ifranji*, *qarabugha* and *shaytani* trebuchets. Al-Yunini and Ibn Taghribirdi, who do not provide a total figure, specify that there were fifteen large *ifranji* trebuchets, some of which were capable of throwing stones weighing a Damascene *qintar* or more, as well as a number of *shaytani* and other types of trebuchets. Al-'Ayni helpfully clarifies that there were fifty-two *shaytani* trebuchets.[128]

The Frankish sources largely agree with their Muslim counterparts. The Templar of Tyre claims that the Mamluk engines were erected eight days after the first Muslim forces arrived and the Amadi account states that it took four days for the Muslims to prepare their siege machines, a similar period to that found in most of the Arabic accounts. According to the *Annales de Terre Sainte*, the Mamluks employed fourteen large engines, only one short of the number of large *ifranji* trebuchets noticed by al-Yunini, and the Templar of Tyre repeats that some trebuchets threw stones weighing a *qintar*. The Templar of Tyre also provides a sense of how these engines were distributed: one notable trebuchet was positioned opposite the Templars' section of Acre's defences, another faced the Hospitallers' section, a third targeted the Accursed Tower and a fourth opposed the Pisans' section. This last engine was called al-Mansur, the same name as that given by Abu'l-Fida' to the trebuchet brought from Crac des Chevaliers, suggesting that these, like those employed against Saone in 1287, had been commissioned by Qalawun after 1277.[129] In addition to these large engines, the Templar of Tyre and John of Villiers, grand master of the Hospitallers, both of whom were among Acre's defenders, as well as Marino Sanudo and the Amadi account, confirm the presence of *carabouha*/*corobonares*/*carabagas*/*caravachani*, which correspond with the *qarabugha* trebuchets found in the Arabic sources. The Templar of Tyre describes these as rapid-shooting hand-operated stone-throwers that posed a greater threat to Acre's defenders than did the larger engines; however, John of Villiers and Marino Sanudo, who do not contradict this description, portray them as being some-

128 Khamisy and Fulton, "Manjaniq Qarabugha," pp. 184-86. See also Little, "The Fall of 'Akka," p. 171 and n. 72; D'Souza, "The Conquest of 'Akka," p. 252.

129 It is possible that these engines were one and the same; however, the engine brought from Crac would not have been that which was operated by the contingent from Hama, as they were positioned on the Muslims' extreme right, opposite the Templars, while the Pisans were positioned close to the Accursed Tower, which anchored Acre's northeast salient.

what more powerful.[130] These, as noted above, were probably traction trebuchets.[131]

Other Christian sources are less helpful. The *Excidium* states that six hundred and sixty-six large and small engines were used, including *petrarie, bibliete, perdicete* and *mangonelli* – probably little more than a list of all the artillery terms known by the author. Three hundred engines are mentioned in the account of Bar Hebraeus, a figure as unlikely as his claim that a thousand miners worked against each of the city's towers. Ludolph of Suchem places the number of engines at a more reasonable sixty but provides no further details.[132]

In the first stage of the siege, while the Muslim forces were still assembling against them, the Franks prepared their own artillery and gathered stones.[133] Some of these trebuchets may have been those built in 1286 under Odo Poilechien, on behalf of Charles of Anjou, which were originally intended for use against Henry II of Cyprus.[134] Little else is known about these engines.

Attacks appear to have begun on 5 April, although artillery continued to arrive until 8 April. The gates of Acre remained open during this early phase as fighting took place outside the city's walls. The Muslims, fighting in shifts, eventually pushed their way to the edge of the fosse that ran in front of the city's outer line of walls. Here, they erected mantlets at the edge of the counterscarp and developed positions behind which the *qarabugha* trebuchets could be placed and operated securely. These traction engines provided close suppressing fire while heavier engines battered the city's fortifications, steadily destroying sections of the city's battlements and limiting the Franks' ability to defend the city from the tops of certain towers and sections of wall.

---

130 *Annales de Terre Sainte*, ed. Röhricht and Raynaud, A pp. 36-37 (460-61); Templar of Tyre, *Gestes des Chiprois* 490-91, ed. Minervini, pp. 206-8, trans. Crawford, pp. 105-6; letter of John of Villiers to William of Villaret, in *Carticulaire général des Hospitaliers*, no. 4157, 3:593; Marino Sanudo, *Liber secretorum* 3.12.21, ed. Bongars, p. 230, trans. Lock, p. 367; *Amadi*, ed. Mas Latrie, pp. 219-20.

131 For different interpretations, see Chevedden, "Black Camels," pp. 251-54; Marshall, *Warfare in the Latin East*, pp. 214, 234.

132 *Excidium Aconis* 2.3, ll. 123-48, ed. Huygens, pp. 67-68; Bar Hebraeus, *Makhtebhanuth Zabhne*, trans. Budge, 1:492-93; Ludolph of Suchem, *De itinere Terrae sanctae liber* 26, ed. Deycks, p. 43, trans. Stewart, p. 56.

133 Templar of Tyre, *Gestes des Chiprois* 485, ed. Minervini, p. 204, trans. Crawford, p. 104; letter of John of Villiers to William of Villaret, in *Carticulaire général des Hospitaliers*, no. 4157, 3:593; *Excidium Aconis* 1.8, ll. 327-34, ed. Huygens, pp. 60-61.

134 Templar of Tyre, *Gestes des Chiprois* 437-38, ed. Minervini, pp. 168-70, trans. Crawford, p. 86.

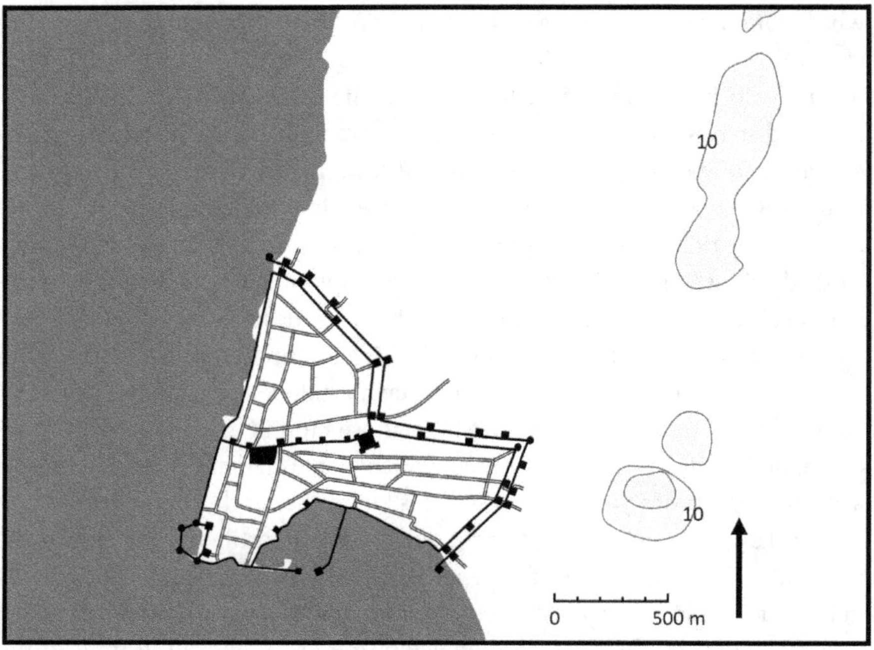

PLAN 35 *Acre, 1291*

Meanwhile, facing less pressure from defenders, sappers began to undermine parts of Acre's outer fortifications.[135]

The Frankish engines that targeted the Muslims' improvised ramparts had little effect. These were probably traction trebuchets, suggesting that the Muslims had advanced sufficiently far to enjoy the protection that the outer wall provided against larger counterweight trebuchets, which had been set up behind the main wall. The Franks' most powerful engines, stockpiled counterweight trebuchets pressed into defensive service, were deployed in certain open areas inside the city and most likely targeted the besiegers' engines of comparable size. A ship-mounted trebuchet enjoyed considerable success applying pressure against the Muslims' right flank, until it was destroyed during a

135 Abu'l-Fida', *al-Mukhtasar*, trans. Holt, pp. 16-17; Templar of Tyre, *Gestes des Chiprois* 489, 491, ed. Minervini, pp. 206, 208-10, trans. Crawford, pp. 105, 106-7; letter of John of Villiers to William of Villaret, in *Carticulaire général des Hospitaliers*, no. 4157, 3:593; Thadeus, *Ystoria* ll. 48-60, ed. Huygens, p. 101; Marino Sanudo, *Liber secretorum* 3.12.21, ed. Bongars, p. 230, trans. Lock, p. 367; *Amadi*, ed. Mas Latrie, p. 220; Ludolph of Suchem, *De itinere Terrae sanctae liber* 26, ed. Deycks, p. 43, trans. Stewart, p. 56; D'Souza, "The Conquest of 'Akka," p. 241.

storm.[136] It may have been the artillery here on the Muslims' extreme right, that of the contingent of Hama, which was targeted in a subsequent sally led by the Templars.[137]

In the mid-twentieth century, about fifty medieval artillery projectiles were found around what had been the northwest limit of Acre's defences.[138] These probably relate to eighteen projectiles excavated *in situ* (and fifty-two found out of context) approximately 100-200 m north of these and the same distance from the present coastline.[139] The stones appear to be around the size of those found at Montfort. These two concentrations represent the positions where stones thrown against the Templars' section of the Montmusard wall ultimately came to rest and roughly where one of the Mamluks' engines, probably operated by the contingent of Hama, was located. While it can be safely assumed that the projectiles found along the line of Acre's walls were thrown there by one or more Mamluk engines, the provenance of those clustered farther back is harder to determine with certainty. They were probably prepared by the besiegers, evidently left unused; however, in the absence of a more detailed study, it seems possible that some may have been thrown by a large Frankish engine targeting the Muslim trebuchet positioned here. In either case, the approximate size of the stones and the range from which they were thrown fits with similar siege scenarios from the previous three decades.

During excavations in the 1970s, a number of large projectiles were discovered around a wall that has been identified as part of the city's medieval eastern wall, near the St Nicholas Gate. Additional projectiles can still be found in the present Christian cemetery, located to the east of where the city's northeast salient would have been in the thirteenth-century.[140] These, approximately 100 m from the medieval curtain and of varying sizes, some quite large,

136 Abu'l-Fida', *al-Mukhtasar*, trans. Holt, pp. 16-17; Templar of Tyre, *Gestes des Chiprois* 491, ed. Minervini, p. 208, trans. Crawford, p. 106.

137 Templar of Tyre, *Gestes des Chiprois* 491, ed. Minervini, pp. 208-10, trans. Crawford, pp. 107-8. Cf. Abu'l-Fida', *al-Mukhtasar*, trans. Holt, p. 17; *Amadi*, ed. Mas Latrie, pp. 220-21. See also Templar of Tyre, *Gestes des Chiprois* 492, ed. Minervini, p. 210, trans. Crawford, p. 108; *Amadi*, ed. Mas Latrie, p. 221.

138 For the trace of Acre's thirteenth-century walls, see Guérin, *Description*, 3:507; Rey, *Les colonies Franques*, pp. 452-54; Kedar, "The Outer Walls," p. 161. See also Frankel, "The North-West Corner," p. 256. Cf. Jacoby, "Crusader Acre," p. 42 and n. 214; Jacoby, "Montmusard," p. 215. For the development of this trace, see Jacoby, "Montmusard," pp. 205-17.

139 Porat, "'Akko, Dov Gruner Street"; Tepper, "Stone Balls."

140 Kedar, "The Outer Walls," pp. 174-75. This cemetery might correspond with the Frankish St Nicholas cemetery, see Pringle, *Churches*, no. 438, 4:150-55.

suggest that at least one large counterweight trebuchet and a number of smaller engines were placed in this general area. Considering that the Mamluks' concentrated their attack on the northeast corner of the city, additional engines were almost certainly located in nearby positions to the north.

As in 1191, the besiegers relied on their sappers to compromise the city's defences. Efforts were focused against the King's Tower, which anchored the forewall ahead of the Accursed Tower at the northeast salient of the inner wall.[141] While larger engines probably bombarded the adjoining sections of curtain and nearby towers, lighter *qarabugha* trebuchets proved critical in providing cover for the sappers, forcing the Franks to countermine.[142] By the time that Henry II of Cyprus arrived on 4 May, a portion of Acre's outer wall had been successfully undermined. The Franks attempted to defend the compromised section of wall but were forced to abandon it by 8 May. As the Franks added improvised defences to face the increasingly exposed salient, the Muslims stepped up their bombardment of the town's defenders.[143]

The King's Tower was eventually evacuated on 15 May, its outer face having fallen away entirely. As the besiegers moved into the tower, it was then targeted by Frankish artillery in a failed last effort to keep it out of Muslim hands.[144] In an anecdote about his own involvement, Baybars al-Mansuri claims that a felt screen was used to help the besiegers approach the damaged section of wall. The effectiveness of this light cover, which proved sufficient to absorb the shock of incoming artillery projectiles and resist arrows, reveals that it was traction trebuchets that were used to assail these assault forces. This screen was then used to shelter workers filling the fosse in preparation for a frontal attack.[145]

At dawn on 18 May 1291, the Mamluks attacked on all fronts. Prevented from concentrating their manpower at the breach in the outer defences, the Franks

141 For the defences along this section of wall, see Pringle, *Churches*, 4:11; Pringle, "Edward I," p. 53.

142 Templar of Tyre, *Gestes des Chiprois* 491, ed. Minervini, p. 208, trans. Crawford, pp. 106-7; Marino Sanudo, *Liber secretorum* 3.12.21, ed. Bongars, p. 230, trans. Lock, p. 367; *Amadi*, ed. Mas Latrie, p. 220.

143 Templar of Tyre, *Gestes des Chiprois* 493, ed. Minervini, p. 212, trans. Crawford, p. 108; Marino Sanudo, *Liber secretorum* 3.12.21, ed. Bongars, pp. 230-31, trans. Lock, pp. 367-68; *Excidium Aconis* 2.3, ll. 148-56, ed. Huygens, pp. 68-69; Ibn Taghribirdi, trans. Gabrieli in *Arab Historians*, p. 347; D'Souza, "The Conquest of 'Akka," p. 243.

144 Templar of Tyre, *Gestes des Chiprois* 494, ed. Minervini, pp. 212-14, trans. Crawford, p. 109; Marino Sanudo, *Liber secretorum* 3.12.21, ed. Bongars, p. 231, trans. Lock, p. 368. See also *Excidium Aconis* 2.3, ll. 165-69, ed. Huygens, pp. 69.

145 Little, "The Fall of 'Akka," pp. 172-73.

were unable to hold off the assault here. Passing through the breach, the attackers rushed both west and south between the inner and outer walls. As panic spread through the town and the first defenders began to flee towards the sea, Muslim forces started to scale the inner walls. A postern in the Accursed Tower was also forced open, leading a party of Mamluks to the Church of St Romanus in the Gardens and the artillery that the Pisans had erected there.[146] While some paused to burn the heaviest engine, almost certainly a large counterweight trebuchet, others continued south along the inner side of the east wall.[147] As the city was steadily overwhelmed, the Templars' complex in the southwest corner of the city, partly secured by the sea, hosted the defenders' last stand until it too was undermined and captured about ten days later. When the last bastion of resistance fell, al-Ashraf ordered Acre demolished.[148]

Artillery does not reappear in the narrative once the Muslims broke into the city: none of the Mamluks' engines were apparently brought in to use against the Templar complex nor were any Frankish engines turned against it. It is unclear whether some of the Frankish engines were disassembled and distributed through Syria or if most were simply destroyed along with the rest of Acre.[149] So thorough was the destruction that the medieval walls have all but disappeared, leaving little indication of the power of the engines that played against them in 1291.

Despite the increasing power of trebuchet technology, it never had the chance to successfully fulfil a breaching role before the end of the Frankish period. As at Acre in 1191 and countless sieges over the following century, mining was the definitive means of bringing down walls. Despite what appears to have been the common use of projectiles weighing 60-100 kg through the

146 For this structure and its location, see Pringle, *Churches*, no. 442, 4:158.

147 Templar of Tyre, *Gestes des Chiprois* 496-99, ed. Minervini, pp. 214-20, trans. Crawford, pp. 110-13; Marino Sanudo, *Liber secretorum* 3.12.21, ed. Bongars, p. 231, trans. Lock, p. 368; letter of John of Villiers to William of Villaret, in *Carticulaire général des Hospitaliers*, no. 4,157, 3:593; Ibn Taghribirdi, trans. Gabrieli in *Arab Historians*, p. 347; Abu'l-Fida', *al-Mukhtasar*, trans. Holt, p. 17; D'Souza, "The Conquest of 'Akka," p. 244. See also al-Maqrizi, *al-Suluk*, trans. Quatremère 2:125-26; *Excidium Aconis* 2.4-12, ll. 180-742, ed. Huygens, pp. 70-93; Thadeus, *Ystoria* ll. 62-77, 709-24, ed. Huygens, pp. 101-2, 127; Ludolph of Suchem, *De itinere Terrae sanctae liber* 26, ed. Deycks, pp. 43-45, trans. Stewart, pp. 56-58.

148 For later testimony of this destruction, see Ibn Battuta, *Tuhfat* 1.1, trans. Gibb, pp. 57-58.

149 Templar of Tyre, *Gestes des Chiprois* 501, ed. Minervini, pp. 220-22, trans. Crawford, p. 114; *Amadi*, ed. Mas Latrie, p. 222; Thadeus, *Ystoria* ll. 252-67, ed. Huygens, pp. 109-10; Ludolph of Suchem, *De itinere Terrae sanctae liber* 26, ed. Deycks, pp. 45-46, trans. Stewart, pp. 59-60; Ibn Taghribirdi, trans. Gabrieli in *Arab Historians*, pp. 348-49; Little, "The Fall of 'Akka," p. 176; D'Souza, "The Conquest of 'Akka," p. 246.

IMAGE 16 *Acre: projectile* (AUTHOR'S PHOTO)

second half of the thirteenth century, the walls of the thirteenth-century tower that anchored the extreme north-west end of Montmusard's defences were only 1 m thick.[150] Thin walls like these should check any drastic overestimations of the destructive power of these engines. Although the size of the largest engines may have increased slightly between the mid-1260s and 1291, it seems the most significant development was the use of increasing numbers of these engines.

Fittingly, it is at this final siege that the largest projectiles clearly datable to the thirteenth century have been found. A visit to the Hospitallers' complex, where many of the biggest stones found over the years are now kept, can leave the impression that only such large projectiles were thrown against the city in 1291, some similar in scale to fourteenth-century stones stockpiled elsewhere by the Mamluks. But the number of these stones instead reflects the correspondingly significant number of large engines that were employed by the Mamluks and their use for more than six weeks. There is every indication that smaller projectiles, both easier to find and throw, would have greatly outnumbered these; however, their less impressive size has allowed them to avoid the same degree of attention over the centuries. The diversity of engine-types found in the sources and the emphasis placed on the use of these smaller trebuchets at various points, would seem to confirm this.

150 Frankel, "The North-Western Corner," p. 256; Kedar, "The Outer Walls," pp. 60-61.

IMAGE 17 *'Ajlun: large projectiles* (PHOTO COURTESY OF RACHEL RABEY)

IMAGE 18 *'Ajlun: small projectiles* (AUTHOR'S PHOTO)

In a fitting parallel, a number of incredibly large stones can be found on display at ‘Ajlun, which remained a part of the Mamluk network of arsenals into the fourteenth century. These stones are stacked and well lit with the clear hope that they impress visitors. In other rooms, significant numbers of smaller stones, perhaps once part of a defensive stockpile, have been collected in darker and out-of-the-way corners. Although less visually impressive, members of the castle's garrison may have placed a higher value on these smaller stones, as they would be more useful if the castle came under siege.

As fell Acre, so fell the Latin presence in the Levant. Tyre, which had never fallen from Frankish hands since it was taken in 1124, was surrendered almost immediately after the capture of Acre by those who had not already fled.[151] The Templars considered defending Sidon but it, like Beirut, was abandoned in July and the garrisons of ‘Atlit and Tortosa similarly abandoned these strongholds in August.[152] The Mamluks then slighted all of these coastal strongholds. The speed with which the Franks were prepared to evacuate their remaining defences after the fall of Acre speaks to the Mamluks' real strength: their comparative numbers. Although these stoutly fortified strongholds could probably have withstood the most formidable artillery for weeks, if not months, al-Ashraf's reserves of manpower would have allowed him to sacrifice waves of men in frontal assaults or to impose near limitless blockades with the knowledge that no hostile army would appear to his rear. Although the Templars continued to hold the island of Ruad off the coast of Tortosa until 1302,[153] and calls for crusades were made well into the fourteenth century, the period of Frankish lordship in the Levant effectively ended with the loss of Acre. With this, almost two hundred years of Frankish-Muslim side-by-side artillery development in the Near East also came to an end.

---

151 Templar of Tyre, *Gestes des Chiprois* 504, ed. Minervini, p. 204, trans. Crawford, p. 504. Marino Sanudo, *Liber secretorum* 3.12.22, ed. Bongars, pp. 231-32, trans. Lock, p. 369; Abu'l-Fida', *al-Mukhtasar*, trans. Holt, p. 17; Ibn Taghribirdi, trans. Gabrieli in *Arab Historians*, pp. 349-50.

152 Templar of Tyre, *Gestes des Chiprois* 509-12, ed. Minervini, pp. 226-28, trans. Crawford, pp. 118-19; Marino Sanudo, *Liber secretorum* 3.12.22, ed. Bongars, p. 232, trans. Lock, p. 369; Abu'l-Fida', *al-Mukhtasar*, trans. Holt, p. 17. John Poloner was impressed by the ruined remains of Sidon's castles in the fifteenth century but Mandrell appears not to even have noticed the sea castle two hundred and fifty years later, John Poloner, *Descriptio Terrae Sanctae*, trans. Stewart, pp. 29-30, 32; Mandrell, *A Journey*, p. 60.

153 Templar of Tyre, *Gestes des Chiprois* 615-23, 635-38, ed. Minervini, pp. 300-4, 308-10, trans. Crawford, pp. 155-58, 160-61; Abu'l-Fida', *al-Mukhtasar*, trans. Holt, p. 40.

Plates

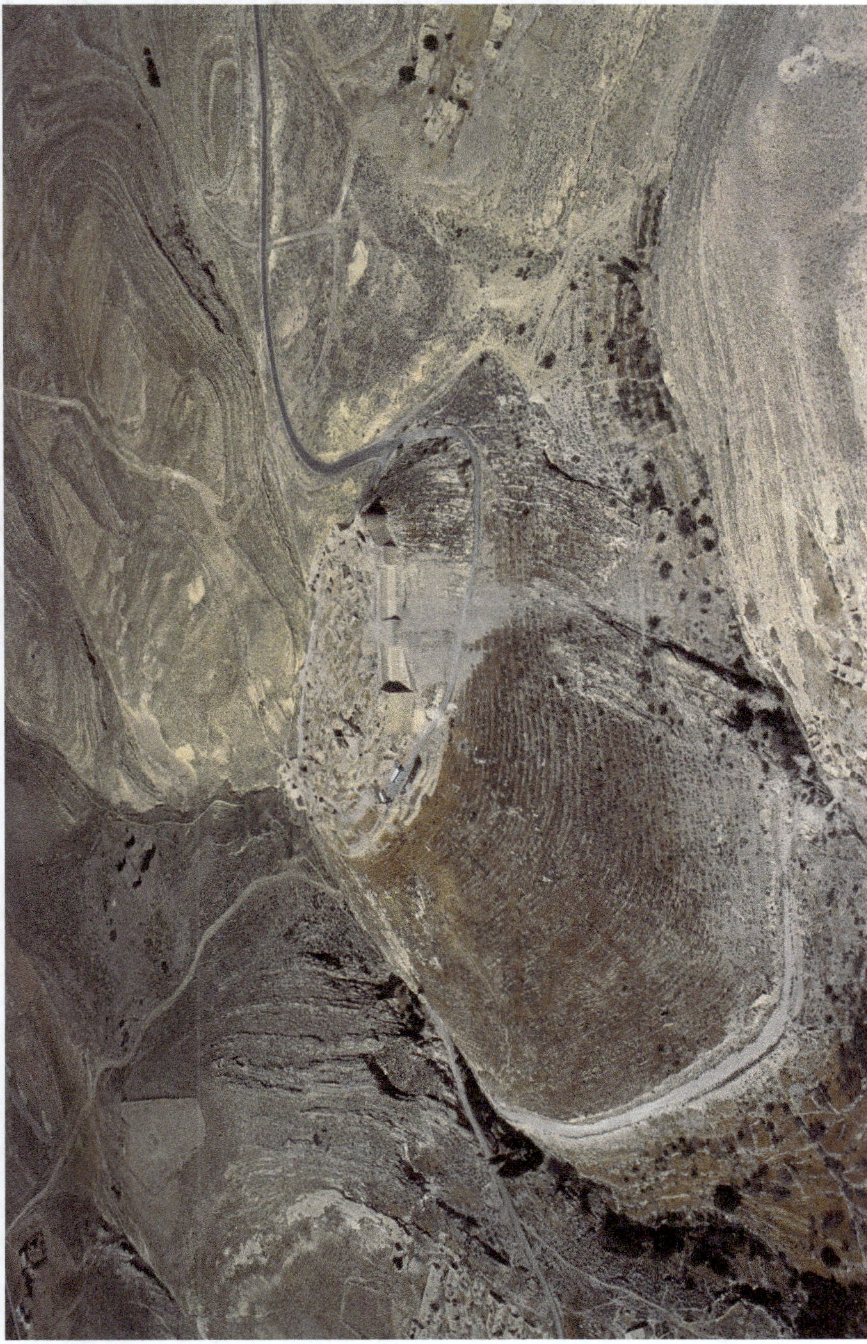

PLATE 1 *Montreal: castle from the northeast* (PHOTO COURTESY OF APAAME)

 | DOI 10.1163/9789004376922_010

PLATE 2 *Montreal: castle from the west* (PHOTO COURTESY OF APAAME)

PLATE 3 *Montreal: tower E from the visitors' centre to the northwest* (AUTHOR'S PHOTO)

PLATE 4 *Kerak: castle from the southeast* (PHOTO COURTESY OF APAAME)

PLATE 5 *Kerak: castle from the northeast* (PHOTO COURTESY OF APAAME)

PLATE 6 *Kerak: Mamluk donjon from the north* (PHOTO COURTESY OF HEATHER CROWLEY)

PLATE 7 *Arsuf: castle from the south* (AUTHOR'S PHOTO)

PLATE 8 *Arsuf: castle from the north* (AUTHOR'S PHOTO)

PLATE 9 *Caesarea: medieval town from the southeast* (PHOTO COURTESY OF MICHAEL EISENBERG)

PLATE 10 *Caesarea: eastern town defences from the south* (AUTHOR'S PHOTO)

PLATE 11 *'Atlit and Le Destroit: strongholds from the northeast* (AUTHOR'S PHOTO)

PLATE 12 *Belvoir: castle from the north* (PHOTO COURTESY OF MICHAEL EISENBERG)

PLATE 13 *Montfort: castle from the north* (AUTHOR'S PHOTO)

PLATE 14 *Montfort: mine dug into the southern wall of the upper castle* (AUTHOR'S PHOTO)

PLATE 15 *Subayba: castle from the south* (AUTHOR'S PHOTO)

PLATE 16 *Subayba: interior of tower 9, southeast firing chamber* (AUTHOR'S PHOTO)

PLATE 17 *Beaufort: castle from the northwest* (PHOTO COURTESY OF JEAN YASMINE)

PLATE 18 *Crac des Chevaliers: castle from the southern outwork* (PHOTO COURTESY OF DENYS PRINGLE)

PLATE 19 *Crac des Chevaliers: outer southern wall and fighting platform from the east* (PHOTO COURTESY OF DENYS PRINGLE)

PLATE 20 *Shayzar: citadel from the north* (PHOTO COURTESY OF DENYS PRINGLE)

PLATE 21 *Shayzar: southern donjon complex from the west* (PHOTO COURTESY OF DENYS PRINGLE)

PLATE 22 *Saone: Frankish donjon* (PHOTO COURTESY OF DENYS PRINGLE)

PLATE 23 *Belvoir: projectiles, most likely dating to 1188* (AUTHOR'S PHOTO)

PLATE 24 *Arsuf: projectiles* (AUTHOR'S PHOTO)

PLATE 25
Arsuf: marble projectile (AUTHOR'S PHOTO)

PLATE 26
Arsuf: rounded columns (AUTHOR'S PHOTO)

PLATE 27 *Montfort: projectile discovered in 2012* (AUTHOR'S PHOTO)

PLATE 28 *Montfort: projectile discovered in 2017* (AUTHOR'S PHOTO)

PLATE 29 *Traction trebuchet at Caerphilly Castle* (AUTHOR'S PHOTO)

PLATE 30 *Small counterweight trebuchet at Cardiff Castle* (AUTHOR'S PHOTO)

CHAPTER 8

Influence of Offensive Artillery

In the nineteenth century, the famed historian, architect and architectural restorer, Eugene Emanuel Viollet-le-Duc remarked, "We cannot doubt that the crusades, during which so many memorable sieges were effected, improved the means of attack, and that consequently important modifications were introduced into the defence of fortified places."[1] As the means of attack and defence advanced, it is widely contested to what degree siege engines were a driving, rather than reciprocally developing, element of this process. In Viollet-le-Duc's opinion, up to the thirteenth century fortifications remained essentially static bulks of masonry, defended by small garrisons. Without expressly stating why or examining the underlying causes, he concluded that it was a direct result of the crusades that such defences were no longer sufficient and more methodically built fortifications, with larger and more active garrisons, were built from the late twelfth century.

Opinions like this reflect noticeable changes in the style and scale of fortifications constructed in Western Europe and the Near East at the end of the twelfth century and start of the thirteenth. This has often led to the assumption that a new style of castle was imported to Europe from the Levant via the crusades. Because the crusades were a natural point of exposure between Latin Christians and Turco-Arabic Muslims, they have been regarded at times as almost a semi-mystical source of cross-cultural technological sharing and are occasionally offered up as a blind explanation for certain developments. However, there appears to be little evidence that the mighty castles built by Richard I of England and Philip II of France were influenced directly by Frankish examples. Belvoir was the most remarkable Frankish-built castle at the time of the Third Crusade, but it had fallen to Saladin before 1191 and, like most other large Frankish castles, was not seen by either monarch. It is often overlooked that most of the iconic 'crusader castles', at least in their recognisable forms, date to the early thirteenth century and thus postdate mighty European castles such as Château Gaillard.

The impressive castles and citadels built or significantly developed in the late twelfth century and, more frequently, early thirteenth century, were often larger than earlier strongholds with thicker walls and more concentric plans. This naturally encourages suspicion that this shift in scale and sophistication

1 Viollet-le-Duc, *Dictionnaire*, 1:341-42, trans. Macdermott, in *An Essay*, p. 31.

 | DOI 10.1163/9789004376922_011

was the result of some acute need to completely replace outdated twelfth-century fortification styles with stronger and more 'scientific' designs. The introduction or development of powerful counterweight trebuchets is often regarded as the impetus behind this change.[2] Neither siege towers nor rams had developed considerably since the early twelfth century and the use of both was sparse in the thirteenth century. Mining remained an effective siege weapon but no technological advance was made until the introduction of gunpowder, long after the Franks had been pushed out of Acre.

Theories Regarding the Influence of Artillery

In the early twentieth century, Sydney Toy attributed the declining use of rams to the superseding power of trebuchets.[3] Taking matters much further, Paul Chevedden has emerged as the champion of those arguing that the power of artillery during the Early and High Middle Ages has been grossly underestimated. The breadth of his studies is commendable and the subsequent impact of his arguments on general interpretations of medieval artillery cannot be understated.

Chevedden claims that the counterweight trebuchet was nothing less than the Big Bertha of the pre-modern era, used to demolish fortifications with projectiles weighing hundreds of kilograms.[4] What began as a PhD thesis addressing the massive towers built by al-'Adil at the citadel of Damascus has developed into a broader theory explaining what he sees as a revolution in fortification planning. Simply put, this transition was necessitated by a need to defend against extremely powerful counterweight trebuchets by mounting similarly large engines on top of towers in defence.[5] In Chevedden's words,

> The counterweight trebuchet was so far superior to any piece of artillery yet invented that its introduction brought about a revolution in siegecraft that rendered existing systems of defence obsolete. This gravity-powered siege-engine could discharge missiles of far greater weight than the

2 The number of historians that have endorse this idea over the last few decades is sufficient to leave any attempt to name them all incomplete. To be fair, most do so in passing, without examining the matter very closely.

3 Toy, *Castles*, p. 142.

4 Chevedden, "King James I," p. 313.

5 See Chevedden, *The Citadel of Damascus*, pp. 283-84; Chevedden, "Fortifications," p. 36; Chevedden, "Invention," p. 71.

> traction or hybrid machines, and it could do so with remarkable accuracy. The machine was thus able to deliver devastating blows against the same spot of masonry time after time, and this made it potentially capable of demolishing the strongest fortified enclosures. The introduction of the counterweight trebuchet led to an increase in the scale of warfare and produced revolutionary changes in military architecture in order to counter the greater destructive power of this new artillery.[6]

This theory was embraced by Hugh Kennedy and widely circulated in his notable work on crusader castles:

> The development of this new and improved artillery fundamentally changed the balance between attack and defence. The engines became so effective that, given uninterrupted firing time, they would reduce any fortress to rubble. It was no use sitting behind castle walls and trusting in their strength: unless the defenders took active measures to neutralise the artillery, the fortress they had built would be destroyed as surely as the incoming tide washes away the children's sandcastle.[7]

It is perhaps revealing that Chevedden's earliest arguments focus on defensive artillery: the power of offensive artillery is emphasised in order to support a theory that defenders were compelled to employ the same engines in defence, elevating them to keep besieging machines out of range. Accordingly, Chevedden concluded that the large mural towers built by al-'Adil were platforms for the heaviest contemporary counterweight trebuchets.[8] This theory will be examined more closely in the next chapter.

Throughout his work Chevedden argues that developments in trebuchet technology took place much earlier than is often accepted. By doing this he is forced to suggest that the century between the invention of the counterweight trebuchet, supposedly by Alexius Comnenus at the time of the First Crusade, and al-'Adil's rebuilding work at Damascus, in the first and second decades of the thirteenth century, was one of development and experimentation.[9] In

---

6 Chevedden, "Invention," p. 76.

7 Kennedy, *Crusader Castles*, p. 108.

8 For his study of the citadel of Damascus and conclusions, see Chevedden, *The Citadel of Damascus*. For the influence of his theory regarding defensive artillery, see also Kennedy, *Crusader Castles*, p. 182; Nicolle, *Crusader Castles*, p. 44.

9 See Chevedden, "Invention," p. 73. Chevedden inaccurately cites a supposed delay between the introduction of effective siege guns and the inception of the bastion system of early modern defences as a comparison. In reality, there was hardly a delay: the earliest

the absence of clear evidence, he has relied upon a selection of isolated examples, pooled from a vast geographical area, to support the notion that counterweight trebuchets were in use at this time. However, the simplicity of the mechanics of a counterweight trebuchet render such a long period of development unlikely: it is enough to see such an engine to reconstruct one and only basic carpentry skills are required to scale up a fairly small model into something far more powerful and impressive.

When considering the relative futility of employing a small counterweight trebuchet, it would have been of great benefit to build ever bigger engines. It seems more likely that this natural and rapid process of building increasingly large machines took place between Saladin's campaigns in the 1180s and the siege of Damietta in 1218-19. The rate at which these engines grew probably peaked in the early thirteenth century, decreasing fairly sharply thereafter.[10] But there is very little evidence to indicate that even the impressive engines used in Egypt during the Fifth Crusade, or those employed by Damascene forces against 'Atlit around the same time, were considered effective breaching engines. Accordingly, the design of al-'Adil's towers was probably influenced by other factors.

Looking at Frankish fortifications in the kingdom of Jerusalem, Ronnie Ellenblum has suggested that the construction of larger more sophisticated Frankish castles and the development of concentric castle plans in the last third of the twelfth century was a response to the Muslims' use of 'heavy artillery'.[11] Concentricity, according to Ellenblum, was the only defence against

---

polygonal bastions in Italy actually predate the effective use of light and mobile siege guns by the French during the Italian Wars from 1494. By the 1530s, polygonal fortifications had been built throughout the Italian peninsula, from the small forts of Nettuno and Barletta to the elaborate urban defences of Ferrera, Lucca, Verona and Florence. See Mallett, "The Transformation of War, 1494-1530," pp. 3-21; Pepper, "The Face of the Siege," pp. 33-56; Black, *European Warfare, 1494-1660*; Parker, "The 'Military Revolution'," pp. 196-214.

10 For the difficulties associated with building ever larger counterweight trebuchets, see Chapter 2.

11 Ellenblum, *Crusader Castles*, pp. 237-38, cf. pp. 177-81. Ellenblum's presumption that many Frankish castles were dramatically developed during this period is largely pinned on a single passage in James of Vitry's history. Writing in the early thirteenth century, James of Vitry places the construction of Montreal, Kerak, Safed and Belvoir with the events of 1168, following Amalric's failed campaign in Egypt. Ellenblum suggests James of Vitry's remarks indicate that these castles were significantly developed around this time, acknowledging that earlier contemporary accounts confirm that most of these castles were built decades earlier, Ellenblum, *Crusader Castles*, pp. 177-81. Cf. James of Vitry, *Historia orientalis* 49, ed. and trans. Donnadieu, pp. 216-17.

this shift to heavy siege equipment, confusingly citing Kerak as an example of a concentric castle.[12] The evidence to support this theory is less forthcoming than Ellenblum would make it out to be. The castle at Jacob's Ford, to which he devotes considerable attention, was captured without the aid of artillery in 1179 and was one of only a few castles in the Kingdom of Jerusalem to have been taken between 1160 and 1186. The walls of this stronghold appear to have been incomplete when it was attacked, in some areas only slightly higher than the earth piled up against them, which probably served as scaffolding.[13] Furthermore, there is no firm evidence to suggest that the stronghold was ever intended to have an outer (concentric) wall; one was certainly not present in 1179.

While no artillery appears to have been used at Jacob's Ford, Ellenblum suggests that the two engines used against Kerak in 1170 were prefabricated counterweight trebuchets.[14] His focus on these engines, rather than those used during the subsequent sieges of 1183 and 1184, where artillery was employed with greater effect, is puzzling. Regardless, at none of the various sieges of Kerak between 1170 and 1188 is there any suggestion that artillery threatened the integrity of the castle's defences.[15] While Ellenblum chalks this up to the greater strength of this 'concentric' castle, he fails to demonstrate that these engines were any more powerful than those used by earlier Muslim armies or that the assailed defences postdate 1160.

When examining the more general shift in the balance of power during the twelfth century, Ellenblum asserts that Muslim sieges began to last longer as a result of the growing resources of Nur al-Din and Saladin. The suggestion that sieges in the region became more protracted is in itself problematic, as the average duration of Muslim sieges with discernable lengths between 1147 and 1186 was about three weeks, while the average length of those from 1099 to 1146 was slightly more than a month. Setting this aside, he attempts to reconcile his belief that drastically more powerful artillery was coming into use with the naturally conflicting notion that sieges were becoming longer by suggesting that the Franks had taken pre-emptive efforts to strengthen their fortifications.[16] Although he discusses certain strongholds, he neglects any examination of the machines that he judges to be responsible for this supposedly

12 Ellenblum, *Crusader Castles*, pp. 282-85.

13 Cf. Ellenblum, *Crusader Castles*, pp. 269-70.

14 Ellenblum, *Crusader Castles*, p. 258.

15 For discussions of the various sieges of Kerak, see Chapter 4.

16 Ellenblum, *Crusader Castles*, p. 237.

drastic shift in fortification design, while he also reveals a tendency to label any large castle 'concentric' and place its construction after 1160.

Emphasising continuity, David Nicolle has highlighted the similarities between late Roman walls and those constructed by the Franks in the twelfth century. Drawing attention to tower spacing and projection as well as wall thickness, he implicitly ties these similarities to the comparable power of Roman and early-twelfth-century artillery.[17] Nicolle, following Ellenblum, interprets the Frankish fortification efforts of the 1160s and 1170s as a response to the effective employment of counterweight trebuchets, decades before al-'Adil's building efforts.[18] Nicolle risks contradicting this, however, when suggesting that this was also a period of continuity in siege warfare, echoing the sentiments of France and Marshall, and ultimately conceding that the great engines of the thirteenth century probably inflicted more psychological than physical damage.[19]

A significant issue with these theories is that they interpret the power of artillery through the hypothesis that it was most responsible for the discernible shifts that may be seen in the design of late-twelfth-century fortifications. A number of other factors contributed to the construction of larger and more sophisticated castles around this time. Not least among these are the greater financial resources of their commissioners, the growing size of Muslim armies, the natural development or enlargement of existing structures, and the opportunities for substantial rebuilding provided by the destruction wrought by earthquakes.

On the other side of this debate, Christopher Marshall has argued that the methods of attack and defence remained largely the same through the thirteenth century as they had been during the twelfth, suggesting that there was little technological change.[20] This tempered assessment of thirteenth-century siege technology echoes sentiments expressed by Viollet-le-Duc, who asserted that even counterweight-powered artillery could do little more than destroy crenellations, clear defenders from parapets and target the machines

17 Nicolle, *Crusader Castles*, p. 57. Chevedden also alludes to these similarities, Chevedden, *The Citadel of Damascus*, pp. 7-8.

18 Nicolle, *Crusader Castles*, p. 31. Michaudel has also accepted the influence of the counterweight trebuchet in driving the development of fortifications, though in a much broader and more subdued Ayyubid-Mamluk context, Michaudel, "Development of Islamic Military Architecture," p. 106.

19 Nicolle, *Crusader Castles*, pp. 139, 142.

20 Marshall, *Warfare in the Latin East*, p. 212.

of the besieged, a view shared by Oman.[21] John France has also supported this notion of continuity and a more obvious balance between attack and defence.

Without having studied trebuchet technology in depth, France appears to appreciate the limitations of this period's artillery, whether traction- or counterweight-powered, better than most.[22] Placing these engines in a more appropriate tactical context, his conclusions align with those of Viollet-le-Duc.[23] Both men accurately appreciated that in the vast majority of instances where artillery is mentioned by the sources, it is not depicted as a kind of super-weapon but merely one of many familiar tools that could be used during a siege.

Siege Length

Perceived trends towards both longer and shorter sieges have been used to argue the increased use of heavy artillery. While Ellenblum has argued the former in the context of Saladin's sieges of the late twelfth century,[24] the latter is often suggested when examining the Mamluk period. Hugh Kennedy and Carole Hillenbrand have both viewed the brevity of certain Mamluk sieges as evidence of their more efficient use of artillery.[25] Such arguments, however, rely heavily on circumstantial evidence, as neither approach sufficiently acknowledges the changing political climate and discrepancy between the resources of the various parties. While no Mamluk siege of a Frankish castle lasted more than six weeks, at no point did Saladin ever besiege a Frankish stronghold, in person, for a period of time any longer.[26]

Arguments focusing on the relative brevity of Mamluk sieges stem largely from a theory proposed by Christopher Marshall, which asserts that Muslim sieges of Latin strongholds in the thirteenth century lasted no more than a few weeks, while Frankish forces were content to besiege an objective for months.[27] What is not emphasised is that all of these protracted Frankish sieges, save that

---

21 Viollet-le-Duc, *Dictionnaire*, 1:344, trans. Macdermott, in *An Essay*, p. 36; Oman, *The Art of War in the Middle Ages*, pp. 57-61; Oman, *A History of the Art of War*, p. 131.

22 His acceptance that al-Zahir Ghazi's artillery breached the western bailey at Saone in 1188 is uncharacteristic of his otherwise critical treatment of artillery.

23 France, *Western Warfare*, pp. 119-26.

24 Ellenblum, *Crusader Castles*, p. 284.

25 Kennedy, *Crusader Castles*, p. 101; Hillenbrand, *the Crusades*, p. 531.

26 The intermittent sieges of Belvoir (1187-89) and Beaufort (1189-90) are the closest to an exception.

27 Marshall, *Warfare in the Latin East*, pp. 241-46.

of Damietta in 1218-19, were directed against the strongholds of fellow Franks.[28] When considering the length of Muslim sieges, if intra-faith engagements in the region are included, the average duration of those with discernable lengths between 1194 and 1258 was seventy-four days, with 57% lasting more than a month. By comparison, Mamluk sieges from 1260 to 1291 lasted an average of just twenty-four days, although half still extended for at least a month.

Although his data might present a somewhat skewed picture of more general siege dynamics, Marshall astutely acknowledges that the most important factor contributing to the brevity of most Muslim sieges of Frankish strongholds was the overwhelming manpower that the Muslims were able to command.[29] Rather than a simple tactical preference, the Muslims' more aggressive approach was a by-product of their ability to commit significant numbers to frontal assaults and endure equally high numerical losses. The cost of maintaining large armies in the field also contributed to some of the more aggressive siege strategies demonstrated by figures such as Zanki, Saladin and Baybars. The prospect of becoming bogged down far from home in a lengthy siege, decreasing potential profits, would also have encouraged opposition from the various contingents within most armies.[30]

When considering Frankish sieges of the twelfth century, most of the longest involved large bodies of European crusaders, such as those of Antioch (1097-98), Harim (1177-78) and Acre (1189-91), similar to the later siege of Damietta (1218-19), while many of the prolonged sieges of coastal strongholds, such as Tyre (1124), involved considerable elements of Italian naval support. The few that remain, such as the lengthy siege of Ascalon (1153), which benefited from the involvement of members of the Second Crusade who remained in the East, can be contrasted with the majority of Frankish sieges, which, like that of Banyas (1140), were concluded in a matter of weeks.

Artillery had little effect on the length of most sieges. Traction trebuchets were useful supporting weapons that could assist sappers and assault troops, but their impact was relative to the contribution of the forces they supported. Although counterweight trebuchets came to target battlements and wreak havoc within defensive enceintes, they too were dependent on miners to bring down large sections of walls and significant bodies of scaling forces to exploit weakened fortifications and the compromised morale of defenders.

---

28 Those sieges directed against fellow Franks lasting more than 90 days include Dieudamor (1229-30), Kantara (1229-30), Beirut (1231-32), and Kyrenia (1232-33).

29 Marshall, *Warfare in the Latin East*, pp. 247-48.

30 Saladin's efforts to keep his army together year to year to oppose the Frankish siege of Acre in 1189-91 speak to these difficulties.

Development, Manpower and Resources

The earthquakes of 1157, 1170 and 1202 necessitated the rebuilding of many Frankish and Muslim strongholds throughout greater Syria.[31] While Nur al-Din's rebuilding efforts might be seen as stylistically 'traditional', a number of Frankish castles rebuilt at the same time are often viewed as radically innovative. Belvoir, built by the Hospitallers from around 1168 in two concentric squares, is the most exceptional. Crac des Chevaliers was also rebuilt by the Hospitallers from around 1170, but it was far from the mighty thirteenth-century castle that it would become in the years following the 1202 earthquake. At the end of the twelfth century, Crac may have had a plan very similar to that of the contemporary Templar castle at Jacob's Ford.[32] In Transjordan, Montreal and Kerak were spared the effects of these earthquakes; their impressive Frankish defences are the product of continuous development by their wealthy lords. Both were developed further by the Ayyubids and Mamluks in the thirteenth century.[33]

As Belvoir would appear to be the only drastically different or considerably stronger Frankish castle built in a single phase in the late twelfth century, there is minimal evidence of a drastic architectural shift before the crisis of 1187. Most Frankish castles, like those elsewhere in the Near East and Europe, were the result of multiple phases of improvement and continual, rather than one-time, investment. Likewise, Belvoir's radical design and obvious use of multiple lines of defences should not be seen as a response to improvements in artillery technology but rather Nur al-Din's expanding reserves of wealth and manpower. Nur al-Din had united Damascus with Aleppo in 1154 and from 1163

31 For these tremors, see Ibn al-Qalanisi, *Dhail*, trans. Gibb, pp. 326, 328-30, 338-40; Matthew of Edessa, *Patmut'iwn*, trans. Dostourian, p. 267; Ibn al-Athir, *al-Kamil*, trans. Richards, 2:87, 89, 185-86; *Anonymous Syriac Chronicle*, trans. Tritton, pp. 302-3; William of Tyre, *Chronicon* 20.18, ed. Huygens, pp. 934-36, trans. Babcock and Krey, 2:370-71; letter from Geoffrey of Donjon, master of the Hospitallers, to Sancho VII of Navarra, and letter from Philip of Plessis, master of the Templars, to the abbot of Cîteaux, ed. Meyer in "Two Unpublished Letters on the Syrian Earthquake of 1202," pp. 306-10.

32 For a study of Crac as it may have stood at the end of the twelfth century, see Biller, "Die erste Burg der Johanniter (nach 1170)," pp. 47-77.

33 For the foundation and development of Montreal, see Fulcher of Chartres, *Historia Hierosolymitana* 2.55, ed. Hagenmeyer, pp. 592-93, trans. Ryan p. 215; Albert of Aachen, *Historia Ierosolimitana* 12.21, ed. and trans. Edgington, pp. 856-57; Faucherre, et al., "La forteresse de Shawbak," pp. 50-64; Pringle, *Churches*, 2:304-14. For the foundation and development of Kerak, see William of Tyre, *Chronicon* 15.21, 22.28, ed. Huygens, 2:692-93, 1,055-56, trans. Babcock and Krey 2:127, 499. See also Deschamps, *Les Châteaux*, 2:35-98.

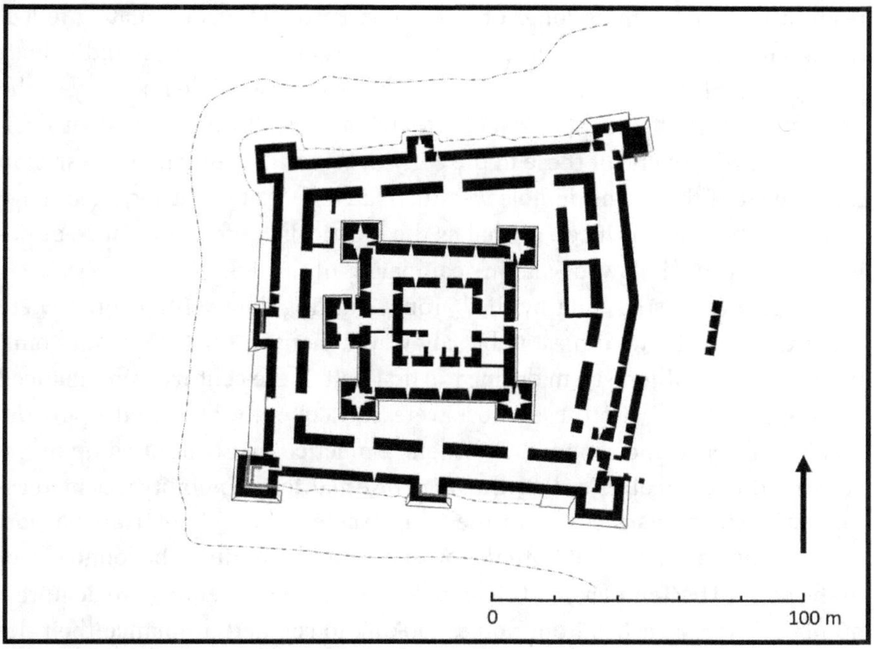

PLAN 36 *Belvoir: castle* (AFTER BILLER AND BAUD)

was engaged in a struggle with Amalric of Jerusalem for control of Egypt, a contest he won by January 1169, just months after the Hospitallers acquired the site of Belvoir. Accordingly, the final plan of the castle may not have been determined until after Cairo, Damascus and Aleppo had come under the same ruler.

It is inconceivable that fortifications built in the last third of the twelfth century would have been designed without an appreciation of contemporary artillery in mind. Likewise, significant existing strongholds that were not developed by proprietors who had the means to do so were probably judged to be sufficiently capable of enduring the effects of contemporary stone-throwers. Critically, both Frankish and Muslim defences dating to the late twelfth century display towers of a similar scale to those built much earlier in the century and walls of a thickness similar to preceding fortification traditions. The architecture, therefore, appears to support the textual evidence: artillery, even if counterweight-powered, was not a significant threat to fortified masonry in the 1170s, nor in the 1180s if Saladin's defences at Cairo are considered. Not until the early thirteenth century is there a notable shift in the thickness of certain walls and the scale of some towers.

When assessing the development of any stronghold, it is necessary to look not only at the broader socio-political situation, but also at the financial resources of the commissioner or developer. Fortifications ultimately reflect the wealth of those who built them and the resources that those individuals were prepared to invest. The military orders were responsible for many of the Franks' significant rural strongholds by the late twelfth century. Their decisions to develop some of these in the early thirteenth century must be seen in the context of their considerable wealth, relative to that of the local baronage and even the ruling princes, as well as the drastically different balance of political power that followed Saladin's campaigns of 1187-88.

The greatest castles built by the baronage during the twelfth century were constructed by the wealthiest nobles; however, many of the large castles built in the kingdom of Jerusalem during the first half of the century were financed by the monarchy.[34] Of these, most were subsequently bestowed upon the secular baronage and the military orders, a practice that continued up to the construction of Jacob's Ford. By the reign of Amalric, the nobility was increasingly reorienting its focus from the countryside to the Mediterranean port towns, taking advantage of both the prosperity and security to be found closer to the coast. The Templars and Hospitallers were wealthy enough to acquire a number of these castles from nobles looking to sell and to finance their defence.

It is no coincidence that most of the strongholds that Saladin failed to take or bypassed in 1187-88, including Chastel Blanc, Tortosa, Crac des Chevaliers and Margat, belonged to the military orders. The extended resistance of the Templars at Safed and Hospitallers at Belvoir can also be highlighted. Until they were taken by Saladin, significant strongholds such as Kerak, Montreal, Beaufort, Chastel Neuf and Toron, as well as Saone and a number of other northern castles in the Syrian Coastal Mountains, had remained in the hands of powerful nobles, who could afford their upkeep.

The loss of Frankish lands in the 1180s further encouraged the baronage to sell their inland properties to the military orders. Meanwhile the impoverished monarchy and visiting crusaders such as Richard I of England, Frederic II of Germany, Richard of Cornwall and Louis IX of France focused their attention and money on developing the citadels and town walls of coastal communities. This was not only where most Franks had congregated but also where Frankish

34 The vast majority of fortified structures built by the Franks during the twelfth century were small towers. Constructed by various landholders, these were centres of administration and even residences as much as defensible strongpoints. In this way, they bear certain similarities to the seventh- through eight-century system of land control established by the Islamic lords of the Near East, see Carver, "Transition to Islam," pp. 192-93. For the Frankish structures see Pringle, *The Red Tower*.

and European commercial interests were based. The Templars may have developed coastal Tortosa and 'Atlit, and eventually Sidon, during the thirteenth century for similar reasons.

Inland, most territories that were reacquired during the Third Crusade or afterward were far more exposed than they had been during most of the twelfth century. The military orders recognised that any besieging force that invested one of their castles could be considerably larger than most of those encountered during the twelfth century, while any friendly relieving force would probably be smaller and slower to muster. This meant that these strongholds might be required to hold out longer than the enemy could afford to besiege them. It was these factors that led to the construction or development of such strong Frankish castles as 'Atlit, Crac des Chevaliers and Margat in the early thirteenth century, not the appearance of some new or drastically more powerful type of artillery.

Concentricity

Although there was continuity in the principles and methods of attack and defence from 1097 to 1291, this was not a period without advances in expertise, organisation and the technologies employed by both attackers and defenders. While artillery may have improved more than any other type of offensive siege weapon, mining remained the most effective means of bringing down walls. It was the threat of mining, coupled with the numerical strength of thirteenth-century Muslim armies, that most likely encouraged the Franks to build their defences with increasing depth. Utilising multiple lines of defences placed as many obstacles as possible between a hostile adversary and a final point of defence, from which a last stand could be made or terms of surrender might be negotiated.

In most instances, concentricity was the product of progressive development rather than a single construction phase: a tendency to add further lines of defences around pre-existing ones. Although outer walls were often added around earlier towers in the twelfth-century,[35] there is a tendency to associate concentricity with only the most impressive strongholds built from the late twelfth century, rather than the concept in and of itself. Many castles in Europe and Near Eastern urban communities enjoyed multiple lines of defences before the participants of the First Crusade set out for the Holy Land. Many ringworks and motte-and-bailey castles contained concentric lines of de-

35 For numerous examples, see Pringle, *Secular Buildings*; Pringle, *The Red Tower*.

fences, while cities like Jerusalem and Acre are described as having forewalls before they were acquired by the Franks. The Theodosian walls of Constantinople, however, are perhaps the most famous example of the use of multiple lines of defences.

When encountering any stronghold, a besieger had to calculate the strength of the obstacles that it presented and the expense in specie and manpower that it would take to overcome them. Factors including morale had to be considered and the costs of assaults and blockades of varying degrees of aggression had to be weighed against the likelihood that the garrison would be relieved before success was achieved. After the battle of Hattin, Saladin's aggressive actions against some strongholds, relatively passive blockades of others, and complete avoidance of certain castles reflect this evaluation process.

Any stronghold could be taken provided the besieger had an endless supply of assault troops and provisions; however, as this was rarely a reality, mining was the most efficient means of compromising a line of fortifications. In appreciation of this, ditches, through-columns and taluses were employed, in addition to multiple lines of walls, to discourage or frustrate sappers.[36] Testament to the effectiveness of medieval miners, it was the very latest defences built by the Franks at Acre that the Mamluks appear to have undermined in 1291. The threats posed by mining and frontal attacks, rather than artillery, most encouraged the construction of these added lines of fortifications.

Wall Thickness

The relatively constant thickness of most defensive walls is perhaps the most definitive evidence of the limited power of this period's artillery. Ellenblum and Raphael have argued the opposite, asserting that the average thickness of fortified walls in the kingdom of Jerusalem increased from about 2-2.5 m in the early and mid-twelfth century to 4-5 m in the 1160s. Having based their theory on only six examples (Bethgibelin, Kerak, Belvoir, Jacob's Ford, 'Atlit and Caesarea), using a single measurement from each to characterise the strength of

36 The former is explained by Ibn 'Abd al-Zahir and Ibn al-Furat in the context of Caesarea, Ibn 'Abd al-Zahir, *al-Rawd*, trans. al-Khowayter, 2:555-56; Ibn al-Furat, *Tarikh*, trans. Lyons and Lyons, 2:70. See also Vitruvius, *De architectura* 2.8.7, ed. Krohn, p. 43, trans. Morgan, pp. 52-53. For a brief introduction to their use in the Levant during the twelfth and thirteenth centuries, see Yovitchitch, *Forteresses*, pp. 84-88.

the entire stronghold, the fallibility of their conclusions is readily apparent.[37] In reality, the average thickness of the walls of most strongholds appears to have increased by only about 0.5 m (at most 1 m) between the twelfth century and the thirteenth.[38]

'Atlit

When wall thickness is discussed, 'Atlit's two inner towers are often referenced. Although its walls are upwards of 5 m thick, the castle's outer wall, which extends about 200 m, is equally impressive.[39] This wall is 6.5 m thick and solid at the first level. A mural gallery, which halves the thickness of the wall at the second level, creates an additional fighting level below the parapet by providing access to casemates.[40] Each casemate is wide enough to accommodate two archers and extends more than 1 m, further reducing the thickness and strength of the curtain wall at this level. It is estimated that as many as seventy archers could have shot simultaneously from this enclosed level of the outer defences alone.[41] The walls of the three outer mural towers mimic the outer curtain: they are close to 6 m thick at the first level but the outward face of the second level is only about 1 m thick, allowing for six embrasures in the eastern walls of the southern and middle towers and four in that of the northern tower.

The designers of this line of defences obviously felt it necessary for the walls to be extremely thick at their base; however, they were quite willing to sacrifice thickness at the second level in order to maximise the number of archers that could shoot outward from within the safety of the towers and the mural gallery, as well as on top of both. This suggests that mining and frontal assaults were

---

37 Ellenblum, *Crusader Castles*, pp. 239-40; Raphael, *Muslim Fortresses*, pp. 42-43; Raphael, "Mighty Towers," pp. 148-49. There would appear to be little evidence to suggest that the thickest section of Kerak's Frankish defences, the northern wall, which is used to 'typify' the castle's defences, was built after 1160.

38 Most walls remained around 2.5 m thick. When considering the much larger scale of most thirteenth-century castles, relative to the smaller towers and simpler enclosures that characterise twelfth century Frankish fortifications, a much larger increase might be expected if only for structural considerations. The thickness of most town walls, where measurable, appears to have remained fairly constant.

39 For excavations and observations at 'Atlit, see Johns *Guide to 'Atlit*; Rey, *Étude sur les monuments*, pp. 97-99; Conder and Kitchener, *Survey of Western Palestine*, 1:293-96.

40 Boas, *Crusader Archaeology*, pp. 111-12.

41 Johns, *Guide to 'Atlit*, pp. 38-39; Johns, "Excavations at Pilgrims' Castle, (1932)," p. 153. This is based on the presumption that the spacing of the surviving casemates was continued along the ruined sections of the mural gallery and that the casemates, 3 m wide, were designed to incorporate two archers.

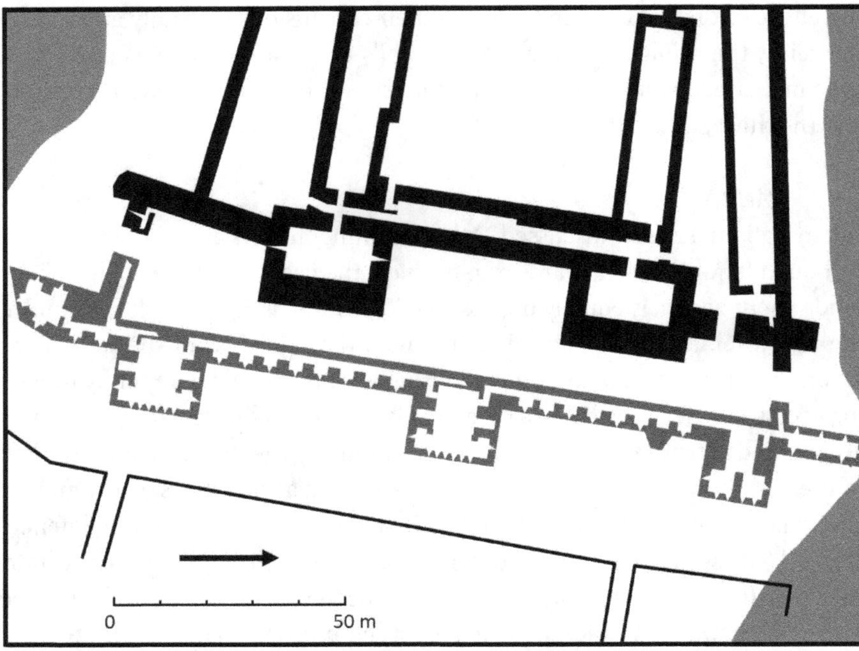

PLAN 37 *'Atlit: second level of outer defences, first level of inner defences* (AFTER JOHNS)

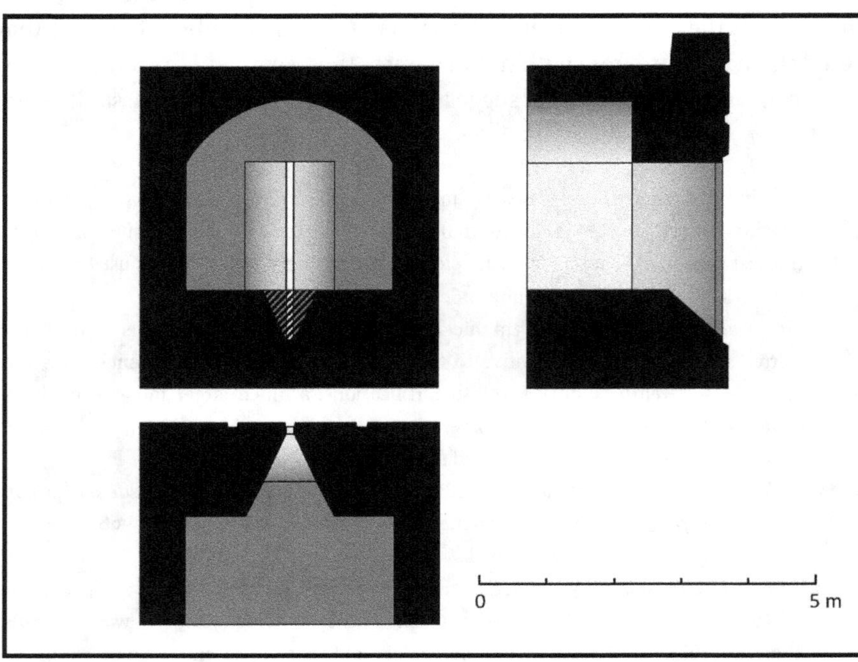

PLAN 38 *'Atlit: casemate and embrasure of the outer wall* (AFTER JOHNS)

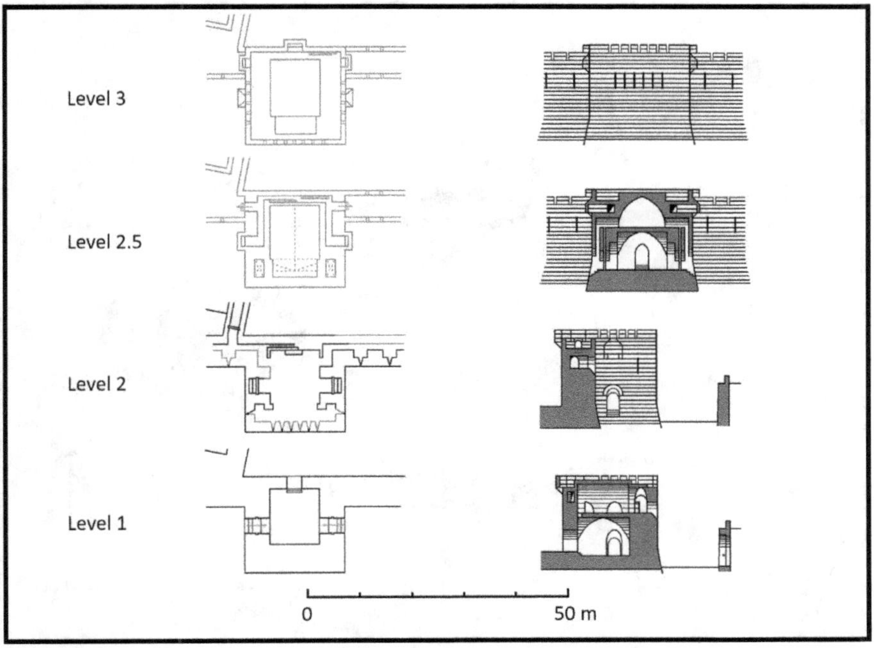

PLAN 39 *'Atlit: south tower of outer eastern line* (AFTER JOHNS)

the threats feared most. The thin outward (eastern) wall of the upper level of each tower would have been that most exposed to stones thrown by artillery and their relative thinness speaks to the perceived threat, or lack thereof, posed by these engines. The designers and their decisions appear to have been vindicated when, following an initial attack while the castle was still under construction, the fortress resisted Mu'azzam 'Isa and his artillery in the autumn of 1220.[42] Although it is possible that these towers postdate the siege of 1220, this would only confirm that it was believed these walls were strong enough to resist just such engines as were employed by Mu'azzam 'Isa.

Purpose and Inconsistency

One of the issues with the simplified measurements of wall thickness provided by Ellenblum and Raphael is that they do not communicate the range of wall thicknesses, let alone the quality of the masonry, employed at most sites. For example, the thickness of the outer walls of Kerak varies considerably, from the stout northern wall to much thinner walls running around the castle's less approachable faces. At Belvoir, the walls of this very regularly planned castle are of a reasonably consistent thickness, yet there is still notable variance. The

42 For the sieges of 1218/19 and 1220, see Chapter 6.

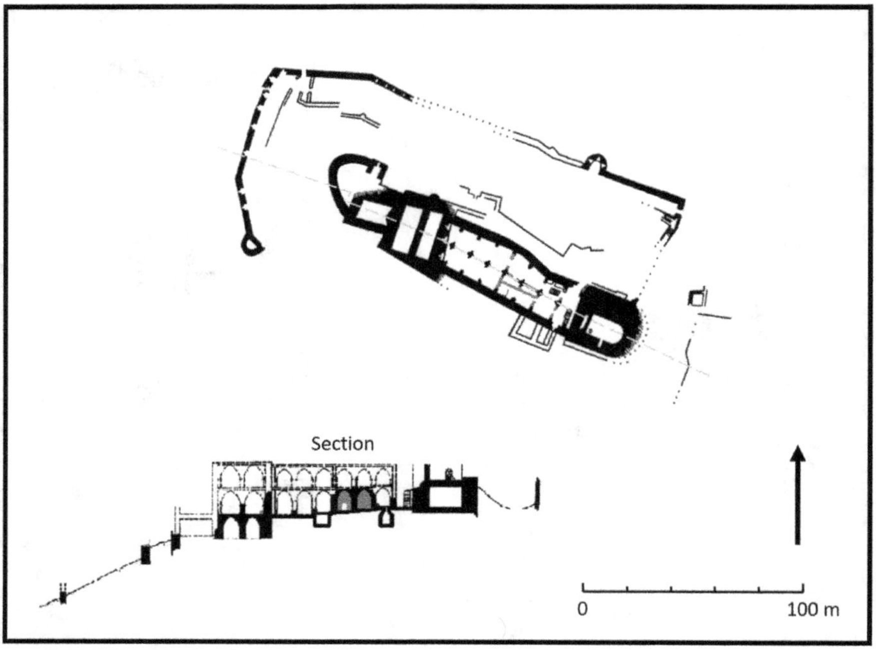

PLAN 40 *Montfort: castle* (AFTER PRINGLE)

curtain and tower walls of the inner enceinte are about 3.5 m thick, while those of the outer enclosure range from about 2.5 m to 3.5 m. At Montfort, where work was completed between about 1226 and 1271, the keep boasts walls about 7 m thick, although pierced by galleries or passages at least at the upper level. The northern and southern walls of the central range that extend to the west of the keep are only 2.3 m and 2 m thick respectively. The curved wall that encloses the western end of the upper castle is 3.1 m thick, but the gate tower at the north end of this curve has walls only 1.1-2.0 m thick. Farther down the slopes of the spur, the walls of the outer bailey are only 1.5-1.9 m thick. The varying thickness of Montfort's defensible walls is mirrored at most other sites. Although extremely thick walls were built in certain parts of some castles in the early thirteenth century, this was not done consistently along all exposed fronts, nor are remarkably thick walls found at all strongholds.

At many sites where there is evidence of efforts to thicken walls in the thirteenth century, this appears to have been done in order to accommodate more defenders on top of the walls. At Subayba, the Ayyubid wall extending north from the southwest tower of the outer bailey (tower 9), originally 1.4 m thick, was thickened considerably in 1240. This seems to have been done to create a broad fighting platform above, as incredibly wide casemates, about 4 m across, were inserted into the bulk of the wall below. To the east of tower 9, a similar platform was created by adding a 0.5 m backing to the wall, allowing for the construction of a barrel vault between the curtain and the nearby reservoir.[43] The thickening of sections of Saone's outer eastern wall after 1188 may have been undertaken with similar objectives.[44] Elsewhere, at sites such as Crac des Chevaliers and Tortosa, otherwise much thicker walls were weakened by creating mural galleries to accommodate additional levels of archers below the parapet, as was done at 'Atlit.

More generally, wall thickness was not something that automatically increased with time. The thickness of the Ayyubid walls at Subayba are comparable to the earlier town walls at nearby Banyas, where exposed sections of the eastern wall are 2.5 m thick and excavations around the southwest tower have revealed the curtain there to be 2.4 m thick.[45] Both are thinner than what appears to be the northern wall of the Frankish citadel of Tiberias, which is 3.4 m thick.[46] Likewise, sections of twelfth-century Frankish masonry may be the thickest medieval walls at Beaufort, despite significant remodelling by both the Ayyubids and Mamluks.[47] The walls of the late-tenth- or eleventh-century tower that anchored the northern defences of the citadel of Shayzar are not much thinner than those of the bridge tower constructed ahead of it by the Mamluks.[48] At Cairo, the northern town defences built under Badr al-Jamali in the late eleventh century, about 3.3-4.0 m thick, are of comparable thickness to the walls of Saladin's citadel constructed a century later.[49] At many smaller Frankish sites, such as Bethgibelin, Belmont and Latrun, the outer walls are thinner than those of the earlier central structures.[50] The same is true of the outer walls of Jubayl and Chastel Rouge, although these castles might have

43 Hartal, *al-Subayba*, pp. 94-99, 103.

44 Michaudel, "Le château de Saône," p. 3. See also Michaudel, "Development of Islamic Military Architecture," pp. 113-16.

45 This tower is 10.6 m × 15.7 m and was subsequently enclosed within a larger tower 18.8 m × 26.0 m, Hartal, "Banyas."

46 Stepansky, "Tiberias, the Courtyard of the Jews "; Stepansky, "The Crusader Castle of Tiberias," p. 179. See also Pringle, *Churches*, 2:352-53 and fig. 99.

47 This is accurate so long as Corvisier's judgement that the southwest circular bastion was built by the Ottomans, Corvisier, "Les campagnes de construction," pp. 243-66.

48 For these structures and the development of the northern entrance system, see Tonghini and Montevecchi, "The Fortification of the Access System," pp. 201-24; Montevecchi "The Northern Area," in Tonghini, *Shayzar I*, pp. 103-77.

49 Creswell, *Muslim Architecture*, 1:183-96.

50 For Bethgibelin, see Cohen, "The Fortification of the Fortress of Gybelin," pp. 67-75. For Belmont, see Harper and Pringle, *Belmont Castle*. For Latrun, see Pringle, *Secular Buildings*, no. 136, pp. 64-65; Boas, *Archaeology of the Military Orders*, p. 111.

been constructed in a single phase.[51] Similar trends can also be seen in Europe, as at Conisbrough where the keep, built around 1170, has walls 4.5 m thick but the later bailey walls are significantly thinner.[52]

The walls of towers constructed in the kingdom of Jerusalem during the early twelfth century average about 2.5 m thick.[53] Most fortified walls built by the Ayyubids during the thirteenth century are 2-3.5 m thick, similar to those built by the Franks. Although walls much thicker than this were constructed at certain sites, such exceptions are not unique to this period. Most classical walls were no more than about 2 m thick; however, the third-century Roman walls of Nicaea are twice as thick as the forewall, 1.6-2 m thick, built around them by the Lascarids in the thirteenth century.[54] The celebrated walls built by Theodosius II around Constantinople in the fifth-century are 5 m thick,[55] while Anna Comnena claims that the walls of Durazzo were so thick that four men could ride abreast along them.[56] The mud-brick main curtain of Raqqa, founded in the eighth century by Caliph al-Mansur, is almost 6 m thick and enclosed by an outer wall 4.5 m thick.[57] These are all dwarfed by the Iron Age walls of Jerusalem, which are 6.4-7.2 m thick.[58]

Although there is a slight rise in the average thickness of fortified walls between the twelfth century and the thirteenth, this appears to have had little to do with artillery. In some instances thick walls were built to provide a broad fighting platform, in others there was a clear effort to deter or combat the work of miners. At some sites the use of much thicker walls was structurally necessary to build towers with more impressive vaulting or higher walls, which will be explored further in the next chapter. It is closer to the end of the thirteenth century when tower walls, built by the Mamluks, become consistently thicker;

---

51 For Jubayl, see Deschamps, *Les Châteaux*, 3:203-15; Rey, *Étude sur les monuments*, pp. 217-19. For Chastel Rouge, see Mesqui, "Qal 'at Yahmur "; Kennedy, *Crusader Castles*, pp. 73-75; Deschamps, *Les Châteaux*, 3:317-19.

52 This castle was built by Hamelin of Warenne, grandson of Fulk of Jerusalem and uncle of Richard I of England. For the castle, see Brindle, "The Keep at Conisbrough Castle," pp. 61-65.

53 Pringle, "Towers in Crusader Palestine," pp. 343-46.

54 A.W. Lawrence, "A Skeletal History," pp. 172-73. The discrepancy here is likely attributable to the height of each wall, the inner wall being twice that of the thinner thirteenth-century outer wall.

55 A.W. Lawrence, "A Skeletal History," p. 180.

56 Anna Comnena, *Alexiad* 13.3, trans. Sewter, p. 403.

57 Creswell, "Fortification in Islam," p. 108.

58 Avigad, "Excavations in the Jewish Quarter," pp. 130-34.

but even as this took place, the thickness of curtain walls remained largely the same.

Mamluk Towers

A discernable trend can be seen towards the construction of stronger towers around the start of the Mamluk period. The Mamluks built few new strongholds but developed many that they acquired, often adding towers or enlarging existing ones. At some sites a finer masonry style was employed, distinguishing the new work, as was done at Subayba, while at others there was a deliberate attempt to match the existing masonry, as can be seen at Shayzar. Although the designs of Mamluk towers varied considerably and an assortment of motivations and influences are evident, they can typically be characterised by their stout plans and emphasis on firepower. They were often sturdier than the large towers built by the Franks and Ayyubids earlier in the thirteenth century and provided numerous positions for archers to shoot from within them. Both aspects would appear to have been more appropriate responses to the threats posed by miners and massed frontal assaults than more powerful artillery.

A process of enlarging certain towers was underway at the citadel of Bosra before it was acquired by the Mamluks. Towers 6 and 8, two of the earliest Ayyubid towers, were encased by al-Salih Ayyub in 1249 and 1251 respectively. Prior to this work, al-Salih Isma'il had wrapped the northwest corner tower, originally constructed by al-'Adil, with a talus in 1240-41. Taluses were then added to the remaining sections of the eastern half of the citadel in the second half of the century. While the taluses were almost certainly used as a defence against mining, the motives behind the enlargement of towers 6 and 8 are less clear.[59]

When Subayba was acquired and then refortified by the Mamluks in the aftermath of the Mongol invasion, towers 9 and 11, located at the west end of the castle, were encased within much bigger towers. The expansion of tower 11 seems to have been done to construct an opulent upper level,[60] rather than to protect it from mining or artillery. The bristling defences given to tower 9 may have been done to inhibit movement around the castle, shielding the small gateway at the base of tower 11. At Kerak, the original Frankish defences at the south end of the castle are largely obscured by the enormous Mamluk tower that now stands there. Whatever the earlier structure looked like, it was

59 The smaller towers may have been enlarged to give the citadel a more uniform appearance from the outside, as a similar masonry style was employed.

60 Hartal, *al-Subayba*, p. 72.

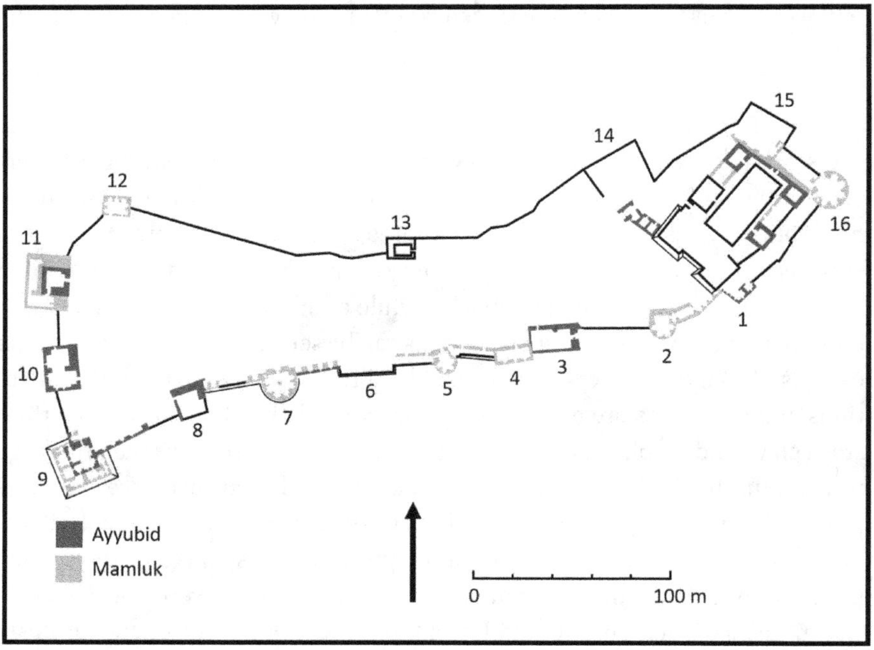

PLAN 41 *Subayba: castle* (AFTER DESCHAMPS AND HARTAL)

replaced by a tower with an exterior southern wall 6.5 m thick.[61] While the scale of this four-level tower indicates that it was built to inspire awe, the thickness of the outer wall and bold exterior inscription appear to defy any attempt to erect artillery on the ridge to the south, from which the castle would have been in range by the end of the thirteenth century.

At Montreal, the impressive elongated northwest tower, round north tower, and quadrangular northeast and south towers (designated E, F, I and L respectively), as well as the two smaller north-eastern towers (G and H), appear to have been constructed around pre-existing Frankish towers.[62] While most of the earlier Frankish embrasures were enlarged into casemates to serve aligning embrasures in the Mamluk masonry, the lower levels of towers F and L were completely enclosed and some of tower E's Frankish embrasures were not carried forward through the encasing Mamluk masonry. This initial loss of potential arrow fire at the lowest level was more than compensated for above. As at Subayba's tower 9, the reason for expanding these towers appears to have been to allow more archers to shoot from within the towers at the level below the battlements; the thickening of the walls, necessary as the perimeter of each

61 Deschamps, *Les Châteaux*, 2:88-89.

62 For a recent study of the castle, see Faucherre, et al., "La forteresse de Shawbak," pp. 43-66.

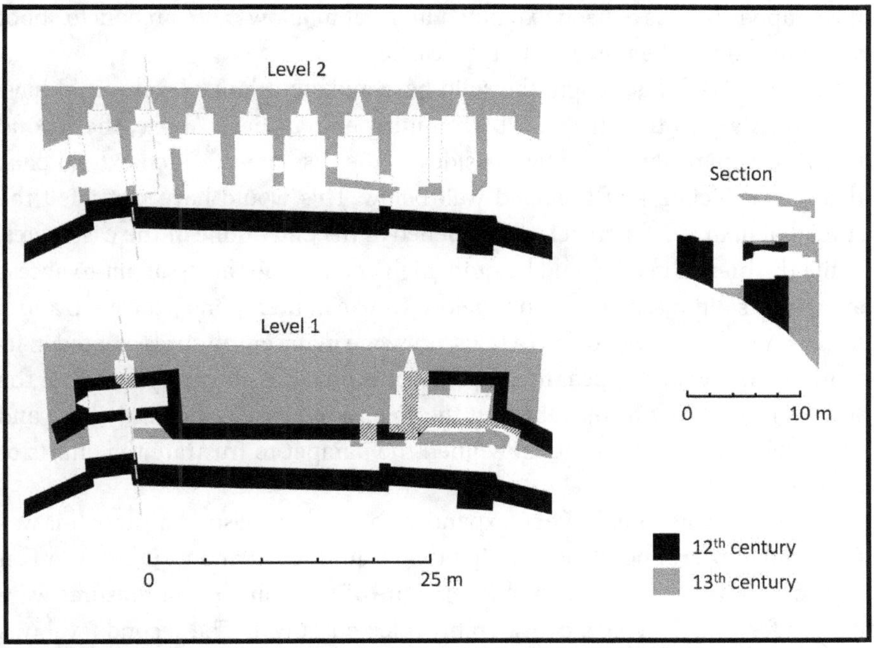

PLAN 42 *Montreal: tower E* (AFTER FAUCHERRE)

tower advanced farther down the slope ahead of it, was simply the means to achieve this. Tower E, the most easily besieged, deserves closer examination: if Montfort was within range of large counterweight trebuchets in 1271, it stands to reason that tower E could have been assailed from the plateau to the north-west of the castle by this period.

Montreal's tower E displays an inscription to Husam al-Din Lajin, dating its completion to the late 1290s. Similar inscriptions on towers F and I indicate that these were part of a larger refortification effort, following the earthquake and destruction that took place during the reign of al-Ashraf Khalil in the early 1290s. It seems reasonable to assume that the designers of tower E, the only one that could be practically bombarded by artillery, took contemporary trebuchet technology into consideration when planning it.

Tower E, almost 40 m wide but projecting less than 10 m, encases two twelfth-century Frankish towers. Its base is solid, built up from the slope in front of the earlier fortifications in order to compensate for the gradient. At the first level, a casemate is accessible from each of the Frankish towers. Between the two earlier towers is a solid mass of masonry, about 8 m thick, which supports a line of eight, or possibly nine, casemates (roughly 3.5 m wide) at the second level. The third level was presumably an open fighting platform. The tower appears to have been expanded in order to allow more archers to shoot at the plateau to the northwest of the castle.

As a result of this design, the solid base and stout lower level would have been resistant to the effects of both mining and artillery. Above, the second level is compartmentalised by the sides of the casemates, which extend back the entire thickness of the solid wall below. This would have restricted the spread of damage if a breach were opened at the end of one of the casemates, while the breach itself would be quite high and hard for an assailant to access above the solid mass of masonry below. Unlike contemporary towers L and I, tower E was not provided with a talus, nor were many embrasures created in its flanks – there would appear to have been only one (possibly two) in each at the second level.[63] The tower relied on the steep topography, its sturdy base and the machicolations that probably lined the parapet to frustrate the efforts of miners.

Tower F seems to have been expanded for similar reasons, although it was reshaped into a rounded tower rather than a quadrangular one. Like tower E, it rests on a solid base. At the first level, none of the Frankish embrasures were carried forward, leaving a considerably thick outer wall. The second level was provided with four embrasures, which enjoyed a good field of fire due to the shape of the tower. The remains of the third level indicate that there were once three box machicolations to protect the base of the tower, suggesting it and others developed at the same time were provided with a double-level parapet. The expansion of tower L similarly increased the number of secure shooting positions at the south end of the castle. No less than six of the new Mamluk embrasures were directed against the north-western plateau, although it was probably out of range. The orientation of so many embrasures in this direction appears to confirm that those who reorganised the castle's defences at the end of the thirteenth century judged this plateau to be the point from which any assault would be organised. Like tower I, it was provided with a large talus, which would have frustrated the work of miners and inhibited movement around the castle.

At Shayzar, the Mamluk enlargement of the Ayyubid donjon complex at the southern end of the citadel provided the structure with additional embrasures, but did little more to strengthen it. Similar to the expansion of certain towers at Subayba and Montreal, it was necessary to set the foundations of the addition down the slope next to the existing tower to the west. As at Subayba, a

63 The length of these two glacises, extending longer at their northern ends, possibly indicate that they were partly intended to obstruct passage around the base of the castle. For these defences see Faucherre, et al., "La forteresse de Shawbak," pp. 43-66.

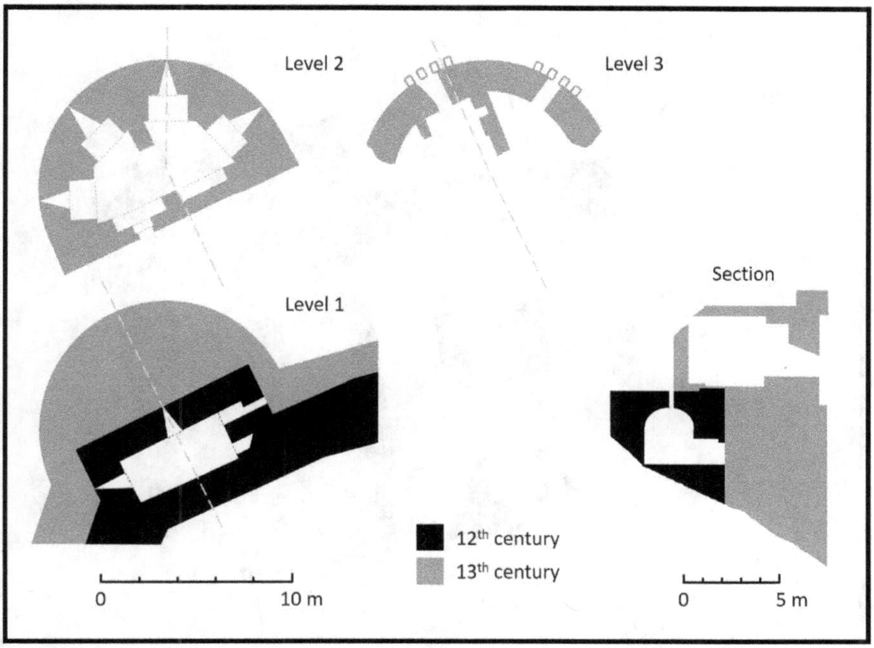

PLAN 43 *Montreal: tower F* (AFTER FAUCHERRE)

lower level was created, provided with six embrasures that were arranged like those of the level above. Both levels were designed to resemble the first level of the Ayyubid tower, which aligns with the second level of the Mamluk addition. Above, a third Mamluk level stretched across both the Mamluk and Ayyubid components below. The plan of this level is unclear, although a number of embrasures are visible in the exterior masonry.

It is noteworthy that although the Mamluk walls of Shayzar's donjon complex are slightly thicker, no effort was made to strengthen the existing Ayyubid walls, which could be bombarded by artillery positioned across the ditch to the south or from the town to the west. Instead, the addition of more embrasures, particularly adding a number that face the river to the east, appears to have been the main defensive consideration. The obvious attempt to blend the new masonry with the existing style, as was similarly done at Bosra, suggests that considerations beyond purely military matters contributed to the design of this work.[64] Perhaps the most obvious examples of stronger Mamluk towers are the solid ones built at Crac des Chevaliers and Margat.

At Crac des Chevaliers, a large quadrangular tower, completed by Qalawun, was built at the centre of the outer southern wall. The tower is solid apart from

64 See Tonghini and Montevecchi, "The Castle of Shayzar," pp. 137-50.

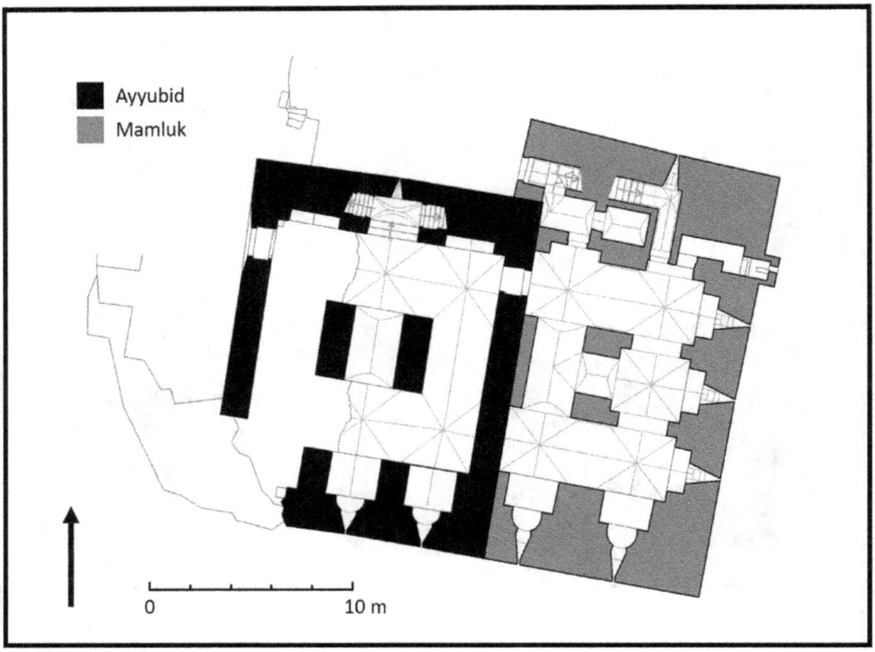

PLAN 44 *Shayzar: Ayyubid level 1 and Mamluk level 2 of the donjon complex* (AFTER TONGHINI AND MONTEVECCHI)

a narrow passage servicing a postern and a single internal level, which is more of a gallery built around a central pier about the level of the parapet of the neighbouring curtain wall. At Margat, a much smaller rounded tower was built to anchor the apex of the castle's southern defences, which had been undermined in 1285. Similar to that at Crac, the tower is solid except for an internal gallery, which wraps around a proportionately larger semi-circular pillar, just below the roof. While it is possible that these towers might have been built solid in response to the increasing power of artillery, the impressive talus at Margat, from which the tower rises, suggests that mining was rather the main threat.[65] A third solid tower can be found at the southwest corner of Beaufort. Although the tower certainly postdates 1268, it is unclear at what point it was added, leaving the possibility that it dates to a period after the introduction of early siege guns. Regardless, these towers were the most solidly constructed in the region during the Frankish period, but they are also unique in this regard. In contrast to the clear defensive planning at certain sites, the Mamluks' development of the lower ward at Beaufort, possibly reveals greater attention to ornamental and residential considerations than purely militaristic ones.

65 For more on these towers, see Chapter 9.

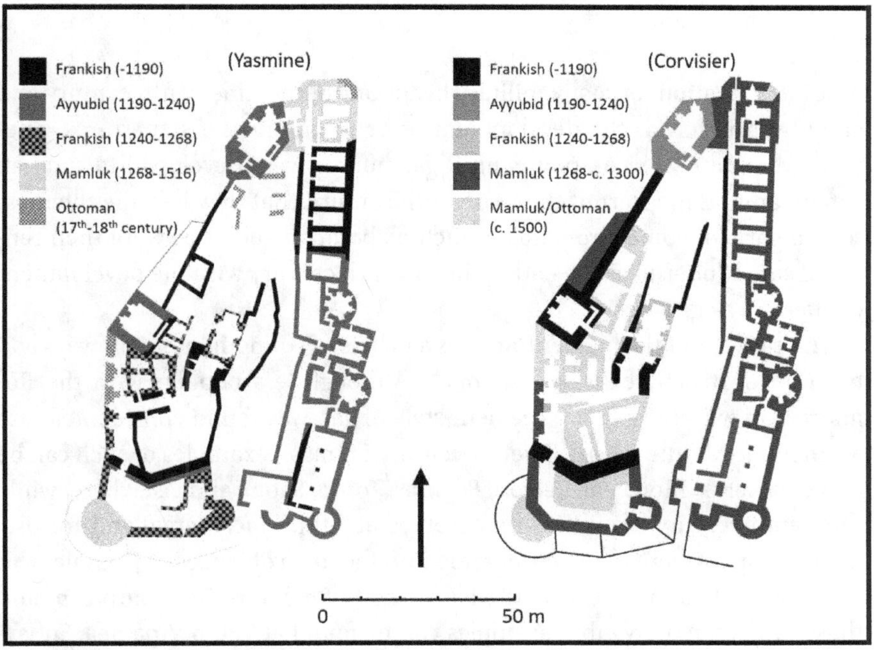

PLAN 45 *Beaufort: interpretations of development* (AFTER YASMINE AND CORVISIER)

While more solid towers appear from the reign of Qalawun, fortifications designed under Baybars, like some of those laid out half a century earlier, placed far greater importance on providing liberal numbers of embrasures along exposed fronts. The degree to which Baybars' architects were influenced by the threat of another Mongol invasion of western Syria, and massed frontal assaults that might accompany this, is unclear. Although there is evidence of increasingly strong counterweight trebuchets through the course of the thirteenth century, it remains an important consideration that not one of these appears to have effected a breach that directly led to the fall of a stronghold.[66] These were not the bombards of the fourteenth and fifteenth centuries nor the early field guns of the late fifteenth and sixteenth centuries. Accordingly, there was no revolution in fortification design, walls were not uniformly thickened and solid towers remained rare. Another indication that artillery was not yet a breaching weapon is the continued development of gateways.

66 Mufaddal ibn Abi'l-Fada'il's account of the siege of 'Akkar would appear to be an exception, Mufaddal, *Kitab*, ed. and trans. Blochet, 1:532.

Entrances

A point of fixation for many military historians of the nineteenth century and early twentieth, was the development of bent entrances. As apertures, gateways were natural weak points until gatehouses were developed to such an extent, around the start of the fourteenth century, that they became the most secure parts of some strongholds. Such elaborate gatehouses were then rendered superfluous near the end of the fifteenth century with the development of effective siege guns.

The concept of the bent entrance as a means of strengthening gateways had been advocated by classical authors.[67] Although less popular with the Romans than with the Greeks, the bent style of gateway found considerable use again in the Middle Ages.[68] Twelfth-century Frankish examples of such can be found at Kerak, Montreal, Belvoir, Baghras, Toron, Saone and elsewhere, while thirteenth-century examples are preserved at ʿAtlit, Crac, Margat and Tortosa. Bent entrances were also incorporated into the town defences of Jerusalem, at the Zion and Tanners' Gates as well as the St Stephen's Gate barbican, and those of Caesarea. Ayyubid examples can be found at ʿAjlun, Apamea, Subayba, al-Jundi, Harim, the citadels of Bosra and Damascus, as well as the town defences of Cairo and Aleppo. The bent entrances at Bourzey and Shughr-Bakas are harder to date, as is that at Qalʿat Najm. As fortifications became more sophisticated, so too did these gateways: elaborate Mamluk examples can be found at Baalbek, Safed, Beaufort and Crac. Among the numerous examples, the two most famous entrances from this period are probably al-Zahir Ghazi's great gatehouse at the citadel of Aleppo and the thirteenth-century Hospitaller/Mamluk entrance ramp at Crac des Chevaliers. Both forced anyone attempting to enter to make a number of 90° or 180° turns, totalling 540° in the case of the gatehouse at Aleppo.

It is difficult to explain the increasing sophistication of gateways during this period if it was, as some have suggested, one dominated by the development of curtain-breaching artillery. Instead, the continued development of gateways is another indication that trebuchet technology was not as powerful as certain literal readings of the sources might suggest.

---

67 For example, Vitruvius, *De architectura* 1.5.2, ed. Krohn, p. 18, trans. Morgan, p. 22.

68 For a discussion of the rediscovery of this means of defence, see Creswell, "Fortification in Islam," pp. 101-8. For gateways and their development in the Levant more generally, see Yovitchitch, *Forteresses du Proche-Orient*, pp. 217-59.

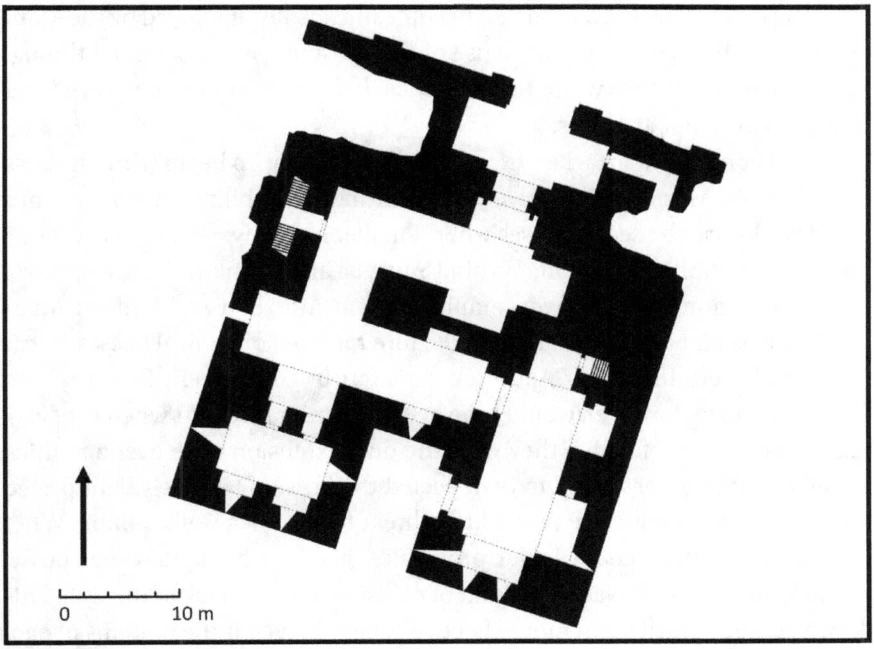

PLAN 46 *Aleppo: citadel gatehouse* (AFTER YOVITCHITCH)

Dressing

Although mechanical artillery was not responsible for radically changing the design of fortifications, some have proposed that it may have had a more subtle influence on the appearance of contemporary defences. Certain historians have suggested that bossed masonry, which was employed heavily through the twelfth and thirteenth centuries, had a defensive purpose.[69] Such arguments assert that by leaving the centre of each block to protrude, it would prevent any incoming projectiles from striking the wall squarely, thus diminishing the percussive force transferred upon collision. But if this was an effective means of defence against such engines, why did this style of masonry, widely employed by Frankish architects during the twelfth century and many Ayyubid

69 For example, Chevedden, *The Citadel of Damascus*, p. 274; Nicolle, *Crusader Castles*, p. 52. For bossed masonry generally, see Viollet-le-Duc, *Dictionnaire*, 2:216-18. Studies of masonry dressing began as attempts to date standing architecture in the nineteenth century, Clermont-Ganneau, *Archaeological Researches*, 1:3-46; Conder and Kitchener, *Survey of Western Palestine*, 3:448. For further thoughts on this topic, see Boase, "Military Architecture," p. 143; Benvenisti, *The Crusaders*, p. 384.

builders in the early thirteenth, not become universally employed by the Mamluk period? If anything, Frankish use of this masonry style decreased through the thirteenth century while the Mamluks, like the Ayyubids, employed it at some sites but not at others.

The theory falls apart when examined more closely. To begin with, the bosses found at most sites are relatively flat, limiting their ability to turn aside projectiles. This can be seen in cases where smaller blocks were used, as at Saone, Jubayl, Cursat, Shayzar, Sidon, Ayyubid Subayba and Mamluk Montreal, as well as where mammoth blocks were employed, as at 'Atlit, Belvoir, Montfort, Mamluk Subayba and the Tower of David.[70] More rounded or conical bosses, as can be found in certain areas at 'Ajlun, are fairly rare by comparison. Even the most rounded bosses, however, would have been of minimal benefit when considering that the walls of which they were the final extension were over 2 m thick. Small rounded bosses could have deflected the force of relatively light projectiles, but these would have posed little threat to the thick walls behind. When considering the effects of larger projectiles, it might be argued that bosses placed more mass between the point of collision and the back of the wall. This, however, could easily, and more effectively, be achieved if the margins of each block were not cut away, preserving a more natural plane and creating an overall thicker wall.

Hopes that a smoother dressing, given the lower surface friction, would better deflect projectiles shot at an oblique angle may have contributed to the declining use of bossed masonry by the Franks in the thirteenth century. Notably, smooth masonry and rounded towers were frequently used together by the Hospitallers, as at Crac, Margat and possibly Arsuf.[71] The Templars continued to employ a range of masonry styles at their large castles, from the smooth masonry at Chastel Blanc to the rough work at Baghras, but appear to have favoured the bossed style found at 'Atlit, Tortosa and Sidon. Like embossed work, if smooth masonry had noticeable defensive benefits, it is hard to explain why it was not employed more broadly and consistently.

The use of marginally-drafted ashlars was probably inspired for reasons other than military necessity. Shaping stones this way was cheaper, easier and

70 Similar bossed masonry can be found at a number of European castles in the Bas-Rhin, such as Ottrott (thirteenth-century), Dambach-la-Ville (thirteenth-century) and Wangenbourg (fourteenth-century), and elsewhere at Montréal in Ardèche (thirteenth-century), Chaudenay-le-Château in Côte-d 'Or (early-fourteenth-century) and the Narbonnaise gate at Carcassonne (late-thirteenth-century).

71 It is possible that the defences of Arsuf predate the castle's acquisition by the Hospitallers; however, they would appear to postdate 1240, see Ewart and Pringle, "Dundonald Castle Excavations," pp. 130-41.

IMAGE 19
'Ajlun: northeast tower
(AUTHOR'S PHOTO)

faster than smoothly dressing each block. The only military consideration might have been one of aesthetics: such rough bosses could disguise the effects of light projectiles (and weathering), shielding the integrity of the smooth margins and an impression of faultlessness.

Tower Shape

Along with smoother masonry, the more frequent use of rounded towers in the thirteenth century has been linked to the increasing power of artillery. The defensive advantages of rounded towers had been recognised in antiquity: Vitruvius had extolled their virtues against mining, rams and other machines and Ammianus identified the weakness of rectangular towers in his account of

Julian's siege of the Assyrian town of Pirisabora.[72] Some modern scholars have amended these sentiments and suggested that the greater use of round towers in the thirteenth century was influenced by their ability to deflect artillery projectiles.

In his attempt to explain the shift towards rounded towers, Hugh Kennedy has suggested that the weakness of quadrangular towers was two-fold: the susceptibility of their corners to mining, and the shelter that was provided at the base of their outward face, which allowed sappers to work in relative safety.[73] While the former would appear to have been inspired by our classical commentators, the latter is suited better to the early modern period and systems of polygonal defences. Although both have merit, neither is sufficient to explain why beaked or pentagonal towers, which were simpler to construct, were not adopted instead.[74]

Rounded towers had been used frequently by the Romans and continued to be a fixture in Byzantine architecture. They were subsequently adopted by Muslim builders at a time when the construction of stone fortifications experienced a hiatus in much of Europe. Before the First Crusade, rounded towers were built by the Armenians in Cilicia and were used to flank the eleventh-century Bab al-Futuh in Cairo.[75] The earliest examples of rounded towers at a Frankish castle appear to be those along the eastern wall of Saone, although these may rest on earlier foundations and are virtually the only examples of such until the early thirteenth century.[76]

Throughout the twelfth century, almost all Frankish towers were quadrangular, similar to those built in Latin Europe.[77] Rectangular towers were simple

---

72 Vitruvius, *De architectura* 1.5.5, ed. Krohn, p. 19, trans. Morgan, p. 23; Ammianus, *Res gestae* 24.2.9-12, ed. Eyssenhardt, pp. 298-99, trans. Younge, pp. 351-52.

73 Kennedy, *Crusader Castles*, p. 114.

74 Towers of this shape have precedent in antiquity, such as those at Salona, and were adopted to a degree by Frederick II in Italy.

75 For the most comprehensive study of Armenian fortifications in Cilicia, see Edwards, *The Fortifications of Armenian Cilicia*. For Bab al-Futuh, see Creswell, *Muslim Architecture*, 1:176-81. See also Creswell, "Fortification in Islam," pp. 113-19.

76 The rounded towers at Baghras may be another rare example of twelfth-century rounded towers built by the Franks; however, the dating of these towers is unclear and it cannot be said with certainty that they were not built after 1188, either by the Armenians or by the Templars.

77 Anglo-French examples of circular or near-circular large-angular towers include the keeps of Arundel, Windsor, Cardiff, Carisbrooke, Tonbridge, Tickhill, Chilham, Odiham, and Château-sur-Epte as well as the enclosure around the keep of Farnham. More developed circular keep-towers date from the reign of Henry III in England and Philip II in France. For some of these early towers, see Colvin, ed., *The History of the King's Works*,

to construct and provided a convenient internal space to inhabit. Although more complicated to build, rounded towers provided a better field of view and distributed structural weight more evenly. Rounded towers became increasingly popular during the thirteenth century but did not replace quadrangular towers completely. The influence of pre-existing forms appears to have been significant. While the Franks seem to have developed earlier rounded towers at Saone, Louis IX, who had built rounded towers at many sites in Europe, such as Aigues-Mortes, appears to have simply developed the existing rectangular towers at Caesarea.[78] Although the castle of Arsuf, dating to the thirteenth century, contains predominantly rounded towers, the towers of the town wall appear to have varied in shape. As at Caesarea, the Franks developed the existing early Islamic town defences, building at least one larger quadrangular tower around one of the smaller solid Islamic towers at the southeast corner of the town. 170 m to the north, the eastern gate, original to the Frankish period and possibly built along a new line of walls, was flanked by a pair of very small rounded towers.[79]

Al-ʿAdil experimented with large-angular tower plans at Harran in the late twelfth century but abandoned these for simpler quadrangular layouts in the early thirteenth century. Although the Mamluks appear to have preferred quadrangular towers in most cases, rounded towers were built at Subayba, Safed, Crac, Margat, Montreal, and elsewhere. The continued and frequent use of quadrangular towers through the thirteenth century, especially in Muslim architecture, is a strong indicator that the threat posed by artillery was not a significant incentive to build round towers.

Kennedy has suggested that the more widespread adoption of round towers in Europe, at least in England and northern France, was in part linked with the crusades.[80]

1:75-78. Experimentation with rounded forms can be found at Provins, Houdan, Etampes and Gisors in France as well as Orford, Conisbrough and Longtown in England. By the thirteenth century, notable circular keeps were built at Coucy, Pembroke, Bothwell, Dourdan, Villeneuve-sur-Yonne, Aigues-Mortes and Flint, some similar to the large keep tower built by Baybars at Safed.

78 For archaeological evidence of this, see Faucherre et al., *Caesarea, 2007*; Faucherre et al., *Caesarea, 2008*; Faucherre et al., *Caesarea, 2009*; Faucherre et al., *Caesarea, 2010*.

79 Roll, "Introduction," pp. 37-41, 44-; Tal, "Apollonia-Arsuf, 2006 and 2009," pp. 108-9.

80 In general, rounded towers became the preference of both Philip II and Richard I by the end of the twelfth century and by the close of the thirteenth most mural towers financed by the French and English crowns, as well as those in Italy by the Holy Roman Emperor, were rounded.

> [I]t is likely that the use of trebuchets and possibly mining were originally learned during the Crusader wars. Certainly Richard I used men from the Crusader states among his artillery men. It is not surprising therefore, that the architects of defensive works adopted some similar designs to thwart these new methods of assault.[81]

However, mining had been undertaken by the Franks since Nicaea (1097), their first siege in the Near East, and the artillery that Richard I employed at Acre in 1191, which he had prepared ahead of time on Sicily, appears to have been at least as powerful as that of Saladin. Furthermore, there is no dramatic change in the way that the chroniclers deal with the artillery employed in Europe in the immediate aftermath of the crusade.[82] Lastly, rounded towers had been built by both Richard I and Philip II in Europe before they went on crusade and examples dating to the Roman period survived in some town defences across Europe and around the Mediterranean, so the idea was nothing new.

There is little to support a direct link between the increased use of rounded towers and the emergence of counterweight artillery, despite the chronological correlation. At best there are circumstantial indications that some rounded towers might have been designed with artillery in mind. The southern faces of the three towers at the south end of the inner defences of Crac are rounded, while their interior levels, except the topmost of the western tower, are rectangular. The towers rise from the great glacis so considerations other than mining or rams would have inspired the shape of these towers. The interior of the circular donjon at Margat is similarly laid out in a rectangle. It may have been hoped that the rounded exteriors of these towers would deflect incoming projectiles, while the rectangular plans within, aligned perpendicular to the direction from which artillery would most likely be positioned, would maximise the thickness of these exposed walls. On the other hand, the exterior plans might have been inspired for other reasons, such as aesthetics, while the quadrangular interiors allowed for the use of simple barrel vaults.

---

81 Kennedy, *Crusader Castles*, pp. 186-87.

82 Although William the Breton's account of Philip's siege of Château Gaillard in 1203-4 is often referenced to support the use of new heavy artillery, the poetic style of the account and lack of corroborating evidence provided by contemporary sources leaves this in doubt, William the Breton, *Philippide* 7, pp. 179-208. Cf. Roger of Wendover, *Flores historiarum*, ed. Hewlett, 1:317-18, 2:8, trans. Giles, 2:207-8, 213; Rigord, *Gesta Philippi Augusti* 141, p. 159. See also Fulton, "Anglo-Norman Artillery," pp. 11-17.

Through the twelfth and thirteenth centuries, fortifications became more sophisticated while the scale and power of counterweight trebuchets similarly increased from the late twelfth century onward. However, these two parallel trends of development were not necessarily linked: there seems to have been little connection between the invention of the counterweight trebuchet and the relatively sudden construction of more impressive fortifications. The more elaborate castles built by the Frankish military orders from the late twelfth century reflect the changing political situation in the Levant: the Latin principalities, which had been the dominant powers in the region in the middle of the century, were now clearly outmatched by the greater resources of the Zankids and later Ayyubids. While the design of Belvoir, built from 1168, signals the declining fortunes of the Franks, about three decades pass before Muslim architecture in the region appears to undergo a similar transition following Saladin's death. As will be addressed in the next chapter, the shift in Ayyubid building styles reflected more a desire to project authority and legitimacy than pure military necessity. Revealingly, the works of al-ʿAdil and al-Zahir, Saladin's brother and son respectively, are the most impressive. Saladin named neither as his primary successor yet these two men emerged as the principal parties in the struggle for control of Ayyubid power and the loyalties of influential figures.

Both Frankish and Muslim fortifications grew stronger during this period as larger sums were invested in the development of certain sites. Artillery, however, does not appear to have been viewed as a considerable threat during the planning of these works. Although wall thickness increased during the thirteenth century, so too did the height of some walls and the areas of internal space that vaulting was required to span. Where military factors were more apparent, the use of thicker walls was often the result of efforts to frustrate miners and increase or maximise the number of archers that could shoot from within the protection of towers and mural galleries. Although the towers built by the Mamluks were often stronger than those of the Ayyubid period, it is not until the end of the thirteenth century that fairly solid towers appear. These, however, seem to have been designed to confront the threat posed by sappers rather than the most powerful trebuchets.

CHAPTER 9

Influence of Defensive Artillery

Throughout the crusades, artillery was used as a weapon of both attack and defence. Latin forces encountered defensive artillery from the time of the First Crusade, but both Franks and Muslims were well aware of the value of utilising artillery in this way centuries before the first wave of crusaders left Europe.[1] Unsurprisingly, the use of defensive artillery continued in the years following. In Britain, Robert of Bellême is said to have strengthened his castles against Henry I in 1102 by supplying them with *machinae*, possibly traction trebuchets; meanwhile, in Persia, the defenders of Isfahan prepared artillery ahead of Sultan Barkyaruq's siege of that place the same year.[2] In the Levant, the effective use of defensive artillery is highlighted at a number of early twelfth-century sieges, including the Muslim defence of Beirut in 1110, the Frankish defence of Jaffa in 1115, and the Muslim defence of Tyre in 1124.[3] Such engines remained important weapons for defenders right up to the siege of Acre in 1291.

In the early twentieth century, Hamilton Thompson, commenting on medieval England, concluded that against the greatest offensive artillery, the besieged were all but powerless as it was impractical to employ the same type of engines in a defensive role.

> The use of such machines upon the walls themselves was as dangerous to the stability of the masonry as their use by the enemy, and hastened the chance of a breach: they could not be employed from the interior of the enclosure, without endangering the defenders on the rampart. The summit of the rectangular keep of the twelfth century was never constructed as a platform for artillery: here, again, engineers probably feared the effect of the constant vibration upon a flat wooden roof, and were content to conceal their ridged roofs within high ramparts.[4]

---

1 For examples of the use of artillery in various regions before the First Crusade, see Chapter 1.

2 John of Worcester, *Chronica chronicarum*, ed. and trans. McGurk, 3:100-1; Roger of Howden, *Chronica*, ed. Stubbs, 1:159, Walter of Coventry, *Memoriale*, ed. Stubbs, pp. 120-21; Ibn al-Athir, *al-Kamil*, trans. Richards, 1:53. See also Fulton, "Anglo-Norman Artillery," p. 27.

3 Ibn al-Qalanisi, *Dhail*, trans. Gibb, pp. 99-100; William of Tyre, *Chronicon* 11.24, 13.6, ed. Huygens, 1:531-32, 593-94, trans. Babcock and Krey, 1:502-3, 2:10-11. For further examples, see Appendix 1.

4 Thompson, *Military Architecture*, pp. 77-78.

 | DOI 10.1163/9789004376922_012

Although apparently unaware of the wealth of source evidence supporting the use of defensive artillery, Thompson keenly identified the impracticality of mounting heavy artillery on top of most medieval towers, a limitation which is chronically overlooked.

While the utility of defensive artillery is emphasised at certain sieges, it is possible to overstate its significance. For example, Ibn al-Athir claims that al-Lawba (in the Biqa' Valley) fell to Shams al-Muluk of Damascus in 1132 because its defenders failed to erect defensive artillery, and Hugh Kennedy has suggested that the defenders of Kerak suffered as a result of their inability to erect any defensive engines in 1183.[5] It is hard to discern the effect that defensive artillery might have had at al-Lawba and it is equally unclear how Kerak suffered by not having any in 1183 – Saladin appears to have been no more successful in 1183 than when both sides employed artillery during the siege of the castle the following year.[6] Taking matters to an extreme, David Nicolle has not only highlighted the potential effectiveness of tower-mounted defensive artillery, but claims that this was how the counterweight trebuchet was most successfully employed.[7]

Analysing the thirteenth century, Christopher Marshall has identified the importance of defensive pressure, including the use of ballistic engines, as Frankish strategies of defence evolved.[8] Artillery appears to have been used relatively effectively at 'Atlit (1220), Beaufort (1268), 'Akkar (1271), Margat (1285) and Acre (1291), despite the ultimate capture of all but 'Atlit. As an important feature of both Muslim and Frankish defensive measures, these engines, their use, and significance warrant closer examination.

---

5 Ibn al-Athir, *al-Kamil*, trans. Richards, 1:296; Kennedy, *Crusader Castles*, pp. 117-18.

6 For the sieges of Kerak in 1183 and 1184, see Chapter 4.

7 Nicolle, *Crusader Castles*, pp. 44, 59.

8 Marshall, *Warfare in the Latin East*, p. 238. The effectiveness of this practice was not confined to the East. Roger of Wendover, for example, describes how the baronial party that invested Mountsorel on behalf of Henry III was restrained by defensive fire from the garrison in the spring of 1217, allowing the castle to be relieved by Prince Louis. He also highlights the prominence of such engines during the two-week defence of Avignon in 1226 before the city fell to treachery, Roger of Wendover, *Flores historiarum*, ed. Hewlett, 2:208-9, 211, 309-10, trans. Giles, 2:389, 391, 478-79. For further instances in which artillery was employed defensively, see Appendix 1.

Ground-mounted Artillery

Most of the defensive artillery employed by the Muslims during the First Crusade may have been situated on the ground. References such as Fulcher of Chartres' to the garrison of Antioch's use of *petrariae* and *fundabula* to throw decapitated heads out of the city, appear to imply that these engines threw their projectiles over the town walls rather than from them.[9] Similarly, Albert of Aachen and William of Tyre seem to distinguish between the positions where defensive engines were placed and where defenders were positioned along the battlements during the Provençal push against Antioch's Bridge Gate.[10] At Jerusalem, Albert of Aachen and Ralph of Caen claim that when the defenders moved their northern artillery, which they were using to target Godfrey of Bouillon's siege tower, they encountered difficulties repositioning these engines and then shooting them in the tight urban constraints of their new locations.[11] Although Albert and Ralph were not at the siege, their assertion that the defending engines were on the ground behind the walls may reflect eyewitness testimony or perhaps a practice familiar to them. William of Tyre similarly suggests that the artillery that defended Jerusalem in 1099, and that which defended Tyre in 1124, was placed behind the walls of each city.[12]

As offensive weaponry developed so too did that of defenders. If counterweight trebuchets were employed by attacking forces by the 1180s, even if they were quite small, it is possible that engines of a similar type and scale were used defensively. Although they shot at a slower rate than traction trebuchets, counterweight engines had a longer range and were more accurate. If placed behind a curtain wall, a counterweight trebuchet could target attacking artillery or approaching siege engines from a position of relative security, while a spotter on the parapet recommended range adjustments. It is possible that engines such as *Mala Cognata/Male Cosine*, which targeted Philip II's notable trebuchet from inside Acre in 1191, and one or more of the nine which defend-

9 Fulcher of Chartres, *Historia Hierosolymitana* 1.15.10, ed. Hagenmeyer, p. 221, trans. Ryan, p. 94.

10 Albert of Aachen, *Historia Ierosolimitana* 3.41, ed. and trans. Edgington, pp. 202-5; William of Tyre, *Chronicon* 4.15, ed. Huygens, 1:254-56, trans. Babcock and Krey, 1:210-11. For the siege of Antioch in 1097-98, see Chapter 3.

11 Albert of Aachen, *Historia Ierosolimitana* 6.17, ed. and trans. Edgington, pp. 424-25; Ralph of Caen, *Gesta Tancredi* 123, RHC Oc 1, p. 691, trans. Bachrach and Bachrach, p. 139. For the siege of Jerusalem in 1099, see Chapter 3.

12 William of Tyre, *Chronicon* 8.13, 13.6, ed. Huygens, 1:403-4, 593-94, trans. Babcock and Krey, 1:361-63, 2:10-11. For the siege of Tyre in 1124, see Chapter 4.

ed Mosul in 1182, were counterweight trebuchets.[13] The comparative exposure of offensive engines is made clear in both instances and is vividly illustrated by Ibn al-Athir in his account of the siege of Bourzey in 1188.[14]

Baybars had the unique opportunity to assail a castle from within another stronghold when he besieged Beaufort in 1268. Having captured the Templar outwork, lighter engines may have been erected on its towers, had there been room, but the heavier engines, much larger and more costly, were probably set up on the ground inside the enclosure, had this been possible. A similar situation may have played out at Crac des Chevaliers a few years later, if indeed Baybars set up artillery within the enclosure of the southern outwork. Counterweight trebuchets may have been used by the Frankish defenders of Margat in 1285 and were almost certainly employed during the defence of Acre in 1291. Although certain defensive engines were damaged at Margat, the ability to repair and redeploy these engines may suggest they were positioned within the castle, rather than on its towers. The defensive artillery at Acre seems to have been well concealed and to have enjoyed relative security throughout the siege. This was certainly true of the Pisans' engine, which was positioned near the church of St Romanus in the Gardens and may have been some distance from the town wall. Such examples appear to confirm that the heaviest offensive engines that were stored in a stronghold could be used to defend it if need be, and that such engines were generally placed on the ground.

Defensive artillery was a reactionary weapon, as attackers possessed the initiative during a siege. Accordingly, the targets of a defensive trebuchet would almost never be right at its maximum potential range. Nevertheless, discussions of how and where defensive artillery was deployed tend to be influenced heavily by impressions of range. Although maximum range is an important consideration, it should not be taken for granted that offensive engines were only ever operated from such distances. There was an obvious advantage to deploying attacking artillery beyond the range of defenders; however, this was not always possible, as in examples where strongholds occupied elevated positions or the topography imposed limits on where engines could be set up. There may also have been a desire to increase accuracy or the amount of energy transferred to a target by advancing an engine closer. It is important to bear in mind that the accuracy of any trebuchet, whether employed by attackers or defenders, decreases with range: environmental and mechanical dis-

13 See Ambroise, *Estoire* ll. 4,737-46, ed. Ailes and Barber, pp. 76-77, trans. Ailes, p. 98; *Itinerarium* 3.7, ed. Stubbs, p. 218, trans. Nicholson, pp. 208-9; Ibn al-Athir, *al-Kamil*, trans. Richards, 2:285-87.

14 Ibn al-Athir, *al-Kamil*, trans. Richards, 2:286, 350.

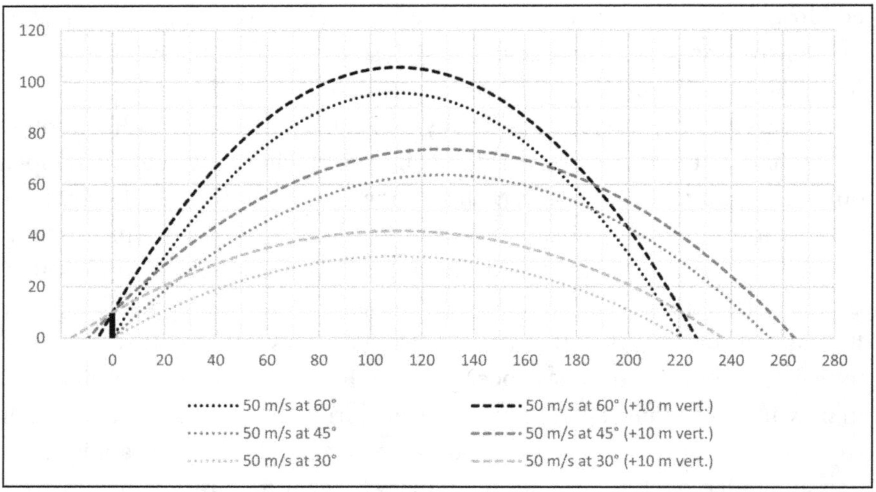

FIGURE 18 *Comparative trajectories: engines on a tower and at the base of a wall*

Position	Horiz. Pos.	Vert. Pos.	Velocity	Release Angle	Max. Range
Ground	0 m	0 m	50 m/s	60°	220.9 m
Tower	0 m	10 m	50 m/s	60°	226.6 m
Ground	0 m	0 m	50 m/s	45°	255.1 m
Tower	0 m	10 m	50 m/s	45°	264.7 m
Ground	0 m	0 m	50 m/s	30°	220.9 m
Tower	0 m	10 m	50 m/s	30°	237.1 m

crepancies from shot to shot are amplified as flight time is extended, and the longer a projectile is in the air, the more time a mobile target has to avoid it.[15]

Due to the implicitly shorter range of an engine placed behind a line of defences, rather than on it, it is often assumed, with little further thought, that artillery was tower-mounted. The horizontal range gained by elevating an engine will always be less than the distance that the engine is raised if it shoots at an angle of 45° or greater (Fig. 18).[16] An engine placed on the ground is also

15 A projectile thrown at about 50 m/s might achieve a theoretical range greater than 250 m; however, it would be subject to a flight time of over seven seconds, during which period environmental factors would considerably slow the projectile, decreasing its actual range, and provide any mobile targets with an opportunity to move.

16 For example, if an engine throws two stones at 45°, but the second starts at an elevation 20 m higher than the first, the second stone will travel only 18.6 m farther if both were thrown at a velocity of 50 m/s. If thrown at 25 m/s, the second stone will land only 16 m farther than the first.

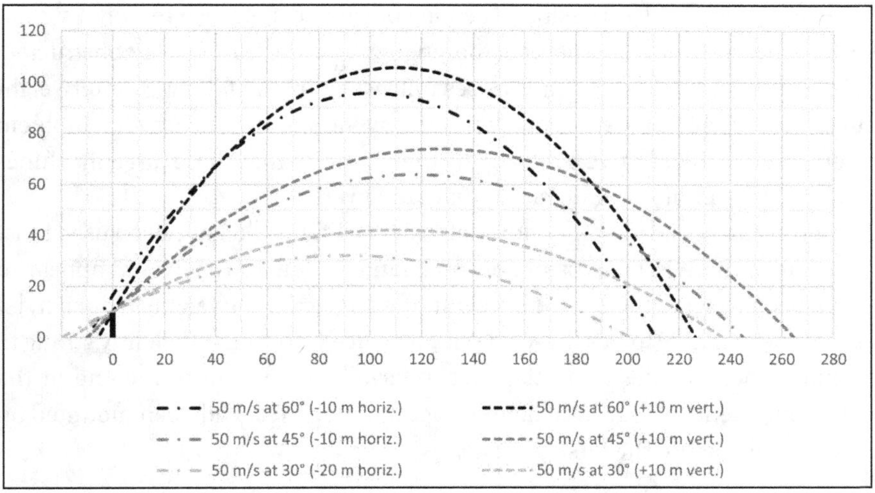

FIGURE 19 *Comparative trajectories: engines on a tower and behind a wall*

Position	Horiz. Pos.	Vert. Pos.	Velocity	Release Angle	Max. Range
Ground	-10 m	10 m	50 m/s	60°	210.9 m
Tower	0 m	0 m	50 m/s	60°	226.6 m
Ground	-10 m	0 m	50 m/s	45°	245.1 m
Tower	0 m	10 m	50 m/s	45°	264.7 m
Ground	-20 m	0 m	50 m/s	30°	200.9 m
Tower	10 m	10 m	50 m/s	30°	237.1 m

subject to a horizontal handicap: it must be backed up from the base of the wall in order to shoot over it (Fig. 19). While this would have encouraged defenders to mount light, rapid shooting engines on top of towers, there are issues of security, strength and space to contend with when considering larger engines.

Besides the risks associated with exposing an expensive but cumbersome, slow-shooting counterweight trebuchet on top of a tower, doing so tied up a valuable fighting platform. Whereas little performance was sacrificed by placing such engines on the ground behind the curtain, more versatile antipersonnel forces, such as archers, slingers and operators of traction trebuchets, who benefited much more with line of sight to the enemy, were less effective in such a position. Any attempt to mix forces would face considerable challenges and risks. Even small counterweight trebuchets had reasonably significant footprints due to the necessity of outriggers, which provided lateral support and ensured that the forces created by the swinging counterweight remained as two-dimensional as possible. The long beam and sling also required significant space in front and behind the engine in order to move unobstructed. There were also the significant forces unleashed by the falling counterweight to consider, which posed a deadly risk to anyone standing nearby if a problem were to arise. Illustrative evidence further suggests that these large machines were placed on the ground, rather than on towers.[17]

Setting aside the matter of how defensive spaces could have been used most efficiently, there is Thompson's observation that few towers would appear to have been designed to bear the weight of a large counterweight trebuchet, let alone the dynamic forces these engines produced. Towers sufficiently strong to support such engines do not appear to have been built until the end of the thirteenth century; however, lighter traction trebuchets had been mounted on towers long before the First Crusade.

Tower-mounted Artillery

The value of mounting defensive artillery on towers was appreciated centuries before the crusades. Evidence dating back to the fourth century BC suggests that some Greek towers were designed specifically to incorporate torsion bows and such machines influenced Vitruvius's theories concerning the optimal spacing of mural towers three centuries later.[18] The engines considered by Vitruvius do not appear to have been significantly heavy, as the weight of defending cohorts is the largest mass that he addresses when dealing with fortifications.[19] *Onagers* are mentioned in a defensive capacity by Ammianus at the Persian siege of Amida in the fourth century AD and by Procopius at the Gothic siege of Rome in the sixth.[20] However, Ammianus warns against using such engines on top of defensive ramparts, as the shock created when shooting an *onager* (*scorpio*/*tormentum*) would threaten the masonry below.[21] With this in mind, Peter Purton has suggested that repeated efforts to strengthen the walls of Rome up to the sixth century may reflect efforts to mount increasingly heavy defensive artillery, rather than attempts to reinforce the walls in order to resist

17 For images of counterweight trebuchets in defensive positions, see Appendix 2.

18 Ober, "Early Artillery Towers," pp. 569-604; Ober, "Towards a Typology," pp. 147-69; Vitruvius, *De architectura* 1.5.4, ed. Krohn, p. 18, trans. Morgan, pp. 22-23.

19 Vitruvius, *de architectura* 1.5.3-8, ed. Krohn, pp. 18-20, trans. Morgan, pp. 22-24

20 Ammianus, *Res gestae* 19.7.7, trans. Younge, p. 197; Procopius, *Wars of Justinian* 5.21, ed. and trans. Dewing, 3:204-7.

21 Ammianus, *Res gestae* 23.4.4-7, ed. Eyssenhardt, pp. 270-71, trans. Younge, pp. 322-23.

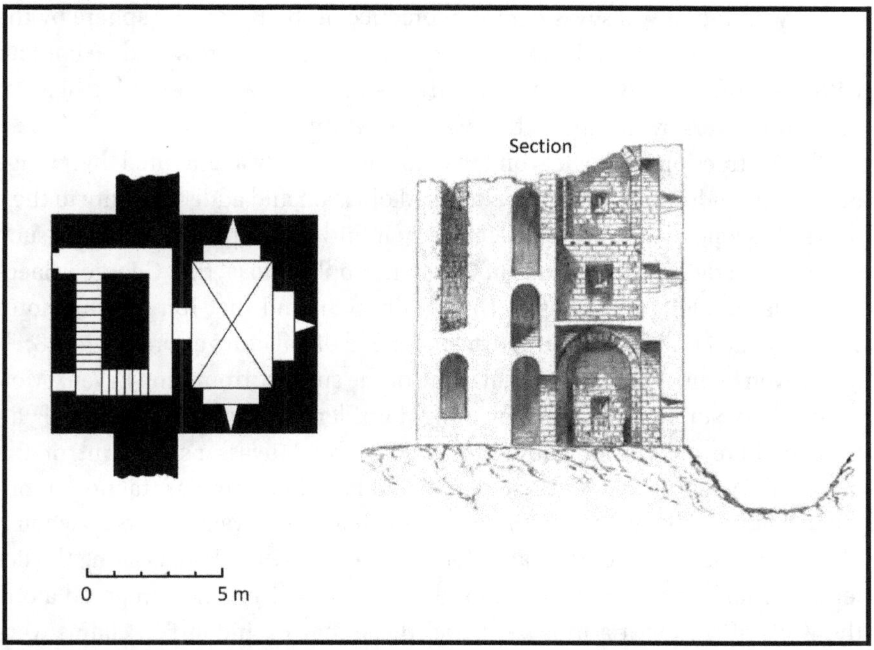

PLAN 47 *Antioch: mural tower* (AFTER REY)

more powerful offensive engines.[22] If this was the case, it stands to reason that the rebuilding work carried out at Antioch by Justinian would have followed a similar model. Perhaps this is why the city's mural towers appear better suited to support a significant mass at their top level than the vast majority of towers built by the Franks and Muslims during the twelfth and thirteenth centuries.[23] Intended or not, this ability came through the solid pier in the back half of these towers, around which the internal staircases were wrapped. But towers capable of bearing such significant loads, as could have been supported by the piers in these towers, were not often necessary.

Although the Muslims may have placed their defensive artillery on the ground during the First Crusade, light traction trebuchets appear to have been elevated during earlier sieges in the general region. As noted in Chapter 1, al-Tabari clearly describes the placement of both *'arradat* and *majaniq* on the gates of Baghdad in the 860s.[24] John Skylitzes gives the impression that placing

22 Purton, *Early Medieval Siege*, p. 16.

23 Towers built by Justinian with a similar plan can be found along the northern and southern walls of Zenobia (Halabiyya), see Lauffray, *Halabiyya-Zenobia*, vol 1. My thanks to Denys Pringle for pointing this out to me.

24 Al-Tabari, *Tarikh*, trans. Saliba, 35:40-41.

artillery on top of walls was common practice in the Byzantine sphere by the eleventh century, as he appears to have assumed his readers would be familiar with this practice.[25] Both Frankish and Muslim forces were mounting light artillery on towers by the end of the twelfth century.

Placing traction trebuchets on top of towers not only augmented their range but also provided crews with a better field of vision and added velocity to their projectiles when striking targets below their initial elevation. It was likely such engines that defended the walls of Constantinople in 1047, that George Palaeologus placed along the walls of Durazzo in 1081, and that Ibn al-Athir notes were installed on the walls of Jerusalem ahead of Saladin's approach in 1187.[26] It has even been hypothesised that some of the spurs surrounding Li Vaux Moise may have supported some type of light artillery in the twelfth century,[27] although there are no indications of this in Ibn Muyassar's account of the eight-day Egyptian attack in 1158.[28] It would have been similar traction trebuchets that Ibn 'Abd al-Zahir claims were erected on the walls of Tarsus, when it revolted in the ninth century, and that were set up by the Franks along the defences of Jaffa in 1266.[29] So too would have been the engines prepared along the walls of Alexandria in 1272-73, and those that Rashid al-Din claims were erected on the walls of Baghdad in response to increasing Mongol pressure.[30]

A number of illustrations confirm that traction trebuchets were mounted on towers.[31] With an eye to these, Creswell postulated that three holes found in the platform over Cairo's Bab al-Nasr may have been used to anchor the legs of a stone-thrower with a tripod base, perhaps similar to those depicted in Peter of Eboli's manuscript.[32] While there is plenty of illustrative evidence to confirm that traction trebuchets were placed on towers, very few illustrations depict towers supporting counterweight trebuchets.

---

25 Hoffmeyer, "Military Equipment," p. 125.

26 Michael Psellus, *Chronographia*, trans. Sewter, p. 210; Anna Comnena, *Alexiad* 4.1, trans. Sewter, p. 135; Ibn al-Athir, *al-Kamil*, trans. Richards, 2:330.

27 Marino, et al., "The Crusader Settlement in Petra," pp. 8-9.

28 Ibn Muyassar, *Akhbar Misr*, RHC Or 3, p. 472.

29 Ibn 'Abd al-Zahir, *al-Rawd*, trans. al-Khowayter, 2:590, 795-96.

30 Ibn 'Abd al-Zahir, *al-Rawd*, trans. al-Khowayter, 2:830; Rashid al-Din, *Ta'rikh-I Ghazani*, trans. Boyle, pp. 190, 232-33.

31 See Appendix 2, **nos. 2-10, 22** and **23**.

32 Creswell, *Muslim Architecture*, 1:166-76. Given the shape, depth and arrangement of these holes, or sockets, there is sufficient reason to doubt Creswell's interpretation of their function; however, there is nothing to rule out the likelihood that defensive artillery was once positioned on this platform.

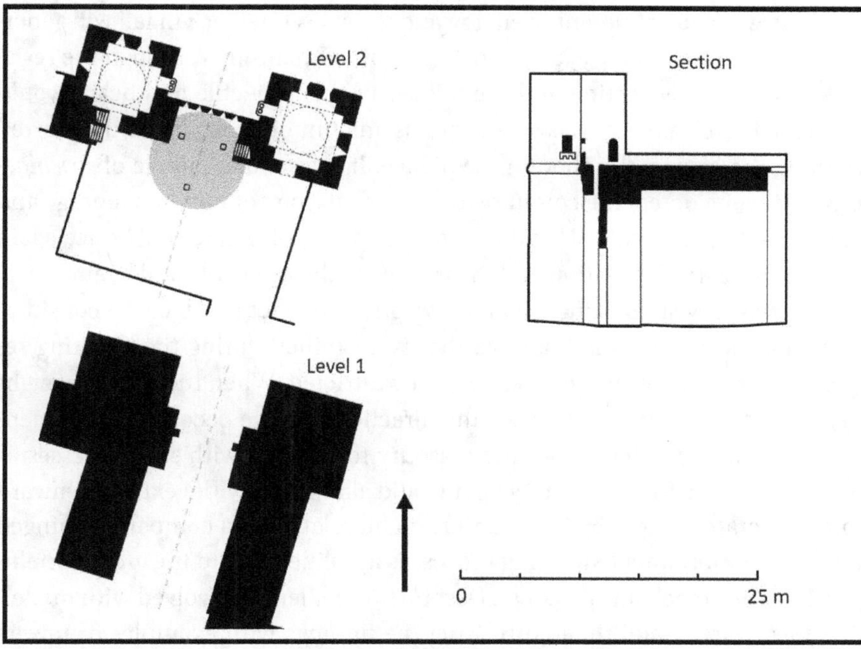

PLAN 48 *Cairo: Bab al-Nasr, with indents highlighted* (AFTER CRESWELL)

It was much more practical and effective to place lighter traction machines on the tops of towers. In addition to being significantly easier to bring up and take down, their ability to yaw with far less difficulty and maintain a higher shooting rate made them more useful against moving targets and attacking troops. Although traction trebuchets were less powerful, they proved capable of disabling siege towers if given enough time, as at Jerusalem (1099) and Beirut (1110). Slow-moving or immobile siege engines, including besieging artillery and siege towers, could also be targeted by counterweight trebuchets, as may have been done at Mansura in early 1250, during the siege of Margat in 1285 and at Acre in 1291.

Elevating Counterweight Artillery

It is unfortunate that no architectural study has been conducted to test the limits of how much additional weight most medieval towers could support. This would be informative but a tower meant to carry a counterweight trebuchet also had to deal with the dynamic stresses that accompanied these engines: there was not only the weight of the machine to support but also the angular forces generated by the falling counterweight and its subsequent swinging. These stresses would have been immense. If the beam of *Arrangement A* is drawn back to an angle of 135°, the massive 10,000 kg counterweight will hold 250,000 J of potential energy and the subsequent 3.4 m fall will generate up to 98,000 N and 333,200 kg·m/s (Ns) of momentum. Although the resistance provided by friction and the masses of the projectile and beam would decrease these values, this is an enormous amount of energy to be transferred through the course of a shooting sequence. If the smaller engine of *Arrangement B* is considered, the counterweight holds 16,700 J of potential energy and falls with a force of 9,800 N and 16,660 kg·m/s (Ns) of momentum, once again slightly mitigated by friction and the masses of the projectile and beam.

The masonry supporting a counterweight trebuchet is subject to considerable uniaxial compression, a force that is magnified during the shooting sequence as the fall of the counterweight is restricted. When the counterweight begins to swing beneath the axle, the direction of force becomes two dimensional, compelling the supporting masonry to contend with shear stresses as well. Although a fixed counterweight would allay some of the extra downward force generated by the falling counterweight, relative to a comparable hinged model, the subsequent swinging process is amplified due to the greater inefficiency of this mechanical system. Despite the challenges involved with modelling these forces, and then correlating the findings with examples of towers built with various materials and different designs, such a study might reveal at what point and in which regions it may have been possible to mount counterweight trebuchets on towers.

Without conducting such a study, Paul Chevedden argued, in his 1986 PhD thesis, that the large towers built by al-ʿAdil at the citadel of Damascus were designed to carry just such engines. Although he addressed issues such as the space required by large counterweight trebuchets, he did not consider the structural strains that these machines would create. It has already been shown that the heaviest counterweight trebuchets were much less powerful at the start of the thirteenth century than Chevedden has suggested; however, it remains to investigate if there is evidence to indicate that any towers were designed to carry counterweight artillery. In the absence of a more quantitative architectural study, indications that towers were designed to withstand considerably greater forces than might otherwise be necessary will be examined.

Possible Ayyubid Artillery Towers

Chevedden has explained the near spontaneous appearance of large towers following Saladin's death, typified by those built by Saladin's brother, al-ʿAdil, and son, al-Zahir Ghazi, as the response to a need to mount heavy counterweight trebuchets on top of towers. This, he argues, was necessary to offset a supposedly enormous advantage held by offensive artillery.[33] With no evidence to indicate the introduction of an outside element, such as a foreign technology, Chevedden has suggested that the appearance of these large towers was a delayed reaction to the more general advancement of offensive weaponry.

The towers built by Saladin at Damascus and Cairo are relatively small, comparable to those constructed by Nur al-Din. This would appear to imply that such towers were judged to be sufficiently strong at a point when counterweight trebuchet technology was known.[34] Descriptions of the siege of Acre suggest that such engines may have been employed there in 1190-91, but confirm that these were not effective breaching engines. The failure of any one of the numerous eyewitness sources to point out any differences between the engines of the Christian besiegers and Muslim defenders, let alone suggest that any new or special type of engine was employed, indicate that neither side held a technological advantage.[35] Thus the artillery that influenced the design of al-'Adil's towers was probably only marginally more powerful than that which would have been considered by Saladin's architects.

Al-'Adil's Early Work

In the mid-1170s, Saladin embarked on an ambitious building campaign to fortify Cairo, constructing a new citadel between the new and old cities. The scale of these defences impressed Ibn Jubayr when he viewed them in the spring of 1183, a year ahead of their completion.[36] The citadel was later developed by al-'Adil, who added a number of large quadrangular towers and encased some of the round ones, notably Burj al-Haddad and Burj al-Ramla, within much larger round towers. The work of the two brothers is easily distinguishable by style and scale. Saladin's towers are faced with smooth masonry and most are of a modest size, typically less than 10 m in diameter. The notable exceptions are Burj al-Mahruq and Burj al-Zafar (both about 16 m in diameter), which were constructed to anchor the south and north ends, respectively, of the new wall Saladin commissioned along the eastern side of Cairo. Al-'Adil's towers, built in the first decade of the thirteenth century, are faced with much rougher

33 Chevedden, *The Citadel of Damascus*, esp. pp. xliii-xliv, 7-8. For his analogy between the supposed architectural response to artillery in the early thirteenth century and that at the start of the early modern period, in response to the early employment of effective siege guns, see Chevedden, "Invention," p. 73 and n. 4.

34 For illustrative proof of this knowledge by the 1180s, see Appendix 2, **no. 34**.

35 For the siege of Acre in 1189-91, see Chapter 5.

36 Ibn Jubayr, *Rihla*, trans. Broadhurst, p. 43. For the dating, an inscription over Bab al-Mudarraj dates it to 570 [1183-84]. Cf. Ibn al-Athir, *al-Kamil*, trans. Richards, 2:249.

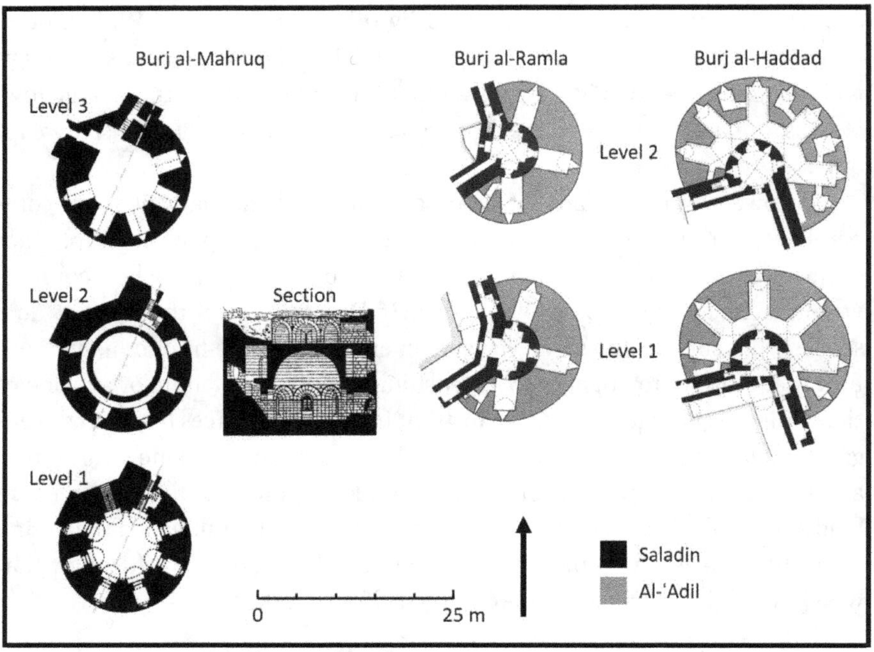

PLAN 49 *Cairo: Ayyubid rounded towers* (AFTER CRESWELL AND YOVITCHITCH)

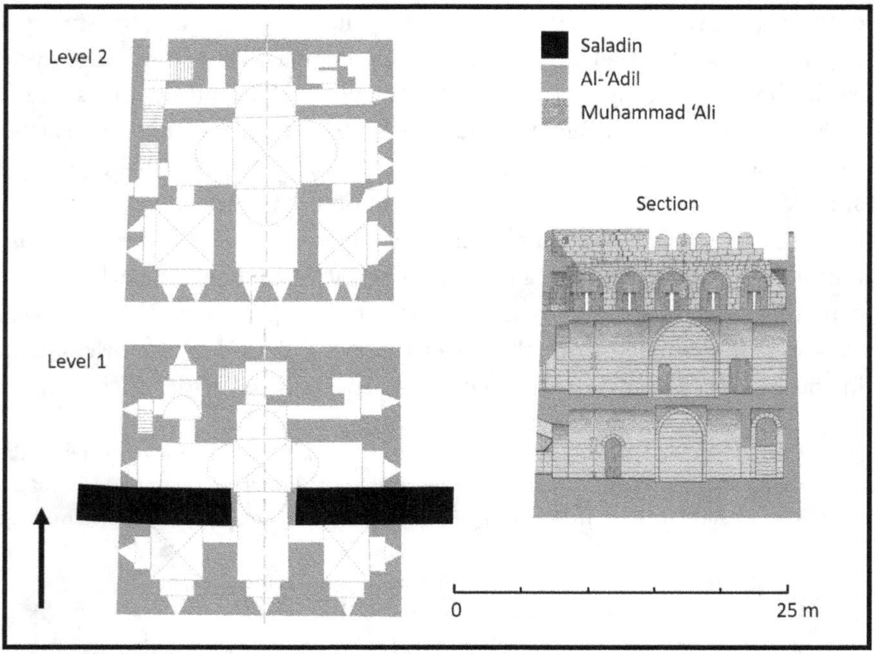

PLAN 50 *Cairo: Burj al-Kerkyalan* (AFTER CRESWELL)

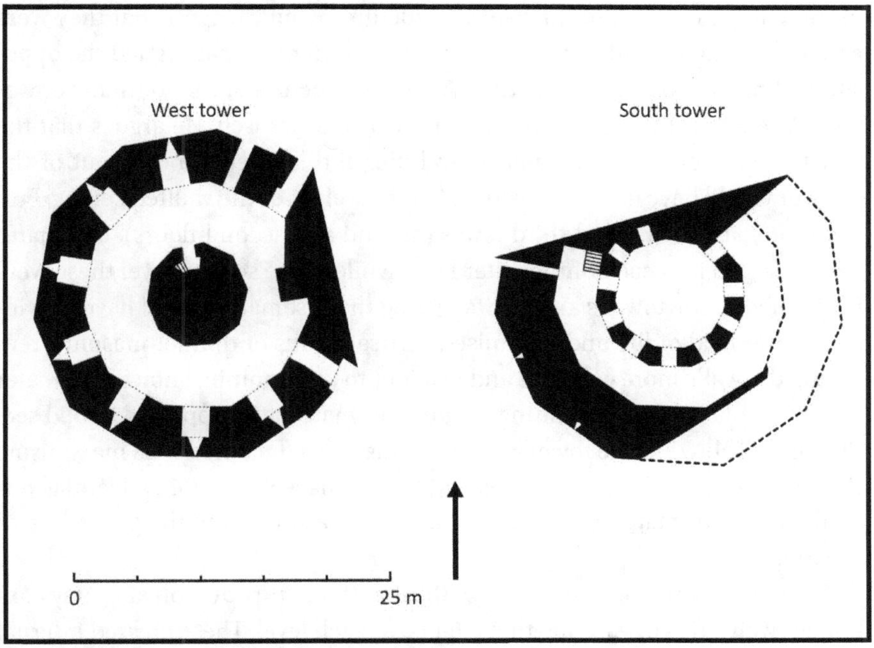

PLAN 51 *Harran: west and south towers of the citadel* (AFTER HANISCH)

masonry and all are comparable in size to the largest of his brother's towers.[37] These elements, bossed masonry and grand scale, also characterise his towers elsewhere.

Late in Saladin's reign, al-'Adil rebuilt the citadel at Harran, constructing an eleven-sided tower, nearly 30 m in diameter, at each corner of the quadrangular stronghold. Each of the three surviving towers has a different internal layout and evidence of what were once three internal levels.[38] The scale of these towers is reminiscent of those built later at Cairo, but their shape is quite different. This, along with their varying internal plans, may suggest that some degree of experimentation was taking place.

The internal structures of the southern and western towers are the clearest. Inside the southern tower, there is an eleven-sided structure laid out to mirror the angles of the outer wall. Open in the centre, there is an aperture in each face of the internal structure at the first level and every other face at the level above. In the centre of the western tower, there is a central ten-sided pillar that contains a winding staircase. The outer walls of the towers appear to have been

37 Creswell, "Fortification in Islam," pp. 121-23; Boase, "Military Architecture," pp. 150-51.

38 Hanisch, "The Works of al-Malik al-'Adil," pp. 168-74. Cf. Lloyd and Brice, "Harran," pp. 77-111.

more than 3 m thick and the central structures would suggest that they were able to bear significant vertical loads; however, Hanisch has argued the opposite.[39] While the open space within the inner structure of the south tower was a weakness, it is the west tower that has worried Hanisch. He argues that the offset created by the decagonal central pier and hendecagonal layout of the interior would have unevenly distributed the weight of the vaulted ceiling. Furthermore, at the first and third levels (ground and second floors), casemates were inserted into each of the outer faces while at the second level these were positioned in the tower's angles. Staggering the casemates would have provided a better field of fire and minimised vertical seams of thinner masonry, rendering the walls more cohesive and resilient to undermining; however, greater stress was placed on the vaulting of each casemate, as it supported a solid section of wall above. If the tower's designer was only concerned with maximising the vertical load that it could support, it might have been more advantageous to align the casemates, creating solid piers of masonry from the ground up to the roof.

By comparison, the similarly large thirteenth-century donjon at Coucy contained twelve ribs of vaulting that aligned at each level. These created internal buttresses that supported the vaulted ceilings. Coucy's smaller outer towers, which were also round but only about 18 m in diameter, had staggered casemates.[40] Although the donjon was large enough to accommodate a significant counterweight trebuchet on its roof, the oculus in the centre, as seen in Viollet-le-Duc's schematics, and vast vaulted internal space would have made this impractical. It seems that neither the towers at Coucy nor those at Harran were designed to support extremely heavy artillery.

The scale of the towers at Harran appears to have been influenced by a desire to impress onlookers rather than to mount or resist sizable counterweight trebuchets. Heidemann has similarly concluded that the citadel of Raqqa, also built by al-ʿAdil to a quadrangular plan in the last decade of the twelfth century, but with round corner towers, was constructed primarily for symbolic and ceremonial purposes, rather than to serve as a seriously defensible strongpoint.[41] But what of al-ʿAdil's later preference for quadrangular towers?

39 Hanisch, "The Works of al-Malik al-ʿAdil," pp. 168-74. Hanisch does not relate these issues to any consideration of artillery.

40 For a study of Coucy predating the castle's destruction, see Viollet-le-Duc, *Description du Château de Coucy*.

41 Heidemann, "The Citadel of al-Raqqa," pp. 136-145.

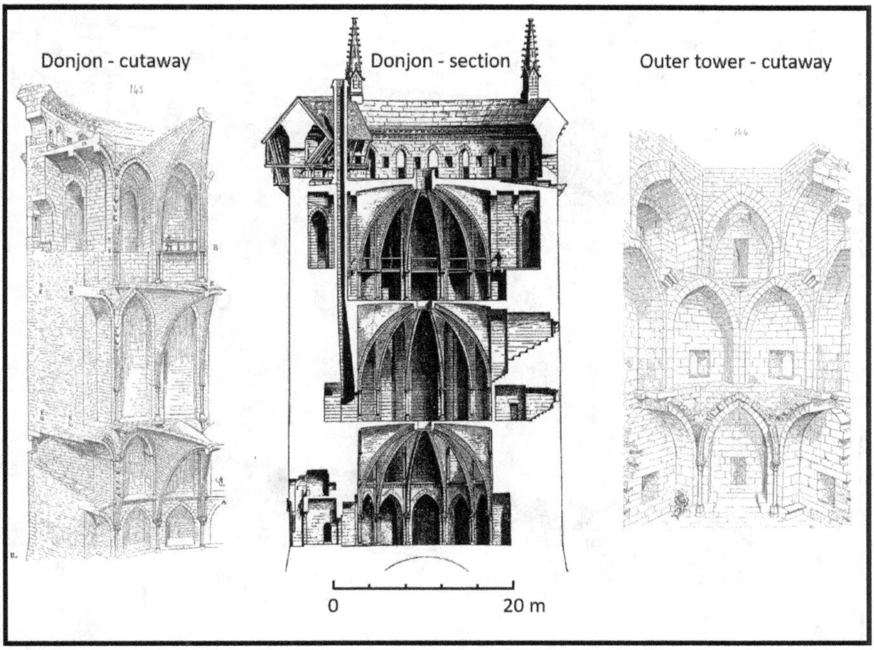

PLAN 52 *Coucy: castle towers* (FROM VIOLLET-LE-DUC)

Damascus

None of al-'Adil's towers have been scrutinised more than those that he built at Damascus.[42] A decade after taking power there, al-'Adil essentially enclosed the earlier Seljuk-Zankid citadel in much the same way as he did certain towers at Cairo. This work, completed between 1207 and 1217, might seem justified when considering that the city had been besieged no less than five times in the decade immediately following Saladin's death in 1193; however, the defences appear to have been adequate. On only one occasion was the city taken and even then the citadel remained a bastion of resistance from which a surrender was negotiated.[43]

42 For studies of the citadel of Damascus, see Sauvaget, "La citadelle de Damas," pp. 59-90; Hartmann-Virnich, "Les portes Ayyoubides de la citadelle de Damas," pp. 287-311; Berthier, "La Citadelle de Damas," pp. 151-64; Hanisch, "Der Langturm als neuer Turmtypus," pp. 183-234; Hanisch, "Die Zitadell von Damaskus," pp. 235-305; Braune, "Die Stadtbefestigung von Damaskus," pp. 202-10. Hanisch has suggested that the large towers of the citadel of Damascus were built to accommodate al-'Adil's personal fighting force, the *halqa*, Hanisch, "The Works of al-Malik al-'Adil," p. 177. Cf. Hanisch "Der Langturm als neuer Turmtypus," pp. 206-20.

43 See Chapter 6.

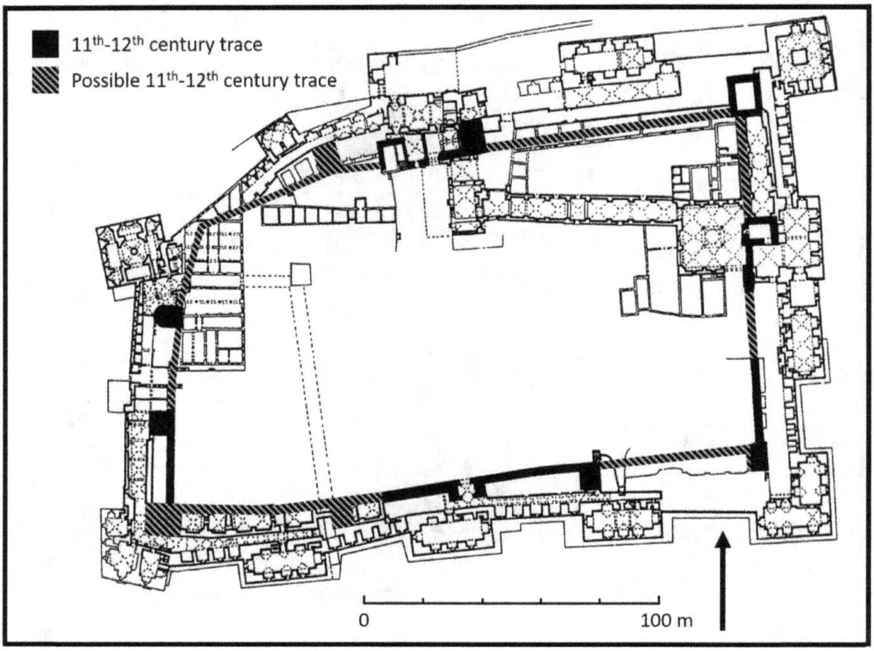

PLAN 53 *Damascus: citadel* (AFTER HANISCH AND BERTHIER)

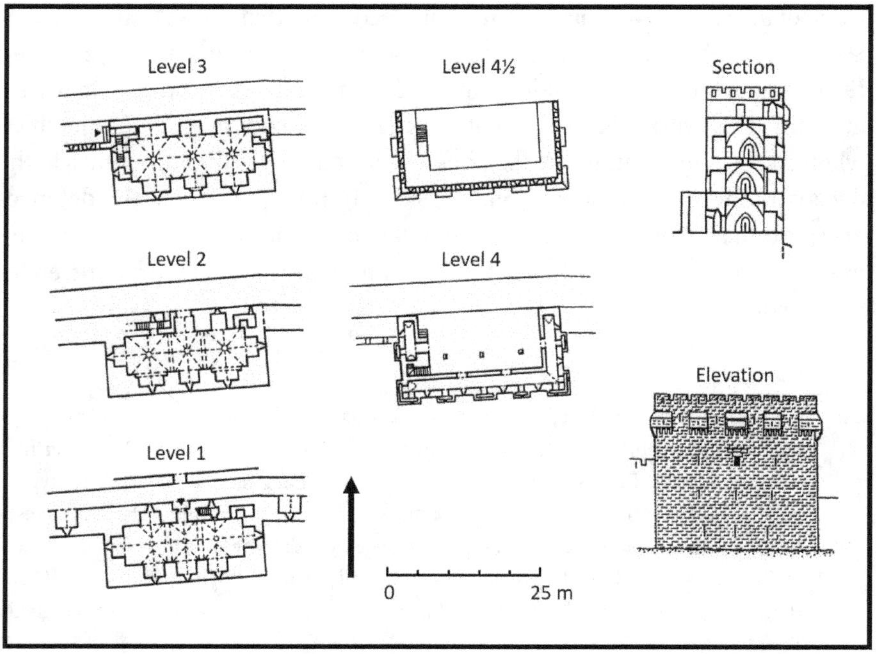

PLAN 54 *Damascus: tower 4 of the citadel* (AFTER BURGER AND KING)

Rather than a response to an acute military threat, the style, timing, and method with which the citadel was developed suggest that the rebuilding was meant to convey an outward display of power while strengthening the city at the same time. Instead of developing the existing defences, al-ʿAdil had a new enceinte constructed just in front of the existing one, erasing any trace of his predecessors' work when viewed from the outside.[44] This, along with the use of much larger towers, was an expression of his authority and preeminent position in the Ayyubid world. It seems this was part of an effort to assert his legitimacy, as Saladin had named his sons, not his brother, as the principal heirs of his empire.

Towers 2, 3 and 4, the mid-wall towers along the southern side of the citadel, which face into the city, display some of the least rebuilding since they were constructed under al-ʿAdil. Each is about 27 m wide, 13 m deep and contains three internal levels. The Ayyubid battlements of these towers, as well as those of others that were rebuilt by the Mamluks, consist of two levels and enclose only the three outward faces of each tower. Chevedden has argued that the use of such battlements, the shape of these towers and their size were dictated by the determinant requirements of mounting large counterweight trebuchets.[45] While the high battlements could have provided good protection and concealment, and the open backs would have allowed the long arm and sling of a trebuchet to rotate freely, such hypothetical considerations are not sufficient in themselves to prove that the design of these towers was dictated by a desire to place heavy artillery on their roofs.

The use of a double-level parapet is not evidence of a desire to protect a valuable counterweight trebuchet. The Franks had built battlements with two levels earlier in the twelfth century, before the invention of the counterweight trebuchet, at sites such as Jubayl and Saone. This was an excellent way of providing an additional level of defensive archers at minimal expense and, when coupled with machicolations, was an effective way of covering the base of a tower. Accordingly, the popularity of such battlements grew from the late twelfth century and can be found at Margat, Crac des Chevaliers, Baalbek, and many other strongholds. Likewise, there is also little to suggest that round towers would have been any less effective platforms for heavy artillery.[46] The curvature of a rounded tower would provide an even better field of fire, if the machine could be yawed, while this shape would distribute the weight of a

44 This is similar to the way in which Saladin surrounded certain parts of Cairo with a new wall that was built ahead of the earlier line.

45 Chevedden, *The Citadel of Damascus*, pp. 277-78, cf. pp. 192-93.

46 Cf. France, *Western Warfare*, p. 124; Kennedy, *Crusader Castles*, p. 114.

FIGURE 20 *Citadel of Damascus: tower and curtain dimensions*[a]

	Face Length	Side Length	Thickness
SW Tower (1)	missing		
Curtain (1-2)	25.0 m	-	3.6-4.9 m
S(w) Tower (2)	26.9 m	12.8 m 8.5 m projection	3.2 m face 3.4 m E side
Curtain (2-3)	29.0 m	-	4.7-4.8 m
S(c) Tower (3)	26.9 m	12.9 m 8.0 m projection	3.4 m
Curtain (3-4)	31.5 m (approx.)	-	4.5 m
S(e) Tower (4)	26.4 m	12.9 m 9.5 m projection, E side	3.4 m face
Curtain (4-5)	31.0 m	-	
SE Tower (5)	25.0 m S face 27.0 m E face		
Curtain (5-6)	24.0 m	-	
E(s) Tower (6)	21.5 m	14.0 m 9.5 m projection, S side	
Curtain (6-7)	10.0 m	-	4.4 m
E Barbican	10 m (approx.)	-	1.9 m
E(n) Tower (7)	27.5 m	11.5 m projection, S side 7.5 m projection, N side	4.4 m S side 4.7 m N side
Curtain (7-8)	32.5 m	-	
NE Tower (8)	24.0 m E face 20.8 m N face	15.0 m projection, S side 8.0 m projection, W side	4.1 m N face 4.4 m S side 3.8 m W side
Curtain (8-9)	36 m (approx.)	-	1.6 m
N(e) Tower 9	31.0 m	14.0 m 9.0 m projection	4.1m face
Curtain (9-10)	35.0 m	-	4.5 m
N(c) Tower (10)	29.8 m	17.3 m 14.0 m projection, E side 17.0 m projection, W side	4.1 m face 4.0 m E side 5.2 m W side
Curtain (10-11)	24.0 m	-	3.6 m
N(w) Tower (11)	13.5 m	10.7 m	2.4 m face 2.5 m side
Curtain (11-12)	43.0 m	-	
NW Tower (12)	20 m		1.8 m E side
Curtain (12-1)		-	5.6 m

a From Chevedden, *The Citadel of Damascus*, Appendix 2, pp. 555-58.

large engine more evenly. The size of these towers is their most spectacular aspect; however, their design and strength are far from exceptional.

The towers are much larger than previous norms, but their walls are not proportionally thicker. Although substantial, roughly 3.4 m thick, this is not dramatically thicker than many dating to the twelfth century.[47] When considering that the towers contain three internal levels, each with a single open room, and that the strength of the walls was compromised by spacious casemates, the thickness of the walls seems to have been a structural necessity. This is further indicated by the side walls of many towers, which are thicker than those facing outward. Finally, when the northern towers of the citadel were rebuilt by the Mamluks, following the Mongol withdrawal from western Syria, their walls were built to a similar scale. Thus, despite evident improvements in artillery technology between 1210 and 1260, the reconstruction of these towers does not appear to have been significantly influenced by considerations of artillery, be it offensive or defensive (Fig. 20).

Following a period of experimentation, al-ʿAdil seems to have settled on a preference for rectangular mural towers and L-shaped corner towers, not because he needed them to support large counterweight trebuchets, but rather because they were imposing and fairly simple to design and construct. The earliest examples of this design may have been built at the citadel of Bosra, though it was at Damascus that they were probably first employed as part of a single, cohesive building strategy. Towers of a similar scale and design characterise refortification efforts completed elsewhere under al-ʿAdil and can be found at Cairo and Jerusalem, where work was overseen by al-ʿAdil's sons al-Kamil and al-Muʿazzam ʿIsa respectively, as well as at Bosra and Mount Tabor. Although the towers vary in size, they are all large, display a similar bossed masonry style, and make use of a double-level parapet and box machicolations.

Bosra and ʿAjlun

Trajan's theatre at Bosra, like that at Caesarea, had been used as a fortress prior to the arrival of the crusaders. The earliest towers built here by al-ʿAdil, predate those he commissioned at Damascus and are typically smaller, in part due to the layout and restrictions imposed by the dimensions of the theatre. Work was carried out in phases, stretching across two decades, and reveals a trend towards the construction of bigger towers with greater spaces inside (Fig. 21).[48]

47 Many of the thickest walls are found along the northern side of the citadel, which was largely rebuilt by the Mamluks. This may have been a structural necessity as the foundations of these defences rest on a riverbank.

48 For the dating of the towers at Bosra, see Yovitchitch, "La citadelle de Bosra," pp. 205-17; Yovitchitch, "Bosra: Eine Zitadelle," pp. 167-77.

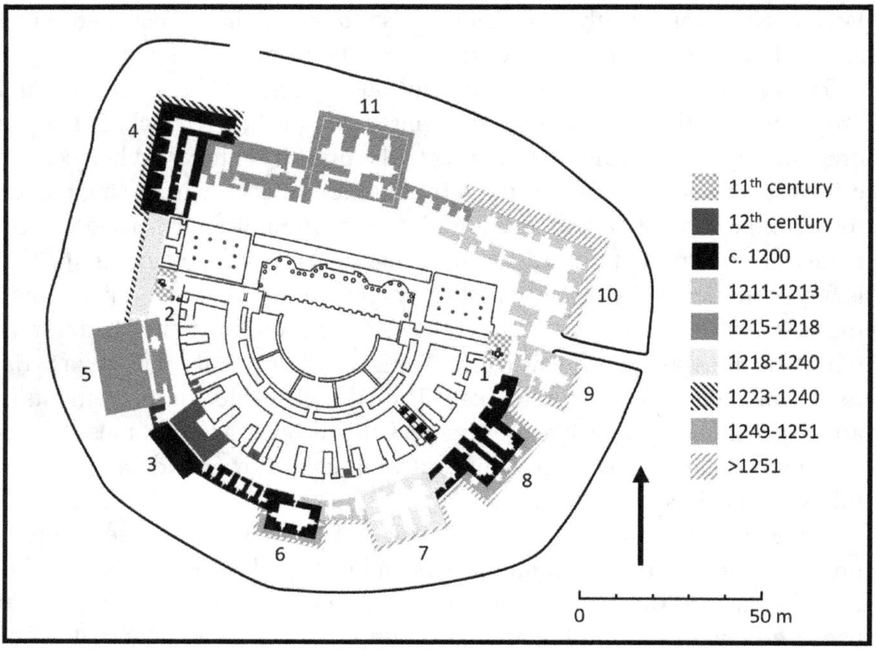

PLAN 55 *Bosra: citadel* (AFTER YOVITCHITCH)

FIGURE 21 *Citadel of Bosra: epigraphic dating*[a]

	Date	Attribution
Tower 1	481 [1089]	*Abu Manur Kumushtekin*
Tower 3	542 [1147-48]	*Mu'in al-Din Unar*
Tower 4	599 [1202-3]	*al-'Adil*
Tower 10	608 [1211]	*al-'Adil*
Tower 9	609-10 [1212-13]	*al-'Adil*
Tower 11	612 [1215]	*al-'Adil*
Tower 5	615 [1218]	*al-'Adil*
Tower 4 (glacis)	638 [1240-41]	*al-Salih Isma'il*
Tower 6	647 [1249]	*al-Salih Ayyub*
Tower 8	659 [1250-51]	*al-Nasir Yusuf*

a From Yovitchitch, "La citadelle de Bosra," p. 216.

Rather than becoming stronger and more solid, structural strength was progressively sacrificed for interior space throughout the first third of the thirteenth century. It might be argued that the later encasing of towers 6 and 8, the

IMAGE 20 *Bosra: citadel from the south* (PHOTO COURTESY OF DENYS PRINGLE)

IMAGE 21 *ʿAjlun: castle from the east* (PHOTO COURTESY OF APAAME)

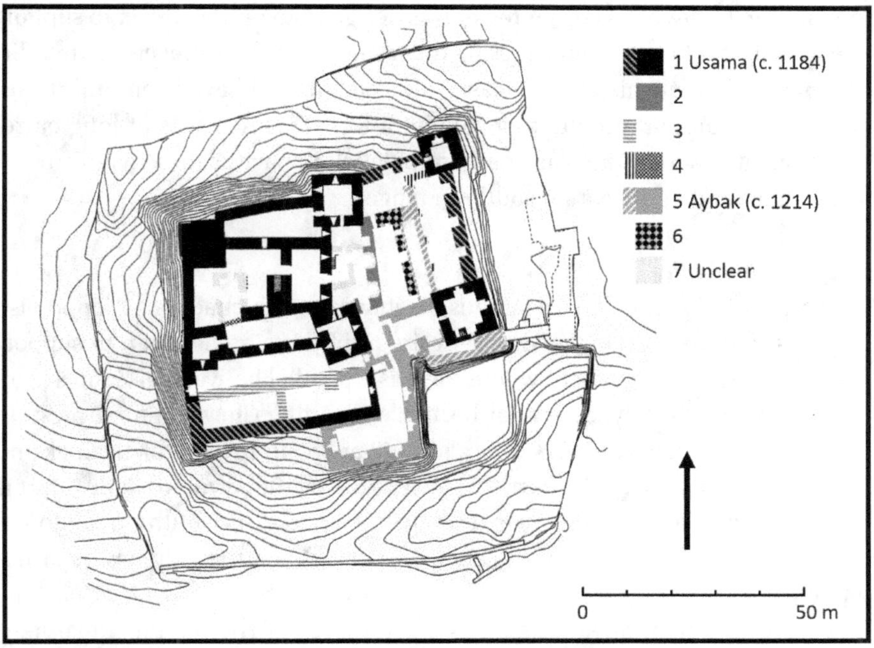

PLAN 56 *'Ajlun: level 2 with exposed sections of level 1* (AFTER YOVITCHITCH AND JOHNS)

earliest and smallest Ayyubid towers, reflected a need for stronger or larger towers; however, this may have been done, at least in part, to give the citadel a more uniform appearance from the outside. This brought them closer to the size of towers 5, 7 and 11, as well as corner towers 4 and 10, and, conspicuously, the same masonry style that had been used earlier was employed once more.

'Izz al-Din Aybak, a mamluk of al-'Adil's son, al-Mu'azzam 'Isa, may have honoured his patron through the addition he built at 'Ajlun. The castle had been founded by one of Saladin's emirs, 'Izz al-Din Usama, in 1184-85. Al-'Adil later became suspicious of Usama and had him arrested, stripping him of 'Ajlun, which was in al-Mu'azzam 'Isa's sphere of influence, and the castle was subsequently bestowed on Aybak. In addition to developing the entrance of the castle, Aybak added a large L-Shaped tower, which appears square from the outside, in 1214. The tower shares certain obvious similarities with those commissioned elsewhere by al-'Adil. Unsurprisingly, Aybak had intimate knowledge of al-'Adil's building style, as an inscription at Mount Tabor states that he had overseen the construction of one of the towers there only two years earlier.

When comparing the work at 'Ajlun completed by Usama with that carried out under Aybak, the exterior walls are all 2-3 m thick and there are few differences in respective masonry techniques apart from the embrasures and other stylistic signatures. Without its uppermost level it is impossible to rule out the possibility that Aybak's corner tower was designed to be a platform to support heavy artillery; however, considering the layout of the lower levels, there is little to suggest that this was the case.[49] Instead, it may have been built to resemble outwardly those built by al-ʿAdil. Internally, it consists of long barrel vaults creating a large internal, perhaps palatial,[50] space as these wrap around the eastern end of the earlier southern range.

Continuity

The scale of al-Zahir Ghazi's famous gatehouse at the citadel of Aleppo also appears to have been inspired by a desire to impress, rather than to support heavy artillery. Here and elsewhere there is a parallel between al-Zahir's projects and those if his dynastic rival, his uncle, al-ʿAdil.[51] Chevedden's suggestion that new large artillery towers were also built at Beaufort, Shayzar, Baalbek and Subayba is confusing:[52] the northern tower built by the Ayyubids at Beaufort is similar in scale to the earlier keep built by the Franks; the southern donjon at Shayzar seems to have been a palatial structure;[53] and the only towers of notable size at Baalbek and Subayba were built by the Mamluks. Looking past the large towers built by al-ʿAdil, as well as a few more built by al-Zahir and others, more traditional building trends continued through the first half of the thirteenth century.

Much smaller Ayyubid towers can be found at sites such as ʿAjlun, Subayba, Baalbek, Apamea, Palmyra, Harim and al-Rahba. Although Chevedden's arguments concerning the need to mount heavy artillery have permeated into studies of even such small towers,[54] these also lack any assessment of the forces that such towers would need to withstand. Among the many towers built in the region, many of those best suited to supporting considerable loads were built by the Franks prior to Saladin's rise to power.

---

49 For archaeological investigations of ʿAjlun, see Johns, "Medieval ʿAjlun," pp. 21-33; Yovitchitch, "Die Aiyubidische Burg ʿAglun," pp. 118-25; Yovitchitch, "The Tower of Aybak," pp. 225-42, see esp. pp. 234-39 for Yovitchitch's similar, and more detailed, conclusions. Cf. Chevedden, "Fortifications," p. 39.

50 See Yovitchitch, *Forteresses du Proche-Orient*, pp. 191-92.

51 See also Michaudel, "Development of Islamic Military Architecture," pp. 112-16; Yovitchitch, *Forteresses du Proche-Orient*, pp. 177-210.

52 Chevedden, *The Citadel of Damascus*, pp. 286-88.

53 See Tonghini and Montevecchi, "The Castle of Shayzar: The Results," p. 145.

54 This is clear in Bylinski's analysis of al-Rahba, Chevedden's arguments having been considered via Kennedy, Bylinski, "Three Minor Fortresses," pp. 157-58, cf. n. 9.

Possible Frankish Artillery Towers

Probably dating to the middle of the twelfth century, the tower built in the centre of the eastern wall at Saone, 24.5 m square, boasts walls 4.4-5.4 m thick.[55] Adding further strength, a solid central pier runs through both internal levels, supporting four bays of groin vaults in each. This combination of strengths make it possibly the most suitable pre-Mamluk tower in the Levant upon which to place a counterweight trebuchet. The accounts of the siege of 1188, however, provide no indication that any such heavy weaponry was positioned here. Given the girth of the castle's eastern defences, which would have protected it from any incoming artillery projectiles, it seems more practical to have placed archers, and possibly a traction trebuchet, here instead. These lighter weapon systems would have been more effective against besieging engines and assault forces massing on the ridge opposite the abyss of the eastern fosse.[56]

Similar to the donjon at Saone, the central keep of Chastel Rouge, which was built by the Hospitallers in the twelfth century and measures approximately 14 m by 16 m, is also built with a strong central pier. This tower has walls over 2 m thick and two internal masonry levels, the upper being divided with a wooden floor to create a third level.[57]

Although the central keep of Jubayl, which is 22 m by 18 m, has no central pier rising through its two levels, its walls are 4 m thick and occupy almost two thirds of the structure's total footprint.[58] Saladin's apparent inability to destroy the keep in 1190 speaks to its strength.[59] Although all three appear to have had two levels of battlements running around them, artillery does not seem to have been used to defend any of them in the late twelfth century.[60]

Two other candidates are the Templar keep at Tortosa and the keep at Toron. The former is 20 m square with walls 3.5-5m thick, vaulted internally by paral-

---

55 Deschamps, *Les Châteaux*, 3:235-36, 239-40; Mesqui, "Saône," pp. 12-13.

56 For Saladin's siege of Saone in 1188, see Chapter 5 and Fulton, "A Ridge Too Far," pp. 33-53.

57 For Chastel Rouge, see Mesqui, "Qal'at Yahmur," pp. 5-6; Kennedy, *Crusader Castles*, pp. 74-75; Pringle, *Red Tower*, pp. 15-17; Müller-Wiener, *Castles of the Crusaders*, p. 52; Deschamps, *Les Châteaux*, 3:317-19.

58 Deschamps, *Les Châteaux*, 3:210-11; Rey, *Étude sur les monuments*, pp. 117-20; Kennedy, *Crusader Castles*, pp. 65-66; Müller-Wiener, *Castles of the Crusaders*, p. 64.

59 Wilbrand of Oldenbourg, *Itinerarium terrae sanctae* 1.6, ed. Pringle, p. 119, trans. Pringle, p. 167. Cf. 'Imad al-Din, *al-Fath*, trans. Massé, p. 231.

60 For Saladin's acquisition of Chastel Rouge, see 'Imad al-Din, trans. Massé, p. 122; Abu Shama, *Kitab al-Raudatain*, RHC Or 4, p. 352. Cf. Ibn al-Athir, *al-Kamil*, trans. Richards, 2:344; Baha' al-Din, *al-Nawadir*, trans. Richards, p. 81. For Jubayl, see Chapter 5.

IMAGE 22 *Saone: interior of the keep* (PHOTO COURTESY OF DENYS PRINGLE)

IMAGE 23 *Chastel Rouge: interior of the keep* (PHOTO COURTESY OF DENYS PRINGLE)

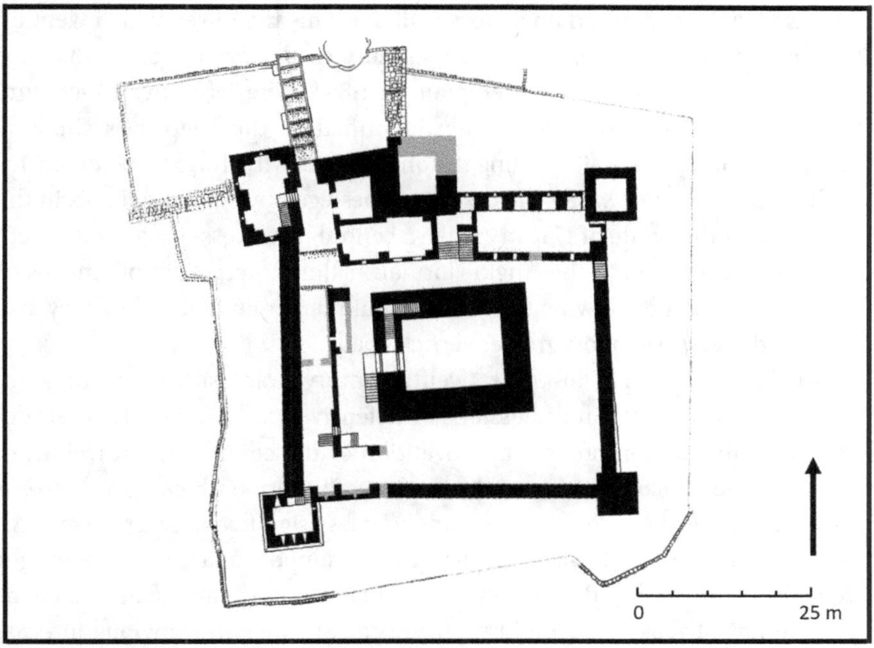

PLAN 57 *Jubayl: castle* (AFTER DESCHAMPS)

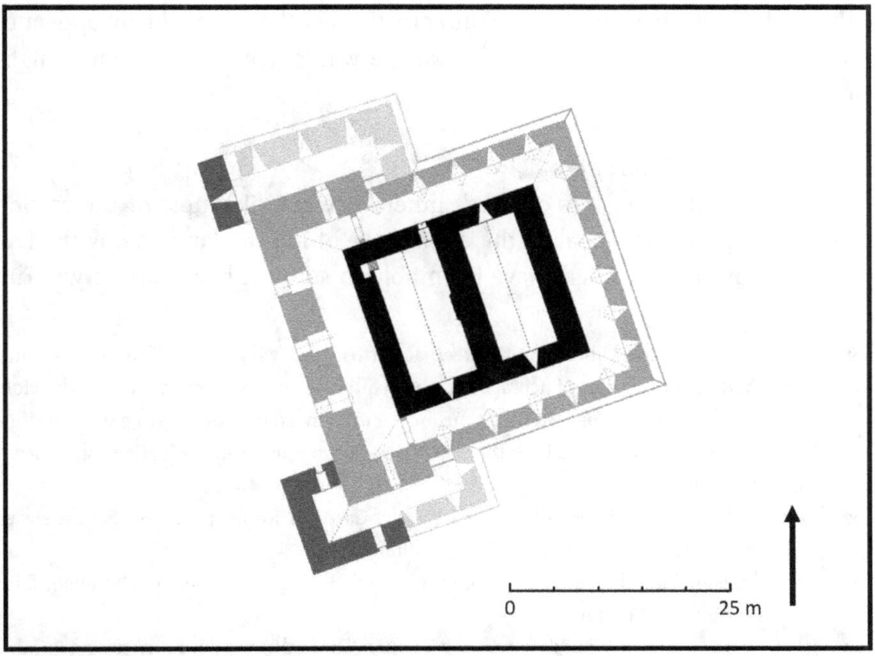

PLAN 58 *Tortosa: keep* (AFTER POSPIESZNY AND BRAUNE)

lel barrel vaults supported on a cross-wall. It seems the tower, which went on to form the nucleus of the thirteenth-century castle, was the same that was successfully defended by the Templars in 1188.[61] The large twelfth-century Frankish tower at Toron, commonly identified as the keep, was similarly planned with a cross-wall dividing its long axis. As with Tortosa, no defensive artillery is mentioned when the castle was besieged by Saladin's forces in the aftermath of the battle of Hattin.[62] Like a central pier, cross-walls, which were commonly employed in the Anglo-Norman realm,[63] were a simple means of adding support and allowing towers to be built larger and taller, but they also permitted them to support greater vertical loads.

Despite the strength of these twelfth-century Frankish towers, they are completely overlooked in discussions of defensive artillery. Because their design and construction predate the invention of the counterweight trebuchet and the appearance of larger castles in the early thirteenth century, it seems their strength is considered to be somehow less significant or irrelevant. Although none of these towers was designed to support a large counterweight trebuchet, they clearly demonstrate that the knowledge and skill existed, as did alternative reasons to build such towers. Significantly, obvious internal supports, such as pillars and cross-walls, are missing from most large towers built in the first half of the thirteenth century, suggesting internal space, rather than the ability to support considerable vertical loads, was what was desired. Although Frankish towers built in the early thirteenth century do not appear to have been planned to support heavy artillery, it is possible that some might have been able to.

Crac des Chevaliers

The three southern towers of Crac's inner enceinte, like most of the castle's impressive elements, postdate the earthquake of 1202 and are among the few Frankish towers that might have been able to support heavy artillery.[64] The

61 Given the similarities between the keep at Tortosa, following the addition of the talus, and Frederick II's keep at Lucera, built c. 1235 after the emperor returned from the Holy Land, Pospieszny has suggested that the former influenced the latter, dating the additions to the original structure no later than the 1220s, Pospieszny, "Tortosa (Syrien) und Lucera (Apulien)," pp. 243-46.

62 For the keep at Toron, see Piana, "The Crusader Castle of Toron," pp. 181-83. For the sieges of these castles by Saladin's forces, see Chapter 5.

63 Examples include the keeps of Portchester, Castle Rising, Rochester, Scarborough, Trim and the Tower of London.

64 For recent studies of these defences, see Häffner, "Die Südtürme der Kernburg im 13. Jahrhundert," pp. 142-84; Mesqui, "Les programmes architecturaux," pp. 3-7.

circular westernmost tower is the smallest. Although more than 10 m in diameter, the encircling parapet would probably have imposed serious restrictions on the sling of even a small counterweight engine. Space would not have been an issue on top of the central and eastern towers because of their rearward extension. Structurally, the walls of the central tower are continuous from their foundations to the parapet, supporting a single barrel vault along the tower's long axis at each level. The eastern tower is more complex as openings were left in its supporting walls to facilitate movement through the castle at the first and second levels (ground and first floors). Despite this apparent weakness, the tower's northern corners and south face were maintained as structural piers as was a portion of the western wall. The tower's eastern wall thins as it rises but a pier, almost 6 m wide, sustains the full 8 m thickness of the wall at its base, dividing a pair of east-facing embrasures at each of the three internal levels. Accordingly, anything sitting on the eastern side of this tower's roof rests on a solid pier of masonry reaching down to the bedrock, similar in concept to the rear half of the much older mural towers of Antioch. It seems possible, albeit far from certain, that the eastern tower, and possibly even the central tower, may have been able to support the weight and stresses of mounting a counterweight trebuchet.

Although a possibility, it seems very unlikely that artillery would have been positioned on these towers. Their height and the fairly confined spaces on top would have made it very difficult to move an engine's components and ammunition up to such positions and then back down to be stored under cover.[65] When considering that the construction of the outer wall would have removed most of the relative range advantage of mounting engines on these towers,[66] it would appear to have been far more sensible to set up an engine on the ground in the open space to the west of the *berquilla*. This would have left the lofty and commanding towers free for other forces (Fig. 22).

Indicative of contemporary priorities, the roof of the stable, which abuts the castle's outer southern wall, created a broad fighting platform 68 m long and 10 m deep along the castle's most assailable front.[67] Although the broad vaulting below probably would not have supported a heavy counterweight engine, this would have been an ideal position to erect lighter defensive traction trebu-

65 The impact of weathering on exposed wood would have made storing the engines under cover necessary, at the very least during the winter months, similar to the storage of sixteenth- through eighteenth-century wooden gun carriages.

66 Although lower, these engines would have been farther forward, between the inner and outer walls, yielding comparable ranges.

67 For this structure, see Biller, et al., *Der Crac*, pp. 249-51.

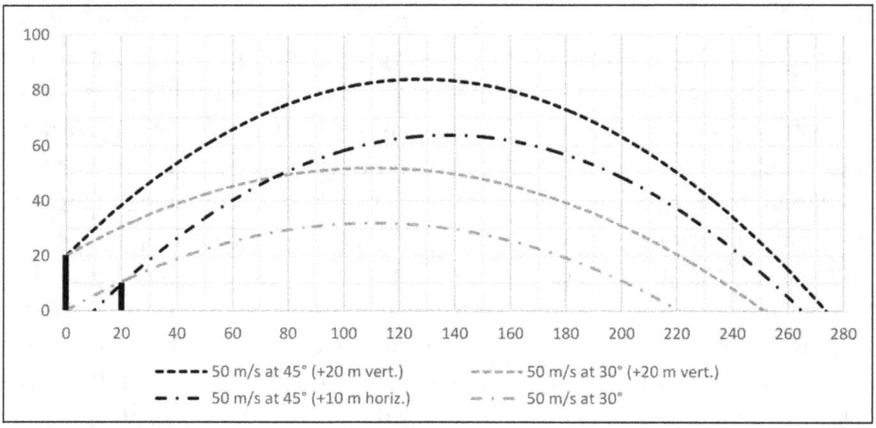

FIGURE 22 *Hypothetical trajectories of defensive engines at Crac des Chevaliers*

Position	Horiz. Pos.	Vert. Pos.	Velocity	Release Angle	Max. Range
Tower	0 m	20 m	50 m/s	45°	273.7 m
Tower	0 m	20 m	50 m/s	30°	251.4 m
Ground	10 m	0 m	50 m/s	45°	265.1 m
Ground	0 m	0 m	50 m/s	30°	220.9 m

chets. The battlements, especially if they were of two levels, as they were during the Mamluk period, would have provided excellent cover while the open space on three sides left plenty of room for personnel to move freely around any engines. The importance and role of this platform continued after the castle fell from Frankish control. Rather than strengthen the vaulting of the stable, the Mamluks appear to have blocked the eastern door and built a staircase, facilitating access from both ends.[68] This suggests that the value of the fighting platform was readily appreciated but that the Mamluks, like the Franks, had no desire to place heavy artillery here. Although it made sense to put large counterweigh trebuchets to use in a defensive role, they were far from the most effective means of defence.

'Atlit

Due to the scale of its towers and walls, it is not uncommon to find suggestions that the towers of 'Atlit were designed to carry the heaviest artillery of the

68 Mesqui, "La troisième enceinte," p. 13.

day.[69] The thick walls of 'Atlit's outer eastern towers appear to have allowed them to bear considerable loads. Although the towers have not survived to their original height, C.N. Johns speculated that the internal vaulting of the second level (first floor) would have projected up through the roof level, thus denying the opportunity to employ any kind of artillery on top of these towers.

Behind the outer line of defences soared two colossal towers, the northern of which still survives to a height of 35 m, almost its original height. The towers were 27 m wide with walls upwards of 5 m thick. Each contained three levels, the highest of which made up nearly half the height of each tower.[70] The towers were built using immense blocks, each weighing more than a tonne, cut with 4 cm margins no less than 7 cm deep, leaving bosses that projected up to 30 cm.[71] Although cut from the same soft stone found elsewhere along the Palestinian coast, the extremely hard Frankish mortar and rubble fill gave these walls considerable strength.

It is the outstanding height of these towers that tends to nominate them as pseudo equivalents to al-'Adil's supposed artillery towers.[72] Despite the unprecedented thickness of their walls, it seems unlikely that they ever supported heavy counterweight trebuchets. There may have been a central pier rising through the middle of the second and third levels, creating bays of groin vaults, but this seems to have been left floating as it does not appear to have been present at the lowest level. While the towers would have provided an impressive vantage point and could very well have supported traction artillery, the vast volume of internal space, especially at the third level, appears to have rendered them unfit to withstand the stresses created by the sudden drop and subsequent swinging of a sizable counterweight. To this can be added the difficulties of moving the components of such an engine up into position.

The height of these towers would have given any artillery placed on them an advantage in range compared to a similar engine situated on the ground in front of them, just behind the outer curtain; however, the latter position provided greater protection and left the tops of the towers free for archers, slingers

69 This can be traced back to Chevedden's original study, which includes the inner towers at 'Atlit, along with the southern inner towers at Crac, Chevedden, *The Citadel of Damascus*, pp. 289-90.

70 Johns, *Guide to 'Atlit*, pp. 36-37, 44; Johns, "Excavations at Pilgrims' Castle, (1932)," pp. 152-53; Boas, *Crusader Archaeology*, pp. 111-12; Conder and Kitchener, *Survey of Western Palestine*, 1:195.

71 Conder and Kitchener, *Survey of Western Palestine*, 1:296. In disagreement with *The Survey of Western Palestine*, T.S.R. Boase suggested that these blocks may have been reused from an earlier Phoenician settlement, Boase, "Military Architecture," pp. 157-58.

72 For example, Nicolle, *Crusader Castles*, p. 67.

and observers, for whom line of sight was more beneficial. Accordingly, the *trabuculum* that defended the castle in 1220 was probably positioned on the ground behind the outermost curtain, regardless of whether the outer line of defences or the third storey of the inner towers had been completed by this date.

Margat

At Margat, some have suggested that artillery could have been placed on the terrace along the eastern face of the citadel,[73] while others have looked to the roof tops of towers. Through his work at the castle, Balázs Major has pointed out the light structure of certain upper-level buildings, which would not be suitable for heavy artillery, but has suggested that the roof of the chapel might have supported a counterweight trebuchet.[74] The chapel, however, faces the steep drop beyond the eastern wall of the castle and is well withdrawn from the exposed southern salient, making the donjon further south a more suitable position.

There would have been plenty of space on top of the donjon, which is just over 20 m in diameter.[75] Despite its thick walls, the open space of the internal levels, each laid out as a single-room under a barrel vault, would have restricted the vertical load and lateral strain that this tower could have supported. If a counterweight trebuchet was brought into action by the defenders in 1285, it seems at least as possible that it was set up on the ground or terrace of the small vaulted structures between the donjon and southern apex of the outer wall.

Arsuf

Arsuf is one of the smaller castles built by the Franks in the thirteenth century, but its design is fairly scientific: built against a cliff overlooking the sea, providing both protection and an avenue of communication, two lines of walls secured its landward fronts. In his assessment of the castle, Jean Mesqui has suggested that the outer walls may have been designed in part to accommodate heavy artillery.[76] Both the north and south protrusions mirror the inner towers, leaving neither a platform nor space on the ground for a sizable engine.

---

73 For this example see, France, *Western Warfare*, p. 105; Nicolle, *Crusader Castles*, p. 79.

74 Major, "Medieval 'light construction buildings'," pp. 165-81; Major, "Medieval Cranes," pp. 1-5.

75 Rey, *Étude sur les monuments*, p. 30; Deschamps, *Les Châteaux*, 3:272; Kennedy, *Crusader Castles*, pp. 173-75.

76 Mesqui, "La fortification des Croisés," p. 8.

There would appear to have been space behind the eastern lobe, directly in front of the inner gate, but erecting a significant engine here would place obvious restrictions on movement. Although it is possible that the defenders employed a counterweight trebuchet in their defence of the castle in 1265, when considering the layout of the castle, it seems far more likely that they would have relied instead on smaller traction trebuchets.

It is difficult to gauge accurately the size of the engines typically employed defensively by the Franks. It seems that more thought was given to defensive artillery than simply using stockpiled engines otherwise intended for offence. Ibn al-Furat hints at this in his characterisation of the defensive engines of Jaffa in 1266.[77] Most defensive engines would have been lighter traction trebuchets; however, heavier counterweight trebuchets appear to have been employed defensively at Acre (1190-91), 'Atlit (1220), and during the Mamluk siege of Acre in 1291. These were probably repurposed offensive weapons and considerations of how best to employ them in a defensive capacity does not appear to have had a significant influence on the design of most fortifications. When dealing with towers, most large and sturdy enough to have possibly supported a counterweight trebuchet date to the Mamluk period.

Possible Mamluk Artillery Towers

It seems wall-mounted artillery was a critical component of the arrangements made to improve the defensibility of Alexandria in 1272-73. In a letter preserved by Ibn 'Abd al-Zahir, the city's governor reported to Baybars that around one hundred machines had been prepared to fulfil this role.[78] Bearing in mind that Baybars was certainly storing heavy artillery in many Mamluk strongholds by this point, it remains to be considered whether a desire to mount the heaviest engines on towers motivated any of his rebuilding efforts, or those of his immediate successors.

Subayba

Under Baybars, efforts were undertaken to refortify and develop the Ayyubid outer defences of Subayba, which had been slighted by the withdrawing Mongols.[79] John France has casually suggested that the round towers built

77 Ibn al-Furat, *Tarikh*, trans. Lyons and Lyons, 2:90-91. Cf. Appendix 2, **no. 23**.

78 Ibn 'Abd al-Zahir, *al-Rawd*, trans. al-Khowayter, 2:795-96; Ibn al-Furat, *Tarikh*, trans. Lyons and Lyons, 2:159.

79 The castle was built by al-'Aziz 'Uthman from about 1227. Despite mistakenly attributing construction to the Franks in the twelfth century, Deschamps' study remains the most

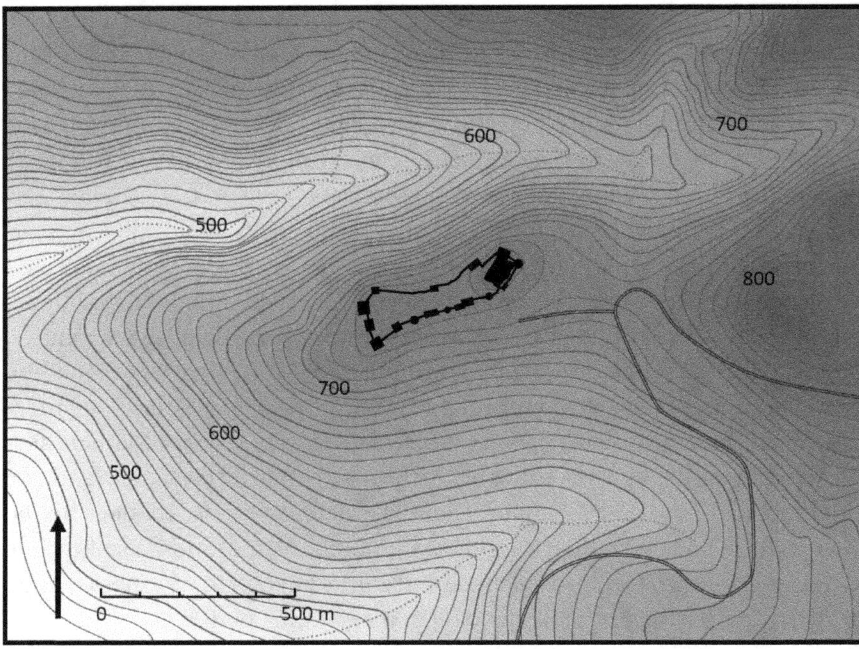

PLAN 59 *Subayba*

along the southern curtain (towers 2, 5, 7 and 16) may have mounted heavy engines.[80] Of these, tower 7, the roof of which is supported by a central pier and an encircling barrel vault 4 m wide, appears to be able to bear the greatest loads. Tower 16, at the castle's eastern end, is quite similar but lacks the support of a central pier. There would have been far greater advantages to placing artillery on tower 16 rather than on tower 7, as the former is the closest to the road running between Banyas and Damascus and the first tower that any besieging force would encounter when attacking from the east. When considering the surrounding topography, there would appear to have been few incentives to place a heavy engine on top of tower 7, suggesting that the tower's designers incorporated its central pier for reasons unrelated to artillery. Although these round towers are large, a few quadrangular towers are even bigger.

---

complete, although Hartal's examination of the castle's outer western defences has revealed how much there is still to learn from this impressive stronghold, Deschamps, *Les Châteaux*, 2:145-74; Hartal, *al-Subayba*. See also Müller-Wiener, *Castles of the Crusaders*, pp. 45-46; Benvenisti, *The Crusaders*, pp. 147-74; Ellenblum, "Who Built Qal'at al-Subayba?" pp. 103-12.

80 France, *Western Warfare*, p. 124.

IMAGE 24
Subayba: interior of tower 7
(AUTHOR'S PHOTO)

Towers 9, 11 and 15 were all developed into quite large structures under the Mamluks. While tower 15 still awaits close study, Hartal has examined towers 9 and 11, which were encased in the 1270s, quite closely. Between them, in the centre of the western wall, tower 10 was not enlarged by the Mamluks and retains its relatively thin Ayyubid walls and open interior space. Tower 11, originally built by al-'Aziz 'Uthman in 1230, had substantial walls, 2.5-3.7 m thick, but was much smaller than those built by his father, al-'Adil.[81] It was enlarged in 1275 using massive blocks, averaging 7 tonnes. Hartal has suggested that the scale of these building stones was necessary to retain the fill behind the walls, as the western wall was advanced down the slope, and to support the two levels above, which reached an estimated height of 30 m.[82] Although the Ayyubid

81 Hartal, *al-Subayba*, pp. 3, 9, 12-14, 17.
82 Hartal, *al-Subayba*, pp. 20, 36.

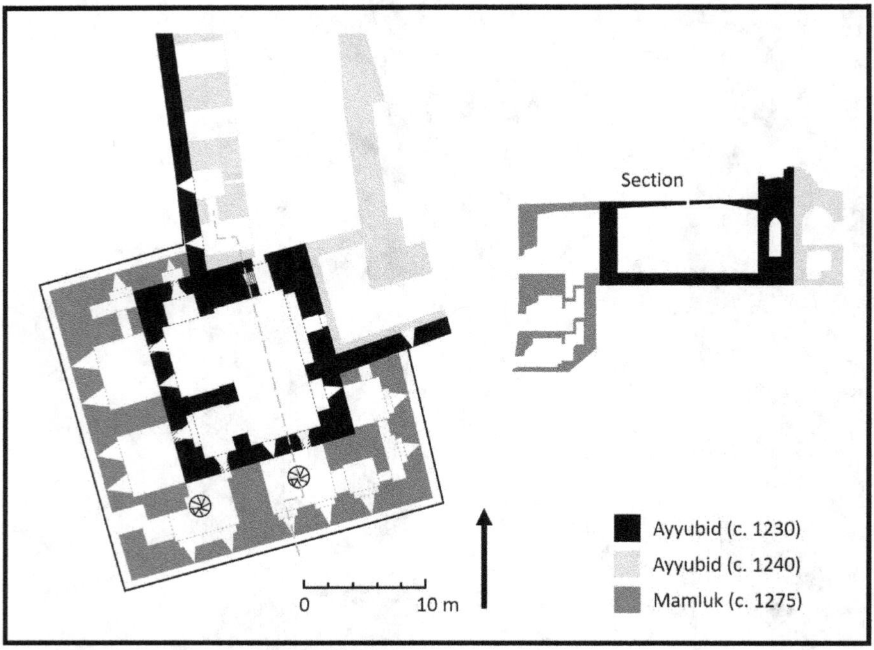

PLAN 60 *Subayba: tower 9* (AFTER HARTAL)

tower and surrounding structure provided quite a strong base, the third level of the Mamluk tower appears to have been a single open room, limiting the vertical load that its vaulting could carry. In Hartal's opinion, the creation of this opulent room was the impetus for the tower's expansion.[83]

The Ayyubid walls of tower 9 were much thinner, about 1.4 m thick. Even at its earliest phase this tower incorporated numerous embrasures, allowing it to control movement around the southwest corner of the castle. The present Mamluk tower has only slightly thicker walls but its construction increased the number of embrasures at the first level from eight to fourteen. The construction of two sub-levels, incorporated into the addition built down the slope in front of the Ayyubid structure, led to the creation of a further five southward, one westward and two eastward-facing casemates at the first of these and another four to the south at the lower sub-level.[84] Intriguingly, the Mamluk extension created firing chambers, essentially a series of connected casemates, leaving large piers of masonry to break up what could otherwise have been a

83 Hartal, *al-Subayba*, pp. 41, 72. The picturesque views from this position and fine internal masonry further suggest that concerns of comfort outweighed those of defence at this upper level.

84 Hartal, *al-Subayba*, pp. 75-79, 68-90.

continuous gallery. These divisions increased the vertical load that each level could support, but their primary purpose appears to have been to increase the cohesive strength of the tower, perhaps as a defence against mining. Above the first level, traces of both Ayyubid and Mamluk building are evident in the northwest corner of the tower, revealing no less than another four Mamluk embrasures in what little remains. Traces of an Ayyubid staircase up to a third level are still discernible, although nothing of this level remains.

While the layout of what has survived of this tower suggests that it may have been able to bear a substantial and perhaps even dynamically shifting load, the plans of the upper levels are unclear. Tactically, it would make little sense to place a large counterweight trebuchet at this corner of the castle, which is already protected from large siege engines by the grade of the topography. The tower appears to have been designed, at least at the first two levels, to maximise the number of archers that could shoot at anyone attempting to encircle the castle or launching a frontal assault up the steep slope to the west.

Crac des Chevaliers

The most obvious example of a tower capable of carrying immense loads is the massive quadrangular tower, about 20 m square, completed by Qalawun at Crac des Chevaliers around 1285.[85] The tower, known as Qalawun's tower, contains a single internal level at the same height as the upper parapet of the adjacent curtain walls. This level is essentially a gallery: eight bays of groin vaults arranged around a central square pillar, slightly larger than the bays themselves. The seven exterior bays provide access to casemates, each extending halfway through the external wall to service an embrasure. The tower is solid below this level with the exception of a narrow passage connecting a postern in the tower's east face with the Frankish stable behind. The roof was surrounded by an open-backed double-level parapet, similar to that found at Damascus.

The solidity of this tower would have made it capable of supporting immense loads and dynamic forces. If it was desirable to elevate a large defensive counterweight trebuchet, this would have been an ideal position. The tower provided both space and protection while there was ample room along the broad fighting platform, created by the roof of the Frankish stable, for lighter artillery, archers and other defenders. Despite the favourable conditions, there is no historical evidence to indicate that this tower ever carried any kind of artillery and it may instead have been a fear of mining that led to the tower's solid construction.

85 Deschamps, *Les Châteaux*, 1:157 n. 2; Biller, et al., *Der Crac*, pp. 251-52.

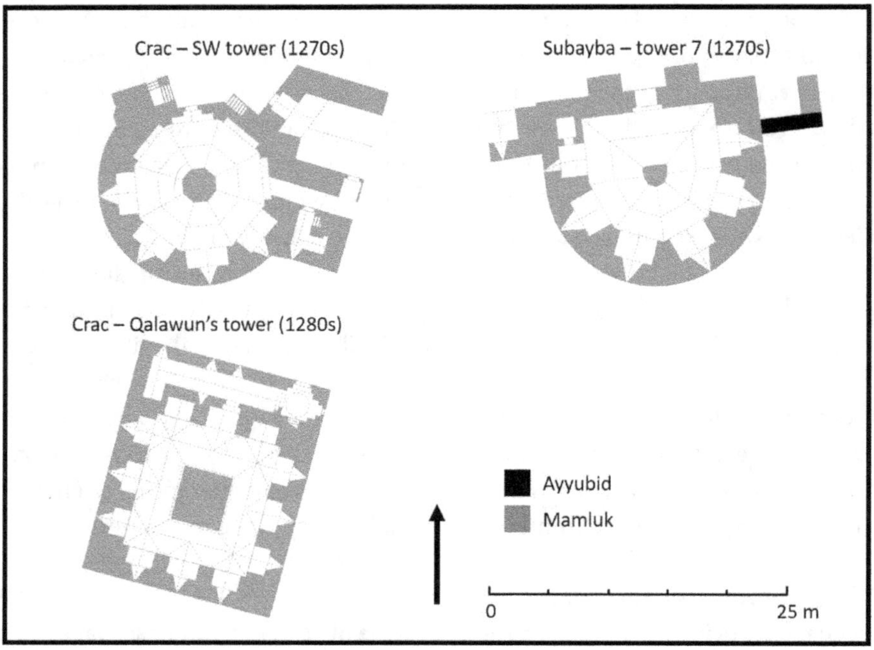

PLAN 61 *Mamluk towers with pillars* (AFTER DESCHAMPS AND YOVITCHITCH)

Just to the west of Qalawun's tower is a slightly earlier rounded tower, which was rebuilt over the remains of a Frankish tower after the siege of 1271.[86] This tower has a similar layout to tower 7 at Subayba. It contains a single internal level, arranged in an octagon around a central pier of the same shape. The diameter of the pier is roughly half the span of the encircling vaulting. Casemates were provided in each of the tower's four external faces. This tower would have been able to support considerable loads, unlike the tower just to the northwest, which is also rounded but was provided with an almost flat ceiling vault at the second level.[87] It is possible that the design of this tower, with its single internal level and central pier, inspired that of the more dramatic quadrangular tower built next to it.

The garrison's ability to defend the castle would not have been seriously compromised if it was denied the ability to use the top of Qalawun's tower or that of the rounded tower to the west. The long line of double-level battlements, which extended the length of the castle's outer southern walls and around the exterior faces of each of its towers, allowed for two levels of archers to defend the castle's most approachable front. This tall parapet would also

86 This tower is dated epigraphically, Deschamps, *Les Châteaux*, 1:170.

87 For a picture of the inside of this tower, see Biller, et al., *Der Crac*, p. 241 fig. 199.

have provided excellent cover for any lighter artillery positioned on top of the Frankish stable, as well as any engines on top of the two Mamluk towers. Regardless of whether Qalawun's tower was designed to support a counterweight trebuchet or not, it was part of an extremely well designed line of defences: sturdy towers were combined with numerous positions for archers, not unlike the earlier outer fortifications of 'Atlit and later work at Montreal.

Margat

The Mamluk rebuilding at Margat clearly displays which section of the castle was undermined in 1285. The apex of the outer southern line of defences was rebuilt with walls 8 m thick. Like the quadrangular southern tower at Crac, the design of the rounded tower that anchors the castle's most exposed section of fortifications is much more solid than earlier ones. The glacis that wraps around this salient, from which the tower rises, suggests that mining, rather than artillery, was still the greatest threat and may have been the sole influence behind the thickness of these walls.

The tower has one internal level, servicing three embrasures and three box machicolations, arranged around a large semi-circular pier much thicker than the passage is wide. While stout, there would not appear to have been enough room on this tower to erect any sizable artillery: space rather than strength is the limiting factor. Similar to the double-level battlements of the Frankish donjon behind this tower, the internal gallery provided a second level from which archers could shoot outwards. More traditional single-level battlements crowned the stout outer tower, keeping it low enough to be commanded from the Frankish keep behind.

Safed

Space would not appear to have been an issue on the large circular donjon built by the Mamluks at Safed, although little remains of it today. The massive tower is similar in some ways to the contemporary donjon built at Coucy. Both towers, about 35 m in diameter, appear to have contained a substructure, reducing the central space to be spanned by the internal vaulting.[88] At Safed, a heliocoidal ramp built into the bulk of the outer wall may have ascended the tower, leaving an inner ring of masonry to support the vaulting over what was likely a central open space; however, neither a detailed description nor illustration of this central space has survived.

---

88 For studies of Safed, see Pringle, "Reconstructing the Castle of Safad," pp. 139-48; Barbé and Damati, "Le château de Safed," 77-93. For Coucy, see Viollet-le-Duc, *Description du Château de Coucy*.

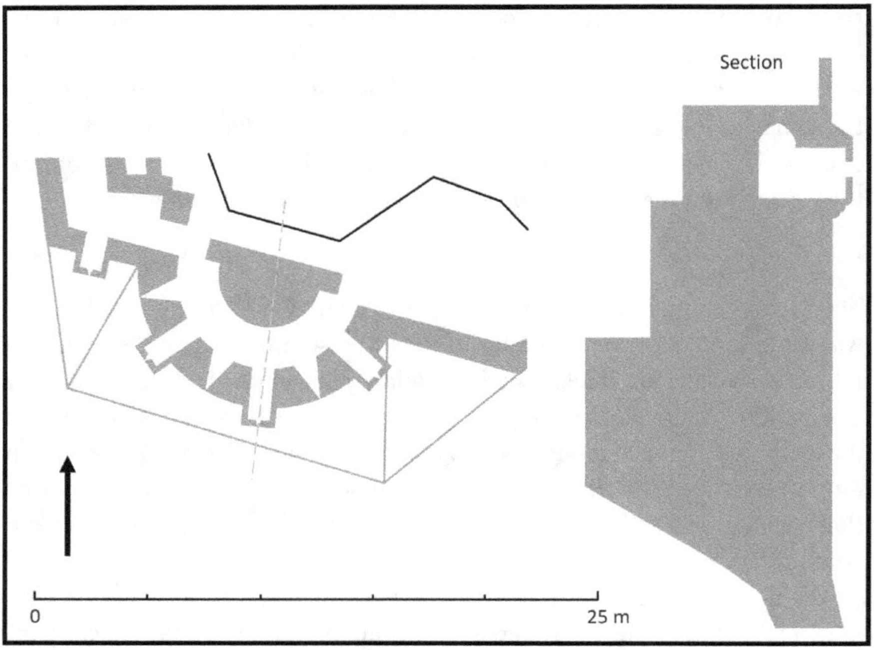

PLAN 62 *Margat: outer southern tower* (AFTER MESQUI)

IMAGE 25 *Margat: outer southern tower from the north* (PHOTO COURTESY OF DENYS PRINGLE)

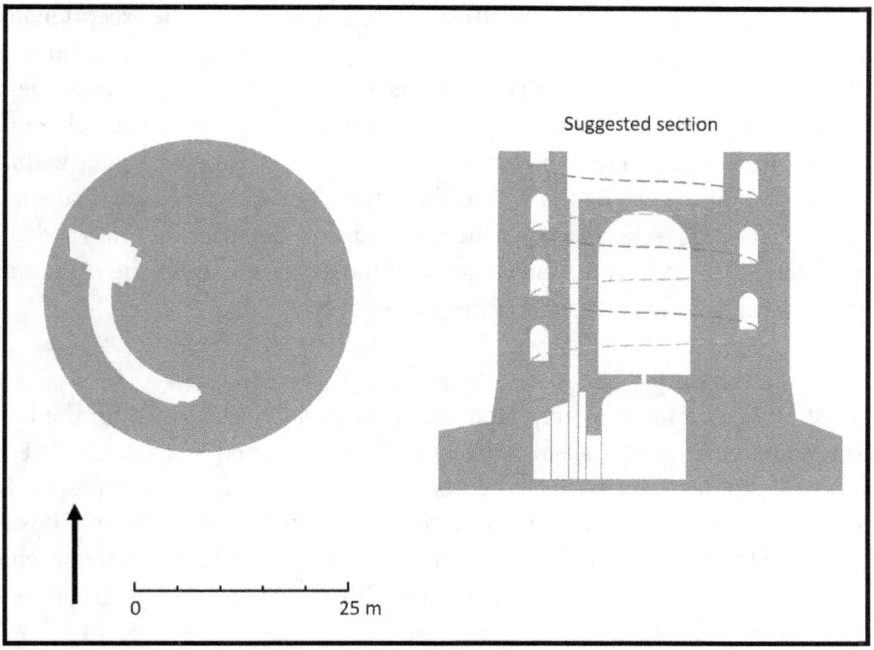

PLAN 63 *Safed: Mamluk donjon* (AFTER BARBÉ AND PRINGLE)

Like that of Coucy,[89] the suggested structure of Safed's Mamluk donjon would have made it capable of carrying relatively large loads if it were not for its enormous scale. The probable volume of open space inside these towers suggests that they were designed to impress viewers rather than to support heavy artillery. Their thick walls, like those of the tall inner towers at 'Atlit, were probably a structural necessity for fortifications of this scale. Had the diameter of these towers been smaller, mounting heavy artillery on each might have been an option. However, at all three – Safed, Coucy and 'Atlit – these massive towers were all positioned behind an outer wall, decreasing the potential range of any artillery that might be set up on their roofs and lowering the probability that they ever supported any sizable engines.

When walls in excess of 5 m become more popular in the fifteenth century, towers tended to be smaller. For example, those built at Clisson, Foix, Dijon

89 Plans of Coucy that predate its destruction in 1917 reveal that a single spiral staircase provided access to three lofty internal levels, each with a partial mezzanine level and sheltered by rib vaulting. At the third level, the mezzanine became a gallery by piercing each of the twelve structural pillars, from which the ribs of each level sprang. This left internal buttresses on the exterior side of the passage and solid pillars on the internal side to carry the third vault, which was also the tower's roof.

and Bourbon-l'Archambault, are all around 15 m in diameter. The exceptionally thick walls of these towers, as well as those of larger towers, such as those at Roumeli Hissar and Ham, reflect a more significant advancement in artillery technology. It had become necessary to build thicker walls to resist much more powerful siege guns and to incorporate similarly heavy defensive guns within and on top of towers. By the end of the fifteenth century, the use of smaller, more mobile guns, which could be grouped into batteries, led to a radical transformation in fortification design and the development of the *trace italienne* in the first three decades of the sixteenth century.

Others

Most other examples of large Mamluk towers built in the Levant in the late thirteenth century also seem unlikely to have supported heavy artillery. For example, the exterior wall of the Mamluk donjon at Kerak is impressively thick, but the tower has an awkward trapezoidal shape, suggesting that it was designed to impress onlookers more than anything else.[90] The numerous embrasures provided for archers appear to be the tower's greatest defensive consideration.

Before Chevedden, Wolfgang Müller-Wiener suggested that the large quadrangular tower built by Qalawun adjoining the temple of Bacchus at Baalbek may have been used as an artillery platform.[91] While there is ample space for this and outwardly the tower resembles that at Crac, a wide cross-vault opens up the inside and additional chambers turn what would otherwise have been solid corners into firing chambers that provide access to casemates. Rather than having been built to support heavy artillery, the tower fits with trends seen elsewhere: it is fairly stout, making it resistant to mining, and provided with numerous embrasures, maximising its ability to repulse frontal attacks. The same motives can clearly be seen at Subayba's tower 9.

The motivations behind the plan of Montreal's tower E have already been addressed, but it is worth noting that its general design was neither exceptional nor new. Rather than the rounded or large quadrangular towers built elsewhere by the Mamluks, which display numerous embrasures in different directions, those of Tower E are stretched out in a line, focusing on the plateau opposite it. The southern donjon at Kerak was designed with a similar focus as was the earlier elongated Ayyubid tower 5 at Apamea.[92] This was an efficient means of deploying numerous archers in protected positions against an area of

90 For this tower, see Deschamps, *Les Châteaux*, 2:88-89.

91 Müller-Wiener, *Castles of the Crusaders*, p. 55.

92 See plan in Dangles, "La refortification d'Afamiyya," p. 195.

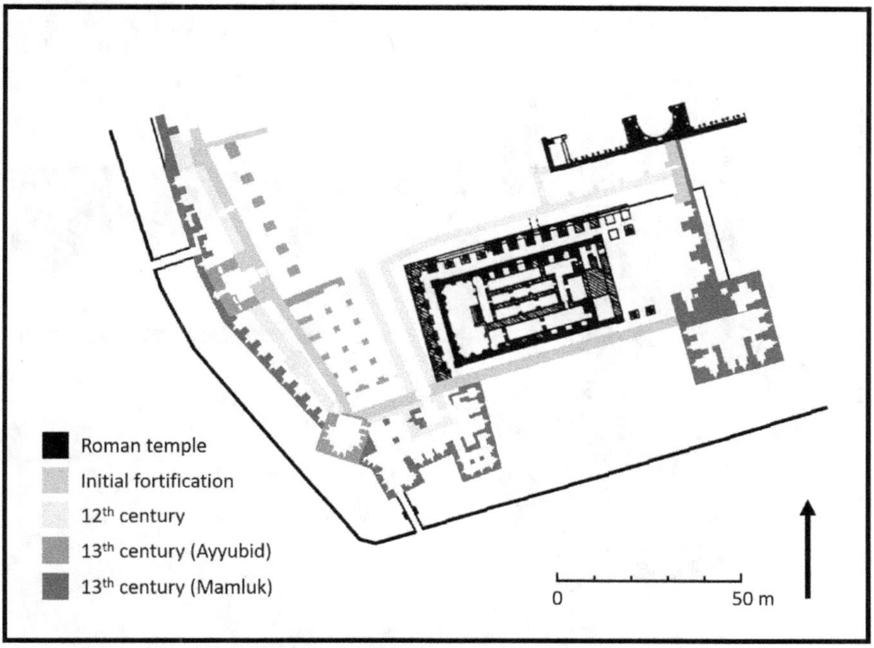

PLAN 64 *Baalbek: south-western area of the citadel* (AFTER YOVITCHITCH AND MÜLLER WIENER)

perceived vulnerability. The Mamluk expansion of the southern complex of Shayzar similarly increased the number of archers that could shoot eastward; however, the internal layout, which mimics the earlier structure, reveals the considerable influence of non-military factors. Although the footprint of the tower was greatly increased, providing enough room for a large counterweight trebuchet on the roof, no efforts appear to have been taken to strengthen the tower to support, or resist, any sizable artillery.

Although some towers would have been big enough to accommodate a large counterweight trebuchet, and a few may have been able to support the stresses that accompanied such engines, few appear to have been designed specifically to do so. When examining the wider use of central pillars by the Mamluks, Yovitchitch has concluded that this was done to provide greater structural cohesion, a countermeasure against mining more than an effort to carry greater loads.[93] This fits with the broader design trends of the century: the construction of stouter towers and focus on maximising the number of positions for archers.

93 Yovitchitch, *Forteresses du Proche-Orient*, pp. 208-10.

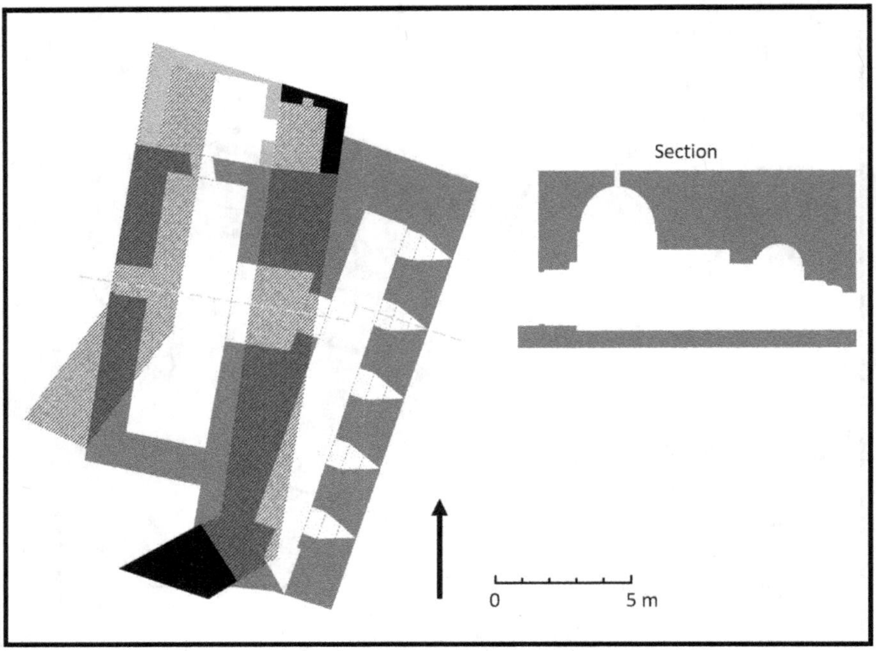

PLAN 65 *Apamea: tower 5 of the citadel* (AFTER DANGLES)

Other Interpretations

If thirteenth-century tower designers desired to support far greater loads on top of their towers, there were a number of quite simple structural solutions: adding interior supports, such as a cross-wall, increasing wall thickness, adding buttressing, etc. Most measures, however, came at a cost to internal space and aesthetic grandeur. While the average thickness of fortified walls increased slightly in the early thirteenth century, so too did vaulted spaces inside towers. Predictably, some of the thickest walls surround the most open internal spaces and, more generally, many of the most impressive towers, whether viewed from within or without, boast thicker than average walls. Examples range from the stout design of al-ʿAdil's encased towers at Cairo and the imposing Mamluk donjon at Kerak, to the far more open and grand spaces inside the upper level of Mamluk tower 11 at Subayba, and the tall inner towers at ʿAtlit. Aside from monumental projects such as these, the thickness of fortified walls continued to average between 2 and 3.5 m.

Most measures undertaken to increase the defensibility of fortifications in the thirteenth century appear to have focused on maximising the number of embrasures and ensuring structural cohesion, rather than elevating heavy defensive artillery. Qalawun's solid tower at Crac, ideally suited to enduring the stresses associated with mounting such engines, is a rare example of a tower both large and strong enough to have accommodated a machine of this type. However, the use of solid towers is not proof in itself that they were meant to support heavy counterweight trebuchets. Completely solid towers had been built across Europe and the Near East by the Romans; more recently, solid towers were built in the southern Levant at Ayla (Aqaba), Kafr Lam, Minat al-Qal'a and elsewhere in the centuries leading up to the arrival of the crusaders, long before the counterweight trebuchet was invented. The rapid ejection of the Franks from the Holy Land and the region's transition from frontier to hinterland, removed much of the incentive to invest in and develop its fortifications. It is therefore a matter of conjecture whether a larger number of solid towers, such as that at Crac, might have been built if the Frankish presence had lasted longer.

It was not until the fifteenth century and the refinement of chemical propellants that a comprehensive shift in tower design was necessitated, encouraging much thicker walls and solid lower levels in order to both support and resist improving gunpowder weapons.[94] The sharpness of this change, relative to any in the thirteenth century, reveals the comparatively limited impact that artillery had on fortification design during the period of the crusades.

Symbolism

The scale of the largest towers was probably intended more to impress onlookers than to perform any particular military function. This is not to say that these buildings were not built to serve a military purpose, only that the size of these towers had more to do with image than a particular aspect of defensibility. The men who built these towers in the early thirteenth century, al-'Adil, al-Zahir Ghazi, the Templars and even the lord of Coucy,[95] used scale as a form of propaganda: each hoped to convey an impression of their authority and to be considered as powerful and grand as their immense towers. Many Ayyubid and Mamluk towers were striped with inscriptions, making it clear for contemporaries and posterity who was responsible for their construction. Despite the grandeur of some of these towers, it is important to bear in mind that most

94 See Mallet, "The Transformation of War," pp. 1-17; Pepper, "The Face of the Siege," pp. 33-52; de la Croix, "Military Architecture," pp. 263-90. See also Black, *European Warfare*, esp. pp. 85-86; Parker, *The Military Revolution*, pp. 7-16, 24-32; Dechert, "The Military Architecture of Francesco di Giorgio," pp. 161-80.

95 It is of little influence, but Enguerrand III of Coucy's father, Ralph I, died at the siege of Acre in 1191, an episode which thus connects all the parties.

towers built during the late Ayyubid and early Mamluk periods were much smaller than these.

Castles and citadels had been used as symbols of power and control in the Near East prior to the arrival of the Franks, much as they had been in Latin Europe. Although some citadels occupied much older tells, many found a new political significance during the Seljuk period.[96] Fortification programmes were carried out as these became symbols of power intended to reinforce notions of the social order as the Turkish elite imposed their rule on local populations of Arabs and other groups. Ayyubid, as well as Frankish, fortification efforts can be regarded similarly. In Saladin's shadow, his successors struggled to secure their authority with more than just the sword. Alongside al-'Adil's building programme, al-Zahir Ghazi launched a parallel construction campaign throughout his northern realm. His work at the citadels of Aleppo, Harim, Qal'at Najm and Apamea is quite similar: each was built on a conical tell faced with masonry and accessed via an impressive ramp and gatehouse.[97] At Harim, al-Zahir appears to have encircled the earlier defences, built by the Franks or Nur al-Din, similar to al-'Adil's encasement of the Seljuk-Zankid citadel of Damascus.[98] Al-'Adil employed his own, perhaps more obvious, stylistic signature: great quadrangular towers faced with a consistent bossed masonry style.

The unprecedented scale of the towers built in the early thirteenth century would remain an influential design option but did not become a new standard. Aside from his impressive gatehouses, al-Zahir Ghazi continued to employ relatively small mural towers, comparable with those constructed a century earlier. Decades later, towers built at Aleppo in the wake of the Mongol conquest were actually smaller than the earlier ones that they replaced.[99] The continued construction of towers with relatively small footprints contradicts the suggestions of Chevedden and Kennedy that the large towers built by al-'Adil were a necessary response to a dramatic improvement in artillery technology.

The development of the counterweight trebuchet seems to have had a far from spectacular genesis. These engines were probably used in the late twelfth century and almost certainly employed in the Levant by the time of the Fifth

96 Rabbat, "The Militarization of Taste in Medieval Bilad al-Sham," pp. 87-92.

97 To a certain extent, these resemble the citadels of Hama and Homs, which were rebuilt by al-Zahir Ghazi's cousins.

98 See Gelichi, "The Citadel of Harim," pp. 191-95. Cf. Berthier, "La Citadelle de Damas," pp. 161-63.

99 Gonnella, "The Citadel of Aleppo," p. 171.

Crusade; however, fortification designs may not have been considerably influenced by artillery until the final decades of the thirteenth century. But even the changes that occurred during the Mamluk period cannot be compared with the more dramatic shift that took place in the fifteenth century and the radical transformation that occurred in the early sixteenth century. Put simply, the counterweight trebuchet did not cause a fortification revolution.[100]

100 Cf. Chevedden, "Invention," p. 73 and n. 4.

Conclusion

In some ways, artillery and the manner in which it was used developed considerably through the twelfth and thirteenth centuries; in others, it changed quite little. Traction trebuchets appear to have developed little, restricted by the nature of their mechanics rather than a lack of skill or interest. These were used in a fairly consistent manner throughout the period: suppressing defenders along a besieged section of parapet, destroying improvised defences that the defenders might employ along their battlements, and bombarding the people and unfortified buildings within an enceinte in an indiscriminate manner at times. In fairly stark contrast, counterweight trebuchets underwent a dynamic period of development. These engines most likely grew quite rapidly through the early thirteenth century, slowing as they reached their peak practical size. As these machines became larger and more powerful, they took on a more destructive role: although they were probably first used much like powerful traction trebuchets, they appear to have been employed to destroy battlements from at least the early thirteenth century. They maintained this function in subsequent decades; although, as they became more powerful, they may also have begun to attack lower sections of wall as well, working their way down from the destroyed parapet. Despite their power, it does not appear as though these engines were ever used successfully as breaching weapons – that was left to the sappers.

Development

It is clear that artillery was familiar to both Frankish and Muslim forces at the time of the First Crusade. Although the descriptions left in the accounts of this episode leave it unclear if these were trebuchets or torsion stone-throwers, the lack of evidence to support the continued use of the latter through the Early Middle Ages suggests that these were almost certainly traction trebuchets. These engines were fairly light, but their frequent use throughout the twelfth century speaks to their value. They were dynamic antipersonnel weapons that threw larger projectiles than the stones of slingers and arrows of archers, tossing them with considerably more energy.

Counterweight trebuchets seem to have appeared in the second half of the twelfth century and were probably used by both Frankish and Muslim forces at certain sieges in the 1180s and 1190s. The earliest of these new engines do not appear to have been significantly more powerful than certain traction trebu-

 | DOI 10.1163/9789004376922_013

chets, receiving neither special notice nor description in the works of contemporary sources. By the second decade of the thirteenth century, these engines were becoming more powerful and new terminology was warranted. Terms such as *trebuchetum* and *manjaniq maghribi* appear in the context of the siege of Damietta (1218-19). Revealingly, the engines associated with these new terms often appear at significant sieges, in limited numbers, and are praised for their destructive power. While such praise was not uncommon earlier, references to structural damage inflicted by engines identified by other terms starts to disappear, implying a clear division of power and of roles.

Physical evidence confirms the increasing scale of the heaviest counterweight trebuchets from the siege of Ascalon in 1247 onwards. In 1271, stones around 100 kg were used at Crac des Chevaliers and, just months later, stones weighing around 70 kg appear to have been thrown at Montfort from at least 200 m. These were considerable engines by this point in time and efforts appear to have shifted from building ever bigger counterweight trebuchets to employing larger numbers of these machines more efficiently.

As the mass of the largest projectiles thrown by counterweight trebuchets rose through the thirteenth century, and such large stones appear with increasing frequency and in greater numbers, these remained vastly outnumbered, at most sieges, by the much smaller stones thrown by traction trebuchets, as was clearly the case at Arsuf in 1265. This is mirrored in the textual evidence: the increasingly specific terminology used to identify engine-types confirms that lighter engines (traction trebuchets) typically made up a significant majority of the engines used at most sieges. Thus while counterweight trebuchets grew larger and more destructive through the Mamluk period, traction trebuchets, probably very similar to those used in the early twelfth century, remained important antipersonnel weapons.

East-West Supremacy

From the First Crusade onward, it is very difficult to discern any differences, let alone a technological advantage, between the trebuchets employed by Muslim forces and those used by their Frankish counterparts. Even during the siege of Acre (1189-91), when individuals from across Europe and the Middle East were drawn together and encountered people from distant regions and different technological traditions for the first time, their accounts provide no suggestion of an advantage or even difference between the engines employed by the Franks and those of the Muslims. While there must have been some regional differences between the machines familiar to someone from London and those built in Mosul, especially during this period when the counterweight trebuchet was in an early and probably quite radical phase of development, we are left to guess what these might have been.

In the past, certain historians have suggested that Muslim artillery was superior to that of their Frankish counterparts in the twelfth century. Such opinions were generally founded on either a belief that siege techniques developed rapidly in Europe during the twelfth century, a result of knowledge brought back from the crusades, or as demonstrated by the successes of the Zankid-Ayyubid conquests of the second half of the twelfth century.[1] Gibb, operating from a Muslim perspective, challenged this by identifying the apparent simplicity of Muslim siege tactics in the Near and Middle East before the First Crusade.[2] Another popular theory holds that the Franks received an education in siege warfare from the Byzantines during the First Crusade. This idea was put forward by Oman and Runciman in much more general terms than Chevedden's more recent artillery-specific theory.[3] From an opposing point of view, Ronnie Ellenblum has argued that the Franks employed more sophisticated siege engines than did the Muslims through most of the twelfth century, a sentiment shared by Lynn White Jr.[4]

The work of John France has been critical in amending these skewed interpretations, pointing out that the Franks and Muslims were largely aware of the other side's technologies but preferred different tactics, which were suited to different technologies.[5] In his study of thirteenth century warfare, Christopher Marshall similarly observed that "the same basic weapons were used by both Muslims and Christians for defence and attack."[6] The present study can thus be seen as the latest instalment supporting this latter school of thought.

The Muslim invasions of southern Europe during the Early Middle Ages, the Reconquista in Iberia, the struggles between Latin and Greek Christians in Eastern Europe and continual conflicts amongst the powers of these various ethnic groups would have naturally spread the relatively simple mechanics of trebuchet technology. As any developments were made, their employment in conflict inadvertently shared this advantage with enemies and allies alike. More distant links would have been made by travellers, including European

1 For examples, see Thompson, *Military Architecture*, p. 66; Benvenisti, *The Crusaders*, p. 284; Kennedy, *Crusader Castles*, p. 102; Nicolle, *Crusader Castles*, p. 31.

2 Gibb, in Ibn al-Qalanisi, *Dhail*, trans. Gibb, pp. 39-40.

3 Oman, *A History of the Art of War*, p. 526; Runciman, *A History of the Crusades*, 1:227-28.

4 Ellenblum, *Crusader Castles*, pp. 203-30; White, "The Crusaders and the Technological Thrust of the West," pp. 102-3.

5 France, "Technology and the Success of the First Crusade," pp. 170-73. See also Rogers, *Latin Siege Warfare*, pp. 62, 244-46.

6 Marshall, *Warfare in the Latin East*, p. 212.

Christians who made pilgrimages to the Levant and Muslims who similarly travelled through the region on their way to Mecca, accelerating the spread of knowledge amongst military elites. But when judging the siege technology used by a particular army or commander at a given siege, context is imperative: the resources of the besieging force, the nature of the stronghold to be assailed, the topography, the availability of local timber and even the weather were all critical factors that influenced what engines might be employed.

Employment

Almost all previous studies of artillery have operated on the premise that artillery was positioned as far as possible from an intended target.[7] Although this would have been ideal in certain circumstances, the evidence suggests that archers could outrange artillery from the start of the twelfth century until at least the early thirteenth. Accordingly, attacking engines were probably positioned so as to optimise their offensive capabilities, taking into consideration the strength of the defenders and their fortifications, the nature of the surrounding topography, and various other factors.

Trajectory

Attacking traction trebuchets were probably positioned within range of defensive archers, but so too were early counterweight engines. Counterweight trebuchets may have been among the engines that Saladin erected close to the north wall of Kerak in 1183 and 1184, while if counterweight trebuchets were employed at Acre in 1189-91, these were also within range of the defenders. As counterweight trebuchets developed, it became possible to set them up beyond the range of the defending archers, possibly even by the siege of Damietta (1218-19). Later, during the Mamluk period, it is possible that counterweight trebuchets were intentionally set up quite close to a target alongside traction engines, as may have occurred in 1265 during Baybars' sieges of the citadel of Caesarea and castle of Arsuf. Engines were certainly set up at a considerable distance at Montfort in 1271, and possibly also at Safed in 1266. At Beaufort (1268), Crac des Chevaliers (1271) and Margat (1285), there appear to have been efforts to advance attacking engines. These may have initially been positioned

7 This includes both historical studies and reconstructions, for example, Payne-Gallwey, *A Summary of the History*, pp. 8-10; Chevedden, "King James I," p. 314; Rogers, *Latin Siege Warfare*, Appendix 3, pp. 254-73; Vemming Hansen, "Experimental Reconstruction," pp. 189-208; Krizek, "Trebuchet Reconstructions," pp. 19-20.

beyond the range of the castles' archers but were moved much closer, after an advantage had been gained, in order to increase the percussive force of their projectiles.[8] Although it is tempting to see these engines acting as breaching weapons, akin to later bombards, such enthusiastic mental images must be checked against considerations of the basic physics involved.

In order to maximise destruction, an engine would need to be as powerful as possible, shooting projectiles as heavy as possible, which strike the target at as high a velocity as possible and at an angle as close to perpendicular as possible. This would ultimately lead to the greatest and most efficient transfer of energy. Although decreasing the distance between an engine and its target would allow for the use of heavier projectiles, and may have led to a more perpendicular angle of impact, moving a large counterweight trebuchet would have been quite difficult. If the engine was built with wheels, considerable time and attention would need to be devoted not only to preparing the ground across which the engine would move but also where it would shoot from. Wheels created pressure points, which would be susceptible to sinking into soft patches of earth or sand, even more so if there was a chance of rain. A base without wheels spreads the weight of the engine across its entire footprint (similar to a snowshoe). But to move an engine without wheels would almost certainly involve going through the time-consuming process of deconstructing the machine and reassembling it in its new position. An engine large enough to inflict structural damage would almost certainly have been too heavy and cumbersome to move under concentrated fire from defenders and it would have been foolhardy to go through the process of advancing such an engine so close to the enemy without first wearing down their numbers and defences. Thus although most trebuchet projectiles were probably shot at angles greater than about 20°, this should not be taken for granted.

Damage

Traction trebuchets were used in an antipersonnel capacity throughout the period of the crusades. When comparing the hypothetical model of an engine at Nicaea (1097), suggested in Chapter 3, with the evidence from the siege of Arsuf (1265), provided in Chapter 7, the power of these engines appears to have increased little over time. Assessing the power and degree of damage that counterweight trebuchets were capable of inflicting is more difficult.

When evaluating the contemporary descriptions, it is important to bear in mind a sense of relativity: it is easy to undervalue the limited appreciation that

8 Cf. the engine used in the final phase of the siege of Château Gaillard in 1204, William the Breton, *Gesta Philippi Augusti* 129, ed. Delaborde, p. 219.

such sources would have had of acute structural destruction compared to modern standards. For example, some sites that are supposed to have been 'destroyed' or 'demolished' were only mildly slighted. Richard I's apparent ability to repair Acre's walls "so they were higher and better than they had been before they were destroyed" in only a month before moving south in 1191, speaks either to the limited degree to which they were 'destroyed' during the siege or an exaggeration of what could practically have been repaired in such a period of time.[9]

In some instances it is clear that authors skewed the reality in order to impress a point, while in others they may simply have been mistaken or misinformed.[10] Reports and estimations of the scale of engines, their power, range and shooting rate can also be seen in this light. Only by avoiding the tendency to fixate on a few vivid accounts and rejecting literal readings of certain details, do more gradual trends of development become apparent. It is critical to evaluate as many sieges as possible, taking into consideration as many accounts from as many different perspectives as are available, in order to appreciate the nuance and gradual development of this technology as it is revealed in the literary sources. For a more quantitative assessment, we must look at the physical evidence.

Datable impact signatures and projectiles provide the most objective measure of the power and scale of trebuchet technology, but these are only parts of a more complex equation. In the absence of timber remains and reliable estimations of scale, the size of a given engine will remain an unknown variable on one side of the equation; however, a sense of this can be deduced by incorporating quantifiable values on the other side. For example, the minimum mechanical energy generated by the engine, the kinetic energy transferred to the projectile (Ekmin), can be represented if the mass of the projectile (Mp) and distance of flight (R), from which the minimum release velocity can be determined (Vmin), are known:

$$\text{Ekmin} = \frac{\text{Mp} \cdot \text{Vmin}^2}{2}$$

9 ... *rex Ricardus muris civitatis operam dabat reparandis in altius et perfectius quam priusquam diruerentur*. Itinerarium, ed. Stubbs, p. 240, trans. Nicholson, p. 227. Cf. letter of Richard I to the abbot of Clairvaux, in Roger of Howden, *Chronica*, ed. Stubbs, 3:131, trans. Riley, 2:222.

10 This is similar to the often wildly inaccurate estimations of combatants at a given battle. For a contemporary understanding of such, see Ibn al-Athir, *al-Kamil*, trans. Richards, 2:242. See also Marshall, *Warfare in the Latin East*, p. 212.

$$\text{where Vmin} = \sqrt{\frac{R \cdot g}{\sin(2 \cdot 45°)}}$$

Similarly, if future investigations are able to quantify the energy required to create a given degree of destruction, this can be used to balance the equation in the absence of other variables:

$$\text{Ekmin} = \text{destructive energy} + \text{energy lost in flight}$$

An approximate idea of the necessary mechanical energy and capabilities of a particular engine can be used to suggest a rough sense of scale.

Most misunderstandings of these engines are due to an inadequate appreciation of the mechanics that govern them. This has allowed for many longstanding overestimations of power and range. While some have misrepresented contemporary testimony, others have looked ahead to a period where there is more evidence – and larger engines – and simply applied this to an earlier context. Among those who have more studiously examined the evidence, Fino has provided a fairly accurate impression of the power of medieval counterweight trebuchets; however, his misunderstanding of the basic operational mechanics of these engines has led to errors relating to accuracy and shooting rate.[11] Unfortunately, those with the best sense of the mathematics and mechanics are often experts in physics or applied engineering and are not sufficiently familiar with the historical material to place the correct engine size in its appropriate historical context. In the absence of a comprehensive understanding, misconceptions have become accepted as fact, something Hollywood and its tendency towards the sensational have reinforced. This has had a knock-on effect and spawned flawed theories regarding the influence of these engines, such as the prominence of artillery as a driving factor behind the development of new fortification styles in the early thirteenth century.

To appreciate the relative ineffectiveness of artillery as a breaching weapon at the end of the twelfth century, consider that Acre was besieged with such engines during 1190 and 1191 but that most damage was reportedly inflicted in the last month of the siege, following Richard I's arrival. Even during this final phase, with a reported nine *perieres/petrariae* and two *mangonels/mangunelli*, commissioned by some of the wealthiest princes of Europe, those who attacked the walls in the last weeks of the siege still required ladders to do so and it was mining that was the greatest threat to the city's walls. A similar scenario played out at Damietta in 1219: although numerous counterweight trebuchets

11 Fino, "Machines de jet médiévales," pp. 36, 38.

played against the city's landward defences for many months, managing to inflict some damage, there are few indications that this directly facilitated the massed frontal attacks that eventually led to the capture of the city. Even at Acre in 1291, where more artillery was gathered than at any other siege of the crusader period, it was mining rather than artillery that compromised the city's outer defences.

The strength of medieval walls and the effect of sappers have been chronically underemphasised by both medieval and modern historians. The difficulties faced by Saladin in attempting to slight the keep of Jubayl echo what took place four and a half centuries later, when the Parliamentarians turned from gunpowder to traditional sapping in order to slight the 3-4 m thick walls of the keep at Raglan Castle.[12] Mining was the favoured means of breaching and slighting most strongholds throughout the medieval period; this was as true in 1097 as it was in 1291.

Coordinated Use

Rather than a breaching weapon, artillery was used in a supportive role, facilitating the progress of other attacking elements while inhibiting the defenders' ability to resist, by inflicting both physical and psychological trauma. If mining alone is considered, artillery can be found providing support at sieges such as Nicaea (1097), Kharput (1123), Jaffa (1123), Edessa (1144), Banyas (1164), Jerusalem (1187), Acre (1191), Jaffa (1192), Beirut (1231), Arsuf (1265), Safed (1266), Montfort (1271), Margat (1285), Tripoli (1289) and Acre (1291). Examples of its use by Frankish forces to support siege towers and Muslim armies to provide cover for massed frontal assaults are equally forthcoming. Topographic conditions usually account for those instances in which artillery was the only type of siege weapon used by attackers. Due to their range, these engines could often apply pressure against hilltop strongholds that were inaccessible to siege towers and left sappers exposed.

The psychological impact of artillery is hard to gauge but it was probably considerable. The besieged had to contend with showers of lighter projectiles, which could be cast at the same rate as an archer could loose arrows and would kill or maim anyone in their way. Heavier projectiles would have sent out shockwaves as they collided with fortifications, destroyed internal buildings and crushed individuals like no other weapon of the period could. Even if they

12 Wilbrand of Oldenburg, *Itinerarium terrae sanctae* 1.6, ed. Pringle, p. 119, trans. Pringle, p. 66. Cf. *Eracles* 25.2, RHC Oc 2, p. 140. See also Conder and Kitchener, *Survey of Western Palestine*, 1:303. The keep of Jubayl still displays cannonballs that were embedded in its walls during the British bombardment of the 1840s.

were incapable of bringing down walls, by creating a hostile environment within a stronghold, they could compromise the defenders' morale, pushing them to seek terms or making them more vulnerable to massed attacks.

It is important to grasp the elemental simplicity of medieval siege warfare. The additional resources that allowed for the construction of newer and stronger defences were the same which allowed traditional ones to be stormed more easily. But as fortifications and siege engines developed over time, the ladder remained the most effective siege weapon throughout the Middle Ages. The advancements made in gunnery and fortifications over the past five hundred years and the dramatic increase in the influence of personal technologies since the end of the nineteenth century, often cause modern historians to overemphasise the importance of complex technologies when examining medieval warfare.

Although the power of the strongest trebuchets increased dramatically between 1097 and 1291, these engines did not have the revolutionary effect that siege guns had from the end of the fifteenth century. As fortifications grew stronger, largely as a result of other factors, their design was governed by the same guiding principles and perceived threats. Iron picks, in the hands of determined and experienced miners, were the greatest threat to the integrity of medieval defences while ladders led to the fall of more strongholds than any other siege weapon. The romanticised notions put forward by hobbyists and imprudent historians that now characterise trebuchets as super weapons, or medieval howitzers, must be discarded.

Appendix 1: Mentions of Artillery

Mentions of Artillery by Select Narrative Sources (1099-1250)[1]

Date	Site & Use (Christian/Muslim)		Employer(s)	Source(s)
First Crusade				
1099	Arsuf	offensive	Godfrey of Bouillon	AA
1100	Haifa	offensive	Tancred	AA
1101	Caesarea	offensive	Baldwin I	FC, AA, WT
1102	Acre	offensive	Raymond of St Gilles	IA
1103	Acre	offensive	Baldwin I	AA
1104	Acre	offensive	Baldwin I	FC, AA, WT
1106	In Acre for Sidon	prepared	Baldwin I	AA
1106	Apamea	offensive	Tancred	AA
1106	Malatya	offensive	Kilij Arslan	MS, BH
1106/7	Castellum Arnaldi	offensive	Egyptians	AA
1108	Sidon	offensive	Baldwin I	AA
1109	Tripoli	offensive	Bertrand of Tripoli	AA, IQ
1110	Baalbek	offensive	Tughtakin	IQ
1110	Beirut	offensive	Baldwin I	AA, WT
		defensive	garrison	IQ
1110	Sidon	offensive	Baldwin I, Norwegians	AA
		defensive	garrison	AA
1110	al-Atharib	offensive	Tancred	AA, AK
1111	Vetula	offensive	Tancred	AA
1111-12	Tyre	offensive	Baldwin I	AA, IQ, IA
		defensive	garrison	AA
1115	Kafartab	defensive	garrison	US
1115	Jaffa	defensive	garrison	WT
1121	Jerash	offensive	Baldwin II	FC
1122	Zardana	offensive	Ilghazi	AK
1122	Balis	offensive	Baldwin II	AK
1123	Kharput	offensive	Balak	ME, MS, AC, BH

1 Chronicles and histories that span several decades have been selected, rather than more specific accounts, with hopes of revealing broader trends. Less specific Latin and Arabic references to 'machines' and 'siege engines' have been omitted unless these are explicitly described as throwing stones larger than those that a slinger might.

 | DOI 10.1163/9789004376922_014

(cont.)

Date	Site & Use (Christian/Muslim)		Employer(s)	Source(s)
1123	Jaffa	offensive	Egyptians	FC
1124	Tyre	offensive	Franks, Venetians	FC, ME, WT, AC
		defensive	garrison	WT
1124	Manbij	offensive	Balak	ME, AK
1125	'Azaz	offensive	al-Bursuqi, Tughtakin	FC, ME, WT, AC
1126	Raffaniya	offensive	Baldwin II, Pons of Tripoli	FC, WT
1137	Antioch	offensive	John Comnenus	WT
1137	Montferrand	offensive	Zanki	WT, AK
1138	Kafartab	offensive	John Comnenus	AK
1138	Shayzar	offensive	John Comnenus	IQ, US, WT, MS, IA, AK, BH, AF
1138	Buza'a	offensive	John Comnenus	IQ, IA
1139	Baalbek	offensive	Zanki	IQ, IA, AK, AC, AF
1140	Banyas	offensive	Mu'in al-Din Unar, Franks	IQ, WT
1144	Li Vaux Moise	offensive	Baldwin III	WT
1144	Edessa	offensive	Zanki	IQ, WT, AK, AC, BH
1145	al-Bira	offensive	Zanki	AC
1145	In Homs for Damascus	prepared	Zanki	IQ
1146	In Damascus for Baalbek	prepared	Mu'in al-Din Unar	IQ
1146	Edessa	offensive	Joscelin II	AC
1147	In Damascus for Sarkhad, Bosra	prepared	Mu'in al-Din Unar	IQ
1150/51	Turbessel	offensive	Mas'ud	ME
1151	Jerusalem	offensive	Baldwin III	WT
		defensive	garrison	WT
1151	In Damascus for Bosra	prepared	summoned by Nur al-Din	IQ
1152	In Damascus for Bosra	prepared	summoned by Mujir al-Din	IQ
1153	Ascalon	offensive	Baldwin III	WT, AC, BH
1157	In Damascus for Banyas	prepared	summoned by Nur al-Din	IQ
1157	Banyas	offensive	Nur al-Din	IQ, WT
1157	Shayzar	offensive	Baldwin III, Thierry of Flanders	WT
1157-58	Casalia*	offensive	Baldwin III, Thierry of Flanders	WT
1158	Harim	offensive	Baldwin III, Thierry of Flanders	IQ

Date	Site & Use (Christian/Muslim)		Employer(s)	Source(s)
1164	Harim	offensive	Nur al-Din	WT, IA, AC
1167	Alexandria	offensive	Amalric	WT
1169	Damietta	offensive	Amalric, Byzantines	WT, BD, MQ
1170	Kerak	offensive	Nur al-Din	ID, IA
1170	Sinjar	offensive	Nur al-Din	IA
1174	Alexandria	offensive	Sicilians	ID, BD, IA, MQ
1175	Homs	offensive	Saladin	MQ
1175	Montferrand	offensive	Saladin	IA
1175	Sinjar	offensive	Sayf al-Din Ghazi II	BD
1176	A'zaz	offensive	Saladin	IA
1176	Masyaf	offensive	Saladin	IA, MQ
1177-78	Harim	offensive	Philip of Flanders, Northern Franks	WT, IA, AC
1179	Jacob's Ford	offensive	Saladin	ID, IA
1182	In Egypt for Syria	prepared	Saladin	MQ
1183	Amid	offensive	Saladin	IA
1183	Tell Khalid	offensive	Saladin	IA
1183	Kerak	offensive	Saladin	WT, IA
1184	Kerak	offensive	Saladin	ID, BD, IA, BH, MQ
1185	Mayyafariqin	offensive	Saladin	BD, IA
Battle of Hattin				
1187	Toron	offensive	Saladin	ID, BD
1187	Ascalon	offensive	Saladin	ID, BD, IA, MQ
1187	Jerusalem	offensive	Saladin	ID, BD, IA, BH, MQ
		defensive	Balian of Ibelin (garrison)	IA
1187	Tyre	offensive	Saladin	ID, BD, IA, BH, MQ
1188	Saone	offensive	Saladin	ID, BD, IA
1188	Shughr-Bakas	offensive	Saladin	ID, BD, IA
1188	Bourzey	offensive	Saladin	ID, BD, IA
		defensive	garrison	IA
1188	Trapessac	offensive	Saladin	ID, BD, IA
1188	Baghras	offensive	Saladin	ID, IA
1188	Safed	offensive	Saladin	ID, BD, IA
1188	Belvoir	offensive	Saladin	ID, IA
1190-91	Acre	offensive	Henry of Champagne	ID, BD, IA
		offensive	Richard I, Philip II, et al.	ID, BD, IA, MQ, BH
		defensive	garrison	ID, BD, IA
1192	Jaffa	offensive	Saladin	BD

(cont.)

Date	Site & Use (Christian/Muslim)		Employer(s)	Source(s)
1199	Montferrand	offensive	al-Mansur Nasir al-Din	AF
1207	Tripoli	offensive	al-'Adil	AF
1217	Mount Tabor	offensive	Crusaders	IA
1218	Tower of the Chain	offensive	Crusaders	ER
1218	Damietta	offensive	Crusaders	ER
1218	Damietta (in the field)	offensive	Crusaders	IA
1220	Caesarea	offensive	al-Mu'azzam 'Isa	ER
1221	Mansura (in the field)	offensive	Crusaders	IA, MQ
1229	Akhlat	offensive	Jalal al-Din	BH, MQ
1231-32	Beirut	offensive	Imperialists	ER
1232	Amida	offensive	al-Kamil	MQ
1239	Jerusalem	offensive	al-Nasir Da'ud	MQ
1245	Damascus	offensive	al-Nasir Da'ud	MQ
1247	Ascalon	offensive	Egyptians	ER
1248-49	Homs	offensive	al-Salih Ayyub	AF, MQ
1249	Mansura	offensive	Loius IX	MQ

Sources, Alphabetically by Abbreviation

Abr.	Source	Period documented
AA	Albert of Aachen	1100-1119
AC	Anonymous Chronicle	1100-1234
AF	Abu'l-Fida	1100-1250
AK	Ibn al-Adim	1100-1144
BD	Baha' al-Din	1163-1193
BH	Bar Hebraeus	1100-1250
ER	*Eracles*	1184-1250
FC	Fulcher of Chartres	1100-1127
IA	Ibn al-Athir	1100-1233
ID	'Imad al-Din	1174-1193
IQ	Ibn al-Qalanisi	1100-1160
ME	Matthew of Edessa	1100-1137
MQ	al-Maqrizi	1174-1250
MS	Michael the Syrian	1100-1200
US	Usama ibn Munqidh	1100-1183
WT	William of Tyre	1100-1184

Sources, Chronologically by Lifetime

Abr.	Source	Approximate life
FC	Fulcher of Chartres	1060-1128
AA	Albert of Aachen	fl. c. 1130s
ME	Matthew of Edessa	d. c. 1140
IQ	Ibn al-Qalanisi	1070-1160
WT	William of Tyre	1130-1186
US	Usama ibn Munqidh	1095-1188
MS	Michael the Syrian	d. c. 1200
ID	'Imad al-Din	1125-1201
BD	Baha' al-Din	1145-1234
IA	Ibn al-Athir	1160-1233
AC	Anonymous Chronicle	c. 1234
AK	Ibn al-Adim	1192-1262
ER	*Eracles*	c. 1277
BH	Bar Hebraeus	1226-1286
AF	Abu'l-Fida	1273-1331
MQ	al-Maqrizi	1364-1442

Use of Specified Types of Artillery by Muslim Forces in the Thirteenth Century

Year	Site	Term	Number	Source
1218	Damietta	*shaytani*		Ibn al-Dawadari
		gharbiya		Ibn al-Dawadari
1229	Baalbek	*maghribi*		al-Ansari
1232-33	Shayzar	*maghribi*	1	Ibn Wasil; Ibn al-'Adim
1236	Harran	*maghribi*	1	*History of the Patriarchs*
1236	Edessa	*maghribi*	1	*History of the Patriarchs*
1248-49	Homs	*shaytani*	12	Ibn Wasil; al-Ansari
		maghribi	1 of 14	Ibn Wasil; al-Ansari
		qarabugha	1 of 14	Ibn Wasil
1265	Caesarea	*ifranji*	4?	Ibn 'Abd al-Zahir; Ibn al-Furat
		maghribi		Ibn 'Abd al-Zahir; Ibn al-Furat
1266	Safed	*maghribi*	>1	Ibn 'Abd al-Zahir; Ibn al-Furat

(cont.)

Year	Site	Term	Number	Source
1275	al-Bira	*ifranji*	23 of 70	Ibn Shaddad; al-Yunini; Shafi' ibn 'Ali
1285	Margat	*shaytani*	4 of 10	Ibn 'Abd al-Zahir
		ifranji	3 of 10	Ibn 'Abd al-Zahir
		qarabugha	3 of 10	Ibn 'Abd al-Zahir
1286	Saone	*ifranji*	3 offensive, 1 defensive	Ibn 'Abd al-Zahir
1289	Tripoli	*ifranji*	6 of 19	Ibn al-Dawadari; al-Yunini; Ibn al-Furat
		qarabugha	13 of 19	Ibn al-Dawadari; al-Yunini; Ibn al-Furat
1291	Acre	*shaytani*	52 of 72	al-Jazari; al-'Ayni; al-Yunini; Ibn al-Furat
		ifranji	15 of 72	al-Jazari; al-'Ayni; al-Yunini; Ibn al-Furat
		qarabugha	(5?) of 72	al-Jazari; al-'Ayni; al-Yunini; Ibn al-Furat

Baybars' Use of Artillery, According to Ibn 'Abd al-Zahir

Date	Site	Intention	Use	Description (if provided)
1265	al-Bira	defensive	prepared	
1265	Caesarea	offensive	prepared, used	*maghribi* and *ifranji*, 4 large, small
1265	Arsuf	offensive	used	
1266	Safed	offensive	prepared, used	1 *maghribi*, others
1266?			prepared	
1268	Beaufort	offensive	prepared, used	26
1271	Crac	offensive	used	
1271	'Akkar	offensive	prepared	some large
1272	'Akkar	defensive	prepared	
1272-73	Alexandria	defensive	prepared	

Appendix 2: Images and Technical Treatises

An analysis of contemporary images has been intentionally omitted from the main body of this book. Because there are so many images, and the cost of reproducing them would be considerable, they warrant their own independent study. For present purposes, it may suffice to provide a list of images featuring artillery, although certainly incomplete, accompanied by a brief discussion of how these images relate to one another and the development of artillery technology. In order to add further context, certain technical works will also be addressed.[1]

Images

Traction Trebuchets: Early West Asian

1. c. 700-99, Piandijikent: Wall-painting from Piandijikent.
   St Petersburg, Hermitage Museum.

Traction Trebuchets: European/Frankish

2. c. 1100-25, Catalonia: Beatus of Liébana, *Commentary on the Apocalypse* – Nebuchadnezzar's siege of Jerusalem.
   Turin, Biblioteca Nazionale Universitaria di Torino, MS I.II.1, fol. 109r.
3. c. 1196, Norman Italy: Peter of Eboli, *Liber ad honorem Augusti* – siege of Naples.
   Bern, Burgerbibliothek, BBB Cod. 120.II, fol. 109r.
4. c. 1196, Norman Italy: Peter of Eboli, *Liber ad honorem Augusti* – defence of Salerno.
   Bern, Burgerbibliothek, BBB Cod. 120.II, fol. 111r.
5. c. 1196, Norman Italy: Peter of Eboli, *Liber ad honorem Augusti*.
   Bern, Burgerbibliothek, BBB Cod. 120.II, fol. 132r.
6. c. 1196, Norman Italy: Peter of Eboli, *Liber ad honorem Augusti* – imprisonment of Roger of Andria.
   Bern, Burgerbibliothek, BBB Cod. 120.II, fol. 104r.
7. c. 1196, Norman Italy: Peter of Eboli, *Liber ad honorem Augusti* – walls of Palermo.
   Bern, Burgerbibliothek, BBB Cod. 120.II, fol. 97r.

---

1 The dating and place of composition listed for the following works are those provided by the curators of each document, unless otherwise stated, with the exception of the William of Tyre manuscripts, which have been dated according to the analysis of Edbury, Folda and Buchthal, see Edbury, "The French Translation of William of Tyre's *Historia*," pp. 69-105; Folda, *Crusader Manuscript Illumination*; Buchthal, *Miniature Painting in the Latin Kingdom of Jerusalem*.

 | DOI 10.1163/9789004376922_015

8. c. 1196, Norman Italy: Peter of Eboli, *Liber ad honorem Augusti.*
Bern, Burgerbibliothek, BBB Cod. 120.II, fol. 96r.
9. c. 1196, Norman Italy: Peter of Eboli, *Liber ad honorem Augusti* – castrum maris, Palermo.
Bern, Burgerbibliothek, BBB Cod. 120.II, fol. 98r.
10. c. 1196, Norman Italy: Peter of Eboli, *Liber ad honorem Augusti* – Monte Cassino.
Bern, Burgerbibliothek, BBB Cod. 120.II, fol. 108r.
11. c. 1196, Norman Italy: Peter of Eboli, *Liber ad honorem Augusti.*
Bern, Burgerbibliothek, BBB Cod. 120.II, fol. 114r.
12. 12th-13th Cs, Sicily: *Skylitzes Chronicle* – siege of Mopsvestia.
Madrid, Biblioteca Nacional de España, MSS Graecus Vitr. 26-2, fol. 151r.
13. 12th-13th Cs, Sicily: *Skylitzes Chronicle* – siege of Preslav.
Madrid, Biblioteca Nacional de España, MSS Graecus Vitr. 26-2, fol. 166r.
14. 12th-13th Cs, Sicily: *Skylitzes Chronicle* – battle of Dorustolon.
Madrid, Biblioteca Nacional de España, MSS Graecus Vitr. 26-2, fol. 169r.
15. c. 1220, Corneto: Graffito in an Etruscan tomb.
Corneto-Tarquinia.
16. c. 1220, south France: Tomb of Simon de Montfort.
Carcassonne, Church of St Nazaire.
17. c. 1225-50, France: *Oxford-Paris-London bible moralisée,* vol. 3a.
London, BL, Harley 1526, fol. 18v.
18. c. 1245-48, Paris: William of Tyre, *Histoire des Croisades* – siege of Nicaea.
Paris, BNF, MS fr. 9081, fol. 26r.
19. c. 1250, France: *Morgan Bible* – Saul's battle against the Ammonites (1 Kings 11:11).
New York, Pierpont Morgan Library, MS M.638, fol. 23v.
20. c. 1250, France: *Morgan Bible* – siege of Able (2 Kings 20:15-22).
New York, Pierpont Morgan Library, MS M.638, fol. 43v.
21. c. 1250, Genoa?: Caffaro, *Annales de Genes.*
Paris, BNF, MS lat. 10136, fol. 107r.
22. c. 1250, Genoa?: Caffaro, *Annales de Genes* – siege of Savona.
Paris, BNF, MS lat. 10136, fol. 141v.
23. c. 1250, England: Matthew Paris, *Chronica maiora* – agreement between the count of Brittany and al-Nasir of Kerak.
Cambridge, Corpus Christi College Lib., MS 16, pt. 2, fol. 139v.
24. c. 1260, c. 1280, Acre: William of Tyre, *Histoire de la guerre sainte* – siege of Tyre.
Paris, BNF, MS fr. 2628, fol. 114r.
25. c. 1278, Acre: William of Tyre, *History of Outremer,* vol. 1 – siege of Antioch.
St. Petersburg, National Library of Russia, MS fr. f° v.IV.5, vol. 1, fol. 18v.
26. c. 1280, Acre: William of Tyre, *Les Estoires d'outremer* – siege of Antioch.
Lyon, Bibliothèque Municipal, MS 828, fol. 33r.

27. c. 1275-99, France: *Arthurian Romances.*
Yale, Beinecke MS 229, fol. 346r.
28. c. 1300-99, France?: *Roman du Chevalier au Cygne et Chanson d'Antioch.*
Paris, BNF, MS fr. 12558, fol. 143v.

Traction Trebuchets: Muslim

29. c. 1180, Egypt: al-Tarsusi, *Tabsira arbab al-albab* – Frankish trebuchet.
Oxford, Bodleian Library, MS Hunt 264, fol. 133r.
30. c. 1180, Egypt: al-Tarsusi, *Tabsira arbab al-albab* – Turkish trebuchet.
Oxford, Bodleian Library, MS Hunt 264, fol. 138r.
31. c. 1180, Egypt: al-Tarsusi, *Tabsira arbab al-albab* – Arab trebuchet.
Oxford, Bodleian Library, MS Hunt 264, fol. 137r.
32. c. 1250-96, Persia: Hasan al-Rammah, *Furusiya.*
Paris, BNF, MS arabe. 2825, fol. 88v.
33. c. 1250-96, Persia: Hasan al-Rammah, *Furusiya.*
Paris, BNF, MS arabe. 2825, fol. 89r.

Counterweight Trebuchets: Al-Tarsusi

34. c. 1180, Egypt: al-Tarsusi, *Tabsirah arbab al-albab* – Persian trebuchet.
Oxford, Bodleian Library, MS Hunt. 264, fols. 134v-135r.

Counterweight Trebuchets: European/Frankish

35. c. 1250, Genoa?: Caffaro, *Annales de Genes* – siege of Savona.
Paris, BNF, MS lat. 10136, fol. 142r.
36. c. 1250, France?: Villard of Honnecourt, *Sketch Book* – base of a trebuchet.
Paris, BNF, MS fr. 19093, fol. 30r.
37. c. 1200-99, France?: *Chevalier au Cygne.*
Paris, BNF, MS fr. 795, fol. 220r.
38. c. 1275-99, France: *Roman de Renart* (Roman de la Rose).
Paris, BNF, MS fr. 1581, fol. 8v.
39. c. 1300, Maastricht: *Book of Hours.*
London, BL, Stowe MS 17, fol. 243v.
40. c. 1280, Castile: *Cantigas de Santa Maria.*
Florence, Biblioteca Nazionale, MS Banco Rari 20, fol. 8r.
41. c. 1250-99, Toulouse: William of Tudela and anonymous continuer, *Chanson de la Croisade contre les Albigeois.*
Paris, BNF, MS fr. 25425, fol. 15r.
42. c. 1291, Acre: William of Tyre, *History of Outremer* – siege of Antioch.
Florence, Biblioteca Medicea-Laurenziana, MS Plu. LXI.10, fol. 42r.

43. c. 1305, Flanders: Courtrai Chest – defeat of the French in 1302.
Oxford, New College.
44. c. 1316, England: Charter to Carlisle by Edward II – siege of Carlisle.
Carlisle, Cumbria Records Office.
45. c. 1320-40, Paris: *Bible Historiale Complétée.*
The Haag, Koninklijke Bibliotheek, KB 71 A, fol. 201v.
46. c. 1300-40, Zürich: *Große Heidelberger Liederhandschrift* (Codex Manesse).
Heidelberg, Universitätsbibliothek, Cod. Pal. germ. 848, fol. 255r.
47. c. 1337, Paris: *Li rommans de Godefroy de Buillon.*
Paris, BNF, MS fr. 22495, fol. 9r.
48. c. 1337, Paris: *Li rommans de Godefroy de Buillon* – siege of Nicaea.
Paris, BNF, MS fr. 22495, fol. 30r.
49. c. 1332-50, Paris: *Chroniques de France ou de St Denis.*
London, BL, Royal 16 G VI, fol. 345v.
50. c. 1332-50, Paris: *Chroniques de France ou de St Denis* – siege of La Rochelle.
London, BL, Royal 16 G VI, fol. 388r.
51. c. 1300-99: William of Tyre, *Les Estoires d'outremer* – siege of Jerusalem.
Paris, BNF, MS fr. 352, fol. 62r.
52. c. 1340, Paris: *Historia de Proeliis* (La vraie ystoire dou bon roi Alixandre) – siege of Saianfu.
London, BL, Royal D I, fol. 111.
53. c. 1340-60, France: *Avis aus Roys* – besieging fortifications.
New York, Pierpont Morgan Library, MS M.456, fol. 127r.
54. c. 1340-60, France: *Avis aus Roys* – defending against siege engines.
New York, Pierpont Morgan Library, MS M.456, fol. 131r.
55. c. 1270, Paris: William of Tyre, *Les Estoires d'Outremer* – siege of Nicaea.
Paris, BNF, MS fr. 2630, fol. 22v.
56. c. 1270-75, Germany: Wolfram von Eschenbach, *Willehalm* (fragments).
Munich, Bayerische Staatsbibliothek, MS Cod. Ger. 193 III, fol. 4v.
57. c. 1280, Castile: *Cantigas de Santa María.*
Madrid: Biblioteca el Escorial, MS T.j.1, fol. 43r.
58. c. 1308-12, London: *Roman de toute chevalière.*
Paris, BNF, MS fr. 24364, fol. 5v.
59. c. 1278, Acre: William of Tyre, *Les Estoires d'outremer*, vol. 1 – siege of Tyre.
St. Petersburg, National Library of Russia, MS fr. f° v.IV.5, vol. 1, fol. 103r.
60. c. 1280, Acre: William of Tyre, *Les Estoires d'outremer* – siege of Tyre.
Lyon, Bibliothèque Municipale, MS 828, fol. 135v.
61. c. 1300, Paris: William of Tyre, *Histoire d'Outre Mer* – siege of Tyre.
Baltimore, Walters Art Gallery, MS W.142, fol. 112r.

62. c. 1300, Paris: William of Tyre, *Histoire d'Outre Mer* – siege of Antioch.
Baltimore, Walters Art Gallery, MS W.142, fol. 28r.
63. c. 1300, France: William of Tyre, *Histoire de la guerre saint* – siege of Nicaea.
Paris, BNF, MS fr. 2824, fol. 15v.
64. c. 1300, France: William of Tyre, *Histoire de la guerre saint* – siege of Jerusalem.
Paris, BNF, MS fr. 2824, fol. 45r.
65. c. 1300, France: William of Tyre, *Histoire de la guerre saint* – siege of Shayzar.
Paris, BNF, MS fr. 2824, fol. 94v.
66. c. 1320, Siena: Simone Martini, *Guidoriccio da Fogliano*.
Siena, Palazzo Pubblico.
67. c. 1325-50, London?: *Book of Hours* (*Neville of Hornby Hours*) – siege of Jerusalem.
London, BL, MS Egerton 2781, fol. 190v.
68. c. 1410, Paris: *Des cad des nobles hommes et femmes*.
Geneva, Bibliothèque de Genève, MS fr. 190-1, fol. 163r.
69. c. 1326, London: Walter of Milemete, *De nobilitatibus, sapientiis et prudentiis regum*.
Oxford, Christ Church, MS 92, fol. 67r.
70. c. 1360, Regensburg: *Weltchronik*.
New York, Pierpont Morgan Library, MS M.769, fol. 193r.
71. c. 1375-85, Swabia: *History Bible* – siege of Ai by Joshua.
New York, Pierpont Morgan Library, MS M.268, fol. 11v.
72. c. 1384-1400, Paris: *Fleurs des chroniques*.
Besançon, Bibliothèque Municipale, MS 677, fol. 45r.
73. c. 1368-1424, Lucca: *Croniche di Giovanni Sercambi* – siege of Nocera.
Lucca, Archivio di Stato, MS 107, fol. 113v.
74. c. 1435, Rottweil: *Ordnung des Hofgerichts zu Rottweil* – Lothar III besieges Conrad III.
Stuttgart: Württembergische Landesbibliothek, HB VI 110, fol. 1r.
75. c. 1400, England: *Alexander and Dindimus*.
Oxford, Bodleian Library, MS 264, fol. 255r.
76. c. 1405, Germany: Konrad Kyeser, *Bellifortis*.
Göttingen, Niedersächsische Staats- und Universitätsbibliothek, Cod. philos. 63, fol. 30r.
77. c. 1405, Germany?: Konrad Kyeser, *Bellifortis*.
Innsbruck, Tiroler Landesmuseum Ferdinandeum, Cod. FB 32009 (olim 16.0.7), fol. 21r.
78. c. 1405, Germany: Konrad Kyeser, *Bellifortis*.
Munich, Bayerische Staatsbibliothek, Clm 30150, tafel 2, fol. 2v.
79. c. 1459, Germany?: Konrad Kyeser, *Bellifortis*, in Talhoffer, *Fechtuch*.
Copenhagen, Det Kongelige Bibliotek, MS Thott. 290.2, fol. 16v.

80. c. 1459, Germany?: Konrad Kyeser, *Bellifortis*, in Talhoffer, *Fechtuch*.
Copenhagen, Det Kongelige Bibliotek, MS Thott. 290.2, fol. 32r.
81. c. 1459, Germany?: Konrad Kyeser, *Bellifortis*, in Talhoffer, *Fechtuch*.
Copenhagen, Det Kongelige Bibliotek, MS Thott. 290.2, fol. 14v.
82. c. 1419-50, Italy: Mariano di Jacopo Taccola, *De ingeneis*, liber secundus.
Munich, Bayerische Staatsbibliothek, Cod. lat. Monacensis 197, fol. 40v.
83. c. 1474-82, Italy: Francesco di Giorgio, *Opusculum de architectura*.
London, British Museum, Cod. 197.b.21 – MS Harley 3281, fol. 3v.
84. c. 1419-49, Italy: Mariano di Jacopo Taccola, *De machinis* – trabocco, for throwing incendiaries.
Paris, BNF, MS lat. 7239, fol. 22v.
85. c. 1433, Italy: Mariano di Jacopo Taccola, *De ingeneis ac edifitiis non usitatis*, liber tertius – manganum.
Florence, Biblioteca Nazionale Centrale, Cod. Palatino 766, fols. 66v-67r.
86. c. 1433, Italy: Mariano di Jacopo Taccola, *De ingeneis ac edifitiis non usitatis*, liber tertius – brichola.
Florence, Biblioteca Nazionale Centrale, Cod. Palatino 766, fols. 67v-68r.
87. c. 1419-50, Italy: Mariano di Jacopo Taccola, *De ingeneis*, liber secundus.
Munich, Bayerische Staatsbibliothek, Cod. lat. Monacensis 197, fol. 39v.
88. c. 1419-50, Italy: Mariano di Jacopo Taccola, *De ingeneis*, liber secundus.
Munich, Bayerische Staatsbibliothek, Cod. lat. Monacensis 197, fol. 95r.
89. c. 1450-1499, Italy?: Anonymous.
Siena, Biblioteca Comunale, Cod. S.IV.5, fol. 25r.
90. c. 1450-1499, Italy?: Anonymous.
Siena, Biblioteca Comunale, Cod. S.IV.5, fol. 57v.
91. c. 1463, Italy: Roberto Valturio, *De re militari libri duodecim*.
Paris, BNF, MS lat. 7236, fol. 148v.
92. c. 1463, Italy: Roberto Valturio, *De re militari libri duodecim*.
Paris, BNF, MS lat. 7236, fol. 149r.
93. c. 1463, Italy: Roberto Valturio, *De re militari libri duodecim*.
Paris, BNF, MS lat. 7236, fol. 149v.
94. c. 1463, Italy: Roberto Valturio, *De re militari libri duodecim*.
Paris, BNF, MS lat. 7236, fol. 150r.
95. c. 1470-75, Hungary?: Paolo Santini, *Tractatus*.
Paris, BNF, MS lat. 7239, fol. 109r.
96. c. 1480-99, Italy: Francesco di Giorgio, *Trattato* – trebuchet.
Turin, Biblioteca Reale, Cod. 148 Saluzzo, fol. 60r.
97. c. 1480-99, Italy: Francesco di Giorgio, *Trattato* – bricola.
Turin, Biblioteca Reale, Cod. 148 Saluzzo, fol. 60r.

98. c. 1480-99, Italy: Francesco di Giorgio, *Trattato*.
Turin, Biblioteca Reale, Cod. 148 Saluzzo, fol. 62r.
99. c. 1480-99, Italy: Francesco di Giorgio, *Trattato*.
Turin, Biblioteca Reale, Cod. 148 Saluzzo, fol. 61v.
100. c. 1478-1519, Italy: Leonardo da Vinci, *Codex Atlanticus* – rotary trebuchet.
Milan, Biblioteca Ambrosiana, fol. 57.
101. c. 1478-1519, Italy: Leonardo da Vinci, *Codex Atlanticus* – tension catapult
Milan, Biblioteca Ambrosiana, fol. 50v.
102. c. 1507, Germany: Jörg Kölderer (woodcut).

Counterweight Trebuchets: Muslim

103. c. 1285, Syria/Persia: Hasan al-Rammah, *Furusiya*.
Paris, BNF, MS arabe. 2825, fol. 87v.
104. c. 1285, Syria/Persia: Hasan al-Rammah, *Furusiya*.
Paris, BNF, MS arabe. 2825, fol. 90r.
105. c. 1306-18, Tabriz: Rashid al-Din, *Jami' al-tawarikh*.
Edinburgh, University of Edinburgh Library, Or. MS 20, fol. 3v.
106. c. 1306-18, Tabriz: Rashid al-Din, *Jami' al-tawarikh*.
Edinburgh, University of Edinburgh Library, Or. MS 20, fol. 124v.
107. c. 1306-18, Tabriz: Rashid al-Din, *Jami' al-tawarikh*.
Edinburgh, University of Edinburgh Library, Or. MS 20, fol. 130v.
108. c. 1306-18, Tabriz: Rashid al-Din, *Jami' al-tawarikh* – siege of Baghdad.
Berlin, Staatsbibliothek, Orientabteilung, Diez A, fol. 70, S. 7 and S. 4.
109. c. 1430, Herat: Rashid al-Din, *Jami' al-tawarikh* – siege of Baghdad.
Paris, BNF, MS pers. 1113, fols. 180v-181r.
110. c. 1373-1462, al-Zardkash, *Kitab al-'aniq fi'l-manajniq*.
Istanbul, Topkapi Museum Library, MS Ahmad III 3469, fol. 20r.
111. c. 1373-1462, al-Zardkash, *Kitab al-'aniq fi'l-manajniq*.
Istanbul, Topkapi Museum Library, MS Ahmad III 3469, fol. 21v.
112. c. 1373-1462, al-Zardkash, *Kitab al-'aniq fi'l-manajniq*.
Istanbul, Topkapi Museum Library, MS Ahmad III 3469, fol. 22r.
113. c. 1373-1462, al-Zardkash, *Kitab al-'aniq fi'l-manajniq*.
Istanbul, Topkapi Museum Library, MS Ahmad III 3469, fol. 24v.
114. c. 1373-1462, al-Zardkash, *Kitab al-'aniq fi'l-manajniq*.
Istanbul, Topkapi Museum Library, MS Ahmad III 3469, fol. 25r.
115. c. 1373-1462, al-Zardkash, *Kitab al-'aniq fi'l-manajniq*.
Istanbul, Topkapi Museum Library, MS Ahmad III 3469, fol. 26v.
116. c. 1373-1462, al-Zardkash, *Kitab al-'aniq fi'l-manajniq*.
Istanbul, Topkapi Museum Library, MS Ahmad III 3469, fol. 27v.

117. c. 1373-1462, al-Zardkash, *Kitab al-'aniq fi'l-manajniq*.
Istanbul, Topkapi Museum Library, MS Ahmad III 3469, fol. 28r.
118. c. 1373-1462, al-Zardkash, *Kitab al-'aniq fi'l-manajniq*.
Istanbul, Topkapi Museum Library, MS Ahmad III 3469, fol. 28v.
119. c. 1373-1462, al-Zardkash, *Kitab al-'aniq fi'l-manajniq*.
Istanbul, Topkapi Museum Library, MS Ahmad III 3469, fol. 29r.
120. c. 1373-1462, al-Zardkash, *Kitab al-'aniq fi'l-manajniq*.
Istanbul, Topkapi Museum Library, MS Ahmad III 3469, fol. 30r.
121. c. 1373-1462, al-Zardkash, *Kitab al-'aniq fi'l-manajniq*.
Istanbul, Topkapi Museum Library, MS Ahmad III 3469, fol. 31r.
122. c. 1373-1462, al-Zardkash, *Kitab al-'aniq fi'l-manajniq*.
Istanbul, Topkapi Museum Library, MS Ahmad III 3469, fol. 32r.
123. c. 1373-1462, al-Zardkash, *Kitab al-'aniq fi'l-manajniq*.
Istanbul, Topkapi Museum Library, MS Ahmad III 3469, fol. 33r.
124. c. 1373-1462, al-Zardkash, *Kitab al-'aniq fi'l-manajniq*.
Istanbul, Topkapi Museum Library, MS Ahmad III 3469, fol. 34r.
125. c. 1373-1462, al-Zardkash, *Kitab al-'aniq fi'l-manajniq*.
Istanbul, Topkapi Museum Library, MS Ahmad III 3469, fol. 37v.
126. c. 1373-1462, al-Zardkash, *Kitab al-'aniq fi'l-manajniq*.
Istanbul, Topkapi Museum Library, MS Ahmad III 3469, fol. 43v.
127. c. 1373-1462, al-Zardkash, *Kitab al-'aniq fi'l-manajniq*.
Istanbul, Topkapi Museum Library, MS Ahmad III 3469, fol. 44r.
128. c. 1373-1462, al-Zardkash, *Kitab al-'aniq fi'l-manajniq*.
Istanbul, Topkapi Museum Library, MS Ahmad III 3469, fol. 45v.

Torsion Engines

129. c. 1326, London: Walter of Milemete, *De nobilitatibus, sapientiis et prudentiis regum*.
Oxford, Christ Church College, MS 92, fol. 78v.

Most images of medieval mechanical artillery postdate the development of the engines that they depict by many decades, if not centuries. Nevertheless, these illustrations serve to confirm many conclusions that can be drawn from the literary sources. These include the light nature of traction trebuchets, the gradual integration of counterweight power systems around the end of the twelfth century, and the development of quite sizable engines by the end of the thirteenth century.

In the following overview, preference, where possible, will be given to images composed in the Near East; however, in the interest of providing a more cohesive impression of the broader development of trebuchet technology, images from Europe to Persia will be included. When considering that individuals from these wider regions travelled to the Levant, were exposed to the engines that were built there, and that some even imported artillery with them, this broader range of images becomes necessary.

Early Frankish Traction Trebuchets

Few illustrations of European and Near Eastern artillery dating earlier than the twelfth century have survived. One of the earliest images of a traction trebuchet found west of the Indus comes from Piandjikent, Transoxania (Panjakent, Tajikistan), **no. 1**.[2] The wall-painting shows the structure of a traction trebuchet and a team of five pullers. The engine is shown to be constructed of slender materials but is sizable and sturdy: the main axle is approximately twice the height of a man and is supported by four legs, each joined to its neighbour at three points below the axle. Although an early example, the engine's proportions appear to be shown very accurately.

One of the earliest surviving illustrations of a traction trebuchet from western Europe is **no. 2**. Two pullers operate a simple beam with a short sling balanced across the fork of a supporting pole. Notably, the engine is depicted in a defensive role. Illustrative evidence from around the Mediterranean becomes more plentiful from the end of the twelfth century.

Peter of Eboli's *Liber ad honorem Augusti* contains a number of images of traction trebuchets. **Nos. 3-11** display engines at a number of sieges, the majority depicted in a defensive capacity. The engines are powered by crews of two to nine and appear to fulfil a function similar to the archers next to whom they are frequently depicted. Each engine is structurally light and shown with an axle housing consisting of two parallel rods, fixed at their ends, which hold the horizontal axle (Fig. 23). Each axle housing is attached to the top of a single vertical support. The tripod bases at the bottom of these supports, clearly shown on the attacking engines in **nos. 3**, **4** and **11**, confirm that these were light machines. In **nos. 3** and **4**, as well as **nos. 16** and **18-20**, a loader is shown holding a projectile prior to release, both securing it in place and providing some initial resistance, as many of the beams are shown bending.

These Italo-Norman illustrations are very similar to those found in John Skylitzes' contemporary Greek manuscript. **No. 12** clearly shows the 'rake' of the short arm, used to secure a broad 'foot', which allows the four pullers to stand directly under the points where their ropes are attached. Similar rakes are also visible in **nos. 3-6** and **11**. It appears that the illustrators of the engines in **nos. 3** and **12**, understood the range of these machines to be less than that of an archer, as attacking archers are shown behind both engines. Furthermore, protective screens can be seen shielding both the attacking and defending trebuchets in **no. 4**.

2 For more on this wall art, see Rice, *Ancient Arts of Central Asia*, pp. 101-10.

FIGURE 23
'Mediterranean' axle arrangement

A simpler engine, a pole trebuchet with no rake, is featured in **no.** 13 and appears to be what is shown in **no.** 14. These images come from the same document as **no.** 12 but display engines more like that in **no.** 2, although with quite different axle arrangements. The similarity of the 'Mediterranean' axles in **nos.** 3-13, as well as that in an Italian graffito from the early thirteenth-century, **no.** 15, suggest that this design was common in Italy and Sicily, although it is possible that none of these were drawn by an ethnically Italian illustrator.[3] A number of illustrations from a more French context, **nos.** 16-20, reveal a similar set of commonalities; in this case, a simplified axle supported by a vertical upright on each side.

No. 16 dates to the early thirteenth century and may represent the trebuchet that killed Simon de Montfort at Toulouse.[4] The beam of this engine, like those in **nos.** 12, 15, 19 and 20, consists of a main spar supported by two others, which connect either end of the foot, at the end of the short arm, to the main beam beyond the axle. This differs from the engines in **nos.** 3-11, 17 and 18 where the supporting rake structure joins the short arm just in front of the main axle. The former would appear to have been a considerably sturdier design (Fig. 24).

In **nos.** 16, 19 and 20 the foot appears to be aligned vertically. While such an arrangement would minimise the stresses incurred across the foot when the pullers suddenly applied force, it would prevent the outer spars from absorbing some of the strain placed on the main beam where it joined the axle. Furthermore, by aligning the foot vertically, fewer pullers could stand directly below it, diminishing the amount of force

3 The *Skylitzes Chronicle* was illuminated in Sicily in a Greek style; Peter of Eboli, a Norman, composed his work in either southern Italy or Sicily; and the graffito, found in Tarquinia, mid-Italy, may have been drawn by a Templar, see Pringle, "A Medieval *Graffito*," pp. 37-46.

4 For the death of Simon de Montfort, see *Chanson de la croisade Albigeoise* 205, ed. Meyer, 1:341-42, trans. Shirley, p. 172; Roger of Wendover, *Flores historiarum*, ed. Hewlett, 2:252, trans. Giles, 2:426.

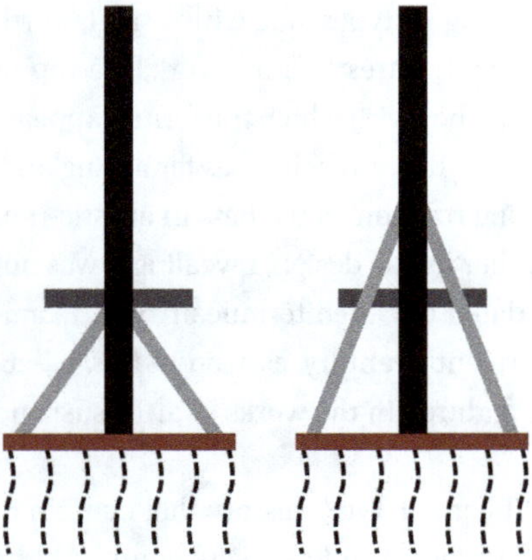

FIGURE 24
Rake designs

that could be communicated when the ropes were pulled. It is possible that the illustrators skewed the design of these engines, representing an otherwise horizontal element vertically, in order to show details that would be hidden when viewed from a horizontal perspective.

Nos. 16, 19 and **20** clearly show that the pulling ropes were attached to rings that extended from the foot of each engine. When reconstructing his traction trebuchet, Tarver had originally fitted his engine with only three rings, inspired by **nos. 19** and **20**; however, this proved too restricting and he later added five more, two more than are shown in **no. 16**.[5]

Returning to an Italian context, albeit a more northern one, **nos. 21** and **22**, from the middle of the thirteenth century, show engines with a similar three-spar arrangement to those in **nos. 12, 15, 16, 19** and **20**. The defensive engine in **no. 22** appears to have a similar axle arrangement and structure to the southern Italian or 'Mediterranean' examples that have been addressed above; however, the assailing engine that is also depicted and the engine in **no. 21** have much sturdier triangular trestle frames. Each of these latter engines also has a substantial foot with pulling ropes attached at three points. Although the foot of the attacking engine in **no. 22** appears to be aligned vertically, the axle, as well as the detailed isometric example in **no. 21**, suggest that it has been skewed in order to display it better. The defensive trebuchet in **no. 23** appears to have a similar axle to that in **no. 22**, as well as many of the earlier 'Mediterranean' examples, and its tripod base is comparable to those of the attacking engines in **nos. 3, 4, 11** and **12**. It is hard to discern this artist's influences, as **no. 23** was composed in England during the middle of the thirteenth century, around the same time as **nos. 17-22** were drawn.

5 Tarver, "The Traction Trebuchet," p. 156.

The images addressed so far display two clear engine types: one with a single vertical support and square axle housing; the other with a trestle framework, the top of each triangular side supporting the axle. The frequency with which the former appears suggests this was a common design in the twelfth century, reaching as far as England, or perhaps became a popular way of representing traction trebuchets in artistic-monastic circles. **No. 1** features a variant of the sturdier trestle design, revealing it was not a later development but instead probably pre-dated the 'Mediterranean' type. Examples of both were drawn in Acre in the late thirteenth century, as seen in **nos. 24-26**, while only engines with trestle frameworks are featured in the works of al-Tarsusi and al-Rammah (**nos. 29-33**).

Three of the earliest surviving editions of William of Tyre's history that contain illuminations of artillery, which were composed in the Latin East, were completed by the early 1280s.[6] **No. 24**, from the earliest and crudest edition, displays a traction trebuchet with a trestle frame in profile. Only one puller is shown but the illustrator may have intended that others were standing beside him, hidden from view. **Nos. 25** and **26** display heavy traction trebuchets, each with a short sling and a particularly large foot. Each engine is shown with a massive projectile, far too large to be practical. The images date to around 1280 so the big stones might reflect the scale of the largest projectiles that were thrown by counterweight trebuchets at that point in time. The illustrator(s) may have scaled up traction engines, which were the type of engines used at the siege of Antioch, the focus of each image, so that a contemporary audience could more easily relate or appreciate the significance of these scenes. Alternatively, ignorance may have been responsible for the images' mechanical impracticalities.

The engine in **no. 25** is shown with at least seven pullers, who are positioned behind the main axle, like those in **no. 16**, while the pullers in **no. 26** are well ahead of their engine. The beam of each machine is clearly bound. This might indicate that the beams were supposed to be composed of a number of spars, but could also have been done to provide extra strength for a single beam. Tarver found that binding the beam of his engine helped protect it from the metal ring at the end of the loose rope of the sling, which had a tendency to swing back against the beam after each shot.[7] The beam of each engine is fitted to an axle that is carried either by a single upright, seated in a base with multiple legs, or by two vertical uprights, each with a number of these legs. The use of two solid uprights, rather than a trestle frame, is also found in **nos. 16** and **18-20**; however, the legs of the engines in **nos. 25** and **26** would have been very impractical, leaving these engines far less stable than even the tripod designs in **nos. 3, 4, 11** and **23**. Why the bases of these engines are represented in this way is unclear given the

6 For the relationship between these manuscripts, see Folda, *Crusader Manuscript Illumination*, pp. 27-36. See also Edbury, "The French Translation," pp. 77-90, 97.

7 Tarver, "The Traction Trebuchet," p. 156.

much more structurally sound precedent established in **no. 24** and that this manuscript appears to have been that from which the illustrator(s) of **nos. 25** and **26** worked.[8] These are not the only images displaying engines with impractical components.

No. 27 contains a traction trebuchet with a single upright and two pullers, similar to **no. 2**, which was produced almost two centuries earlier. Although the engine has been implausibly simplified, featuring a cup at the end of the long arm rather than a sling, it may suggest that very simple engines remained in use at the end of the thirteenth century. On the other hand, machines could be operated by as many as a dozen pullers by this point, as seen in **no. 28**, where, despite the added power, the use of shields implies that the range of these engines was still less than that of the defenders. The illustrative issues with both images, the lack of a sling in **no. 27** and obscure axle arrangement in **no. 28**, along with those found in other illustrations, urge a degree of caution when drawing any interpretations or conclusions based on these images.

Certain exaggerations can lead to misguided conclusions regarding the power of traction trebuchets. At first glance the heavy foot seen in some illustrations, such as those in **nos. 16, 19-22, 25, 26** and possibly **24**, appear to suggest the use of a small counterpoise, while that in **no. 53** leaves it unclear if the engine is meant to be a traction trebuchet or heavier counterweight engine. Through his trials, Tarver found that his engine, built according to the specifications of al-Tarsusi and using the images found in the Morgan Bible (**nos. 19** and **20**), benefited from the addition of a block of wood to reinforce the foot of the machine. This simple addition provided both structural support to an area of high strain and a small amount of extra mass to help balance the beam.[9] This is the extent to which any theory of a hybrid model can be taken. Hypotheses that a heavier beam could be balanced by a proportionally greater mass and operated by correspondingly large crews ignore the fundamental mechanics that govern these engines.[10]

Muslim Traction Trebuchets

Al-Tarsusi, an Egyptian who may have been of Armenian heritage, provides the most detailed descriptions of Muslim trebuchets during the period of the crusades. Around 1180, he presented a book to Saladin that included a number of known and proposed weapon designs, including five types of trebuchets. The simplest was the *lu'ab*, a pole trebuchet that could be easily aimed in different directions and used by a single opera-

8 For the relationship between these manuscripts, see Folda, *Crusader Manuscript Illumination*, pp. 27-36.

9 Tarver, "The Traction Trebuchet," p. 156.

10 See Chapter 2.

tor, perhaps similar to those in **nos. 13** and **27**. This model was so common, according to al-Tarsusi, that it did not warrant illustration. **Nos. 29, 30** and **31**, which all have trestle frames, he labels the 'Frankish', 'Turkish' and 'Arab' trebuchets respectively.

The beam of the Frankish model is shown with a ratio of about 6.5:1. At the end of the short arm, a foot, about 2 cubits long, is supported by two braces of a similar length. These join each end of the foot to a point ⅓ along the length of the beam and are fixed to the axle for greater stability. The beam is similar to those in nos. **17** and **18**, although the point at which the supports join the beam is more akin to those in nos. **15, 16, 19** and **20-22**. At the end of the long arm there is a hook, like those in nos. **3-11, 13, 25** and **26**, which has been skewed to one side in the illustration to distinguish it. The sturdy trestle structure of this engine, similar to those in nos. **21** and **24**, is shown in profile and al-Tarsusi recommends that the height of the axle be ⅗ the length of the beam.

The Turkish model, **no. 30**, also has a triangular structure in profile; however, the axle is supported independently by the larger of the two struts on each side. It is recommended that the axle should be ¾ as high as the beam is long. This is the lightest of the engines with a trestle frame and al-Tarsusi praises it for its simplicity.

The Arab model, no. **31**, has a beam with a ratio of 6:1 and a length 4/3 the height of the axle. The iron hook, to which the sling would be attached, is shown some distance from the tip of the long arm. This seems to be an artistic error, as having a length of beam extend beyond the hook risks it interfering with the sling. The engine's framework is similar to that of the Frankish model, but it was provided with an additional wooden cover to protect the pulling crew, indicating that the engine's effective range was less than that of the opposing defenders.[11] A sling 1 cubit (0.54 m) long appears to be recommended for each type of engine.[12]

Al-Tarsusi also gives more general advice regarding the working of traction trebuchets:

> If [the loader] stands directly under the pouch [of the sling], the stone will be very high and [the range] will be short, and it may possibly fall on the [pulling crew]. If [the loader] moves out from the pouch toward the end of the beam by a distance of one span [of a hand, 22-24 cm], the launch will be farther. The most one should move out from the beam is two spans [44-48 cm], [and] no more, for, if one goes beyond this, the launch will be short. The longest distance which the stone can reach is 60 *ba'* [c. 120 m], and the shortest is 40 *ba'* [c. 80 m]. Another

11 Similar protective screens appear to have been used at the siege of Thessalonica in the sixth century, see *Miracles of St Demetrius*, ed. Lemerle, p. 154.

12 For the Arabic of al-Tarsusi's discussion of traction trebuchets, see al-Tarsusi, *Tabsirat Arbab al-Albab*, ed. Sader, pp. 165-70; Cahen, "Un traité d'armurerie," pp. 118-20. For English translations, see Lewis, *Islam*, 1:218-21; Chevedden et al., "The Traction Trebuchet," pp. 460-61.

> principle which determines the farness or the shortness of the distance [of the shot] is the flexibility or dryness of the beam. When the beam is flexible, but not excessively so, it has a farther range and is more effective. When it is dry, it is less so. The shooter should have his feet wide apart, grasp the pouch with his hands, and sit down while he pulls the pouch each time. The best and most suitable wood to make the beam is cherry wood. If there is none of this kind, it must be of a closely-knotted wood of intermediate [quality] such as cedar or the like.[13]

This description provides important details to complement the illustrations, notably the position of the loader and a sense of range. These factors appear to support the power models that have been proposed above – a velocity of at least 34.5 m/s, negating air resistance, is necessary to yield a range of 120 m. This suggests that the engines found in certain images, such as **nos. 1, 3, 12, 19** and **20**, are relatively proportionate to the men shown operating them.

The suggestion that cherry wood was optimal for the beam is sensible but intriguing. Cherry trees rarely have straight branches or even trunks. Writing in Egypt, al-Tarsusi probably implied that the cherry, or even cedar, wood would be imported, possibly from Galilee, Lebanon, western Syria, Anatolia or Cyprus, otherwise it would need to be scavenged on campaign. If he meant the former, most of these areas were under Frankish or Greek control when he was writing, speaking to the ongoing trade in materials of war. In the early fourteenth century, al-'Umari notes that pine, which would have been an appropriate substitute, was imported to Egypt by sea from Anatolia.[14]

Traction trebuchets are also featured in Hasan al-Rammah's *Furusiya* (**nos. 32** and **33**). Although composed a century after al-Tarsusi's work, the engines seem to be about the same size, each with what appear to be four pulling ropes hanging from the end of the short arm. Both engines are shown loaded with a pot of incendiaries. Projectiles such as these may have been used at sieges such as Jerusalem (1099) and Acre (1189-91).

Al-Tarsusi's Counterweight Trebuchet

Al-Tarsusi's most famous illustration is that of the 'Persian' trebuchet (**no. 34**), the earliest known image of a counterweight trebuchet.[15] All of the fundamental components

13 Adapted from the partial translation by Paul Chevedden in Tarver, "The Traction Trebuchet," pp. 148-49. For other translations, see Lewis, *Islam*, 1:218-19; Chevedden et al., "The Traction Trebuchet," p. 460. See the above note for editions of the Arabic.

14 Al-'Umari, *Masalik al-absar*, trans. Richards, p. 17.

15 For translations of the description, see Appendix 5. For the Arabic, see Cahen, "Un traité d'armurerie," pp. 119-20.

of the engine are shown, including the beam, counterweight, sling, and even a winch and a trigger. The advantage of this design, according to al-Tarsusi, is that it could produce as much power as a traction trebuchet but required only one person to operate it. He states that the power of the engine is 50 *artal*; however, it is unclear which *ratl* is meant.[16] If the Damascene measure is intended (1.85 kg), this may refer to the mass of the counterweight (92.5 kg), about the weight of a solidly built man. If it was instead the Egyptian measure (0.44 kg), as might be assumed given al-Tarsusi was based in Egypt, it almost certainly refers to the mass of the projectile, since a winch or pulley system would not be needed to lift 22 kg.

Although al-Tarsusi states that an engine of this type had been built before, the immaturity of the design is clear. Its scale and basic components are comparable to those of a traction trebuchet, if the crossbow is used as a reference. The beam is fixed at a ratio of 3.5:1, mechanically sensible for an engine of this size, and a small counterweight, suspended in netting, is shown to be connected to the end of the short arm by three ropes.

The machine's principal design flaw is its low axle, so low that it required a hole be dug into which the counterweight could fall. This would have prevented the beam from rotating more than 90° before the projectile was released. The hole would need to be deep enough to allow the counterweight to fall as long as possible before the projectile was released but shallow enough so that the counterweight would hit the bottom before it pulled against the axle with all of the momentum it had built up. Thus, the counterweight would need to reach the bottom of the hole just as, or slightly before, the beam came into line with it, or as al-Tarsusi less precisely states, the ropes connecting the counterweight to the beam should be the same length as the distance between the axle and the bottom of the hole.[17] This design flaw was subsequently addressed by simply raising the height of the axle. This allowed the beam to be drawn back to a greater angle and the counterweight to swing freely above the ground.

Another issue with al-Tarsusi's design is the arrangement of the trigger, which connects the back of the sling pouch to a point directly below the tip of the long arm. This causes the sling to start in a vertical position at the beginning of each shooting sequence, compromising the engine's potential release velocity. If the sling is instead arranged horizontally, extending back below the beam, the projectile must travel farther, and thus faster, before release. The greater distance that the projectile must travel

16 Bradbury's simplistic interpretation of 50 lbs is inexact at best, Bradbury, *The Medieval Siege*, p. 255. This is the same issue confronting efforts to determine the power of Tughrul's engine at Manzikert in 1054-55, see Chapter 1.

17 If the ropes attaching the counterweight to the beam were the same length as the distance between the axle and the bottom of the hole, the counterweight would always rest on the bottom of the hole.

and resulting higher rate of acceleration, also allow for the use of a longer sling, which further increases the projectile's distance of travel, rate of acceleration and ultimate release velocity. The use of a short sling, arranged vertically, and a horizontally cocked beam, probably reflect al-Tarsusi's better understanding of traction trebuchets, which make use of both. Alternatively, if this is an accurate description and illustration of a historical technology, it is essentially a traction trebuchet with a dead load as the power source. The trigger of this counterweight engine has simply replaced the loader who would otherwise hold the sling pouch of a traction trebuchet in exactly the same way before it was pulled out of his hands.

The most mysterious aspect of al-Tarsusi's design is the incorporation of the crossbow. The text seems to indicate that the tension of the crossbow was used to provide extra downward force when the engine was shot;[18] however, the amount of force to be gained in this way would have been almost negligible. It is also unclear why the crossbow should be loaded before firing, as is stated in the text, if it is to be shot downwards. The crossbow is also linked with the trigger somehow, which is clearly labelled at the opposite end of the engine. In his translation of the accompanying text, Chevedden portrays the crossbow as independent of the trebuchet's shooting sequence: after using the winch to draw back the beam, which also draws back the string of the crossbow, the operator could then shoot the crossbow before shooting the engine. What appears to be a more accurate translation has been provided by Shihab al-Sarraf.[19]

It seems likely that the crossbow is far less significant than some have suggested. A considerable portion of al-Tarsusi's work is dedicated to archery and most of the innovative weapons that he presents are traditional weapons that incorporate a crossbow in their design.[20] Accordingly, the presence of the crossbow is possibly an extension of this exercise: it has been added to an engine that would not otherwise require it. Although al-Tarsusi explicitly states that the engine he describes represents an actual historical weapon, he makes similar claims to this effect when discussing other unlikely weapons.

Despite the shortfalls of al-Tarsusi's design, the drawing remains the earliest known illustration of a counterweight trebuchet. It is noteworthy that al-Tarsusi emphasises

18 Fino has supported this apparently literal interpretation, Fino, "Machines de jet médiévales," p. 34.

19 Chevedden, "Invention," pp. 115-16; Shihab al-Sarraf in Nicolle, "Early Trebuchet," pp. 275-76. Bernard Lewis has provided a third English translation, but it is based on Claude Cahen's edition and translation, which mistakenly reads the text, which is written across two pages, one page at a time, Lewis, *Islam*, 1:221-22; Claude Cahen, "Un traité d'armurerie," pp. 119-20, 142-43.

20 For the archery components, see al-Tarsusi, ed. and trans. Boudot-Lamotte, *Contribution à l'étude de l'archerie musulmane*. For some of the innovative designs, see also Cahen, "Un traité d'armurerie," pp. 103-63.

the accuracy of this engine and its ability to be operated by a single man rather than its power. This appears to confirm that knowledge of counterweight trebuchets had spread through the Near East by about 1180 but that these engines may have been only marginally more powerful than contemporary traction trebuchets at this time.

Villard of Honnecourt, Egidio Colonna and Marino Sanudo

Determining the earliest European illustration of a counterweight trebuchet is difficult. The vertical beam of the central engine in **no. 35** might suggest that this, and perhaps its neighbours, were counterweight engines; however, no counterweights are shown and the structure of these engines is the same as the traction trebuchet in **no. 22**, which is found on the previous folio of the same manuscript. A drawing by Villard of Honnecourt, with an accompanying description, appears to be the first conclusive drawing of a European counterweight trebuchet, even if the most essential components are missing.

Villard of Honnecourt

Among the various drawings in Villard of Honnecourt's 'sketch book' is the footprint of a counterweight trebuchet (**no. 36**); unfortunately, the accompanying elevation has not survived. The description reads:

> If you want to make the strong engine that is called a trebuchet, pay close attention here. Here is the base as it rests on the ground. Here in front are the two windlasses and the double rope with which one draws back the shaft [beam] as you can see on the other page. There is a great weight to pull back, for the counter-poise is very heavy, being a hopper full of earth which is two large *toizes* [roughly 'two fathoms' or about 12 feet] long and nine feet across and twelve feet deep. Remember the arc of the arrow when discharged and take great care, because it must be placed against the stanchion in front.[21]

21 *Se ù voles faire le fort engieng con apiele trebucet prendes ci gard. Ves ent ci les soles si com il siet sor tierre. Ves la devant les .II. windas et le corde ploie a coi on ravale le verge, veir le poes en cele autre pagene. Il i a grant fais al ravaler, car li cõtrepoise est mlt pezans. Car il i a une huge plainne d'tierre. Ki .II. grans toizes a d'lonc et .VIIII. pies de le, et .VII. pies de pfont. Et al descocier de le fleke penses, et si ù en dones en dones gard. Car il le doit etre atenue a cel estancon la devant.* Translation adapted from those of Lassus, *Album*, p. 203 and Contamine, *War in the Middle Ages*, pp. 103-4.

The size of the counterweight box has attracted considerable attention. This is the best indication of the engine's size, but the lack of standardised measurements, the absence of a clear sense of scale and missing illustration have left this open to interpretation. Chevedden has argued that the counterweight box would have had a volume of about 18 m^3, allowing it to hold up to 30 tons. This, according to Chevedden, would allow it to throw a 100 kg projectile more than 400 m or one 250 kg more than 160 m; if the mass of the counterweight was halved, it would still be sufficient to hurl a 100 kg projectile over 200 m or one 60 kg over 350 m.[22] Similarly, Fino believed that the counterweight box might have been able to hold 26,000 kg of earth – suggesting that this might correspond with a beam 10-12 m long, with a ratio of 3:1. With this in mind, he calculated that two tread-wheels, about 4 m in diameter, with accompanying pulleys would be sufficient to lift the massive counterweight.[23] Using the modern 12 inch foot, Bradbury has interpreted the capacity of the counterweight box to have been an enormous 1,296 ft^3 (36.7 m^3).[24] Whereas Chevedden and Fino appear to have assumed that the counterweight box was a right triangular prism, Bradbury's calculations reveal he believed it instead to be a right rectangular prism, or cuboid.

An engine capable of both producing and withstanding the forces involved with a counterweight of over 25,000 kg would have been huge. I know of no modern reconstructed engine with a counterweight even half this size. Considering the troubles encountered by Louis-Napoleon and Favé with their engine in the nineteenth-century, the counterweight of which was only 20% as heavy, it seems unlikely that such a large engine was ever built and practically employed during the thirteenth century.

The date of Honnecourt's work is debated. Chevedden places its production between 1220 and 1240, Bradbury and France date it to around 1270, and Fino attributes it more generally to the mid-thirteenth century.[25] Although the accompanying sketch of the elevation has been lost, the base plan has been the inspiration for Viollet-le-Duc's nineteenth-century interpretations[26] and many modern reconstructions.

Egidio Colonna

Egidio Colonna (Giles of Rome) composed his *De regimine principum* for Philip IV of France around 1280. Within, he describes four types of trebuchets: traction models, which are praised for their shooting rate but noted as being incapable of throwing

---

22 Chevedden, "Invention," p. 72, see also n. 3.

23 Fino, "Machines de jet médiévales," p. 34.

24 Bradbury, *The Medieval Siege*, p. 263.

25 Chevedden, "Invention," p. 72; Bradbury, *The Medieval Siege*, p. 263; France, *Western Warfare*, p. 121; Fino, "Machines de jet," pp. 33-34.

26 Viollet-le-Duc, *Dictionnaire*, 5:227-28, figs. 9, 10.

heavy stones; the *trabutium*, which had a fixed counterweight and is celebrated for its accuracy; the *biffa*, which had a mobile counterweight and exceptional range; and the *tripantum*, which had both a hinged and a fixed counterweight, allowing it to be more accurate than the *biffa* with a range greater than the *trabutium*.[27] Although Egidio Colonna correctly observed that a trebuchet with a hinged counterweight would have had a greater range than a similar engine with a fixed counterweight, neither would have enjoyed an advantage in accuracy.

The description of the *tripantum*'s movable counterweight has left it open to interpretation.[28] Bradbury, possibly influenced by Oman, has suggested that this mass could be moved along the length of the short arm, so as to adjust the engine's range.[29] Hill judged this instead to describe a traditional hinged counterweight that could be rotated (essentially propped) to start above the end of the short arm, possibly inspired by images such as **nos.** 77, 79 and **91**.[30] Hill acknowledged that his interpretation of a rotating counterweight would be impractical, due to the additional stresses that such an arrangement would introduce. Cathcart King simply suggested that Egidio's descriptions were fanciful.[31] It is possible, however, that Egidio overcomplicated his description of what was otherwise an engine that had both a weighted short arm (a fixed counterweight) and a hanging counterweight. Like that of any other trebuchet with a hanging counterweight, the hinged component would rotate the better part of 180° relative to the axis of the beam through the course of a natural shooting sequence. Counterweight trebuchets with a weighted short arm and hinged counterweight can be seen in **nos. 40, 41, 73, 74, 95** and **102**. Marino Sanudo's early fourteenth-century description of contemporary artillery provides another perspective.

Marino Sanudo

In his suggestions for how the Holy Land might be retaken for Latin Christendom, Marino Sanudo provides a fairly detailed description of artillery, as he understood it. He distinguishes between two types of engines, the common variant and a long-range model. He describes the former as the following:

> The hips of a machine ought to be as broad on the ground as the height of the machine at its tip and underneath the said machine ought to be open on the

---

27 Egidio Colonna, *De regimine principum* 3.3, ed. Samaritanium, pp. 604-6, ed. Hahn (as chap. 18), pp. 50-52. For a French translation of this section, see Bonaparte, 2:29-30.

28 *Aliud genus machinarum habet contrapondus mobiliter adhaerens circa flagellum, vel circa virgam ipsius machinae, vertens se circa huiusmondi virgam*. Egidio Colonna, *De regimine principum* 3.3, ed. Samaritanium, p. 605.

29 Bradbury, *The Medieval Siege*, p. 267; Oman, *A History of the Art of War*, pp. 543-44.

30 Hill, "Trebuchets," p. 105.

31 King, "The Trebuchet," pp. 463-64.

> ground, between 2 hips minus a third part [of its height]: that is, if the aforesaid machine is 24 feet at its highest point, it ought to be 16 feet on the ground. The masters divide the main pole [beam] of the machine from the weighted bottom to the tip into 5 or 6 parts and place the sow [axle] between the fifth and the sixth, so that if the pole is 30 feet long from top to bottom, that is 5 or 6 feet, and the fifth is 6 feet, and the sixth above-mentioned as is seen will be 5 feet. Concerning which thing, the cross-bar [axle] of the catapult ought to be placed five and a half feet from the cross-bar of the container mentioned above.[32]

What he appears to be describing is a counterweight trebuchet with a beam, at a ratio of either 5:1 or 4:1, mounted on a trestle frame.[33]

The long-range trebuchet is similar: it also has a beam with a ratio of 5:1 and hypothetical length of 30 feet. However, a container, attached to the end of the short arm, is also clearly described. It is possible that the long-range variety, which had only a hinged counterweight, was distinguished from the 'common' type that made use of either a fixed counterweight or both fixed and hinged components. The 'common' engine might thus correspond with Egidio Colonna's *trabutium* or *tripantum* and the long-range type with the *biffa*.

Although Sanudo recommends that the container of the long-range model be as big as possible, so as to allow the engine to throw projectiles as large and as far as possible, he asserts that the scale of the machine determined the appropriate mass of the counterweight, which in turn determined the size of the projectiles. Revealing his familiarity with these machines, he adds that the range of an engine could be altered by adjusting the iron 'finger' at the end of the long arm.[34] This detail suggests that he, or someone close to him, understood how a counterweight trebuchet operated in practice.

Although Sanudo gives the impression that these engines were quite large, he claims that they were deployed on ships and even seems to imply that their use on land was secondary. This view was probably influenced by the Venetian naval tradition that

---

32 Marino Sanudo, *Liber secretorum* 2.4.22, ed. Bongars, pp. 79-80, trans. Lock, p. 135.

33 If the height of the axle is meant to correspond with the height of the frame (24 feet in this example), it is impractically high relative to the length of the beam. With a 4:1 beam, the length of the long arm would be the same as the height of the axle; with a ratio of 5:1, if the engine was loaded with the tip of the long arm touching the ground, the beam would be at an angle of almost 164°. The added strain incurred as the counterweight was pulled outward, as the short arm rotated downward, and then inwards, as it continued to rotate, would make this engine relatively unstable. Few contemporary illustrations show trebuchets loaded beyond about 140° and I know of no sizable reconstructed engines that are loaded to an angle greater than this.

34 Marino Sanudo, *Liber secretorum* 2.4.22, ed. Bongars, p. 80, trans. Lock, pp. 135-36.

would have been prevalent around him, but also seems to confirm Abu'l-Fida's contemporary claim that an engine mounted on a ship assailed his forces at the siege of Acre in 1291.[35] Sanudo states that the counterweight would hang over the side of the vessel when the engine was loaded and would then swing down through a channel across the middle of the boat as it fell through a shooting sequence.[36] The scale of such engines would necessarily have been limited, due to the shortage of available space on board these vessels and the dynamic forces that accompanied the fall and swing of a counterweight. Drawn the better part of two centuries later, **no. 90** features a ship-borne trebuchet used to throw incendiaries. Marino Sanudo admits that he was not a military man but his descriptions, along with those of Egidio Colonna, clearly convey that counterweight trebuchets were far from novel at the end of the thirteenth century, something confirmed by the illustrative evidence from this period.

Frankish Counterweight Trebuchets

Returning to illustrative sources, simple counterweight trebuchets are found in **nos. 37** and **38**. The frames of these engines are similar to those in **nos. 16** and **18-20**. Each has a beam with a ratio of about 4:1, a counterweight that resembles a barrel and a short sling, which is drawing the projectile out from under the engine.[37] Although the framework and sling of each machine appears more suited to a traction engine, the basic mechanics of a counterweight trebuchet are obviously known to the illustrators of these images. Sturdier trestle frames and more familiar triangular counterweight boxes are found in **nos. 39-42**. The slings of these engines are more proportionate, although the trebuchet in **no. 42** has no short arm. The beams in **nos. 39** and **42** appear to be composed of multiple spars and all three engines have a trough along which their projectiles would slide before being lifted into the air.

The framework of the engine in **no. 42**, which was drawn in Acre only months before the city fell in 1291, is very similar to that of al-Tarsusi's 'Frankish' engine (**no. 29**), while that of the engine in **no. 40** is very similar to some of those found in editions of Konrad Kyeser's *Bellifortis*, composed well over a century later (see **nos. 76** and **77**). **No. 41** contains only a sketch of a counterweight trebuchet, left unilluminated; however, it is particularly precise and features a long arm composed of two components lashed together, possibly indicating that it was meant to be adjustable. The engine, drawn in

35 Abu'l-Fida', *al-Mukhtasar*, trans. Holt, pp. 16-17

36 Marino Sanudo, *Liber secretorum* 2.4.22, ed. Bongars, p. 80, trans. Lock, p. 136, cf. 2.4.6, ed. Bongars, pp. 57-58, trans. Lock, p. 104.

37 A similar reconstructed trebuchet, with a barrel as a counterweight, can be found at King John's castle, Limerick.

Toulouse, has a framework very similar to those found in **no. 57**, illustrated not far away in Castile. The engine in **no. 43**, which is far less realistic, has a scale similar to those of **nos. 39** and **41**, while the trigger man's use of a large mallet to shoot the engine is quite similar to contemporary **no. 107**, drawn more than 3,000 km away in Tabriz.

Although one of the reconstructed trebuchets at Castelnaud is similar in scale to the engines in **nos. 39**, **41** and **43**,[38] the engine on the Courtrai Chest (**no. 44**), seen peaking up from behind a crenelated parapet, would appear to represent more accurately the scale of the largest counterweight trebuchets at the end of the thirteenth century. The vast majority of illustrations featuring trebuchets, however, depict much smaller engines relative to the people around them. Although this might suggest that most counterweight trebuchets had axles no more than about 3 m high in the thirteenth century, it seems far more likely that they are simply reduced in scale in order to fit them into various scenes that include many other elements. One example is the engine found on Edward II's charter to Carlisle (**no. 45**). The engine is no taller than the Scott behind it, impaled by a spear thrown from the city wall. Like **nos. 43** and **107**, the trigger man is shown using a hammer to knock loose a pin in order to shoot the engine. The document dates to the second decade of the fourteenth century, postdating not only the use of significant engines against Acre (1291) but also some Scottish castles, such as Stirling (1304).[39]

Extremely light trebuchets are found in **nos. 46-48**. The Parisian artists of these machines appear to have been familiar with the basic concept of a counterweight trebuchet but neither the broader mechanics nor typical framework. The single supporting strut of each engine leaves it unclear if these are meant to be pole trebuchets or if a second strut is hidden from view due to the profile perspective. **Nos. 49** and **50**, from a contemporary Parisian manuscript, reveal an even poorer understanding of trebuchet technology. They too are counterweight trebuchets with a single support, although the axle consists of a fork and the ropes, from which solid counterweights hang, are so long that the counterweight would hit the ground before the beam could rotate anywhere near a desired release angle. Further issues with proportions reinforce the notion of how few monastic illustrators would have actually seen one of these engines.

Similar issues can be found with the engine in **no. 51** and even more problems can be seen in **nos. 52**, **53** and **54**. The engine in **no. 52** does not appear to have a counterweight and the diameter of the immense projectile in its sling is more than half the height of the axle. The apparent 'hybrid' trebuchet in **no. 53** should not be considered

38 For photos of this reconstructed engine, see Norris, *Siege Warfare*, p. 200, figs. 104, 105.

39 For the siege of Stirling, see Peter of Langtoft, *Chronicle*, ed. and trans. Wright, pp. 356-59; Freeman, "Wall-Breakers and River-Bridgers," pp. 13-15. Cf. Bain, ed., *Calendar of Documents Relating to Scotland*, nos. 1,491, 1,498-500, 1,504, 1,510, 1,519, 1,539, 1,560, 1,599, pp. 387, 388-89, 390, 395, 400, 405, 419-20.

any more historically accurate than the engines in **no. 54**. One of these latter engines uses a rooted tree as its supporting frame and the ropes of the sling and counterweight of the other, which has a similar pole frame and forked axle but is found on board a ship (possibly suggesting it is the mast), are simply looped over either end of the beam.

Although images of trebuchets with fixed counterweights are less common than those with hinged or hanging ones, they too appear more frequently from the late thirteenth century. One of the earliest surviving European illustrations of a fixed counterweight trebuchet (**no. 55**) is found in an edition of William of Tyre's history that was composed around 1270. The illumination appears to show two pole-framed trebuchets throwing heads towards the defenders of Nicaea. The beams have a ratio of about 1:1 and at the end of the short arms there is a bulge, large enough to suggest that these might be counterweights but small enough to leave it open to possibility that the engines were meant to be traction trebuchets. Whatever the illustrator's intention, his lack of understanding is clear. By contrast, the *tripochen* in **no. 56** reveals greater appreciation of the mechanics. Although it is drawn with a more primitive sense of perspective, it has a framework similar to that in **no. 42**.[40]

More accurate illustrations of fixed counterweight trebuchets are found in **no. 57**: two miniatures show an engine in the process of being set up and then complete and loaded. The engine has a sturdy trestle frame and a beam with a 2:1 ratio. Part of a traction pole trebuchet can be seen in the background of one of the miniatures – a rare example of an image that includes both engine types. Far less accuracy is apparent in **no. 58**. Unlike the reasonable beam ratio and sturdy frame seen in **no. 57**, the engine has an unrealistic beam ratio of 1:2 and a frail frame similar to those in **nos. 37**, **38** and **45**. In other examples, illustrative issues go beyond the structural strength of the framework and impractical beam ratios.

Illustrative Confusion

The engines in **nos. 59** and **60** appear to have beams with no short arm section: in **no. 59** the beam ends at the axle while in **no. 60** the end of the beam appears to rest on top of the axle. Perplexingly, these engines were drawn by the same hand(s) as those in **nos. 25** and **26**. It is unclear what the mysterious circle shown ahead of the engine in **no. 59** is supposed to be: it may be a winch, although against what this would have drawn the long arm back is unclear; it may be a strange interpretation of a counterweight; or it may be some kind of a defensive screen to protect the engine. Given its numerous issues, it is even possible that the illustrator combined what he had read

40 Cf. Marino Sanudo's description of a *tripantum* above and the *tribok* and *tribracho* noted respectfully in *Cronica Reinhardsbrunnensis*, ed. Holder-Egger, p. 58, and *Chronica S Petri Erfordensis*, ed. Holder-Egger, p. 383. See also Chapter 1.

about classical torsion engines with a basic understanding of the trestle framework of a trebuchet. **No. 60** appears to postdate **no. 59** and the illustrator has not included the mysterious circle. Due to the current state of the image, it is hard to tell whether the trebuchet in **no. 60** was meant to have a counterweight hanging below the axle similar to that in **no. 42**, which was also completed in Acre but about ten years later. A different set of peculiar engines are found in a slightly later manuscript of William of Tyre's chronicle.

No. 61 seems to show a traction trebuchet, similar to those in **nos. 17** and **18**, with a beam like that in **no. 29**; however, the ropes of the pullers are not connected to the foot at the end of the short arm but instead appear to be restraining the long arm. **No. 62**, found in the same manuscript, contains a very similar engine, but the team of pullers, again appearing to restrain the long arm, are positioned under the main axle and what seems to be a small weight has been hung from one end of the foot at the end of the short arm. This weight might be interpreted as balancing the beam if it were not for the obvious efforts of the three men to pull down the long arm. The illustrator would appear to have drawn inspiration from images he had seen of traction trebuchets and mixed in what he understood or had heard about counterweight trebuchets.

Like **nos. 61** and **62**, **nos. 63-65** were completed around 1300. The images contain engines similar to that in **no. 45** but each has a cup at the end of its long arm rather than a sling. In theory, this would have been the best mechanical arrangement to throw an extremely heavy projectile, relative to the mass of the counterweight, a short distance. As noted in Chapter 2, it is very unlikely that engines without slings were ever used; accordingly, these images seem to be further examples of the ignorance of most illustrators.

Defensive Counterweight Trebuchets

All of the illustrations of counterweight trebuchets introduced thus far have depicted engines in an offensive capacity, with the possible exception of **no. 43**. If the trebuchet in **no. 43** is meant to be within a fortified enceinte, rather than behind one, it is clearly positioned on the ground rather than on top of a tower or section of wall. The engine in **no. 66**, of which only the long arm of the beam and sling are visible above the battlements, is similarly positioned. Both of these early fourteenth-century images reveal engines that are much larger than most of those found in manuscript illuminations. The engines appear to be portrayed in rough proportion to the architecture around them and clearly reveal the issues associated with attempting to mount a counterweight trebuchet on top of a tower.[41] If anything, the impractically small trebuchet in

41 For the positioning of defensive engines, see Chapter 9.

no. 67 supports this. This image depicts the Roman siege of Jerusalem and the counterweight trebuchet, which is either positioned on top of the walls or behind them, represents the classical engines used in defence.[42] The engine is so small, relative to the people around it, that the man who is being thrown is depicted smaller than the rest in order to fit him in the sling. A very large defensive counterweight trebuchet is shown in **no. 68**, which dates to the early fifteenth century. The machine appears to have been designed, or is at least presented, so that the struts of the engine abut the curtain, allowing the counterweight to hang over the wall when the beam is cocked.

Although there are relatively few images of counterweight trebuchets in defensive roles, the illustrative evidence supports the notion that these engines were deployed on the ground. There are numerous images of traction trebuchets on the tops of towers; however, there are few indications in the illustrative evidence that counterweight trebuchets were placed in such positions.

Late Medieval and Renaissance Illustrations

European illustrations of trebuchets become more accurate, in general, from the second half of the fourteenth century. The engine in **no. 69**, which slightly predates this point, has a number of mechanical issues but nevertheless seems to show the engine far closer to scale with the operator/observer than most previous images. The number of mechanical and structural issues decrease with time as knowledge of these engines spread beyond the military elements of society and permeated deeper into artistic circles. **Nos. 70-73** come from various parts of Europe and although each has certain simplifications and slight inaccuracies, such as the chronically undersized counterweights, the trestle framing and general proportions of the various components reveal that their creators had a greater degree of understanding than most earlier illustrators. Although each is relatively accurately drawn, they, like the strongholds also depicted, are shown out of scale relative to the people in each image.

The scale of the engine in **no. 74**, like the engine itself, appears to have been drawn quite accurately. Dating to the early fifteenth century, it represents the popular element of a new artistic movement. Although examples of fairly inaccurate or simplistic engines can still be found, such as that in **no. 75**, images dating to the fifteenth century tend to be far more precise, proportional and detailed than earlier illustrations. In no small part, this was because most of the responsible artists were secular figures with interests in the applied sciences rather than monks.

As the significance of trebuchets was ebbing, broader social changes, greater communication and advances in mechanical science led to increasingly accurate images in

42 See Josephus, *The Jewish Wars* 5.6, trans. Whiston, pp. 810-11.

the fifteenth century. **Nos. 76-79**, from various copies of Konrad Kyeser's *Bellifortis*, epitomise this. A number of measurements are found on the engine in **no. 76**, including 46 and 8 for the long and short arms of the beam respectively. Chevedden has interpreted these measurements to be in 'workfeet', each 0.288 m, giving the beam a total length of about 15.5 m and a ratio of 5.75:1.[43] Despite these schematic details, the illustration has a number of significant errors: the trough has a lip, as in **no. 71**, and both ends of the sling appear to be fixed to the beam, which is not provided with a finger.

A major design flaw that runs through Kyeser's plans is the use of a cuboid counterweight box, similar to those in **nos. 70**, **71** and **74**. These require that the struts of an engine be set widely apart in order to allow the box to swing between them. This would be exacerbated by the need to minimise strain on the axle by limiting the space between the beam and housing on each side. The steep angling of the struts is evident in Kyeser's measurement of 23 (6.6 m according to Chevedden) between them at ground level. Steeply angled struts are subject to torque during a shooting sequence, placing considerable strain on the framework. By using a narrower counterweight, allowing the struts to stand vertically, the framework can more easily withstand the strain as the forces are kept relatively two-dimensional.

No. 77 presents a more schematic view. The beam is two units longer, providing an even ratio of 6:1 and a finger, albeit of wood, has been added. Although the counterweight is shown in a more familiar triangular shape, the outline makes it clear that this is just a cross-section of the same cuboid design shown in **no. 76**. The counterweight is also given a prop to raise it higher as the beam is rotated into a cocked position. Although this would prevent the counterweight from falling as straight during a shooting sequence, the added height would provide greater power as long as the projectile was released before the beam reached about 30° to the vertical, at which point the counterweight would retard the rotation of the beam given the proportions of the short arm and counterweight.[44] The degree of detail found in this image, such as the composite beam, which is pinned and lashed together, stands in contrast to those in other editions of *Bellifortis*, such as **nos. 78** and **79**.

Despite the apparent accuracy shown **nos. 76** and 77, *Bellifortis*, like al-Tarsusi's treatise, contains a series of novel and impractical weapons. Although the counterweight trebuchet was certainly a familiar siege engine across Europe by the fourteenth century, Kyeser may have exaggerated the scale of his engines and added certain elements that he believed might improve an already familiar design. A lifting machine, similar to

43 Chevedden, "Invention," fig. 3.

44 For a mathematical representation of a propped counterweight, see Siano, *Trebuchet Mechanics*, pp. 32-34.

Vegetius's *tollenno*,[45] appears in **no. 81**, provided as a means to cross a river. This engine has a beam almost identical to that of a trebuchet, as highlighted in **no. 80**. This theme of experimentation can also be seen in **no. 84**, where a counterweight trebuchet, similar to those in **nos. 82**, **83** and **85**, is shown among other methods of delivering incendiaries.

The broader interest of Renaissance figures, such as Mariano di Jacopo Taccola, Roberto Valturio and Francesco di Giorgio, in practical and imagined trebuchet designs can be seen in **nos. 82-101**. Although accuracy continued to increase into the late fifteenth century, some depictions still include simple design flaws, such as the insufficiently high axle in Kolderer's famous woodcut (**no. 102**).[46]

The engines in **nos. 91-94**, from Roberto Valturio's 1463 edition of *De re militari*, have formed the essential basis of our modern conceptions of the various types of counterweight trebuchets. There was little development technologically beyond this point as the use of trebuchets declined through the fifteenth century, eclipsed by gunpowder artillery. Despite the waning importance of these weapons, even Leonardo da Vinci experimented with new designs, as seen in **nos. 100** and **101**, presumably drawn during his period of service with Ludovico Maria Sforza (1483-99) or Cesare Borgia (1502-3). But while there is relatively plentiful illustrative evidence to trace the ways in which counterweight trebuchets were drawn in Western Europe from the thirteenth century, comparatively few Muslim images have come to light.

Muslim Counterweight Trebuchets

Following al-Tarsusi's illustration of a counterweight trebuchet in the late twelfth century, evidence from Persia confirms that trebuchet technology in the Islamic world had been significantly refined by the end of the thirteenth century. **Nos. 103** and **104** clearly show counterweight trebuchets with trestle frames and beams with ratios of 10:1 and 4:1 respectively. The counterweight of each engine appears to be a solid mass hung with ropes from the end of the short arm. Like the traction trebuchets in **nos. 32** and **33**, these engines appear to be loaded with incendiaries rather than stone projectiles.

More familiar container counterweights can be found in **nos. 105-107**, which date to the early fourteenth century. The detailed engine in **no. 105** appears to have a slightly heavier framework than the other two, but all appear to be drawn extremely accurately. Whether intended or not, the large beam ratios, 10:1 and 7:1, and long slings of

45 Vegetius, *Epitoma Rei Militaris* 4.21, ed. Lang, pp. 142-43. See also Roberto Valturio, *De Re Militari* (Paris: BNF, MS lat. 7236), fol. 133r; Francesco di Giorgio, *Trattato* (Turin, Biblioteca Reale, Cod. 148 Saluzzo), fol. 60r.

46 For Cathcart King's interpretation of this engine, see King, "The Trebuchet," p. 467.

the engines in **nos. 106** and **107** would have made them particularly well suited to throwing light projectiles considerable distances. By employing a small short arm, a significant counterweight is necessary, as the distance that it can fall is limited. With the counterweight relatively close to the axle, strain is limited, allowing the use of a slender beam. A long but light beam, when fitted with an equally long sling, maximises the distance that the projectile must travel before release, and thus maximises release velocity. These proportions are ideal so long as the projectile is extremely light: the retarding force of any mass on the opposite side of the axle from the counterweight is amplified in proportion to the relative shortness of the short arm. Accordingly, there is no issue with the projectile in **no. 107** but those in **no. 106** are far too large for the engine.

Similar engines can be seen in the two-page depiction of the Mongol siege of Baghdad in **no. 108**. Of the three counterweight trebuchets, the engine in the bottom left corner seems to be heavier than the other two. The framework of this trebuchet is more like that of the engine in **no. 105**, while the other two are more like those in **nos. 106** and **107**. The heavy engine also has a winch, like those in **nos. 105-107**, while the other two were presumably light enough that their beams were drawn back by hand, sacrificing power for a higher potential shooting rate.

A similar scene (**no. 109**) is displayed in an edition of the same work composed more than a century later. The two engines are again counterweight trebuchets but appear to be lighter than any of those in **no. 108**. These trebuchets also appear to be cocked by hand as the machine on the left seems to have three men pulling the beam back and the one on the right has a single man drawing back the beam while his colleague prepares to load a projectile. The wheels at the rear of each engine's trusses might be there to help manoeuvre the machines or they are meant to be winches, which the artist seems to have misunderstood when adapting them from an earlier illustration. While these images help to confirm the widespread use of counterweight engines, a late fourteenth-century source is perhaps the most informative medieval work on trebuchets.

Yusuf ibn Aranbugha al-Zardkash's treatise on trebuchets can be seen, in some ways, as representing a middle point between al-Tarsusi's work and that of Roberto Valturio in their treatment of these engines. Although it presents a number of practical designs with a reasonable degree of accuracy, so too does it appear to include a degree of imagination. **Nos. 111** and **121** each display a traction trebuchet with a single pole support and two short arms that diverge as they extend back from the axle. The two short arms spread the pulling ropes and provide more room for the pullers, acting like a raking structure. The single vertical support allows the beam to yaw and aim at different targets without rotating the whole machine. The engines in **nos. 110** and **112** act on the same principle but make use of counterweights rather than traction power, similar to **nos. 86, 87, 89, 92, 94** and **97**. Slightly lighter traction trebuchets appear in **nos. 123**,

126 and **127**, the square axle housings of which are reminiscent of many of the twelfth- and thirteenth-century 'Mediterranean' examples addressed above.

The counterweight trebuchets in **nos. 115, 116, 118** and **120** have more traditional trestle frames, which are quite similar to those of the lighter engines in **no. 108**. Isolated components suited to these engines can also be seen in **nos. 113, 114, 117** and **119**. The hanging counterweights appear to be solid and are similar to those in **nos. 103** and **104**. The beams of all of al-Zardkash's trebuchets highlighted so far have ratios of around 4:1 or 5:1, each bound with six to eight tight rings of lashing. A straight finger can also be seen extending from the end of each long arm. Indicative of the power of these engines, many are shown with winches, similar to those in **nos. 105-109**, and possibly those shown in **nos. 103** and **104**, as well as contemporary European examples, such as those in **nos. 74, 76, 79, 92** and **95**.

The engines in **nos. 122** and **124** are more obscure. They appear to be imagined fixed-weight trebuchets designed to throw bolts, perhaps similar to the trebuchet in **no. 99**. Neither appears to have an axle and each has some kind of a platform on top of the framework – that in **no. 122** appears to be crenelated. The degree of confusion increases when considering that certain component labels refer to the use of hair, suggesting that these were supposed to be quasi-torsion engines.[47]

A number of al-Zardkash's images display engines on top of towers. **Nos. 123, 125** and **128** feature counterweight engines in such a position while **nos. 123, 126** and **127** also include traction engines. As there is no obvious sense of scale in these images, they may be more a creative exercise than an accurate portrayal of how these engines might have been deployed.[48] Although there are issues with some of the images in al-Zardkash's work, its importance is obvious; however, it should be remembered that it dates to a period more than a century after the sieges of Crac, 'Akkar and Montfort.

Revisiting the Arabic Terminology

Al-Zardkash, like al-Tarsusi two centuries earlier, clearly named some of the engines that he included in his work. This is significant when considering that his terms, unlike most of those of al-Tarsusi, appear to correspond with those used by narrative sources when classifying the engines employed at certain thirteenth-century sieges.

47 Chevedden has suggested that these images depict more than one type of engine mixed together, Chevedden, "Black Camels," pp. 237-38.

48 Chevedden calls the left engine in **no. 116** a *ziyar*, which he believes to have been a classical two-armed torsion catapult, Chevedden, "Black Camels," p. 237 n. 21. Ibn al-Athir mentions the use of such engines by al-Zahir Ghazi at Saone in 1188, Ibn al-Athir, *al-Kamil*, trans. Richards, 2:347-48.

Trebuchets of the *shaytani* ('devilish') type are found in **nos. 111** and **121**. These traction trebuchets (occasionally rendered as *sultani*) appear to match the vague qualities attributed to such engines in certain narrative accounts (Fig. 25). The lighter traction trebuchet in **no. 127**, similar to those in **nos. 123** and **126**, is identified as the *'arrada*, which also seems to fit with portrayals of these as the lightest trebuchets.

FIGURE 25 *Mentions of the* manjaniq shaytani/sultani[a]

Year	Site	Engines	Source
1218	Damietta	multiple	Ibn al-Dawadari
1248-49	Homs	12	Ibn Wasil; al-Ansari
1285	Margat	4	Ibn 'Abd al-Zahir
1291	Acre	52	al-Jazari; al-'Ayni; al-Yunini; Ibn al-Furat

a See Chevedden, *The Citadel of Damascus*, pp. 282-83.

Al-Zardkash does not use the term *manjaniq maghribi*; however, al-Rammah clearly labels the trestle-framed counterweight trebuchet in **no. 104** as such. Engines of this type are mentioned at sieges in the mid-thirteenth century and the presence of only one at a number of these suggests that they were relatively exceptional. Chevedden has suggested that a technological change accompanied the appearance of the *manjaniq maghribi* in the sources, postulating that this might correlate with the development of the hinged counterweight, supplementing previous fixed types.[49] This theory may have been inspired by a similar one put forward by Fino.[50] Although a convenient explanation, it seems unlikely that the hinged counterweight developed so late. Consider that al-Tarsusi's primitive design in the late twelfth-century made use of a hanging counterweight.[51] Thorau has suggested that '*maghribi*' in this context might be understood as European, rather than 'western Islamic' (or 'North African'), and may refer to a particular design originally developed in Western Europe.[52] How he sees this relating to the oft-mentioned *manjaniq ifranji* is unclear.

The earliest mention of *maghribi* trebuchets (appearing in a slightly corrupt form), like those of the *shaytani* type, is found in Ibn al-Dawadari's account of the siege of Damietta in 1218-19. Writing in the fourteenth century, he appears to have applied these terms retroactively, inferring it was appropriate to do so given the assortment of

49 Chevedden, "King James I," p. 317 and n. 10.

50 Fino, "Machines de jet médiévales," p. 32.

51 See **no. 34**.

52 Thorau, *The Lion of Egypt*, p. 160 n. 4 (on p. 179).

engines employed at the siege. Considering that the technology was probably developed before the terminology with which it later came to be associated,[53] he seems justified in doing this. Both terms then appear before 1250 in Ibn Wasil's contemporary account of events (Fig. 26).

FIGURE 26 *Mentions of the* manjaniq maghribi[a]

Year	Site	Engines	Source
1218	Damietta	multiple	Ibn al-Dawadari
1232-33	Shayzar	1	Ibn Wasil; Ibn al-'Adim
1236	Harran	1	*History of the Patriarchs*
1236	Edessa	1	*History of the Patriarchs*
1248-49	Homs	1 of 14	Ibn Wasil; al-Ansari
1265	Caesarea	multiple	Ibn 'Abd al-Zahir; Ibn al-Furat
1266	Safed	multiple	Ibn 'Abd al-Zahir; Ibn al-Furat

a See Chevedden, "Invention," p. 106 n. 137; Chevedden, *The Citadel of Damascus*, p. 280 and ns. 38-45 (on pp. 303-4).

Al-Zardkash calls the counterweight trebuchet with a forked short arm, seen in **nos. 110** and **112**, the Frankish trebuchet. This type of engine is comparable to what is commonly identified as the *bricola* in European contexts, featured in **nos. 86, 87, 89, 92, 94** and **97**. When the *manjaniq ifranji* is distinguished in accounts of sieges during the late thirteenth century, however, it is described as being the heaviest or most powerful type of trebuchet. This is not at all an accurate portrayal of the type of engine presented by al-Zardkash, which necessarily sacrifices power for versatility – the strength of a more complex and rigid framework is exchanged for the ability to yaw.

Accepting al-Zardkash's identification of the *manjaniq ifranji* as a trebuchet with a pole frame and searching for the early use of such an engine in Europe, given its designation as 'Frankish', Chevedden has concluded that the first discernible use of such a machine dates to the Norman siege of Thessalonica in 1185. His argument rests on the account of Eustathius, which states that the engines that besieged the city from the west were of a new form.[54] Concluding that the small and large engines of a 'traditional' type, which attacked the city from the east, were traction and counterweight trebuchets respectively, he asserts that the new engines to the west were a type of

53 For this trend, see Fulton, "Diffusion of Artillery Terminology," pp. 1-27.

54 See Eustathius, *Capture of Thessalonica*, trans. Melville Jones, pp. 92-105. Cf. Niketas Choniates, *Historia*, trans. Magoulias, pp. 164-65.

bricola. By identifying this as the earliest known reference to a *bricola*, a term which does not come into use until the thirteenth century, Chevedden is able to suggest that this type of engine was developed in Europe in the late twelfth century and can thus be equated with the *manjaniq ifranji*, which is first mentioned in the second half of the thirteenth century.[55]

Contrary to Chevedden's theory, the only description of these 'new' engines provided by Eustathius is that they were cumbersome and ineffective. It is more likely that these were early attempts to build sizable counterweight trebuchets, as it is hard to qualify the necessarily light *bricola* as cumbersome. The larger of the 'traditional' engines to the east were probably smaller counterweight trebuchets or simply sizable traction engines. The *bricola* was the counterweight variety of the pole trebuchet, a light and versatile engine but not a particularly powerful one. Accordingly, the *manjaniq ifranji*, when identified by early Mamluk historians, rather than al-Zardkash, was an impressive trestle-framed counterweight trebuchet. Whereas all of the stresses incurred during a *bricola*'s shooting sequence were placed on the single vertical support, the more substantial framework of a typical counterweight engine was far better equipped to absorb the strain associated with employing a large counterweight.

It is significant that references to the *maghribi* trebuchet decline with the appearance of the *ifranji* trebuchet. As both appear to refer to the heaviest stone-throwing engines of their day, it is possible that the two terms refer to the same, or very similar, type of engine: *maghribi* ('western') being replaced with *ifranji* ('Frankish') in the 1260s. When found together at the siege of Caesarea in 1265, there is nothing to help distinguish any differences between them. It seems there may have been a terminological shift but not necessarily a technological one; perhaps the increasing scale of these engines was judged sufficient to warrant a new term or simply 'Frankish' became more fashionable for some reason (Fig. 27).

Probably writing in the late fourteenth century, al-Zardkash may have used terms that were appropriate to the engines of his own day or, requiring two terms for two types of counterweight trebuchet, may even have arbitrarily assigned the labels *maghribi* and *ifranji* without understanding what distinguished them in the thirteenth century. It is perhaps noteworthy that Muslim sources rarely mention Frankish forces making use of a *manjaniq ifranji*, although there are few instances when Latin forces would have had the chance to employ such an engine from the point when the new vocabulary emerged.

55 Chevedden, "King James I," p. 318.

FIGURE 27 *Mentions of the* manjaniq ifranji[a]

Year	Site	Engines	Source
1265	Caesarea	multiple	Ibn 'Abd al-Zahir; Ibn al-Furat
1275	al-Bira	23 of 70	Ibn Shaddad; al-Yunini; Shafi' ibn 'Ali
1285	Margat	3 of 10	Ibn 'Abd al-Zahir
1286	Saone	4	Ibn 'Abd al-Zahir
1289	Tripoli	6 of 19	Ibn al-Dawadari; al-Yunini; Ibn al-Furat
1291	Acre	15 of 72	al-Jazari; al-'Ayni; al-Yunini; Ibn al-Furat

a See Chevedden, *The Citadel of Damascus*, pp. 281-82; Chevedden, "Invention," p. 107 n. 9.

Al-Zardkash's treatment of the *manjaniq qarabughra/qarabugha* ('black camel'/ 'black bull' trebuchet) is peculiar. While he might have associated this term with the bolt-throwing engines in **nos. 114** and **115**, he may also have used the term far more generally when discussing artillery. Despite being unable to find any contemporary evidence to suggest that the *manjaniq qarabugha* was a bolt-throwing engine in the thirteenth century, Chevedden has suggested not only that this was the case, but also that this technology was transmitted from the Muslim world to Europe.[56]

References to bolt-throwing trebuchets are rare and the physics involved render the practical use of such an engine questionable. The natural strength of the counterweight trebuchet was its ability to transfer the potential energy within a suspended counterweight to a relatively heavy projectile. Because these engines rely on the accelerating force of gravity, the rate of acceleration and final release velocity are fairly slow; however, the amount of force and efficiency of the system can be quite high. Once up to speed, spherical projectiles of a considerable mass are less susceptible to the decelerating force of drag than are lighter projectiles with more aerodynamic shapes. Although a heavy arrow, perhaps one filled with incendiaries, could hypothetically be thrown along a parabolic path, there would have been few incentives to do so when the same materials could otherwise be hurled in a spherical container, as seen in **nos. 103** and **104**.

Spheres are easier to produce and, given the necessary mass involved, lose little comparable velocity through flight despite being less aerodynamic. Furthermore, throwing an arrow would require a unique type of sling to balance the weight of the projectile and ensure that it was released along a vector perfectly aligning with the arrow's long axis at the point of release. When considering that arrows are most effective when they are light, shot along a relatively level trajectory and at a reasonably high

56 See Chevedden, "Black Camels," pp. 227-77, esp. pp. 259-60.

rate, it seems unlikely that a bolt-throwing trebuchet was ever practically employed. Illustrations of such engines should perhaps be viewed as little more than theoretical exercises. Unsurprisingly, most European examples, such as **no. 99**, were drawn by Renaissance figures.[57]

When examined more closely, al-Zardkash does not appear to associate the *manjaniq qarabugha* with the illustrations of bolt-throwing engines. Instead, he includes the term at the end of a list of engines, perhaps intending the term to encompass all types of trebuchets. When examining the sources, the *manjaniq qarabugha* is portrayed as a relatively light stone-throwing engine, similar to, and frequently mentioned alongside, the *manjaniq shaytani*. It is unclear what differentiated the *qarabugha* and *shaytani* trebuchets, but both were almost certainly traction trebuchets (Fig. 28).[58]

FIGURE 28 *Mentions of the* manjaniq qarabughra/qarabugha[a]

Year	Site	Engines	Source
1248-49	Homs	1	Ibn Wasil
1285	Margat	3 of 10	Ibn 'Abd al-Zahir
1289	Tripoli	13 of 19	Ibn al-Dawadari; al-Yunini; Ibn al-Furat
1291	Acre	multiple	al-Jazari; al-'Ayni; al-Yunini; Ibn al-Furat

a See Chevedden, "Black Camels," p. 247.

Modern Representations and Torsion Engines

Modern interpretations of trebuchets have been considerably influenced by those of certain Renaissance figures. Louis-Napoleon used a variety of Renaissance drawings, adapted from works by Roberto Valturio, Paolo Santini and others, to illustrate what he believed to be types of medieval artillery.[59] A step forward was taken by Viollet-le-Duc. As a military man and architect, he injected practicality and precision into his

57 For additional European examples, all of which date to the Renaissance period, see Mariano di Jacopo Taccola, *De ingeneis*, liber secundus (Munich: Bayerische Staatsbibliothek, Cod. lat. Monacensis 197), fol. 68v; Francesco di Giorgio Martini, *Opusculum de architectura* (London: British Museum, Cod. 197.b.21 – MS Harley 3281), fol. 40r; Francesco di Giorgio, *Codicetto* (Vatican: Biblioteca Apostolica, MS Vat. Urb. lat. 1757), fol. 99v. See also Chevedden, "Black Camels," pp. 239-42.

58 For a closer examination of these engines, see Khamisy and Fulton, "Manjaniq Qarabugha," pp. 179-201.

59 Bonaparte, *Études sur le passé et l'avenir de l'artillerie*, vol. 2, pl. 1 (after p. 26).

drawings. Images from his *Dictionnaire* have gone on to form the basis of most 'historically accurate' reconstructed counterweight trebuchets.[60] The footprint of these engines is derived from Honnecourt's plan but the superstructure is a conglomeration of inspiration drawn from Renaissance examples and his own imagination. His guesswork is revealed in figure 14: a counterweight trebuchet complete with tread-wheels and a triggerman but with the addition of a pulling team to provide extra traction power.[61] In reality, the efforts of the sixteen pullers would have added little force given the size of the counterweight and apparent necessity of tread-wheels to lift it.

Unlike images of trebuchets, there are very few surviving medieval illustrations of torsion engines. That in **no. 129**, dating to the fourteenth century, appears to be the earliest and is a relative anomaly. The clear inaccuracies reveal that the illustrator was unfamiliar with this type of engine. It appears to be an interpretation of an *onager*, which the illustrator might have read about in certain classical texts. Although the lower coil appears to provide resistance and the projectile is held in a cup, the beam seems better suited to a trebuchet and the second, superfluous coil looks conspicuously like the axle of such an engine.

Illustrations like this serve as a reminder that monks, most of whom lived sequestered lives in monasteries, were responsible for composing the vast majority of surviving images of medieval artillery. Many of these men probably never saw a trebuchet. In lieu of personal experience, they would have looked to classical texts, which may have influenced their illustrations as they did their vocabulary. Muslim artists of the thirteenth and fourteenth centuries tended to be more worldly. They travelled and interacted with different social and cultural groups to a greater degree than did their European counterparts. This exposure probably contributed to the greater degree of accuracy apparent in their illustrations.

---

60 Viollet-le-Duc, *Dictionnaire*, 5:227-28, figs. 9, 10.

61 Viollet-le-Duc, *Dictionnaire*, 5:237, fig. 14.

Appendix 3: Mathematical Scenarios

The following data sets have been calculated by Mark Denny using variables recommended to him based on the findings of this study. The formula that he has developed calculates the ideal mechanical ratios, i.e. the lengths of the long and short arms (Lla and Lsa), for given masses and beam dimensions, as well as sling lengths (Ls) and the corresponding release velocities (V) and angles of release (α), in order to effect a maximum horizontal range (Rh).[1] Beam lengths (Lb) have been selected between 4 m and 10 m, counterweight masses (Mcw) up to 4,000 kg and projectile masses (Mp) between 10 kg and 70 kg. For each scenario the load angle of the beam (θ) is 135°, a reasonable balance between maximising the effective distance that the counterweight can fall and a practical axle height relative to the other components of the engine. A trough for the projectiles to slide along before being lifted into the air has also been factored in. The small ratios, between 2:1 and 3.4:1, reflect the reasonable beam lengths and counterweight masses, as well as consistent shape of the beam from end to end. The beam in each case has a density of 750 kg/m3 and an originally circular cross-section, 0.5 m in diameter, but which has had the left and right sides shaved down so as to be no more than 0.3 m wide, shedding mass but retaining its vertical thickness and strength. The mass of the beam significantly decreases velocity when coupled with a relatively light counterweight, hence a beam no longer than 6 m has been paired with a counterweight of 1,000 kg.

70 kg Projectiles

Mcw	*Mp*	*Lb*	Lla	Lsa	Ls	V	~α	Rh
4,000 kg	*70 kg*	*10 m*	7.1 m	2.9 m	6.4 m	51.7 m/s	45°	250 m
4,000 kg	*70 kg*	*8 m*	5.7 m	2.3 m	5.1 m	48.7 m/s	45°	220 m
4,000 kg	*70 kg*	*6 m*	4.3 m	1.7 m	3.9 m	44.4 m/s	45°	190 m
4,000 kg	*70 kg*	*4 m*	2.9 m	1.1 m	2.6 m	38.0 m/s	45°	140 m
3,000 kg	*70 kg*	*10 m*	6.9 m	3.1 m	6.1 m	43.7 m/s	45°	180 m
3,000 kg	*70 kg*	*8 m*	5.6 m	2.4 m	5.0 m	41.5 m/s	45°	170 m
3,000 kg	*70 kg*	*6 m*	4.2 m	1.8 m	3.7 m	38.2 m/s	45°	140 m
3,000 kg	*70 kg*	*4 m*	2.8 m	1.2 m	2.5 m	33.0 m/s	45°	105 m
2,000 kg	*70 kg*	*10 m*	6.7 m	3.3 m	5.6 m	32.7 m/s	45°	105 m
2,000 kg	*70 kg*	*8 m*	5.4 m	2.6 m	4.6 m	31.9 m/s	45°	100 m
2,000 kg	*70 kg*	*6 m*	4.1 m	1.9 m	3.5 m	29.9 m/s	45°	90 m
2,000 kg	*70 kg*	*4 m*	2.7 m	1.3 m	2.4 m	26.4 m/s	45°	70 m

1 See Denny, "Siege Engine Dynamics," pp. 563-76; Denny, *Ingenium*, pp. 70-90.

 | DOI 10.1163/9789004376922_016

50 kg Projectiles

Mcw	*Mp*	*Lb*	Lla	Lsa	Ls	V	~α	Rh
4,000 kg	*50 kg*	*10 m*	7.2 m	2.9 m	6.5 m	56.3 m/s	45°	290 m
4,000 kg	*50 kg*	*8 m*	5.8 m	2.3 m	5.2 m	53.9 m/s	45°	270 m
4,000 kg	*50 kg*	*6 m*	4.4 m	1.7 m	4.0 m	49.6 m/s	45°	235 m
4,000 kg	*50 kg*	*4 m*	2.9 m	1.1 m	2.6 m	43.1 m/s	45°	180 m
3,000 kg	*50 kg*	*10 m*	7.0 m	3.0 m	6.2 m	48.0 m/s	45°	220 m
3,000 kg	*50 kg*	*8 m*	5.7 m	2.3 m	5.1 m	46.0 m/s	45°	200 m
3,000 kg	*50 kg*	*6 m*	4.3 m	1.7 m	3.9 m	42.8 m/s	45°	175 m
3,000 kg	*50 kg*	*4 m*	2.9 m	1.1 m	2.5 m	37.4 m/s	45°	135 m
2,000 kg	*50 kg*	*10 m*	6.7 m	3.3 m	5.7 m	36.7 m/s	45°	130 m
2,000 kg	*50 kg*	*8 m*	5.5 m	2.5 m	4.8 m	36.0 m/s	45°	125 m
2,000 kg	*50 kg*	*6 m*	4.2 m	1.8 m	3.7 m	34.0 m/s	45°	115 m
2,000 kg	*50 kg*	*4 m*	2.8 m	1.2 m	2.5 m	30.4 m/s	45°	95 m

30 kg Projectiles

Mcw	*Mp*	*Lb*	Lla	Lsa	Ls	V	~α	Rh
4,000 kg	*30 kg*	*10 m*	7.3 m	2.7 m	6.6 m	62.5 m/s	40°	345 m
4,000 kg	*30 kg*	*8 m*	5.9 m	2.1 m	5.3 m	60.2 m/s	40°	330 m
4,000 kg	*30 kg*	*6 m*	4.5 m	1.5 m	4.1 m	56.4 m/s	40°	290 m
4,000 kg	*30 kg*	*4 m*	3.0 m	1.0 m	2.7 m	50.1 m/s	40°	230 m
3,000 kg	*30 kg*	*10 m*	7.0 m	3.0 m	6.2 m	53.4 m/s	40°	260 m
3,000 kg	*30 kg*	*8 m*	5.8 m	2.2 m	5.2 m	52.1 m/s	40°	250 m
3,000 kg	*30 kg*	*6 m*	4.4 m	1.6 m	4.0 m	49.2 m/s	40°	130 m
3,000 kg	*30 kg*	*4 m*	2.9 m	1.1 m	2.6 m	43.9 m/s	40°	180 m
2,000 kg	*30 kg*	*10 m*	6.8 m	3.2 m	5.9 m	41.6 m/s	40°	160 m
2,000 kg	*30 kg*	*8 m*	5.5 m	2.5 m	4.9 m	41.2 m/s	40°	160 m
2,000 kg	*30 kg*	*6 m*	4.3 m	1.7 m	3.9 m	39.7 m/s	40°	150 m
2,000 kg	*30 kg*	*4 m*	2.9 m	1.1 m	2.6 m	36.2 m/s	40°	130 m

10 kg Projectiles

Mcw	*Mp*	*Lb*	Lla	Lsa	Ls	V	~α	Rh
3,000 kg	*10 kg*	*10 m*	7.2 m	2.8 m	6.5 m	59.5 m/s	40°	320 m
3,000 kg	*10 kg*	*8 m*	5.9 m	2.1 m	5.4 m	59.5 m/s	40°	320 m
3,000 kg	*10 kg*	*6 m*	4.5 m	1.5 m	4.1 m	57.8 m/s	40°	300 m
3,000 kg	*10 kg*	*4 m*	3.1 m	0.9 m	2.8 m	54.1 m/s	40°	270 m
2,000 kg	*10 kg*	*10 m*	7.0 m	3.0 m	6.2 m	47.0 m/s	45°	210 m
2,000 kg	*10 kg*	*8 m*	5.7 m	2.3 m	5.1 m	47.4 m/s	45°	210 m
2,000 kg	*10 kg*	*6 m*	4.4 m	1.6 m	4.0 m	46.9 m/s	45°	210 m
2,000 kg	*10 kg*	*4 m*	3.0 m	1.0 m	2.7 m	44.7 m/s	45°	190 m
1,000 kg	*10 kg*	*6 m*	4.1 m	1.9 m	3.5 m	30.7 m/s	45°	90 m

Appendix 4: Reconstructed Engines

Most reconstructed trebuchets are of the counterweight type. The majority of those built to be 'historically accurate' use Honnecourt's illustration as a plan for the base, while Viollet-le-Duc's nineteenth century hypothetical drawings have influenced the superstructures of many.[1] Among those drawing inspiration from these sources include those built by Favé at Vincennes, Radim Zepletal in Czechoslovakia, Peter Vemming Hansen in Denmark, the engines at Cardiff and Caerphilly in Wales and Beffeyte's hinged engine built at Urquhart Castle on Loch Ness, Scotland. Tarver's traction trebuchet is largely based on the illustrations and descriptions of al-Tarsusi, supplemented by the illuminations in the Morgan Bible.[2] The design of the much heavier Danish traction trebuchet was inspired by the engines found in BNF MS 10136.[3] The following abbreviations will once more be used:

Mcw	Mass of the counterweight (kg)
Mp	Mass of the projectile (kg)
Mb	Mass of the Beam (kg)
Lb	Length of the beam (m)
Lla	Length of the long arm (m)
Lsa	Length of the short arm (m)
Ls	Length of the sling (m)
Lcw	Length of the counterweight's centre of mass from the beam (m)
Lax	Height of the main axle (m)
Rmax	Maximum theoretical horizontal range (m)
Rh	Horizontal range (m)
V	True velocity of the projectile at release (m/s)
θ	Angle of the short arm to the vertical below the beam
α	Angle of release to the horizontal

1 See Appendix 2, **no. 36**, and Viollet-le-Duc, *Dictionnaire*, 5:227, 228, figs. 9 and 10.

2 See Appendix 2, **nos. 19, 20** and **29-31**.

3 See Appendix 2, **nos. 21** and **22**.

 | DOI 10.1163/9789004376922_017

Reconstructed Counterweight Trebuchets

Engine	Lb	Lla	Lsa	Lcw	Ls	Lax	Mcw	Mp	Mb	Rh	V
Vincennes (1850)	10.3 m	8.6 m	1.7 m				4,500 kg	11 kg		175 m	(>42 m/s)
Helfstyn (1986)	11.5 m					~8 m	6,000 kg	100 kg		200 m	(>45 m/s)
Nykøbing (1989)	6.6 m	5.5 m	1.2 m	1.4 m	5 m	3.5 m	2,000 kg	15 kg [47 kg]		168 m [~85 m]	(>41 m/s) [>29 m/s]
Caerphilly / Cardiff (1990s)	6.5 m	5.2 m	1.3 m			4.2 m	2,000 kg	15 kg		<100 m	(~31 m/s)
Loch Ness, fixed (1999)	8.0 m					7.0 m	6,000 kg	136 kg		200 m	56 m/s
Loch Ness, hinged (1999)	15 m					7.2 m	6,000 kg	136 kg	2,700 kg	200 m	52 m/s
Warwick (2005)							5,500 kg	15 kg		~200 m	(>45 m/s)

Reconstructed Traction Trebuchets

Engine	Lb	Lla	Lsa	Ls	Lax	Pullers	Mp	Rmax
Tarver A (1989)	4.9 m	4.2 m	0.7 m		3.7 m	8-25	1-8 kg	137 m (1.9 kg, 20 pullers)
Tarver B [refit] (1991)	4.9 m	4.2 m	0.7 m	1.0 m	3.7 m	14-16	3.9-5 kg	145 m (3.1 kg, 14 pullers)
Nykøbing hybrid (1991)	~8 m	6 m	1.7 m		4.3 m	20		
Caerphilly (1990s)						6-9	5-15 kg	~100 m

Tarver B- Trials[a]

Trail	Pullers	Mp	1	2	3	4	5	6	Avg.	Time
1	16	4.7 kg	81 m	79 m	76 m	77 m	76 m	76 m	78 m	70 s (1/11.7 s, 5.1/min)
2	16	3.9 kg	77 m	89 m	94 m	89 m	94 m	89 m	89 m	70 s (1/11.7 s, 5.1/min)
3	15	4.7 kg	100 m	90 m	105 m	100 m	105 m	93 m	99 m	
4	4	3.1 kg	65 m	68 m	52 m	55 m	55 m	69 m	63 m	75 s (1/12.7 s, 4.8/min)

a Tarver, "Traction Trebuchet," p. 162. A further trial threw a 5 kg stone from the citadel of Damascus 83 m with 12 pullers.

Bibliography

Manuscripts

Alexander and Dindimus. Oxford: Bodleian Library, MS 264.

Anonymous. Siena: Biblioteca Comunale, Cod. S.IV.5.

Arsenal Bible. Paris: Bibliothèque de l'Arsenal, MS 5211.

Arthurian Romances. Yale: Beinecke, MS 229.

Avis aus Roys. New York: Pierpont Morgan Library, MS M. 456.

Beatus of Liébana, *Commentary on the Apocalypse*. Turin: Biblioteca Nazionale Universitaria di Torino, MS I.II.1.

Bible Historiale Complétée. The Haag: Koninklijke Bibliotheek, KB 71 A.

Biblia Sancti Petri Rodensis (*Roda Bible*). Paris: BNF, MS lat. 6 (3).

Book of Hours (*Neville of Hornby Hours*). London: BL, MS Egerton 2781.

Des cas des nobles hommes et femmes. Geneva: Bibliothèque de Genève, MS fr. 190-1.

Caffaro, *Annales de Gênes*. Paris: BNF, MS lat. 10136.

Cantigas de Santa María. Escorial: Biblioteca de El Escorial, MS T.j.1.

Cantigas de Santa Maria. Florence: Biblioteca Nazionale, MS Banco Rari 20.

Chevalier au Cygne. Paris: BNF, MS fr. 795.

Chroniques de France ou de St Denis. London: BL, Royal 16 G VI.

Croniche di Giovanni Sercambi. Lucca: Archivio di Stato, MS 107.

Fleurs des chroniques. Besançon: Bibliothèque Municipale, MS 677.

Francesco di Giorgio Martini, *Codicetto*. Vatican: Biblioteca Apostolica, MS Vat. Urb. lat. 1757.

Francesco di Giorgio Martini, *Opusculum de architectura*. London: British Museum, Cod. 197.b.21 - MS Harley 3281.

Francesco di Giorgio Martini, *Trattato I*. Turin: Biblioteca Reale, Cod. 148 Saluzzo.

Hasan al-Rammah, *Furusiya*. Paris: BNF, MS arabe 2825.

Histoire Universalle. Dijon: Bibliothèque Municipal, MS. 562.

History Bible. New York: Pierpont Morgan Library, MS M.268.

Ibn Wasil, *Mufarrij al-kurub fi akhbar Bani Ayyub*, Paris: BNF, MS arabe 1703.

Konrad Kyeser, *Bellifortis*. Göttingen: Niedersächsische Staats- und Universitätsbibliothek, Cod. philos. 63.

Konrad Kyeser, *Bellifortis*. Innsbruck: Tiroler Landesmuseum Ferdinandeum, MS 16.0.7, FB 32009.

Konrad Kyeser, *Bellifortis*. Munich: Bayerische Staatsbibliothek, Clm 30150.

Leonardo da Vinci, *Codex Atlanticus*. Milan: Biblioteca Ambrosiana.

Mariano di Jacopo Taccola, *De ingeneis*, liber secundus. Munich: Bayerische Staatsbibliothek, Cod. lat. Monacensis 197.

Mariano di Jacopo Taccola, *De ingeneis ac edifitiis non usitatis*, liber tertius. Florence: Biblioteca Nazionale Centrale, Cod. Palatino 766.
Mariano di Jacopo Taccola, *De machinis*. Paris: BNF, MS lat. 7239.
Matthew Paris, *Chronica maiora*. Cambridge: Corpus Christi College Lib., MS 16, pt. 2.
Morgan (*Maciejowski*) *Bible*. New York: Pierpont Morgan Library, MS M.638.
Ordnung des Hofgerichts zu Rottweil. Stuttgart: Württembergische Landesbibliothek, HB VI 110.
Oxford-Paris-London bible moralisée. Vol. 3a. London: BL, Harley 1526.
Paolo Santini, *Tractatus*. Paris: BNF, MS lat. 7239.
Peter of Eboli, *Liber ad honorem Augusti*. Bern: Burgerbibliothek, BBB Cod. 120.II.
Rashid al-Din, *Jami al-Tawarikh*. Edinburgh: University of Edinburgh Library, Or. MS 20.
Rashid al-Din, *Jami al-Tawarikh*. Paris: BNF, MS pers. 1113.
Rashid al-Din, *Jami al-Tawarikh*. Berlin: Staatsbibliothek, Orientabteilung, Diez A.
Roberto Valturio, *De re militari libri duodecim*. Paris: BNF, MS lat. 7236.
Roman du Chevalier au Cygne et Chanson d'Antioche. Paris: BNF, MS fr. 12558.
Roman de Renart (*Roman de la Rose*). Paris: BNF, MS fr. 1581.
Roman du Saint Graal. London: British Library, MS Add. 10292.
Roman de toute chevalière. Paris: BNF, MS fr. 24364.
Li rommans de Godefroy de Buillon et de Salehadin et de tous lez autres roys qui ont esté outre mer jusques a saint Loys qui darrenierement y fu. Paris: BNF, MS fr. 22495.
Skylitzes Chronicle. Madrid: Biblioteca Nacional de España, MSS Graecus Vitr. 26-2.
Al-Tarsusi, Mardi ibn 'Ali, *Tabsirat Arbab al-Albab fi Kayfiyat al-Najah fi al-Hurub*. Oxford: Bodleian Library, MS Hunt. 264.
Talhoffer, *Fechtuch*. Copenhagen: Det Kongelige Bibliotek, MS Thott. 290.2.
Villard of Honnecourt, *Sketchbook*. Paris: BNF, MS fr. 19093.
Walter of Milemete, *De nobilitatibus, sapientiis et prudentiis regum*. Oxford: Christ Church, MS 92.
Weltchronik. New York: Pierpont Morgan Library, MS M.769.
William of Tyre, *Les Estoires d'outremer*. Bern: Burgerbibliothek, MS 112.
William of Tyre, *Les Estoires d'outremer*. Brussels: Bibliothèque Royale, MS 9492-3.
William of Tyre, *Les Estoires d'outremer*. Florence: Biblioteca Medicea-Laurenziana, MS Plu. LXI.10.
William of Tyre, *Les Estoires d'outremer*. Lyons: Bibliothèque Municipale, MS 828.
William of Tyre, *Les Estoires d'outremer*. Paris: BNF, MS fr. 352.
William of Tyre, *Les Estoires d'outremer*. Paris: BNF, MS fr. 2630.
William of Tyre, *Les Estoires d'outremer*. Vol. 1. St. Petersburg: National Library of Russia, MS fr. f° v.IV.5.
William of Tyre, *Histoire des Croisades; traduction française en XXII livres*. Paris: BNF, MS fr. 9081.
William of Tyre, *Histoire de la guerre sainte*. Paris: BNF, MS fr. 2628.

William of Tyre, *Histoire de la guerre sainte*. Paris: BNF, MS fr. 2824.
William of Tyre, *Histoire d'Outre Mer*. Baltimore: Walters Art Gallery, MS W.137.
William of Tyre, *Histoire d'Outre Mer*. Baltimore: Walters Art Gallery, MS W.142.
Wolfram von Eschenbach, *Willehalm* (fragments). Munich: Bayerische Staatsbibliothek, MS Cod. Ger. 193 III.
Al-Zardkash, Ibn Aranbugha, *Kitab al-'aniq fi'l-manajniq*. Istanbul: Topkapi Museum Library, MS Ahmad III 3469.

Primary Sources

Abbo of Saint-Germain-des-Prés, *Bella Parisiacae urbis*
Ed. Georg Heinrich Pertz. *Abbonis de bello Parisiaco libti III*. MGH SrG 1. Hanover, 1871.
Abu'l-Fida', *al-Mukhtasar fi Akhbar al-Bashar*
Ed. and trans. as *Résumé des Histoire des Croisades tiré des Annales d'Abou'l-Feda*. RHC Or 1. Paris: Imprimerie Nationale, 1872.
Trans. P.M. Holt. *The Memoirs of a Syrian Prince: Abu'l-Fida', Sultan of Hamah (672-732/1273-1331)*. Wiesbaden: Steiner, 1983.
Abu Shama, *Kitab al-Raudatain*
Ed. and trans. as *Livre des deux Jardins*. RHC Or 4-5. Paris: Imprimerie Nationale, 1898-1906.
Albert of Aachen, *Historia Ierosolimitana*
Ed. as *Historia Hierosolymitana*. RHC Oc 4. Paris: Imprimerie Nationale, 1879.
Ed. and trans. Susan B Edgington. *Historia Ierosolimitana: History of the Journey to Jerusalem*. Oxford: Oxford University Press, 2007.
Alexander of Telese, *Ystoria Rogerii regis Sicilie Calabrie atque Apulie*
Ed. Ludovica de Nava. *Ystoria Rogerii regis Sicilie Calabrie atque Apulie*. SI 112. Rome: Istituto Storico Italiano, 1991.
Trans. Graham A. Loud. *The Deeds Done by King Roger of Sicily*. www.leeds.ac.uk/history/weblearning/MedievalHistoryTextCentre/medievalTexts.htm [last accessed 5 January 2010].
(*Chronique d'*) *Amadi*
Ed. René de Mas Latrie. In *Chroniques d'Amadi et de Strambaldi*. Vol. 1. Paris: Imprimerie Nationale, 1891.
Amatus of Montecassino, *Ystoire de li Normant*
Ed. Vincenzo de Bartholomaeis. *Storia de' Normanni di Amato di Montecassino*. SI 76. Rome: Istituto Storico Italiano, 1935.
Trans. Prescott N. Dunbar. *The History of the Normans*. Woodbridge: Boydell, 2004.

Ambroise, *Estoire de la Guerre Saint*

Ed. Gaston Paris. *L'Estoire de la Guerre Saint*. Paris: Imprimerie Nationale, 1897.

Ed. Marianne Ailes and Malcolm Barber. *The History of the Holy War: Ambroise's* Estoire de la Guerre Saint. Woodbridge: Boydell, 2003.

Trans. Marianne Ailes. *The History of the Holy War: Ambroise's* Estoire de la Guerre Saint. Ed. Marianne Ailes and Malcolm Barber. Woodbridge: Boydell, 2003.

Trans. Merton Jerome Hubert. *The Crusade of Richard Lion-Heart*. Columbia University Press, 1941 reprinted New York: Octagon Books, 1976.

Ammianus Marcellinus, *Res gestae* (*Rerum gestarum libri XXXI*)

Ed. Francis Eyssenhardt. *Rerum gestarum libri*. Berlin: F. Vahlen, 1871.

Trans. C.D. Yonge. *The Roman History of Ammianus Marcellinus: During the Reigns of the Emperors Constantius, Julian, Jovianus, Valentinian, and Valens*. London: George Bell & Sons, 1902.

Anna Comnena, *Alexiad*

Trans. Elizabeth A. Dawes. *The Alexiad*. London: Routledge, Kegan Paul, 1928.

Trans. E.R.A. Sewter. *The Alexiad of Anna Comnena*. London: Penguin, 1969.

Annales de Burton

Ed. Henry Richards Luard. In *Annales Monastici*. Vol. 1. London: Longman, 1864.

Annales Colonienses Maximi

Ed. Georg Heinrich Pertz. *Annales Colonienses Maximi*. MGH SS 17. Hanover, 1861.

Annales Marbacenses

Ed. R. Wilmans. *Annales Marbacenses*. MGH SS 17. Hanover, 1861.

Annales regni Francorum

Ed. Friedrich Kurze. *Annales regni Francorum inde ab a. 741 usque ad a. 829*. MGH SrG 6. Hanover, 1895.

Annals of St Bertin and St Vaast

Ed. Chrétien César Auguste Dehaisnes. *Les Annales de Saint-Bertin et de Saint-Vaast*. Paris: SHF, 1871.

Annales sancti Rudberti Salisburgenses

Ed. D. Wilhelmus Wattenbach. *Annales Sancti Rudberti Salisburgenses*. MGH SS 9. Hanover, 1851.

Annales de Terre Sainte

Ed. Peter W. Edbury. "A New Text of the *Annales de Terre Sainte*." In *In Laudem Hierosolymitani*. Ed. Iris Shagrir, Ronnie Ellenblum and Jonathan Riley-Smith. Aldershot: Ashgate, 2007, pp. 145-61.

Ed. Reinhold Röhricht and Gaston Raynaud. *Annales de Terre Sainte*. Paris: Ernest Leroux, 1884.

Anonymous Syriac Chronicle

Ed. J.-B. Chabot. *Anonymi auctoris chronicon ad annum christi 1234 pertinens*. 2 vols. CSCO Scr. Syri, ser. 3, vols. 14 and 15. Paris, 1916-20.

Part trans. Arthur S. Tritton with notes by H.A.R. Gibb. "The First and Second Crusade from an Anonymous Syriac Chronicle." *Journal of the Royal Asiatic Society* 65 (1933), pp. 69-101, 273-305.

Ansbert, *Historia de expeditione Friderici Imperatoris*

Ed. Anton Chroust. *Quellen zur Geschichte des Kreuzzuges Kaiser Friedrichs I*. MGH SrG NS 5. Berlin, 1928.

Ed. as *Ystroia de expeditione Friderici Imperatoris edita a quodam Austriensi cleric, qui eidem interfuit, nomine Ansbertus*. FRA (1 Abt.), vol. 5. Vienna: K.K. Hof- und Staatsdruckerei, 1863.

Trans. Graham Loud. *The Crusade of Frederick Barbarossa*. Farnham: Ashgate, 2010.

Anselm of Ribemont, *Letters*

Ed. as *Epistyolae Stephani, comitis Carnotensis, atque Anselm de Ribodi Monte*. RHC Oc 3. Paris: Imprimerie Impériale, 1866.

Part. trans. August C. Krey. In *The First Crusade: The Accounts of Eye-Witnesses and Participants*. Princeton: Princeton University Press, 1921.

Aristakes Lastiverts'i, *Patmut'iwn Aristakisi Lastiverts'woy*

Ed. K.N. Yuzbashyan. *Patmut'iwn Aristakisi Lastiverts'woy*. Erevan, 1963.

Trans. Robert Bedrosian. *Aristakes Lastiverc'i's* History. New York, 1985.

Arnold of Lubek, *Chronica Slavorum*

Ed. Georg Heinrich Pertz. *Chronica Slavorum*. MGH SS 21. Hanover, 1869.

Auctarium Affligemense

Ed. D.L.C. Bethmann. *Auctarium Affligemense*. MGH SS 6. Hanover, 1844.

Auctarium Aquicinense

Ed. D.L.C. Bethmann. *Auctarium Aquicinense*. MGH SS 6. Hanover, 1844.

Aymar the Monk (Haymarus Monachus), *Expeditione Ierosolimitana*

Ed. William Stubbs. In Roger of Howden, *Chronica*. RS 51. Vol. 3. London: Longman, 1870, pp. cv-cxxxvi.

Ed. and Trans. Sascha Falk and Antonio Placanica. *Der "Rithmus de expedition Ierosolimitana" des sogenanntes Haymarus Monachus Florentinus*. Florence: SISMEL Edizioni del Galluzzo, 2006.

Al-'Ayni, Badr al-Din, *'Iqd al-Juman*

Ed. and trans. as *Le collier de perles*. RHC Or 2a. Paris: Imprimerie Nationale, 1887.

Baha' al-Din Yusuf ibn Shaddad. *al-Nawadir al-Sultaniyya wa'l-Mahasin al-Yusufiyya*

Ed. and trans. as *Anecdotes et Beau Traits de la Vie du Sultan Youssof*. RHC Or 3. Paris: Imprimerie Nationale, 1884.

Trans. C.R. Conder with notes by C.W. Wilson. *Beha ed-Din. The Life of Saladin*. London: PPTS, 1897.

Trans. D.S. Richards. *The Rare and Excellent History of Saladin*. Aldershot: Ashgate, 2001.

Al-Baladhuri, 'Ahmad Ibn Yahya, *Kitab Futuh*

Trans. Philip Hitti and Francis Murgotten. *The Origins of the Islamic State*. 2 vols. New York: Columbia University Press, 1916-24.

Bar Hebraeus, *Makhtebhanuth Zabhne*

Ed. Paul Bedjan. *Makhtebhanuth Zabhne*. Paris, 1890.

Ed. and trans. Ernest A. Wallis Budge. *The Chronography of Gregory Abu'l Faraj*. 2 vols. London: Oxford University Press, 1932, reprinted Amsterdam: Philo Press, 1976.

Caesar, Julius, *De bello civili*

Ed. Albert Doberenz. *De Bello Civili Commentarii Tres*. Leipzig: B.G. Teubner, 1876.

Caesar, Julius, *De bello Gallico*

Ed. Albert Doberenz. *Commentariorum de Bello Gallico*. Leipzig: B G. Teubner, 1877.

Ed. and trans. H.J. Edwards. *The Gallic War.* Loeb. Cambridge: Harvard University Press, 1917, reprinted 1986.

Caffaro (and cont.), *Annales*

Ed. Luigi Tommaso Belgrano and Cesare Imperiale. In *Annali Genovesi di Caffaro e de' suoi continuatori*. 5 vols. SI. Genoa: Istituto Storico Italiano, 1890-1901, Rome: Tipografia del Senato, 1923-29.

Trans. Martin Hall and Jonathan Phillips. In *Caffaro, Genoa and the Twelfth-Century Crusades*. Farnham: Ashgate, 2013.

Caffaro, *De liberatione civitatum Orientis*

Ed. Luigi Tommaso Belgrano. In *Annali Genovesi di Caffaro e de' suoi continuatori*. Vol. 1. SI. Genoa: Istituto Storico Italiano, 1890.

Trans. Martin Hall and Jonathan Phillips. In *Caffaro, Genoa and the Twelfth-Century Crusades*. Farnham: Ashgate, 2013.

Caffaro, *Ystoria captionis Almarie et Turtuose*

Ed. Luigi Tommaso Belgrano. In *Annali Genovesi di Caffaro e de' suoi continuatori*. Vol. 1. SI. Genoa: Istituto Storico Italiano, 1890.

Trans. Martin Hall and Jonathan Phillips. In *Caffaro, Genoa and the Twelfth-Century Crusades*. Farnham: Ashgate, 2013.

Carmen de bello Saxonico

Ed. Osward Holder-Egger. *Carmen de bello Saxonico*. MGH SrG 17. Hanover, 1889.

Chanson d'Antioche

Trans. Susan B. Edgington and Carol Sweetenham. *The* Chanson d'Antioche*: An Old French Account of the First Crusade*. Farnham: Ashgate, 2011.

Chanson de la croisade Albigeoise

Ed. and trans. Paul Meyer. *La chanson de la croisade contre les Albigeois*. 2 vols. Paris: SHF, 1875-79.

Trans. Janet Shirley. *The Song of the Cathar Wars*. Aldershot: Ashgate, 1996.

Chronicle of Lanercost

Ed. and trans. Sir Herbert Maxwell. *Chronicle of Lanercost, 1272-1346*. Glasgow: James Maclehose and Sons, 1913.

Chronicle of Melrose

Ed. Joseph Stevenson. *Chronica de Mailros*. Edinburgh, 1835.

Trans. Joseph Stevenson. In *The Church Historians of England* 4.1. London: Seeleys, 1856.

Chronico Sancti Martini Turonensi

Ed. O. Holder-Egger. *Chronico S. Martini Turonensi*. MGH SS 26. Leipzig, 1925.

Chronicon Salernitanum

Ed. Ulla Westerbergh. *Chronicon Salernitanum*. Stockholm: Almquist & Wiksell, 1956.

De constructione castri Saphet

Ed. R.B.C. Huygens. In "Un nouveau texte du traité 'De constructione castri Saphet'." *Studi medievali* 3.6 (1965), pp. 355-87.

Trans. Hugh Kennedy. In *Crusader Castles*. Cambridge: Cambridge University Press, 1994, pp. 190-98.

Cronica Reinhardsbrunnensis

Ed. O. Holder-Egger. *Cronica Reinhardsbrunnensis*. MGH SS 30, pt. 1. Hanover, 1896.

Cronica S Petri Erfordensis moderna

Ed. O. Holder-Egger. *Cronica S. Petri Erfordensis Moderna*. MGH SS 30, pt. 1. Hanover, 1896.

Egidio Colonna (Giles of Rome), *De regimine principum*

Ed. Girolamo Samaritano (Hieronymum Samaritanium). *De Regimine Principum, Lib. III*. Rome: Bartholomeum Zannettum, 1607.

Ed. D. Simon Frederick Hahn. As *De re militari veterum*, in *Collectio monumentarum veterum et recentium*. Vol. 1. Brunswick: Frid. Wilh. Meyer, 1723.

Ekkehard of Aura, *Hierosolymita*

Ed. Heinrich Hagenmeyer. *Hierosolymita*. Tubingen, 1877.

Eracles Continuation of William of Tyre

Ed. as *L'Estoire de Eracles empereur et la conqueste de la terre d'Outremer: c'est la continuation de l'estoire de Guillaime arcevesque de Sur*. RHC Oc 2. Paris: Imprimerie Impériale, 1859.

Part. trans. Janet Shirley. In *Crusader Syria in the Thirteenth Century: The* Rothelin *Continuation of the* History *of William of Tyre with part of the* Eracles *or* Acre *text*. Aldershot: Ashgate, 1999.

Ernoul, *Chronique*

Ed. Louis de Mas Latrie. *Chronique d'Ernoul et de Bernard le Trésorier*. Paris: SHF, 1871.

Eustathius of Thessalonica, *Capture of Thessalonica*

Ed. and trans. John R. Melville Jones. *Eustathios of Thessaloniki, The Capture of Thessaloniki*. Canberra: Australian Association for Byzantine Studies, 1988.

Evagrius Scholasticus, *Ecclesiastical History*

Ed. Henri de. Valois. *Evagrii Scholastici epiphaniensis et ex praefectis, Ecclesiasticae histriae, libri sex*. London, 1844.

Trans. as *Evagrius's Ecclesiastical History, From 431 to 594 A.D.* Greek Ecclesiastical Historians of the First Six Centuries of the Christian Era. London: Samuel Bagster, 1846.

Excidium Aconis

Ed. R.B.C. Huygens. As *Excidii Acconis Gestorum Collectio*, in *The Fall of Acre 1291*. CCCM 202. Turnhout: Brepols, 2004.

De expugnatione Lyxbonensi

Ed. and trans. Charles Wendell David. *De Expugnatione Lyxbonensi: The Conquest of Lisbon*. New York: Columbia University Press, 1936.

De expugnatione terrae sanctae per Saladinum.

Ed. Joseph Stevenson. *De Expugnatione Terrae Sanctae per Saladinum*. RS 66. London: Longman, 1875.

Part. trans. James Brundage. In *The Crusades: A Documentary History*. Milwaukee: Marquette University Press, 1962, reprinted 1976.

Florence of Worcester, *Chronicon ex chronicis*

Ed. Benjamin Thorpe. *Florentii Wigorniensis Monachi Chronicon ex Chronicis*. 2 vols. London: Sumptibus Societatis, 1848-49.

Trans. Thomas Forester. *The Chronicle of Florence of Worcester with the Two Continuations*. London: Henry G. Bohn, 1854.

Frontinus, Sextus Julius, *Strategemata*

Ed. Mary B. McElwain, trans. Charles E. Bennett. In *The Stratagems and the Aqueducts of Rome*. Loeb. London: William Heinemann, 1925.

Froissart, John, *Chroniques*

Ed. Siméon Luce and G. Raynaud. *Chroniques*. 11 vols. Paris: SHF, 1869-99.

Fulcher of Chartres, *Historia Hierosolymitana*

Ed. Heinrich Hagenmeyer. *Historia Hierosolymitana*. Heidelberg: Carl Winter's Universitätsbuchhandlung, 1913.

Ed. as *Historia Iherosolymitana, Gesta Francorum Iherusalem Peregrinatium, Ab Anno Domini MXCV ad Annum MCXXVII*. RHC Oc 3. Paris: Imprimerie Impériale, 1866.

Trans. Frances Rita Ryan and ed. Harold S. Fink. *A History of the Expedition to Jerusalem, 1095-1127*. Knoxville: The University of Tennessee Press, 1969.

Geoffrey of Malaterra, *De rebus gestis Rogerii*

Ed. Ernesto Pontieri. *De rebus gestis Rogerii Calabriae et Sicliae comitis et Roberti Guiscardi ducis fratris eius*. Rerum Italicarum scriptores [n.s.] 5.1. Bologna: Nicola Zanichelli, 1925-28.

Trans. Kenneth Baxter Wolf. *The Deeds of Count Roger of Calabria and Sicily and His Brother Duke Robert Guiscard*. Ann Arbor: University of Michigan Press, 2005.

Geoffrey of Villehardouin, *Conquête de Constantinople*

Ed. and trans. M. Natalis de Wailly. *Conquête de Constantinople, avec la continuation de Henri de Valenciennes*. 3rd. ed. Paris: Firmin-Didot, 1882.

Trans. M.R.B. Shaw. In *Chronicles of the Crusades*. London: Penguin, 1963.

Gesta Francorum

Ed. and trans. Rosalind Hill. *The Deeds of the Franks and the other Pilgrims*. Edinburgh: Thomas Nelson and Sons, 1962, reprinted Oxford: Oxford University Press, 1972.

Giovanni Sercambi

Ed. Salvatore Bongi. *Le Croniche di Giovanni Sercambi Lucchese*. 3 vols. SI. Lucca, Tipografia Giusti, 1892.

Guibert of Nogent, *Die gesta per Francos*

Ed. as *Historia quae dicitur Gesta Dei per Francos*. RHC Oc 4. Paris: Imprimerie Nationale, 1879.

Ed. R.B.C. Huygens. *Die Gesta per Francos*. CCCM 127a. Turnhout: Brepols, 1996.

Trans. Robert Levine. *The Deeds of God through the Franks*. Woodbridge: Boydell, 1997.

Guidebook to Palestine

Trans. J.H. Bernard. *Guide-Book to Palestine (Circ. A.D. 1350)*. London: PPTS, 1894.

Hayton, *La flor des estoires de la terre d'orient* (*Flos historiarum terre orientis*)

Ed. and trans. as *La flor des estoires de la terre d'orient*. RHC Ar 2. Paris: Imprimerie Nationale, 1906.

Henry of Huntingdon, *Historia Anglorum*

Ed. Thomas Arnold. *Historia Anglorum*. RS 74. London: Longman, 1879.

Trans. Thomas Forester. *The Chronicle of Henry of Huntingdon*. London: Henry G. Bohn, 1853.

Henry of Livonia, *Chronicon Lyvoniae*

Ed. Leonid Arbusow and Albert Bauer. *Heinrichs Livändische Chronik*. MGH SrG 31. Hanover, 1955.

Trans. James A. Brundage. *Chronicle of Henry of Livonia*. Madison: University of Wisconsin Press, 1961, new edition New York: Columbia University Press, 2003.

Histoire des ducs de Normandie et des rois d'Angleterre

Ed. Francisque Michel. *Histoire des Ducs de Normandie et des Rois d'Angleterre*. Paris: SHF, 1840.

Historia de Leone Bardae Armenii filio

Ed. Immanuel Bekker. In *Leonis Grammatici Chronographia*. Corpus scriptorum historiae Byzantinae. Bonn, 1842.

Historia Nicaena vel Antiochena

Ed. as *Historia Nicaena vel Antiochena*. RHC Oc 5. Paris: Imprimerie Nationale, 1886.

History of the Patriarchs of the Egyptian Church

Ed. and trans. O.H.E. Khs-Burmester, et al. *History of the Patriarchs of the Egyptian Church*. 4 vols. Cairo: Publications de la Sociétés d'Archéologie Copte, Textes et Documents, 1942-74.

Hugo Falcandus, *Liber de regno Sicilie*

Ed. Giovanni Battista Siragusa. *La historia o liber de regno Siciliae e la Epistola ad Petrum Panormitanae ecclesiae thesaurarium di Ugo Falcando*. SI 22. Rome: Forzani, 1897.

Trans. Graham A. Loud and Thomas Wiedemann. *The History of the Tyrants of Sicily by 'Hugo Falcandus' 1154-69*. Manchester: Manchester University Press, 1998.

Iacobi de Guisia (Jaques of Guise), *Annales historiae illustrium principum Hanoniae*

Ed. Ernest. Sackur. *Annales Historiae Illustrium Principum Hanoniae*. MGH SS 30, pt. 1. Hanover, 1896.

Ibn 'Abd al-Zahir, Muhi al-Din, *al-Rawd al-Zahir fi sirat al-Malik al-Zahir*

Ed. and trans. Abdul Aziz al-Khowayter. *A Critical Edition of an Unknown Arabic Source for the Life of al-Malik al-Zahir Baibars, with Introduction, Translation, and Notes*. 3 vols. S.O.A.S., unpublished PhD thesis, 1960.

Ed. and trans. Syedah Fatima Sadeque. *Baybars I of Egypt*. Dacca: Oxford University Press, 1956.

Ibn al-'Adim, Kamal al-Din, *Zubdat al-Halab fi tarikh Halab*

Ed. and trans. as *Extraits de la Chronique d'Alep*. RHC Or 3. Paris: Imprimerie Nationale, 1884.

Ibn al-'Amid, al-Makin, *al-Majmu al-Mubarak*

Trans. Anne-Marie Eddé and Françoise Micheau. *Chronique des Ayyoubides (602-658/1205-6-1259-60)*. DRHC 16. Paris: L'Académie des Inscriptions et Belles-Lettres, 1994.

Ibn al-Athir, *al-Kamil fi'l-ta'rikh*

Trans. D.S. Richards. *The Chronicle of Ibn al-Athir for the Crusading Period from al-Kamil fi'l-ta'rikh*. 3 vols. Aldershot: Ashgate, 2008.

Ibn al-Azraq, *Tarikh Mayyafariqin wa Amid*

Trans. Carole Hillenbrand. In *A Muslim Principality in Crusader Times*. Istanbul: Nederlands Historisch-Archeologisch Instituut, 1990.

Ibn Battuta, *Tuhfat an-Nuzzar fi Ghara'ib al-Amsar wa 'Aja'ib al-Asfar*

Trans. H.A.R. Gibb. *Travels in Asia and Africa, 1325-54*. London: George Routledge & Sons, 1929.

Ibn al-Furat, *Tarikh al-Duwal wa'l-Muluk*

Ed. and trans. U. and M.C. Lyons. *Ayyubids, Mamlukes and Crusaders: Selections from the Tarikh al-Duwal wa'l-Muluk*. 2 vols. Cambridge: W. Heffer and Sons, 1971.

Ibn Ishaq, *Sirat Rasul Allah*

Trans. A. Guillaume. *The Life of Muhammad*. Oxford: Oxford University Press, 1955, reprinted 2004.

Ibn Jubayr, *Rihla*

Trans. R.J.C. Broadhurst. *The Travels of Ibn Jubayr*. New Delhi: Goodword Books, 2001.

Ibn Muyassar, *Akhbar Misr*

Ed. and trans. as *Extraits d'Ibn Moyesser*. RHC Or 3. Paris: Imprimerie Nationale, 1884.

Ibn al-Qalanisi, *Dhail Tarikh Dimashq*

Trans. H.A.R. Gibb. *The Damascus Chronicle of the Crusades*. London: Luzac & Company, 1932, reprinted 1967.

Ibn Shaddad, 'Izz al-Din, *Al-A'laq al-Khatira fi Dhikr Umara' al-Sham wa al-Jazira*

Trans. Anne-Marie Eddé-Terrasse. *Description de la Syrie du nord*. Damascus: Institut Français de Damas, 1984.

Ibn al-Suqa'i, *Tali Kitab Wafayat al-A'yan*

Ed. and trans. Jacqueline Sublet. *Tali Kitab Wafayat al-A'yan*. Damascus: Institut Français de Damas, 1974.

Ibn Wasil, Jamal al-Din Muhammad, *Mufarrij al-kurub fi akhbar Bani Ayyub*

Part. trans. Peter Jackson. In *The Seventh Crusade, 1244-1254*. Aldershot: Ashgate, 1988.

'Imad al-Din al-Isfahani, *al-Fath al-Qussi fi'l-Fath al-Qudsi*

Trans. Henri Massé. *Conquête de la Syrie et de la Palestine par Saladin*. DRHC 10. Paris: Académie des inscriptions et belles-lettres, 1972.

Iohannes Codagnellus (Giovanni Codagnelli), *Annales Placentini*

Ed. Georg Heinrich Pertz. *Annales Plancentini Guelfi*. MGH SS 18. Hanover, 1863, reprinted 1925.

Ed. Oswald Holder-Egger. *Iohannis Codagnelli Annales Placentini*. MGH SrG 23. Hanover, 1901.

Itinerarium peregrinorum et gesta regis Ricardi

Ed. William Stubbs. *Itinerarium Peregrinorum et Gesta Regis Ricardi*. RS 38. Vol. 1. London: Longman, 1864.

Trans. Helen J. Nicholson. *Chronicle of the Third Crusade: A Translation of the Itinerarium Peregrinorum et Gesta Regis Ricardi*. Aldershot: Ashgate, 1997.

James of Aragon, *Llibre dels fets*

Trans. Damian Smith and Helena Buffery. *The Book of Deeds of James I of Aragon: A Translation of the Medieval Catalan* Llibre dels Fets. Aldershot: Ashgate, 2003.

James of Vitry, *Historia orientalis*

Ed. and Trans. Jean Donnadieu. *Histoire Orientale / Historia Orientalis*. Turnhout: Brepols, 2008.

Trans. Aubrey Stewart. *The History of Jerusalem*. London: PPTS, 1896.

James of Vitry, *Lettres*

Ed. R.B.C. Huygens. *Lettres de Jacques de Vitry*. Leiden: Brill, 1960.

John of Joinville, *Vie de Saint Louis*

Ed. and trans. Jacques Monfrin. *Vie de Saint Louis*. Paris: Classiques Garnier Multimédia, 1998.

Trans. M.R.B. Shaw. In *Chronicles of the Crusades*. London: Penguin, 1963.

John Kinnamos, *Deeds of John and Manuel Comnenus*

Trans. Charles M. Brand. *Deeds of John and Manuel Comnenus.* New York: Columbia University Press, 1976.

John of Marmoutier, *Historia Gaufredi ducis Normanorum et comitis Andegavorum*

Ed. Louis Halphen and Rene Poupardin. In *Chroniques des Comtes d'Anjou et des Seigneurs d'Amboise.* Paris: Auguste Picard, 1913.

John Poloner, *Descriptio Terrae Sanctae*

Trans. Aubrey Stewart. *John Poloner's Description of the Holy Land (Circa 1421 A.D.).* London: PPTS, 1894.

John of Worcester, *Chronica chronicarum*

Ed. R.R. Darlington and P. McGurk, trans. Jennifer Bray and P. McGurk. *The Chronicle of John of Worcester.* 3 vols. Oxford: Oxford University Press, 1995-98.

Jordan Fantosme, *Chronique*

Ed. and trans. Richard Howlett. In *Chronicles of the Reigns of Stephen, Henry II, and Richard I.* RS 82. Vol. 3. London: Longman, 1886.

Ed. and trans. Francisque Michel. *Chronicle of the War Between the English and the Scots in 1173 and 1174.* London: J.B. Nichols and Son, 1840.

Josephus, Flavius, *The Jewish War*

Trans. William Whiston. In *The Works of Flavius Josephus.* Auburn: John E. Beardsley, 1895.

Joshua the Stylite, *Chronicle*

Trans. W. Wright, *The Chronicle of Joshua the Stylite, Composed in Syriac A.D. 507.* Cambridge: Cambridge University Press, 1882.

Ludolph of Suchem, *De itinere Terrae sanctae liber*

Ed. Ferdinand Deycks. *Ludolphi, rectoris ecclesiae parochialis in Suchem De itinere Terrae sanctae liber.* Stuttgart: Litterarischen Vereins, 1851.

Trans. Aubrey Stewart. *Ludolph von Suchem's Description of the Holy Land and the Way Thither.* London: PPTS, 1895.

Al-Maqrizi, Ahmad ibn 'Ali Taqi al-Din, *Kitab al-Suluk li-Ma'rifat Duwal al-Muluk*

Trans. R.J.C. Broadhurst. *A History of the Ayyubid Sultans of Egypt.* Library of Classical Arabic Literature 5. Boston: Twayne Publishers, 1980.

Trans. M. Quatremère. *Histoire des Sultans mamlouks.* 2 Vols. Paris, 1837-45.

Al-Maqrizi, Ahmad ibn 'Ali Taqi al-Din, *Mawaiz wa al-'Athar*

Trans. Urbain Bouriant and M. Paul Casanova. *Description Topographique et Historique de l'Egypte.* 4 vols. Mémoires publies par les Membres de la Mission Archéologique Française au Caire 17. Paris: Libraire de la Société Asiatique, 1895-1920.

Marino Sanudo Torsello, *Liber secretorum fidelium crucis*

Ed. J. Bongars. *Liber Secretorum Fidelium Crucis.* Hanover, 1611, reprinted with an introduction by Joshua Prawer, Jerusalem: Massada Press, 1972.

Trans. Peter Lock. *The Book of the Secrets of the Faithful of the Cross - Liber Secretorum Fidelium Crucis*. Farnham: Ashgate, 2011.

Maronite Chronicle

Part trans. Andrew Palmer. "Extract from the *Maronite Chronicle* (AD 664+)." In *The Seventh Century in the West-Syrian Chronicles*. Liverpool: Liverpool University Press, 1993, pp. 29-35.

Matthew of Edessa, *Patmut'iwn*

Trans. Ara Edmond Dostourian. *The Chronicle of Matthew of Edessa*. Lanham: University Press of America, 1993.

Matthew Paris, *Chronica maiora*

Ed. Henry Richards Luard. *Chronica Majora*. RS 57. 7 vols. London: Longman, 1872-83.

Trans. J.A. Giles. *Matthew Paris's English History: From the Year 1235 to 1273*. 3 vols. London: Henry G. Bohn, 1852-54.

Menkonis Chronicon

Ed. L. Weiland. *Menkonis Chronicon*. MGH SS 23. Leipzig, 1925.

Michael Attaleiates, *History*

Ed. and trans. Anthony Kaldellis and Dimitris Krallis. Dumbarton Oaks Medieval Library. Cambridge: Harvard University Press, 2012.

Michael Psellus, *Chronographia*

Trans. E.R.A. Sewter. *Fourteen Byzantine Rulers: The Chronographia of Michael Psellus*. Harmondsworth: Penguin, 1966.

Michael the Syrian, *Chronique*

Ed. and trans. Jean B. Chabot. *Chronique de Michel le Syrien, Patriarche Jacobite d'Antioche (1166-1199)*. 4 vols. Paris, 1899-1910.

Ed. and trans. as *La Chronique de Michel le Syrien*. RHC Ar 1. Paris: Imprimerie Impériale, 1869.

Miracles of St Demetrius

Ed. Paul Lemerle. *Les Plus Anciens Recueils des Miracles de Saint Démétrius*. Vol 1. Paris: Editions du Centre National de le Recherche Scientifique, 1979.

Movses Dasxuranci, *Patmut'iwn Aluanic*

Trans. C.J.F. Dowsett. *The History of the Caucasian Albanians*. London: Oxford University Press, 1961.

Mufaddal ibn Abi'l-Fada'il, *Kitab al-Nahj al-Sadid wa'l-Durr al-Farid*

Ed. and trans. E. Blochet. *Histoire des Sultans Mamlouks*. 3 fascs. PO 12.3 (1919), pp. 345-550; 14.3 (1920), pp. 375-672; 20.1 (1929), pp. 3-270, reprinted 3 vols. Brepols, 2003.

Nerses Shnorhali, *Voghb Yedesyo*

Ed. and trans. as *Elegie sur la prise d'Edesse*. RHC Ar 1. Paris: Imprimerie Impériale, 1869.

Niketas Choniates, *Historia*

Trans. Harry J. Magoulias. *O City of Byzantium, Annals of Niketas Choniates*. Detroit: Wayne State University Press, 1984.

Nishapuri, Zahir al-Din, *Jami' al-Tawarikh*

Trans. Kenneth Alin Luther. *The History of the Seljuks*. Ed. C. Edmund Bosworth. Richmond: Routledge, 2001.

Oliver of Paderborn, *Historia Damiatina*

Ed. Hermann Hoogeweg. In *Die Schriften des Kölner Domscholasters, späteren Bischofs von Paderborn und Kardinal-Bischofs von S. Sabina*. Tubingen: Litterarischen Verein in Stuttgart, 1894.

Trans. John J. Gavigan. In *Christian Society and the Crusades*. Ed. Edward Peters. Philadelphia: University of Pennsylvania Press, 1971.

Orderic Vitalis, *Historiae ecclesiasticae*

Ed. and trans. Marjorie Chibnall. *The Ecclesiastical History of Orderic Vitalis*. 6 vols. Oxford: Oxford University Press, 1969-80.

Ed. Auguste Le Prevost. Orderic Vitalis. *Historiae Ecclesiasticae*. 5 vols. Paris: SHF, 1838-55.

Trans. Thomas Forester. *The Ecclesiastical History of England and Normandy*. 4 vols. London, Henry G. Bohn, 1853-56.

Otto of Freising and Rahewin, *Gesta Friderici I imperatoris*

Ed. Georg Waitz. *Ottonis et Rahewini Gesta Friderici I. imperatoris*. 3rd ed. MGH SrG 46. Hanover, 1912.

Trans. Charles Christopher Mierow. *The Deeds of Frederick Barbarossa*. New York: Columbia University Press, 1953, republished Toronto: University of Toronto Press, 1994.

Paul the Deacon, *Historia Langobardorum*

Ed. Georg Waitz. *Pauli Historia Langobardorum*. MGH SrG 48. Hanover, 1878.

Trans. William Dudley Foulke. *History of the Langobards*. Philadelphia: University of Pennsylvania, 1907.

Peter Tudebode, *Historia de Hierosolymitano itinere*

Ed. John Hugh Hill and Laurita L. Hill. *Historia de Hierosolymitano Itinere*. DRHC 12. Paris: L'Académie des Inscriptions et Belles-Lettres, 1977.

Ed. as *Historia de Hierosolymitano Itinere*. RHC Oc 3. Paris: Imprimerie Impériale, 1866.

Trans. John Hugh Hill and Laurita Hill. *Historia de Hierosolymitano Itinere by Peter Tudebode*. Memoirs of the American Philosophical Society 101. Philadelphia: American Philosophical Society, 1974.

Peter of Langtoft, *Chronicle*

Ed. and trans. Tomas Wright. *The Chronicle of Pierre de Langtoft, in French Verse, from the Earlier Period to the Death of King Edward I*. RS 47. Vol. 2. London: Longman, 1868.

Philip of Novara, *Gestes des Chiprois*

Ed. Gaston Raynaud. In *Les Gestes des Chiprois*. Recueil de Chroniques Françaises. Paris: Société de l'Orient Latin, 1887, reprinted Osnabruck: Otto Zeller, 1968.

Trans. John L. La Monte. *The Wars of Frederick II Against the Ibelins in Syria and Cyprus*. New York: Columbia University Press, 1936.

Pipe Roll 36 (1190)

Ed. Doris M. Stenton. *The Great Roll of the Pipe for the Second Year of the Reign of King Richard the First, Michaelmas 1190 (Pipe Roll 36)*. Vol. 39 (new series, vol. 1). London: Pipe Roll Society, 1925.

Pipe Rolls 37 and 38 (1191 and 1192)

Ed. Doris M. Stenton. *The Great Rolls of the Pipe of the Third and Fourth Years of the Reign of King Richard the First, Mich. 1191 and Mich. 1192 (Pipe Rolls 37 and 38)*. Vol. 40 (new series, vol. 2). London: Pipe Roll Society, 1926.

Pipe Roll 40 (1194)

Ed. Doris M. Stenton. *The Great Roll of the Pipe for the Sixth Year of the Reign of King Richard the First, Michaelmas 1194 (Pipe Roll 40)*. Vol. 43 (new series, vol. 5). London: Pipe Roll Society, 1928.

Procopius, *Wars of Justinian*

Ed. and trans. H.B. Dewing. *History of the Wars*. 7 vols. Loeb. London: William Heinemann, 1914-40, reprinted 1962.

Ralph of Caen, *Gesta Tancredi*

Ed. as *Gesta Tancredi*. RHC Oc. 3. Paris: Imprimerie Impériale, 1866.

Trans. Bernard S. Bachrach and David S. Bachrach. *The* Gesta Tancredi *of Ralph of Caen, A History of the Normans on the First Crusade*. Aldershot: Ashgate, 2005.

Ralph of Coggeshall, *Chronicon Anglicanum*

Ed. Joseph Stevenson. *Chronicon Anglicanum*. RS 66. London: Longman, 1875.

Ralph of Diceto, *Ymagines historiarum*

Ed. William Stubbs. In *Opera Historica: The Historical Works of Ralph de Diceto, Dean of London*. RS 68. 2 vols. London: Longman, 1876.

Rashid al-Din Hamadani, *Ta'rikh-i Ghazani*, in *Jami' al-Tawarikh*

Trans. John Andrew Boyle. *The Successors of Genghis Khan*. New York: Columbia University Press, 1971.

Raymond of Aguilers, *Historia Francorum qui ceperunt Iherusalem*

Ed. as *Historia Francorum qui Ceperunt Iherusalem*. RHC Oc. 3. Paris: Imprimerie Impériale, 1866.

Trans. John Hill and Laurita Hill. *Historia Francorum qui Ceperunt Iherusalem*. Philadelphia: The American Philosophical Society, 1968.

Regino of Prum, *Chronicon*

Ed. Frederick Kurze. *Regionis abbatis Prumiensis Chronicon cum continuatione Treverensi.* MGH SrG 50. Hannover, 1890.

Richard of Devizes, *Chronicon de rebus gestis Ricardi Primi*

Ed. Richard Howlett. In *Chronicles of the Reigns of Stephen, Henry II, and Richard I.* RS 82. Vol. 3. London: Longman, 1886.

Trans. in *Chronicles of the Crusades, being Contemporary Narratives of the Crusade of Richard Coeur de Lion, by Richard of Devizes and Geoffrey de Vinsauf; and of the Crusade of Saint Louis, by Lord John de Joinville.* London: Henry G. Bohn, 1848.

Richard of Morins, *Annales prioratus de Dunstaplia*

Ed. Henry Richards Luard. In *Annales Monastici.* RS 36. Vol. 3. London: Longman, 1866.

Rigord, *Gesta Philippi Augusti*

Ed. H.-Francois Delaborde. In *Oeuvres de Rigord et de Guillaume le Breton.* Vol. 1. Paris: SHF, 1882.

Robert the Monk, *Historia Iherosolimitana*

Ed. as *Historia Iherosolimitana.* RHC Oc 3. Paris: Imprimerie Impériale, 1866.

Trans. Carol Sweetenham. *Robert the Monk's History of the Crusade: Historia Iherosolimitana.* Aldershot: Ashgate, 2005.

Robert of Torigni, *Chronique*

Ed. Richard Howlett. In *Chronicles of the Reigns of Stephen, Henry II, and Richard I.* RS 82. Vol. 4. London: Her Majesty's Stationary Office, 1889.

Roger of Howden, *Chronica*

Ed. William Stubbs. *Chronica.* RS. 51. 4 vols. London: Longman, 1868-71.

Trans. Henry T. Riley. *The Annals of Roger de Hoveden: Comprising the History of England and of Other Countries of Europe, from A.D. 732 to A.D. 1201.* 2 vols. London: H.G. Bohn, 1853.

Roger of Wendover, *Flores historiarum*

Ed. Henry Coxe. *Chronica, sive Flores Historiarum: in qua lectionum varietas additionesque, quibus chronicon istud ampliavit et instruxit Matthaeus Parisiensis.* 4 vols. London: Sumptibus Societatis, 1841-44.

Ed. Henry G. Hewlett. *Rogeri de Wendover Liber qui Dicitur Flores Historiarum ab Anno Domini MCLIV. Annoque Henrici Anglorum Regis Secundi Primo / The Flowers of History by Roger de Wendover: From the Year of Our Lord 1154, and the First Year of Henry the Second, King of the English.* RS 84. 3 vols. London: Her Majesty's Stationary Office, 1886-89.

Trans. J.A. Giles. *Roger of Wendover's Flowers of History, comprising the History of England from the descent of the Saxons to A.D. 1235.* 2 vols. London: Henry G. Bohn, 1849.

Rothelin Continuation of William of Tyre

Ed. as *La continuation de Guillaume de Tyre de 1229 a 1261, dite du manuscrit de Rothelin.* RHC Oc 2. Paris: Imprimerie Impériale, 1859.

Trans. Janet Shirley. *Crusader Syria in the Thirteenth Century: The* Rothelin *Continuation of the* History *of William of Tyre with part of the* Eracles *or* Acre *text.* Aldershot: Ashgate, 1999.

Sebeos, *History*

Trans. R.W. Thomson, with commentary by James Howard-Johnston and assistance from Tim Greenwood. *The Armenia History attributed to Sebeos.* 2 vols. Liverpool: Liverpool University Press, 1999.

Trans. Robert Bedrosian. *History.* New York: Sources of Armenian Tradition, 1985.

Seneca, Lucius Annaeus, *Qaestiones naturales*

Ed. Giovanni Monfrè. In *Philosophi opera.* Vol. 2. Padua: Typographia Seminarii, 1728.

Trans. John Clarke. *Physical Science in the Time of Nero.* London: MacMillan, 1910.

Stephen of Blois, *Letters*

Ed. as *Epistyolae Stephani, Comitis Carnotensis, atque Anselm de Ribodi Monte.* RHC Oc 3. Paris: Imprimerie Impériale, 1866.

Trans. August C. Krey. In *The First Crusade: The Accounts of Eye-Witnesses and Participants.* Princeton: Princeton University Press, 1921.

Al-Tabari, *Tarikh al-Rusul wa al-Muluk*

Trans. various. *The History of al-Tabari.* Ed. Ehsan Yarshater. 40 vols. Albany: State University Press of New York, 1985-2007.

Al-Tarsusi, Mardi ibn 'Ali, *Tabsirat Arbab al-Albab fi Kayfiyat al-Najah fi al-Hurub*

Ed. Karen Sader. *Dictionary of Ancient Weapons.* Beirut: Dar Sader, 1998.

Part. ed. and trans. Antoine Boudot-Lamotte. *Contribution à l'étude de l'archerie musulmane.* Damascus: Institut Français de Damas, 1968.

Templar of Tyre, *Gestes des Chiprois*

Ed. Gaston Raynaud. In *Les Gestes des Chiprois.* Recueil de Chroniques Françaises. Paris: Sociétés de l'Orient Latin, 1887, reprinted Osnabruck: Otto Zeller, 1968.

Ed. and trans. Laura Minervini. *Cronaca del Templare di Tiro (1243-1314).* Naples: Liguori Editore, 2000.

Trans. Paul Crawford. *The 'Templar of Tyre': Part III of the Deeds of the Cypriots.* Aldershot: Ashgate, 2003.

Thadeus, *Ystoria de desolatione et conculcatione civitatis Acconensis et tocius terre sancte*

Ed. R.B.C. Huygens. In *The Fall of Acre 1291.* CCCM 202. Turnhout: Brepols, 2004.

Theodosius the Deacon, *Expugnatio Cretae*

Ed. and trans. J.-P. Migne. In *Sapientissimi Imperatoris Constantini Porphyrogeniti.* Patrologiae Grecae 113. Paris, 1864.

Theophylact Simocatta, *History*

Trans. Michael Whitby and Mary Whitby. *The* History *of Theophylact Simocatta*. Oxford: Oxford University Press, 1986.

Translatio et Miracula Sanctorum Senesii et Theopontii

Ed. P.E. Schramm. *Translatio et Miracula Sanctorum Senesii et Theopontii*. MGH SS 30, pt. 2. Leipzig, 1934.

Al-'Umari, *Masalik al-absar*

Trans. D.S. Richards. *Egypt and Syria in the Early Mamluk Period*. Abingdon: Routledge, 2017.

Usama ibn Munqidh, *Kitab al-I'tibar*

Trans. Paul M. Cobb. *The Book of Contemplation: Islam and the Crusades*. London: Penguin, 2008.

Trans. Philip K. Hitti. *An Arab-Syrian Gentleman and Warrior in the Period of the Crusades: Memoirs of Usamah ibn Munqidh*. New York: Columbia University Press, 1929. Reprinted 2000.

Vahram, *Chronicle*

Trans. Charles Fried Neumann. *Vahram's Chronicle of the Armenian Kingdom in Cilicia*. London: Oriental Translation Fund, 1831.

Vegetius Renatus, Flavius. *Epitoma rei militaris*

Ed. Carl Lang. *Epitoma Rei Militaris*. Leipzig, 1885.

Vincent of Beauvaus, *Speculum historiale*

Ed. Antoine Koburger. *Speculum historiale*. Nuremberg, 1483.

Part. trans. Peter Jackson. In *The Seventh Crusade, 1244-1254*. Aldershot: Ashgate, 1988.

Vitruvius Pollio, Marcus, *De architectura libri decem*

Ed. F. Krohn. *De Architectura libri decem*. Leipzig: B.G. Teubneri, 1912.

Trans. Morris Hicky Morgan. *The Ten Books on Architecture*. Cambridge: Harvard University Press, 1914.

Walter the Chancellor, *Bella Antiochena*

Ed. Heinrich Hagenmeyer, *Galterii Cancellarii Bella Antiochena*. Innsbruck: Wagner'schen universitäts-buchhandlung, 1896.

Trans. Thomas S. Asbridge and Susan B. Edgington, *Walter the Chancellor's* The Antiochene Wars. Aldershot: Ashgate, 1999.

Walter of Coventry, *Memoriale*

Ed. William Stubbs. *Memoriale fratris Walteri de Coventria*. RS 58. 2 vols. London: Longman, 1872-73.

Wilbrand of Oldenburg, *Itinerarium terrae sanctae*

Ed. Denys Pringle. In "Wilbrand of Oldenburg's Journey to Syria, Lesser Armenia, Cyprus, and the Holy Land (1211-1212): A New Edition." *Crusades* 11 (2012), pp. 109-37.

Trans. Denys Pringle. In *Pilgrimage to Jerusalem and the Holy Land*. Ashgate, 2012.

William of Apulia, *Gesta Roberti Wiscadi*

Ed. Marguerite Mathieu. *La geste de Robert Guiscard.* Palermo: Istituto Siciliano di studi Bizantini e Neoellenici, 1961.

Trans. Graham A. Loud. *The Deeds Done by King Roger of Sicily.* www.leeds.ac.uk/arts/download/1049/the_deeds_of_robert_guiscard_by_william_of_apulia [last accessed 28 April 2014].

William the Breton, *Gesta Philippi Augusti*

Ed. H.-François Delaborde. In *Oeuvres de Rigord et de Guillaume le Breton.* Vol. 1. Paris: SHF, 1882.

William the Breton, *Philippide*

Ed. H.-François Delaborde. In *Oeuvres de Rigord et de Guillaume le Breton.* Vol. 2. Paris: SHF, 1885.

William of Malmesbury, *De gestis regum Anglorum*

Ed. Thomas Duffy Hardy. *Gesta Regum Anglorum.* 2 vols. London: Sumptibus Societatis, 1840.

Ed. William Stubbs. *De Gestis Regum Anglorum.* RS 90. 2 vols. London: Her Majesty's Stationary Office, 1887-89.

Trans. J.A. Giles. *Chronicle of the Kings of England.* London: George Bell and Sons, 1904.

William of Newburgh, *Historia rerum Anglicarum*

Ed. Richard Howlett. In *Chronicles of the Reigns of Stephen, Henry II, and Richard I.* RS 82. Vols. 1 & 2. London: Longman, 1884.

Trans. Joseph Stevenson. In *The Church Historians of England.* Vol. 4, pt. 2. London: Seeleys, 1856.

William of Tripoli, *Tractatus de statu Sarracenoroum*

Ed. Hans Prutz. In *Kulturgeschichte der Kreuzzuge.* Berlin: E.S. Mittler und Sohn, 1883.

William of Tyre, *Cronique*

Ed. as *Historia Rerum in Partibus Transmarinis.* RHC Oc 1. Paris: Imprimerie Royale, 1844.

Ed. R.B.C. Huygens. *Cronique.* CCCM 63. 2 vols. London: Brepols, 1986.

Trans. Emily Babcock and A.C. Krey. *A History of Deeds Done Beyond the Sea.* 2 vols. New York: Octagon Books, 1976.

Al-Yunini, Qutb al-Din Musa, *Dhayl Mir'at al-zaman*

Ed. and trans. Li Guo. *Early Mamluk Syrian Historiography: Al-Yunini's Dhayl Mir'at al-zaman.* 2 vols. Leiden: Brill, 1998.

Al-Zardkash, Ibn Aranbugha, *Kitab al-'aniq fi'l-manajniq*

Ed. Fuat Sezgin. *Armoury Manuel.* Frankfurt: Institute for the History of Arabic-Islamic Science, 1883.

Edited Collections of Primary Sources

Bain, Joseph, ed. *Calendar of Documents Relating to Scotland, Preserved in Her Majesty's Public Record Office, London*. Vol. 2, A.D. 1272-1307. Edinburgh: H.M. General Register House, 1884.

Barber, Malcolm and Keith Bate, eds. *Letters from the East: Crusaders, Pilgrims and Settlers in the 12th-13th Centuries*. Farnham: Ashgate, 2010.

Belgrano, Luigi Tommaso, et al., eds. *Annali Genovesi di Caffaro e de' suoi continuatori*. 5 vols. SI. Genoa: Istituto Storico Italiano, 1890-1929.

Brundage, James, ed. *The Crusades: A Documentary History*. Milwaukee: Marquette University Press, 1962, reprinted 1976.

Delaville le Roulx, Joseph, ed. *Cartulaire Général de l'ordre des Hospitaliers de S. Jean de Jérusalem (1100-1310)*. 3 vols. Paris: Ernest Leroux, 1894-99.

Duchesne, André, ed. *Historiae Francorum Scriptores*. 5 vols. Paris, 1636-49.

Edwards, J.G., ed. *Calendar of Ancient Correspondence Concerning Wales*. Cardiff: University of Wales Press, 1935.

Eickelman, Dale F. and James Piscatori, eds. *Muslim Travellers: Pilgrimage, Migration, and the Religious Imagination*. Berkeley, CA: University of California Press, 1990.

Gabrieli, Francesco, ed. *Storici arabi delle Crociate*. Turin: Giulio Einaudi, 1957. Trans. E.J. Costello as *Arab Historians of the Crusades*. New York: Dorset Press, 1989.

Hallam, Elizabeth, ed. *Chronicles of the Crusades: Eyewitness Accounts of the Wars Between Christianity and Islam*. London: George Weidenfeld and Nicolson, 1989.

Halphen, Louis and Rene Poupardin, eds. *Chroniques des Comtes d'Anjou et des Seigneurs d'Amboise*. Paris: Auguste Picard, 1913.

Hoogeweg, Hermann, ed. *Die Schriften des Kölner Domscholasters, späteren Bischofs von Paderborn und Kardinal-Bischofs von S. Sabina*. Tubingen: Litterarischen Verein in Stuttgart, 1894.

Howlett, Richard, ed. *Chronicles of the Reigns of Stephen, Henry II, and Richard I*. RS 82. 4 vols. London: Her Majesty's Stationary Office, 1884-89.

Huygens, R.B.C., ed. *The Fall of Acre 1291*. Turnhout: Brepols, 2004.

Jackson, Peter, ed. *The Seventh Crusade, 1244-54: Sources and Documents*. Aldershot: Ashgate, 2007.

Krey, August C., ed. *The First Crusade: The Accounts of Eye-Witnesses and Participants*. Princeton: Princeton University Press, 1921.

Lewis, Bernard, ed. *Islam: From the Prophet Mohammad to the Capture of Constantinople*. 2 vols. Oxford: Oxford University Press, 1974, reprinted 1987.

Luard, Henry Richards, ed. *Annales Monastici*. RS 36. Vol. 3. London: Longman, 1866.

Munro, Dana Carleton, ed. "Letters of the Crusaders." In *Translations and Reprints from the Original Sources of European History*. Vol. 1.4. Philadelphia: Department of History of the University of Pennsylvania, 1897.

Palmer, Andrew, ed. *The Seventh Century in the West-Syrian Chronicles*. Liverpool: Liverpool University Press, 1993.

Peters, Edward, ed. *The First Crusade: The Chronicle of Fulcher of Chartres and Other Source Material*. 2nd ed. Philadelphia: University of Pennsylvania Press, 1998.

Pringle, Denys, ed. *Pilgrimage to Jerusalem and the Holy Land, 1187-1291*. Farnham: Ashgate, 2012.

Prutz, Hans, ed. *Kulturgeschichte der Kreuzzuge*. Berlin: E.S. Mittler und Sohn, 1883.

Röchricht, Reinhold, ed. *Regesta Regni Hierosolymitani*. Innsbruck: Libraria Academica Wagneriana, 1893.

Shaw, M.R.B, ed. *Chronicles of the Crusades*. London: Penguin, 1963.

Verci, Giambatista, ed. *Storia degli Ecelini*. 3 vols. Bassano: Stamperia Remondini, 1779.

Wilkinson, John, ed., with Joyce Hill and W.F. Ryan. *Jerusalem Pilgrimage, 1099-1185*. London: The Hakluyt Society, 1988.

Secondary Sources

Adan-Bayewitz, David and Mordechai Aviam. "Iotapata, Josephus, and the Siege of 67: Preliminary Report on the 1992-94 Seasons." *Journal of Roman Archaeology* 10 (1997), pp. 131-65.

Amitai, Reuven. "The Conquest of Arsuf by Baybars: Political and Military Aspects." *Mamluk Studies Review* 9.1 (2005), pp. 61-83.

Amitai, Reuven. "Foot Soldiers, Militiamen and Volunteers." In *Texts, Documents and Artefacts: Islamic Studies in Honour of D. S. Richards*. Ed. Chase. F. Robinson. Leiden: Brill, 2003.

Amitai, Reuven. "Mongol Provincial Administration: Syria in 1260 as a Case-study." In *In Laudem Hierosolymitani*. Ed. Iris Shagrir, Ronnie Ellenblum and Jonathan Riley-Smith. Aldershot: Ashgate, 2007, pp. 117-43.

Amitai, Reuven. *Mongols and Mamluks: The Mamluk-Ilkhanid War, 1260-1281*. Cambridge: Cambridge University Press, 1995.

Amitai, Reuven. "Notes on the Ayyubid Inscriptions at al-Subayba (Qal'at Nimrud)." *Dumbarton Oaks Papers* 43 (1989), pp. 113-19.

Amt, Emilie. "Besieging Bedford: Military Logistics in 1224." *Journal of Medieval Military History* 1 (2002), pp. 101-24.

Asbridge, Thomas. *The Crusades: The Authoritative History of the War for the Holy Land*. New York: Harper Collins Publishers, 2010.

Asbridge, Thomas. *The First Crusade: A New History*. Oxford: Oxford University Press, 2004.

Aviam, Mordechai. "The Archaeological Illumination of Josephus, Narrative of the Battles at Yodefat and Gamla." In *Making History*. Ed. Zuleika Rodgers. Leiden: Brill, 2006, pp. 372-84.

Avigad, N. "Excavations in the Jewish Quarter of the Old City of Jerusalem, 1970 (Second Preliminary Report." *Israel Exploration Journal* 20 (1970), pp. 129-40.

Baatz, Dietwulf. "Recent Finds of Ancient Artillery." *Britannia* 9 (1978), pp. 1-17.

Bachrach, Bernard S. "Medieval Siege Warfare: A Reconnaissance." *The Journal of Military History* 58:1 (Jan., 1994), pp. 119-33.

Bachrach, David S. "English Artillery 1189-1307: The Implications of Terminology." *English Historical Review* 121.494 (Dec., 2006), pp. 1,408-30.

Barbé, Hervé. *Le Château de Safed et son Territoire à l'Epoque des Croisades*. Hebrew University of Jerusalem, unpublished PhD thesis, 2010.

Barbé, Hervé. "Zefat, Hativat Yiftah Way: Final Report." *Hadashot Arkheologiyot* 118 (2006), 29/3/2006.

Barbé, Hervé and Emanuel Damati. "Le château de Safed: sources Historiques, problématique et premiers résultats des recherches archéologiques." In *La fortification au temps des croisades*. Ed. Nicolas Faucherre, Jean Mesqui and Nicolas Prouteau. Rennes: Presses Universitaires des Rennes, 2004, pp. 77-93.

Barber, Malcolm. "Frontier Warfare in the Latin Kingdom of Jerusalem: The Campaign of Jacob's Ford, 1178-79." In *The Crusades and Their Sources: Essays Presented to Bernard Hamilton*. Ed. John France and William Zajac. Aldershot: Ashgate, 1998, pp. 9-22.

Barthet, Laure. "La prise de la barbacane de Montségur (Ariège) en février 1244: une introduction à l'archéologie de la poliorcétique." In *Artillerie et fortification, 1200-1600*. Ed. Nicolas Prouteau, Emmanuel de Crouy-Chanel and Nicolas Faucherre. Rennes: Presses Universitaires de Rennes, 2011, pp. 41-48.

Bartlett, W.B. *Islam's War against the Crusaders*. Stroud: The History Press, 2008.

Bauden, Frédéric. "Taqi al-Din Ahmad ibn 'Ali al-Maqrizi." In *Medieval Muslim Historians and the Franks of the Levant*. Ed. Alex Mallet. Leiden: Brill, 2015, pp. 161-200.

Beffeyte, Renaud. *Les machines de guerre au Moyen Age*. Rennes: Ouest France, 2008.

Benvenisti, Meron. *The Crusaders in the Holy Land*. New York: The Macmillan Company, 1972.

Berkovich, Ilya. "The Battle of Forbie and the Second Frankish Kingdom of Jerusalem." *Journal of Military History* 75 (2011), pp. 9-44.

Berthier, Sophie. "La Citadelle de Damas: Les Apports d'une Etude Archéologique." In *Muslim Military Architecture in Greater Syria: From the Coming of Islam to the Ottoman Period*. Ed. Hugh Kennedy. Leiden: Brill, 2006, pp. 151-64.

Biller, Thomas. "Die erste Burg der Johanniter (nach 1170). In *Der Crac des Chevaliers: Die Baugeschichte einer Ordensburg der Kreuzfahrerzeit*. Ed. Thomas Biller, et al. Regensburg: Schnell & Steiner, 2006, pp. 47-77.

Biller, Thomas, et al. "The Crac des Chevaliers - once again, comments on the state of research." (Internet publication, version March 14, 2013).

Bishop, M.C. and J.C.N. Coulston. *Roman Military Equipment: From the Punic Wars to the Fall of Rome*. 2nd. ed. Oxford: Oxbow Books, 2006.

Black, Jeremy. *European Warfare, 1494-1660*. London: Routledge, 2002.

Blary, Francois. "Aux origines d'une place forte des Vermandois au IXe-Xe siècle: Château-Thierry (Aisne, France)." *Château Gaillard* 25 (2012), pp. 39-60.

Boas, Adrian J. "Archaeological Evidence for the Mamluk Sieges and Dismantling of Montfort: A Preliminary Discussion." In *Montfort*. Ed. Adrian Boas. Leiden: Brill, 2017, pp. 41-55.

Boas, Adrian J. "Archaeological Sources for the History of Palestine: The Frankish Period: A Unique Medieval Society Emerges." *Near Eastern Archaeology*, 61.3 (Sep., 1998), pp. 138-73.

Boas, Adrian J. *Archaeology of the Military Orders: A Survey of the Urban Centres, Rural Settlement and Castles of the Military Orders in the Latin East (c. 1120-1291)*. Abingdon: Routledge, 2006.

Boas, Adrian J. *Crusader Archaeology: The Material Culture of the Latin East*. London: Routledge, 1999.

Boas, Adrian J. "Interpretation of the Parts." In *Montfort*. Ed. Adrian Boas. Leiden: Brill, 2017, pp. 102-19.

Boas, Adrian J. *Jerusalem in the Time of the Crusades: Society, Landscape and Art in the Holy City under Frankish Rule*. London: Routledge, 2001.

Boas, Adrian J. "Stone, Metal, Wood and Worked Bone Finds from the 1926 Expedition." In *Montfort*. Ed. Adrian Boas. Leiden: Brill, 2017, pp. 195-220.

Boas, Adrian J. and Rabei G. Khamisy. "Initial Thoughts on the Architectural Development of the Castle. In *Montfort*. Ed. Adrian Boas. Leiden: Brill, 2017, pp. 95-101.

Boase, T.S.R. *Kingdoms and Strongholds of the Crusaders*. London: Thames & Hudson, 1971.

Boase, T.S.R. "Military Architecture in the Crusader States in Palestine." In *A History of the Crusades. Vol. 4: The Art and Architecture of the Crusader States*. Ed. Harry W. Hazard. Madison: University of Wisconsin Press, 1977, pp. 140-64.

Bonaparte, Louis-Napoléon. *Études sur le passé et l'avenir de l'artillerie*. 6 vols. Paris: Libraire Militaire de J. Dumaine, 1848-71.

Booth, Kevin and Paul Roberts. "Recording the Keep, Dover Castle." *Château Gaillard* 19 (2000), pp. 21-23.

Bosworth, C. Edmund. "The City of Tarsus and the Arab-Byzantine Frontiers in Early and Middle 'Abbasid Times." *Oriens* 33 (1992), pp. 268-86.

Bradbury, Jim. *The Medieval Siege*. Woodbridge: Boydell, 1992.

Braune, Michael. "Die Stadtbefestigung von Damaskus." In *Burgen und Städte der Kreuzzugszeit*. Ed. Mathias Piana. Petersburg: Michael Imhof Verlag, 2008, pp. 202-10.

Brindle, Steven. "The Keep at Conisbrough Castle, Yorkshire." *Château Gaillard* 25 (2012), pp. 61-74.

Broshi, Magen "Mount Zion." *Israel Exploration Journal* 24 (1974), p. 285.

Broshi, Magen and Shimon Gibson. "Excavations along the Western and Southern Walls of the Old City of Jerusalem." In *Ancient Jerusalem Revealed*. Ed. Hillel Geva. Jerusalem: Israel Exploration Society, 1994, pp. 147-55.

Brown, Robin M. "Excavations in the 14th Century A.D. Mamluk Palace at Kerak." *Annual of the Department of Antiquities of Jordan* 33 (1989), pp. 287-304.

Brown, Robin M. "Summary Report of the 1986 Excavations: Late Islamic Shobak." *Annual of the Department of Antiquities of Jordan* 32 (1988), pp. 225-46.

Buchthal, Hugo. *Miniature Painting in the Latin Kingdom of Jerusalem*. Oxford: Oxford University Press, 1957, reprinted London: Pindar Press, 1986.

Burton, Richard Francis and Charles. F. Tyrwhitt Drake. *Unexplored Syria*. 2 vols. London: Tinsley Brothers, 1872.

Butler, Lawrence. "The Siege of Dolforwyn Castle in 1277." *Château Gaillard* 19 (2000), pp. 25-26.

Bylinski, Janusz "Three Minor Fortresses in the Realm of Ayyubid Rulers of Homs in Syria: Shumaimis, Tadmur (Palmyra) and al-Rahba." In *La fortification au temps des croisades*. Ed. Nicolas Faucherre, Jean Mesqui and Nicolas Prouteau. Rennes: Presses Universitaires des Rennes, 2004, pp. 151-64.

Cahen, Claude. *La Syrie du nord a l'époque des croisades et la principauté franque d'Antioche*. Paris: Geuthner, 1940.

Cahen, Claude. "Un traité d'armurerie composé pour Saladin." *Bulletin d'Etudes Orientales* 12 (1947-48), pp. 103-63.

Campbell, Duncan B. "Ancient Catapults: Some Hypotheses Reexamined." *Hesperia* 80 (2011), pp. 677-700.

Campbell, Duncan B. *Greek and Roman Artillery, 399 BC-AD 363*. Oxford: Osprey, 2003.

Campbell, Duncan B. *Greek and Roman Siege Machinery, 399 BC-AD 363*. Oxford: Osprey, 2003.

Carver, M.O.H. "Transition to Islam: Urban Roles in the East and South Mediterranean, Fifth to Tenth Centuries AD." In *Towns in Transition: Urban Evolution in Late Antiquity and the Early Middle Ages*. Ed. N. Christie and S.T. Loseby. Aldershot: Scolar Press, 1996, pp. 184-212.

Chakrabarti, Birbhushan, Tim Yates and Andrew Lewry. "Effect of Fire Damage on Natural Stonework in Buildings." *Construction and Building Materials* 10.7 (Oct., 1996), pp. 539-44.

Chevedden, Paul E. "Artillery in Late Antiquity: Prelude to the Middle Ages." In *The Medieval City Under Siege*. Ed. Ivy A. Corfis and Michael Wolfe. Woodbridge: Boydell, 1995, pp. 131-73.

Chevedden, Paul E. "Artillery of King James I." In *Iberia and the Mediterranean World of the Middle Ages*. Ed. Larry J. Simon, et al. Vol. 2. Leiden: Brill, 1996, pp. 47-94.

Chevedden, Paul E. "Black Camels and Blazing Bolts: The Bolt-Projecting Trebuchet in the Mamluk Army." *Mamluk Studies Review* 8.1 (2004), pp. 227-77.

Chevedden, Paul E. *The Citadel of Damascus*. University of California, Los Angeles, unpublished PhD thesis, 1986.

Chevedden, Paul E. "Fortifications and the Development of Defensive Planning in the Latin East." In *The Circle of War in the Middle Ages*. Ed. Donald J. Kagay and L.J. Andrew Villalon. Woodbridge: Boydell, 1999 pp. 33-43.

Chevedden, Paul E. "The Hybrid Trebuchet: The Halfway Step to the Counterweight Trebuchet." In *On the Social Origins of Medieval Institutions*. Ed. Donald J. Kagay and Teresa M. Vann. Leiden: Brill, 1998, pp. 179-222.

Chevedden, Paul E. "The Invention of the Counterweight Trebuchet: A Study in Cultural Diffusion." *Dumbarton Oaks Papers* 54 (2000), pp. 71-116.

Chevedden, Paul E. "King James I the Conqueror and the Artillery Revolution of the Middle Ages." In *Jaume I: Commemoració del VIII centenari del naixement de Jaume I*. Ed. María Teresa Ferrer i Mallol. Vol. 1. Memòries de la Secció Històrico-Arqueològica 91 (Barcelona: Institut d'Estudis Catalans, 2011), pp. 313-39.

Chevedden, Paul E., et al. "The Traction Trebuchet: A Triumph of Four Civilizations." *Viator* 31 (2000), pp. 433-86.

Chevedden, Paul E., et al. "The Trebuchet: Recent Reconstructions and Computer Simulations Reveal the Operating Principles of the Most Powerful Weapon of its Time." *Scientific American* (Jul., 1995), pp. 66-71.

Christie, Niall. "Ibn al-Qalanisi." In *Medieval Muslim Historians and the Franks of the Levant*. Ed. Alex Mallet. Leiden: Brill, 2015, pp. 7-28.

Clermont-Ganneau, Charles. *Archaeological Researches in Palestine During the Years 1873-1874*. Trans. Aubrey Stewart. Vol. 1. London, Published for the Palestine Exploration Fund, 1899.

Cohen, Michael. "The Fortification of the Fortress of Gybelin." In *La fortification au temps des croisades*. Ed. Nicolas Faucherre, Jean Mesqui and Nicolas Prouteau. Rennes: Presses Universitaires des Rennes, 2004, pp. 67-75.

Coldstream, Nicola. "Architects, Advisers and Design at Edward I's Castles in Wales." *Architectural History* 46 (2003), pp. 19-36.

Colvin, Howard, gen. ed. *The History of the King's Works*. 6 vols. London: Her Majesty's Stationary Office, 1963-82.

Conder, C.R. and H.H. Kitchener. *The Survey of Western Palestine: Memoirs of the Topography, Orography, Hydrography, and Archaeology*. Ed. E.H. Palmer and Walter Besant. 3 vols. London: The Committee of the Palestine Exploration Fund, 1881-83.

Contamine, Philippe. *La Guerre au moyen âge*. Paris: Presses Universitaires de France, 1980. Trans. Michael Jones as *War in the Middle Ages*. Basil Blackwell Publishers, 1984.

Cooper, John P. "'Fear God; Fear the *Bohaze*': The Nile Mouths and the Navigational Landscapse of the Medieval Nile Delta, Egypt." *Al-Masaq* 24.1 (2012), pp. 53-73.

Corvisier, Christian. "Les campagnes de construction du château de Beaufort (Qal'at as-Sharqif): une relecture." In *La fortification au temps des croisades*. Ed. Nicolas Faucherre, Jean Mesqui and Nicolas Prouteau. Rennes: Presses Universitaires des Rennes, 2004, pp. 243-66.

Crandell, Otis. "'Fire' Cracked Rocks - An Archaeological Experiment." *Corviniana* 10 (2007), pp. 1-10.

Creswell, K.A.C. "Fortification in Islam Before A.D. 1250." *Proceedings of the British Academy*. (1952), pp. 89-125.

Creswell, K.A.C. *The Muslim Architecture of Egypt*. 2 vols. Oxford: Oxford University Press, 1952-59, reprinted New York: Hacker Art Books, 1978.

Dangles, Philippe. "La refortification d'Afamiyya - Qal'at al-Mudiq sous le sultanat ayyoubide d'Alep (fin XIIe-mi XIIIe siècle)." In *La fortification au temps des croisades*. Ed. Nicolas Faucherre, Jean Mesqui and Nicolas Prouteau. Rennes: Presses Universitaires des Rennes, 2004, pp. 189-204.

De la Croix, Horst. *Military Considerations in City Planning: Fortifications*. New York: George Braziller, 1972.

De la Croix, Horst. "Military Architecture and the Radial City Plan in Sixteenth Century Italy." *The Art Bulletin* 42.4 (Dec., 1960), pp. 263-90.

De Meulemeester, Johnny and Denys Pringle. "Al-'Aqaba Castle, Jordan." *Château Gaillard* 22 (2006), pp. 97-102.

De Waha, Michel. "Les traces du bombardement de Beersel en 1489." *Château Gaillard* 19 (2000), pp. 305-13.

Dean, Bashford. "The Exploration of a Crusaders' Fortress (Montfort) in Palestine." *The Metropolitan Museum of Art Bulletin* 22.9 (Sep., 1927), pp. 5-46.

Dechert, Michael S.A. "The Military Architecture of Francesco di Giorgio in Southern Italy." *The Journal of the Society of Architectural Historians* 49:2 (Jun., 1990), pp. 161-80.

Dennis, George. "Byzantine Heavy Artillery: The Helepolis." *Greek, Roman, and Byzantine Studies* 39 (1998), pp. 99-115.

Denny, Mark. *Ingenium*. Baltimore: The John Hopkins University Press, 2007.

Denny, Mark. "Siege Engine Dynamics." *European Journal of Physics* 26 (2005), pp. 561-77.

Deschamps, Paul. *Les Châteaux des Croisés en Terre-Sainte*. 3 vols. Bibliothèque Archéologique et historique 19, 34, 90.

1. *Le Crac des Chevaliers*. Paris: Geuthner, 1934.
2. *La Défense du royaume de Jérusalem*. Paris: Geuthner, 1939.
3. *La Défense du comté de Tripoli et de la Principauté d'Antioche*. Paris: Geuthner, 1973.

Description de l'Égypte. 2nd ed. 37 vols. Paris: C.L.F. Panckoucke, 1821-26.

Desideri, Andrea Vanni. "Un contrappeso da *manjaniq* dalle fasi mamelucche del castello di Mont Réal (Shawbak). Considerazioni tra archeologia e iconografia / A Manjaniq Counterweight from the Mamluk Period in the Mont Réal Castle (Shawbak). Comments on Archaeology and Iconography." In *Da Petra a Shawbak, Archeologia di una Frontiera*. Ed. Guido Vannini and Michele Nucciotti. Florence: Giunti Editore, 2009, pp. 122-25.

D'Souza, Andreas. "The Conquest of 'Akka (690/1291): A Comparative Analysis of Christian and Muslim Sources." *The Muslim World* 80 (1990), pp. 234-50.

DeVries, Kelly and Robert Douglas Smith. *Medieval Military Technology*. 2nd ed. Toronto: University of Toronto Press, 2012.

Dothan, Moshe. "'Akko, 1976." *Israel Exploration Journal* 26 (1976), pp. 207-8.

Drap, Pierre, et al. "Going to Shawbak (Jordan) and Getting the Data Back: Towards a 3D GIS Dedicated to Medieval Archaeology." *Citeseer* (Jan., 2010).

Druks, Adam. "Akko, Fortifications." *Excavations and Surveys in Israel* 3 (1984), pp. 2-4.

Dufour, Guillaume. *Mémoire sur l'artillerie des anciens et sur celle du moyen age*. Paris: Cherbuliez, 1840.

Durdik, Tomas. "Anfange des Koniglichen Burgenbaus in Bohem." *Château Gaillard* 25 (2012), pp. 125-38.

Edbury, Peter W. "*Ernoul, Eracles* and the Fifth Crusade." In *The Fifth Crusade in Context: The Crusading Movement in the Early Thirteenth Century*. Ed. E.J. Mylod et al. Abingdon: Routledge, 2017, pp. 163-74.

Edbury, Peter W. "The French Translation of William of Tyre's *Historia*: the Manuscript Tradition." *Crusades* 6 (2007), pp. 69-105.

Edbury, Peter W. *The Kingdom of Cyprus and the Crusades, 1191-1374*. Cambridge: Cambridge University Press, 1991.

Edbury, Peter W. "Making Sense of the *Annales de Terre Sainte*: Thirteenth-century Vernacular Narratives from the Latin East." *Crusader Landscapes in the Medieval Levant: The Archaeology and History of the Latin East*. Ed. Micaela Sinibaldi et al. Cardiff: University of Wales Press, 2016, pp. 403-13.

Eddé, Anne-Marie. "Kamal al-Din 'Umar Ibn al-'Adim." In *Medieval Muslim Historians and the Franks of the Levant*. Ed. Alex Mallet. Leiden: Brill, 2015, pp. 109-35.

Edwards, Robert W. *The Fortifications of Armenian Cilicia*. Washington: Dumbarton Oaks Research Library and Collection, 1987.

El-Hakim, Sabry, et al. "Detailed 3D Modelling of Castles." *International Journal of Architectural Computing* 5.2 (2007), pp. 200-20.

Ellenblum, Ronnie. *Crusader Castles and Modern Histories*. Cambridge: Cambridge University Press, 2007.

Ellenblum, Ronnie. "Frankish Castles, Muslim Castles, and the Medieval Citadel of Jerusalem." In *In Laudem Hierosolymitani*. Ed. Iris Shagrir, Ronnie Ellenblum and Jonathan Riley Smith. Aldershot: Ashgate, 2007, pp. 93-109.

Ellenblum, Ronnie. *Frankish Rural Settlement in the Latin Kingdom of Jerusalem*. Cambridge: Cambridge University Press, 1998.

Ellenblum, Ronnie. "Who Built Qal'at al-Subayba?" *Dumbarton Oaks Papers* 43 (1989), pp. 103-12.

Ellenblum, Ronnie, et al. "Crusader Castle Torn Apart by Earthquake at Dawn, 20 May 1202." *Geology* 26.4 (Apr., 1998), pp. 303-6.

Ewart, Gordon and Denys Pringle. "Dundonald Castle Excavations 1986-93." *Scottish Archaeological Journal* 26.1-2 (2004).

Faucherre, Nicolas, et al. [*Caesarea, 2007*] *Mission archéologique a Césarée*. Ministère des Affaires Etrangères Français, 2007.

Faucherre, Nicolas, et al. [*Caesarea, 2008*] *Césarée: Fortifications Médiévales*. Ministère des Affaires Etrangères Français, 2008.

Faucherre, Nicolas, et al. [*Caesarea, 2009*] *Caesarea Maritima: Fortifications Médiévales*. Ministère des Affaires Etrangères Français, 2009.

Faucherre, Nicolas, et al. [*Caesarea, 2010*] *Caesarea Maritima: Fortifications Médiévales*. Ministère des Affaires Etrangères Français, 2010.

Faucherre, Nicolas, et al. "La forteresse de Shawbak (Crac de Montréal), une des premières forteresses Franques sous son corset mamelouk." In *La fortification au temps des croisades*. Ed. Nicolas Faucherre, Jean Mesqui and Nicolas Prouteau. Rennes: Presses Universitaires des Rennes, 2004, pp. 43-66.

Faucherre, Nicolas, Bruno Merlin and Jean Mesqui. "Césarée (Israël) – Le château et l'enceinte urbaine." *Bulletin Monumental - Actualités* (2007), pp. 1-4.

Fedden, Robin and John Thomson. *Crusader Castles*. London: John Murray Publishers, 1957.

Feld, Istvan. "Die Burgen des Konigreichs Ungarn im 11-12. Jahrhundert." *Château Gaillard* 25 (2012), pp. 159-69.

Finan, Thomas. "3D Castle Reconstruction as Interpretive Modelling: the Medieval Lough Ce Project." *Château Gaillard* 25 (2012), pp. 171-76.

Fino, J.F. "Machines de jet médiévales." *Gladius* 10 (1972), pp. 25-43.

Folda, Jaroslav. *Crusader Manuscript Illumination at Saint-Jean d'Acre, 1275-1291*. Princeton: Princeton University Press, 1976.

Folda, Jaroslav. *Crusader Art in the Holy Land, From the Third Crusade to the Fall of Acre, 1187-1291*. Cambridge: Cambridge University Press, 2005.

Fowke, Frank Rede. *The Bayeux Tapestry*. London: Arundel Society, 1875.

Fournet, Thibaud and Jean-Claude Voisin. "Le Château de Aakkar al-Aatiqa (Nord-Liban)." *Bulletin d'archéologie et d'architecture libanaises* 4 (2000), pp. 149-63.

France, John. *The Crusades and the Expansion of Catholic Christendom, 1000-1714*. Abingdon: Routledge, 2005.

France, John. "Fortifications East and West." In *Muslim Military Architecture in Greater Syria: From the Coming of Islam to the Ottoman Period*. Ed. Hugh Kennedy. Leiden: Brill, 2006, pp. 281-94.

France, John. *Hattin*. Oxford: Oxford University Press, 2015.

France, John. "Technology and the Success of the First Crusade." In *War and Society in the Eastern Mediterranean, 7th-15th Centuries*. Ed. Yaacov Lev. Leiden: Brill, 1997, pp. 163-76.

France, John. *Victory in the East: a Military History of the First Crusade*. Cambridge: Cambridge University Press, 1994.

France, John. *Western Warfare in the Age of the Crusades, 1000-1300*. Ithaca: Cornell University Press, 1999.

Frankel, Rafael. "The North-Western Corner of Crusader Acre." *Israel Exploration Journal* 37 (1987), pp. 256-61.

Freeman, A.Z. "Wall-Breakers and River-Bridgers: Military Engineers in the Scottish Wars of Edward I." *Journal of British Studies* 10.2 (May, 1971), pp. 1-16.

Frova, A. "Caesarea." *Israel Exploration Journal* 12 (1962), pp. 150-51.

Fulton, Michael S. "Anglo-Norman Artillery in Narrative Histories, from the Reign of William I to the Minority of Henry III." *Journal of Medieval Military History* 14 (2016), pp. 1-31.

Fulton, Michael S. *Artillery in and around the Latin East, 1097-1291*. Cardiff University, unpublished PhD thesis.

Fulton, Michael S. "Development of Prefabricated Artillery during the Crusades." *Journal of Medieval Military History* 13 (2015), pp. 51-72.

Fulton, Michael S. "Diffusion of Artillery Terminology." *SHARE* 3.1 (2016), pp. 1-27.

Fulton, Michael S. "The Myth of the Hybrid Trebuchet." *Viator* 48.2 (forthcoming).

Fulton, Michael S. "A Ridge Too Far: The siege of Saone/Sahyun in 1188 and Contemporary Trebuchet Technology." *Crusades* 16 (2017), pp. 33-53.

Gelichi, Sauro. "Die Burg Harim." In *Burgen und Städte der Kreuzzugszeit*. Ed. Mathias Piana. Petersburg: Michael Imhof Verlag, 2008, pp. 211-20.

Gelichi, Sauro. "The Citadel of Harim." In *Muslim Military Architecture in Greater Syria: From the Coming of Islam to the Ottoman Period*. Ed. Hugh Kennedy. Leiden: Brill, 2006, pp. 184-200.

Genequand, Denis. "Umayyad Castles; The Shift from Late Antique Military Architecture to Early Islamic Palatial Building." In *Muslim Military Architecture in Greater Syria:*

From the Coming of Islam to the Ottoman Period. Ed. Hugh Kennedy. Leiden: Brill, 2006, pp. 3-25.

Gillmor, Carroll. "The Introduction of the Traction Trebuchet into the Latin West." *Viator* 12 (1981), pp. 1-8.

Goldsworthy, Adrian. *In the Name of Rome*. Weidenfeld & Nicolson, 2003, reprinted London: Phoenix, 2004.

Gonnella, Julia. "The Citadel of Aleppo: Recent Studies." In *Muslim Military Architecture in Greater Syria: From the Coming of Islam to the Ottoman Period*. Ed. Hugh Kennedy. Leiden: Brill, 2006, pp. 165-75.

Goodall, John. "Dover Castle and the Great Siege of 1216." *Château Gaillard* 19 (2000), pp. 91-102.

Grose, Francis. *Military Antiquities Respecting a History of the English Army from the Conquest to the Present Time*. New Edition. 2 vols. London: T. Egerton & G. Kearsley, 1801.

Guérin, M.V. *Description géographique, historique et archéologique de la Palestine: Troisième Partie - Galilée*. 3 vols. Paris: Imprimerie Nationale, 1880.

Guo, Li. "Mamluk Historiographic Studies: The State of the Art," *Mamluk Studies Review* 1 (1997), pp. 15-43.

Haarmann, Ulrich. "L'édition de la chronique mamelouke syrienne de Sams ad-Din Muhammad al-Gazari." *Bulletin d'études orientales* 27 (1974), pp. 195-203.

Hack, Robert and Marco Huisman, "Estimating the Intact Rock Strength of a Rock Mass by Simple Means." *Engineering Geology for Developing Countries - Proceedings of the 9th Congress of the International Association for Engineering Geology and the Environment*, Durban, South Africa, 16-20 September 2002. Ed. J.L. van Rooy and C.A. Jermy. pp. 1,971-77. Pretoria: South African Institute of Engineering and Environmental Geologists, 2002.

Häffner, Hans-Heinrich. "Die Südtürme der Kernburg im 13. Jahrhundert." In *Der Crac des Chevaliers: Die Baugeschichte einer Ordensburg der Kreuzfahrerzeit*. Ed. Thomas Biller, et al. Regensburg: Schnell & Steiner, 2006, pp. 142-84.

Hamblin, William. *The Fatimid Army During the Early Crusades*. University of Michigan, unpublished PhD thesis, 1985.

Hamblin, William. "The Fatimid Navy During the Early Crusades: 1099-1224." *The American Neptune* 46 (1986), pp. 77-83.

Hamilton, Bernard. "The Elephant of Christ: Reynald of Châtillon." *Studies in Church History*, 15 (1978), pp. 97-108.

Hamilton, Bernard. *The Leper King and His Heirs: Baldwin IV and the Crusader Kingdom of Jerusalem*. Cambridge: Cambridge University Press, 2000.

Hanisch, Hanspeter. "Der Langturm als neuer Turmtypus im aiyubidischen Wehrbau, dargestellt an einem Linienturm der Zitadell von Damaskus." In *Egypt and Syria in the Fatimid, Ayyubid and Mamluk Eras*. Ed. U. Vermeulen and K. D'Hulster. Leuven: Peeters, 2007, pp. 183-234.

Hanisch, Hanspeter. "The Works of al-Malik al-'Adil in the Citadel of Harrân." In *La fortification au temps des croisades*. Ed. Nicolas Faucherre, Jean Mesqui and Nicolas Prouteau. Rennes: Presses Universitaires des Rennes, 2004, pp. 165-78.

Hanisch, Hanspeter. "Die Zitadell von Damaskus, 14/635 – 1410/1995." In *Egypt and Syria in the Fatimid, Ayyubid and Mamluk Eras*. Ed. U. Vermeulen and K. D'Hulster. Leuven: Peeters, 2007, pp. 235-305.

Hansen, Peter Vemming. "Experimental Reconstruction of a Medieval Trebuchet." *Acta Archaeologica* 63 (1992), pp. 189-208.

Hansen, Peter Vemming. "The Witch with Ropes for Hair." *Military Illustrated* 47 (1992), pp. 15-18.

Harper, Richard P. and Denys Pringle. *Belmont Castle: The Excavation of a Crusader Stronghold in the Kingdom of Jerusalem*. Oxford: Oxford University Press, 2000.

Hartal, Moshe. "Banyas, the Southwestern Tower." *Hadashot Arkheologiyot* 119 (2007), 16/4/2007.

Hartal, Moshe. *The al-Subayba (Nimrod) Fortress: Towers 11 and 9*. IAA Reports 11. Jerusalem: Israel Antiquities Authority, 2001.

Hartmann-Virnich, Andreas. "Les portes Ayyoubides de la citadelle de Damas: le regard de l'archéologie du bâti." In *La fortification au temps des croisades*. Ed. Nicolas Faucherre, Jean Mesqui and Nicolas Prouteau. Rennes: Presses Universitaires des Rennes, 2004, pp. 287-311.

Hawari, Mahmoud K. *Ayyubid Jerusalem (1187-1250): An Architectural and Archaeological Study*. Oxford: British Archaeological Reports, 2007.

Heidemann, Stefan. "The Citadel of al-Raqqa and Fortifications in the Middle Euphrates Area." In *Muslim Military Architecture in Greater Syria: From the Coming of Islam to the Ottoman Period*. Ed. Hugh Kennedy. Leiden: Brill, 2006, pp. 122-50.

Hill, Donald. "Military Technology." In *The Different Aspects of Islamic Culture, Vol. 4: Science and Technology in Islam, Part 2: Technology and Applied Sciences*. Ed. A.Y. al-Hassan. Paris: UNESCO, 2001, pp. 247-70.

Hill, Donald. "Trebuchets." *Viator* 4 (1973), pp. 99-114.

Hillenbrand, Carole. *The Crusades: Islamic Perspectives*. Edinburgh: Edinburgh University Press, 1999.

Hillenbrand, Carole. *A Muslim Principality in Crusader Times*. Istanbul: Nederlands Historisch-Archaeologisch Instituut, 1990.

Hinz, Walther. *Islamische Masse und Gewichte*. Trans. M. Ismail Marcinkowski as *Measures and Weights in the Islamic World*. Kuala Lumpur: International Islamic University of Malaysia, 2003.

Hirschler, Konrad. "Ibn Wasil: An Ayyubid Perspective on Frankish Lordships and Crusades." In *Medieval Muslim Historians and the Franks of the Levant*. Ed. Alex Mallet. Leiden: Brill, 2015, pp. 136-60.

Hirschler, Konrad. *Medieval Arabic Historiography: Authors as Actors*. London: Routledge, 2006.

Hoffmeyer, Ada Bruhn. "Military Equipment in the Byzantine Manuscript of Scylitzes in Biblioteca Nacional in Madrid." *Gladius* 5 (1966), pp. 1-194.

Holley, Andrew E. "The *Ballista* Balls from Masada." In *Masada IV: The Yigael Yadin Excavations 1963-1965 Final Reports*. Ed. Joseph Aviram, Gideon Foerster and Ehud Netzer. Jerusalem: Israel Exploration Society, 1994, pp. 349-65.

Holley, Andrew E. "Stone Projectiles and the Use of Artillery in the Siege of Gamla." In *Gamla III*, pt. 1. Ed. Danny Syon. Jerusalem: Israel Antiquities Authority, 2004, pp. 35-55.

Holmes, Urban Tignor. "Life among the Europeans in Palestine and Syria in the Twelfth and Thirteenth Centuries." In *A History of the Crusades. Vol. 4: The Art and Architecture of the Crusader States*. Ed. Harry W. Hazard. Madison: University of Wisconsin Press, 1977, pp. 3-35.

Holt, P.M. "Qalawun's Treaty with Acre in 1283." *English Historical Review* 91 (1976), pp. 802-12.

Humphreys, R. Stephen. *From Saladin to the Mongols: The Ayyubids of Damascus, 1193-1260*. Albany: State University of New York Press, 1977.

Humphries, Peter H. *Engines of War: Replica Medieval Siege Weapons at Caerphilly Castle*. Cardiff: CADW, 1992.

Huuri, Kalervo. *Zur Geschichte des mittelalterlichen Geschützwesens aus orientalischen Quellen*. Studia Orientalia 9.3. Helsinki: Druckerei A.G. der Finnischen Literaturgesellschaft, 1941.

Huygens, R.B.C. "Un nouveau texte du traité 'De constructione castri Saphet'." *Studi medievali* 3.6 (1965), pp. 355-87.

Ingham, Jeremy. "Forensic engineering of fire-damaged structures." *Proceedings of the Institute of Civil Engineers – Civil Engineering* 162 (May, 2009), pp. 12-17.

Irwin, Robert. "The Mamluk Conquest of the County of Tripoli." In *Crusade and Settlement*. Ed. Peter W. Edbury. Cardiff: University College Cardiff Press, 1985, pp. 246-50.

Irwin, Robert. *The Middle East in the Middle Ages: The Early Mamluk Sultanate 1250-1382*. Beckenham: Croom Helm, 1986.

Irwin, Robert. "Usamah ibn Munqidh: An Arab-Syrian Gentleman at the Time of the Crusades Reconsidered." In *The Crusades and Their Sources: Essays Presented to Bernard Hamilton*. Ed. John France and William Zajac. Aldershot: Ashgate, 1998, pp. 71-87.

Jackson, Peter. "Crisis in the Holy Land in 1260." *English Historical Review* 95.376 (Jul., 1980), pp. 481-513.

Jacoby, David. "Crusader Acre in the Thirteenth Century: Urban Layout and Topography." *Studi medievali* 3.20 (1979), pp. 1-45.

Jacoby, David. "Montmusard, Suburb of Acre: The First Stage of its Development." In *Outremer: Studies in the History of the Crusading Kingdom of Jerusalem*. Ed. B.Z. Kedar, et al. Jerusalem: Izhak Ben-Zvi Institute, 1982, pp. 205-17.

Jacoby, David. "Three Notes on Crusader Acre." *Zeitschrift des Deutchen Palästina-Vereins* 109.1 (1993), pp. 83-96.

Johns, C.N. "The Citadel, Jerusalem: A Summary of Work since 1934." *Quarterly of the Department of Antiquities in Palestine* 14 (1950), pp. 121-90.

Johns, C.N. "Excavations at Pilgrims' Castle ('Atlit): The Faubourg and its Defences." *Quarterly of the Department of Antiquities in Palestine* 1 (1931), pp. 111-29.

Johns, C.N. "Excavations at Pilgrims' Castle, 'Atlit, (1932): The Ancient Tell and the Outer Defences." *Quarterly of the Department of Antiquities in Palestine* 3 (1934), pp. 145-64.

Johns, C.N. *Guide to 'Atlit: The Crusader Castle, Town and Surroundings*. Jerusalem: Government of Palestine, Department of Antiquities, 1947, in *Pilgrims' Castle ('Atlit), David's Tower (Jerusalem) and Qal'at ar-Rabad ('Ajlun): Three Middle Eastern Castles from the Time of the Crusades*. Ed. Denys Pringle. Aldershot: Ashgate, 1997.

Johns, C.N. "Medieval 'Ajlun: I. The Castle (Qal'at ar-Rabad)." *Quarterly of the Department of Antiquities in Palestine* 1 (1931), pp. 21-33.

Kalayan, Haroutune. "The Sea Castle of Sidon." *Bulletin du Musée de Beyrouth* 26 (1973), pp. 81-90.

Kane, James H. "Wolf's Hair, Exposed Digits, and Muslim Holy Men: the *Libellus de expugnatione Terrae Sanctae per Saladinum* and the *Conte* of Eracles." *Viator* 47.2 (2016), pp. 95-112.

Kedar, Benjamin Z. "The Battle of Hattin Revisited." In *The Horns of Hattin*. Ed. Benjamin Kedar. (Jerusalem: Yad Izhak Ben-Zvi, 1992), pp. 190-207.

Kedar, Benjamin Z. "The Outer Walls of Frankish Acre." *'Atiqot* 31 (1997), pp. 157-80.

Kennedy, Hugh. *Crusader Castles*. Cambridge: Cambridge University Press, 1994.

King, D.J. Cathcart. "The Taking of Le Krak des Chevaliers in 1271." *Antiquity* 23.90 (Jun., 1949), pp. 83-92.

King, D.J. Cathcart. "The Trebuchet and other Siege-Engines." *Château Gaillard* 9-10 (1982), pp. 457-69.

Khamisy, Rabei G. "History and Archaeology of the Frankish Village of Tarphile." In *Montfort*. Ed. Adrian Boas. Leiden: Brill, 2017, pp. 128-36.

Khamisy, Rabei G. "Montfort Castle (Qal'at Al-Qurayn) in Mamluk Sources." In *Montfort*. Ed. Adrian Boas. Leiden: Brill, 2017, pp. 28-40.

Khamisy, Rabei G. "Some Notes on Ayyubid and Mamluk Military Terms." *Journal of Medieval Military History* 13 (2015), pp. 73-92.

Khamisy, Rabei G. and Michael S. Fulton, "Manjaniq Qarabugha and Thirteenth-Century Trebuchet Nomenclature." *Studia Islamica* 111.2 (2016), pp. 179-201.

Khan, Geoffrey. *Arabic Legal and Administrative Documents in the Cambridge Genizah Collection*. Cambridge: Cambridge University Press, 1993.

Köhler, G. *Die Entwickelung des Kriegswesens und der Kriegführung in der Ritterzeit von Mitte des 11. Jahrhunderts bis zu den Hussitenkriegen*. 3 vols. Breslau: Wilhelm Koebner, 1887-89.

Kölzer, Theo and Marlis Stähli, eds. *Liber ad honorem Augusti sive de rebus Siculis, Codex 120 II der Burgerbibliothek Bern*. Sigmaringen: Jan Thorbecke Verlag GmbH & Co., 1994.

Kostick, Conor. *The Siege of Jerusalem: Crusade and Conquest in 1099*. London: Continuum, 2009.

Kotzer, Hans -Jürgen, ed. *Die Kreuzzüge: Kein Kreig ist heilig*. Mainz: Philipp von Zabern, 2004.

Krizek, Leonid. "Trebuchet Reconstructions in Czechoslovakia." *Military Illustrated* 47 (1992), pp. 19-20.

La Viere Leiser, Gary. "The Crusader Raid in the Red Sea in 578/1182-83." *Journal of the American Research Centre in Egypt* 14 (1977), pp. 87-100.

Lassus, J.B.A., ed. *Album de Villard de Honnecourt, architecte du XIIIe siècle*. Paris: Imprimerie Impériale, 1858.

Lauffray, J. *Halabiyya-Zenobia place forte du limes oriental et la Haute-Mésopotamie au VIe siècle*. 2 vols. Bibliothèque archéologique et historique 119, 138. Paris: Geuthner, 1983-91.

Lawrence, A.W. "The Castle of Baghras." In *The Cilician Kingdom of Armenia*. Ed. T.S.R. Boase. Edinburgh: Scottish Academic Press, 1978, pp. 34-83.

Lawrence, A.W. "A Skeletal History of Byzantine Fortification." *The Annual of the British School at Athens* 78 (1983), pp. 171-227.

Lawrence, T.E. *Crusader Castles*. London: Golden Cockerel Press, 1936. Ed. Denys Pringle. *Crusader Castles*. Oxford: Oxford University Press, 1988.

Le Maho, Jacques. "Fortifications de siège et 'contre-châteaux' en Normandie (XIe-XIIe s.)." *Château Gaillard* 19 (2000), pp. 181-89.

Le Strange, Guy. *Palestine under the Muslims: A Description of Syria and the Holy Land from A.D. 650 to 1500*. Boston: Houghton, Mifflin and Company, 1890.

Leroy, Jules. *Les manuscrits syriaques à peintures*. 2 vols. Bibliothèque archéologique et historique 77. Paris: Geuthner, 1964.

Lewis, Raphael. "How Strong was Strong Mountain? Preliminary Remarks on the Possible Location of the Mamluk Siege Position at Montfort Castle." In *Montfort*. Ed. Adrian Boas. Leiden: Brill, 2017, pp. 282-86.

Lindsay, Lord. *Letters on Egypt, Edom and the Holy Land.* 5th ed. London: Henry G. Bohn, 1858.

Lithgow, William. *The Rare Adventures and Painful Peregrinations of William Lithgow.* Ed. Gilbert Phelps. London: The Folio Society, 1974.

Little, Donald P. "The Fall of 'Akka in 690/1291: the Muslim Version." In *Studies in Islamic History and Civilization in Honour of Professor David Ayalon.* Ed. M. Sharon. Jerusalem: Cana, 1986, pp. 159-81.

Little, Donald P. "Historiography of the Ayyubid and Mamluk Epochs." In *The Cambridge History of Egypt, Volume 1: Islamic Egypt, 640-1517.* Ed. Carl F. Petry. Cambridge: Cambridge University Press, 1998, pp. 412-44.

Little, Donald P. *An Introduction to Mamluk Historiography: An Analysis of Arabic Annalistic and Biographical Sources for the Reign of Al-Malik An-Nasir Muhammad Ibn Qala'un.* Montreal: McGill-Queen's University Press, 1970.

Lloyd, Seton and William Brice. "Harran." *Anatolian Studies* 1 (1951), pp. 77-111.

Lot, F. *L'Art Militaire et les Armées au Moyen Age en Europe et dans le Proche Orient.* Vol. 1. Paris: Payot, 1946.

Maalouf, Amin. *Les croisades vues par les Arabes.* Paris: Jean-Claude Lattès, 1983. Trans. Jon Rothschild as *The Crusades Through Arab Eyes.* London: Al Saqi Books, 1984.

Major, Balázs. "Al-Malik al-Mujahid, Ruler of Homs, and the Hospitallers (The Evidence in the Chronicle of Ibn Wasil)." In *The Crusades and the Military Orders: Expanding the Frontiers of Medieval Latin Christianity.* Ed. Zaolt Hunyadi and Jozsef Laszlovsky. Budapest: Department of Medieval Studies, Central European University, 2001, pp. 61-75.

Major, Balázs. "Medieval Cave Fortifications in the Upper Orontes Valley (a preliminary report)." In *Muslim Military Architecture in Greater Syria: From the Coming of Islam to the Ottoman Period.* Ed. Hugh Kennedy. Leiden: Brill, 2006, pp. 251-68.

Major, Balázs. "Medieval Cranes in Qal'at al-Marqab." *Hungarian Archaeology E-Journal* (Winter 2012), pp. 1-5.

Major, Balázs. "Medieval 'light construction buildings' on top of the vaulted halls of al-Marqab Citadel (Syria)." In *Das Ordens Militares: Atas do Encontro Internacional.* Ed. Isabel Cristina Ferreira Fernandes. Lisbon: Direção-Geral do Património Cultural, 2014, 165-81.

Mallett, Alex. "Sibt Ibn al-Jawzi." In *Medieval Muslim Historians and the Franks of the Levant.* Ed. Alex Mallett. Leiden: Brill, 2015, pp. 84-108.

Mallett, Alex. "A Trip Down the Red Sea with Reynald of Châtillon." *Journal of the Royal Asiatic Society of Great Britain & Ireland* 18.2 (2008), pp. 141-53.

Mallett, Michael. "The Transformation of War, 1494-1530." In *Italy and the European Powers: The Impact of War, 1500-1530.* Ed. Christine Shaw. Leiden: Brill, 2006, pp. 3-21.

Mandrell, Henry. *A Journey from Aleppo to Jerusalem in 1697.* Beirut: Khayats Oriental Reprints, 1963.

Marino, Luigi, et al. "The Crusader Settlement in Petra." *Fortress* 7 (Nov., 1990), pp. 3-13.

Marriner, Nick, Christopher Morhange and Samuel Meulé. "Holocene morphogenesis of Alexander the Great's isthmus at Tyre in Lebanon." *Proceedings of the National Academy of Sciences* 104.22 (2007), pp. 9,218-23.

Marsden, E.W. *Greek and Roman Artillery: Historical Development.* Oxford: Oxford University Press, 1969.

Marsden, E.W. *Greek and Roman Artillery: Technical Treatises.* Oxford: Oxford University Press, 1971.

Marshall, Christopher. *Warfare in the Latin East, 1192-1291.* Cambridge: Cambridge University Press, 1992.

Masterman, E.W.G. "A Visit to the Ruined Castles of the Teutonic Knights." *Palestine Exploration Fund, Quarterly Statement* (1919), pp. 71-75.

Mayer, Hans Eberhard. "The Origins of the Lordships of Ramla and Lydda in the Kingdom of Jerusalem." *Speculum* 60.3 (Jul., 1985), pp. 537-52.

Mazor, Gaby. "The Northern City Wall of Jerusalem on the Eve of the First Crusade (Excavations at the 'Jordanian Garden')." *'Atiqot* 51 (Mar., 2006), pp. 49-53.

Mayer, Hans Eberhard. "Two Unpublished Letters on the Syrian Earthquake of 1202." In *Medieval and Middle Eastern Studies in Honour of Aziz Suryal Atiya.* Ed. Sami A. Hanna. Leiden: Brill, 1972, pp. 295-310.

McGeer, Eric. "Byzantine Siege Warfare in Theory and Practice." In *The Medieval City Under Siege.* Ed. Ivy A. Corfis and Michael Wolfe. Woodbridge: Boydell, 1995, pp. 123-29.

Meckseper, Cord. "Frankreich und das Reich: Die Europaisch Burg im Vergleich." *Château Gaillard* 23 (2008), pp. 279-88.

Mesqui, Jean. "Bourzeÿ, une forteresse anonyme de l'Oronte." In *La fortification au temps des croisades.* Ed. Nicolas Faucherre, Jean Mesqui and Nicolas Prouteau. Rennes: Presses Universitaires des Rennes, 2004, pp. 95-133.

Mesqui, Jean. "L'enceinte médiévale de Césarée." In *L'architecture en Terre sainte au temps de Saint Louis.* Ed. Nicolas Faucherre, Benjamin Kedar and Jean Mesqui. *Bulletin monumental* 164.1. Paris: SFA, 2006, pp. 83-94.

Mesqui, Jean. "La fortification des Croisés au temps de Saint Louis au Proche-Orient." In *L'architecture en Terre sainte au temps de Saint Louis.* Ed. Nicolas Faucherre, Benjamin Kedar and Jean Mesqui. *Bulletin monumental* 164.1. Paris: SFA, 2006, pp. 5-29.

Mesqui, Jean. "Les programmes architecturaux." Ed. Philippe Dangles, et al. www.castelloreint.fr (version 14/07/03 12:16).

Mesqui, Jean. "Qal'at Yahmur (*Castrum Rubrum*), description archéologique." Ed. Philippe Dangles, et al. www.castelloreint.fr (version 14/07/03 12:33).

Mesqui, Jean. "Saône / Sahyoun / Qalaat Salah ad-Din: Rapport préliminaire de la mission effectuée du 15 au 20 mai 2002." Ed. Philippe Dangles, et al. www.castelloreint.fr (version 15/02/2004).

Mesqui, Jean. "La troisième enceinte." Ed. Philippe Dangles, et al. www.castellorient.fr (version 13/07/03 11:59).

Mesqui, Jean, et al. "The Portcullis in Medieval Great Towers: An Impression" *Château Gaillard* 23 (2008), pp. 289-95.

Mesqui, Jean and Benjamin Michaudel. "Bourzey - Hisn Burzaih / Burzuih / Qal'at Mirze: rapport de la mission effectuée du 27/05/2001 au 02/06/2001." Ed. Philippe Dangles, et al. www.castelloreint.fr (version 19/06/03).

Meyer, Werner et al. "Le Krak des Chevaliers (République Arabe Syrienne): rapport préliminaire sur les recherches 2005/2006." *Château Gaillard* 23 (2008), pp. 297-314.

Michaudel, Benjamin. "Burzaih." In *Burgen und Städte der Kreuzzugszeit*. Ed. Mathias Piana. Petersburg: Michael Imhof Verlag, 2008, pp. 178-87.

Michaudel, Benjamin. "The Development of Islamic Military Architecture during the Ayyubid and Mamluk Reconquest of Frankish Syria." In *Muslim Military Architecture in Greater Syria: From the Coming of Islam to the Ottoman Period*. Ed. Hugh Kennedy. Leiden: Brill, 2006, pp. 106-21.

Michaudel, Benjamin. "Le château de Saône / Qal'at Salah al-Din." (Unpublished paper, 2007).

Michaudel, Benjamin. "Étude historique de Qal'at Salah al-Din (Sahyun - Château de Saône)." Ed. Philippe Dangles, et al. www.castellorient.fr (version automne, 2002).

Michaudel, Benjamin. "Les refortifications Ayyoubides et Mameloukes en Syrie du nord (fin XIIe-début XIVe siècle)." In *La fortification au temps des croisades*. Ed. Nicolas Faucherre, Jean Mesqui and Nicolas Prouteau. Rennes: Presses Universitaires des Rennes, 2004, pp. 179-88.

Micheau, Françoise. "Ibn al-Athir." In *Medieval Muslim Historians and the Franks of the Levant*. Ed. Alex Mallet. Leiden: Brill, 2015, pp. 52-83.

Milwright, Marcus. *The Fortress of the Raven: Karak in the Middle Islamic Period (1100-1650)*. Leiden: Brill, 2008.

Mokary, Abdalla and Moshe Hartal. "Tiberias." *Hadashot Arkheologiyot* 122 (2010), 17/11/2010.

Morris, John E. *The Welsh Wars of Edward I: A Contribution to Medieval Military History, Based on Original Documents*. Oxford: Oxford University Press, 1901.

Müller-Wiener, Wolfgang. *Burgen der Kreuzritter*. Munich, Deutscher Kunstverlag, 1966. Trans. J. Maxwell Brownjohn as *Castles of the Crusaders*. London: Thames and Hudson, 1966.

Needham, Joseph. "China's Trebuchets, Manned and Counterweighted." In *On Pre-Modern Technology and Science*. Ed. Bert S. Hall and Delno C. West. Malibu: Undena Publications, 1976, pp. 107-45.

Needham, Joseph and Robin Yates. *Science and Civilisation in China*. Vol. 5, pt. 6 (Cambridge: Cambridge University Press, 1994.

Negev, A. "Caesarea." *Israel Exploration Journal* 10 (1960), pp. 264-65.

Negev, A. "Caesarea." *Israel Exploration Journal* 11 (1961), pp. 81-83.

Nicolle, David. *Arms and Armour in the Crusading Era, 1050-1350*. 2 vols. White Plains: Kraus International Publications, 1988.

Nicolle, David. *Crusader Castles in the Holy Land: An Illustrated History of the Crusader Fortifications of the Middle East and Mediterranean*. Oxford: Osprey, 2008.

Nicolle, David. "The Early Trebuchet: Documentary and Archaeological Evidence." In *La fortification au temps des croisades*. Ed. Nicolas Faucherre, Jean Mesqui and Nicolas Prouteau. Rennes: Presses Universitaires des Rennes, 2004, pp. 269-78.

Nicolle, David. "The Vocabulary of Medieval Warfare: Arms, Armour and Siege Weaponry Terminology in the *Excidium Aconis*." In *The Fall of Acre 1291*. Ed. R.B.C. Huygens. CCCM 202. Turnhout: Brepols, 2004, pp. 165-87.

Niebuhr, M. *Travels through Arabia, and Other Countries in the East*. Trans. Robert Heron. 2 vols. Edinburgh, 1792.

Norris, John. *Medieval Siege Warfare*. Stroud: Tempus, 2007.

Northrup, Linda S. *From Slave to Sultan: The Career of al-Mansur Qalawun and the Consolidation of Mamluk Rule in Egypt and Syria (678-689 A.H./1279-1290 A.D.)*. Stuttgart: Steiner, 1998.

Nossov, Konstantin. *Ancient and Medieval Siege Weapons*. Staplehurst: Spellmount, 2005.

Ober, Josiah. "Early Artillery Towers: Messenia, Boiotia, Attica, Megarid." *American Journal of Archaeology* 91.4 (Oct., 1987), pp. 569-604.

Oman, Sir Charles William Chadwick. *The Art of War in the Middle Ages: A.D. 378-1515*. Oxford: B.H. Blackwell, 1885.

Oman, Sir Charles William Chadwick. *A History of the Art of War: The Middle Ages from the Fourth to the Fourteenth Century*. London: Methuen & Co., 1889.

Omran, Mahmud Said. "King Amalric and the Siege of Alexandria, 1167." In *Crusade and Settlement*. Ed. Peter W. Edbury. Cardiff: University of Cardiff Press, 1985, pp. 191-96.

Payne-Gallwey, Sir Ralph. *A Summary of the History, Construction and Effects in Warfare of the Projectile - Throwing Engines of the Ancients*. London: Longman, 1907.

Parker, Geoffrey. *The Military Revolution: Military Innovation and the Rise of the West, 1500-1800*. 2nd ed. Cambridge: Cambridge University Press, 1996.

Parker, Geoffrey. "The 'Military Revolution,' 1560-1660 - a Myth?" *The Journal of Modern History* 42:2 (Jun., 1976), pp. 196-214.

Pepper, Simon. "The Face of the Siege: Fortification, Tactics and Strategy in the Early Italian Wars." In *Italy and the European Powers: The Impact of War, 1500-1530*. Ed. Christine Shaw. Leiden: Brill, 2006, pp. 33-56.

Piana, Mathias. "The Crusader Castle of Toron: First Results of its Investigation." *Crusades* 5 (2006), pp. 173-91.

Piana, Mathias. "Die Kreuzfahrerstadt Tortosa (Tartus) in Syrien." In *Burgen und Städte der Kreuzzugszeit*. Ed. Mathias Piana. Petersburg: Michael Imhof Verlag, 2008, pp. 408-21.

Piana, Mathias. "Die Templerburg Chastel Blanc (Burg as Safita)." In *Burgen und Städte der Kreuzzugszeit*. Ed. Mathias Piana. Petersburg: Michael Imhof Verlag, 2008, pp. 293-301.

Popplow, Marcus. "The Concept of *Machina* in the Roman Period." In *Homo Faber: Studies on Nature, Technology, and Science at the Time of Pompeii*. Ed. Jürgen Renn and Giuseppe Castagnetti. Rome: "L'erma" di Bretscheider, 2002, pp. 83-90.

Porat, Leea. "'Akko, Dov Gruner Street: Final Report." *Hadashot Arkheologiyot* 117 (2005), 29/9/2005.

Pospieszny, Kazimierz. "Tortosa (Syrien) und Lucera (Apulien), Die Idealen Residenz-Kastelle der Kreuzzugezeit: Eine Untersuchung nach Militärischen Aspekten." *Château Gaillard* 19 (2000), pp. 243-46.

Powell, James M. *Anatomy of a Crusade, 1213-1221*. Philadelphia: University of Pennsylvania Press, 1986.

Prager, Frank D. and Gustina Scaglia. *Mariano Taccola and his Book* de ingeneis. Cambridge: MIT Press, 1972.

Prawer, Joshua. *The Crusaders' Kingdom: European Colonialism in the Middle Ages*. London: Phoenix Press, 1972.

Prawer, Joshua. "The Jerusalem the Crusaders Captured: a Contribution to the Medieval Topography of the City." In *Crusade and Settlement*. Ed. Peter W. Edbury. Cardiff: University of Cardiff Press, 1985, pp. 1-14.

Pringle, Denys. "Castellology in the Latin East: An Overview." *Château Gaillard* 23 (2008), pp. 361-77.

Pringle, Denys. "A Castle in the Sand: Mottes in the Crusader East." *Château Gaillard* 18 (1998), pp. 187-91.

Pringle, Denys. "The Castle of Ayla (al-'Aqaba) in the Crusader, Ayyubid and Mamluk Periods." In *Egypt and Syria in the Fatimid, Ayyubid and Mamluk Eras* 4. Ed. U. Vermeulen and J. Van Steenbergen. Leuven-Dudley: Peeters, 2005, pp. 333-53, reprinted in Pringle, Denys. *Churches, Castles and Landscapes in the Frankish East*. Farnham: Ashgate, 2013.

Pringle, Denys. *The Churches of the Crusader Kingdom of Jerusalem*. 4 vols. Cambridge: Cambridge University Press, 1993-2009.

Pringle, Denys. "Crusader Castles: The First Generation." *Fortress* 1 (May, 1989).

Pringle, Denys. "Crusader Castles and Fortifications: The Armenian Connection." In *La Méditerranée des Arméniens, XIIe-XVe siècle*. Ed. Claude Mutafian. Paris: Geuthner, 2014, pp. 353-72.

Pringle, Denys. "Edward I, Castle-building and the Tower of the English in Acre." In *A Fresh Approach: Essays Presented to Colin Platt in Celebration of his Eightieth*

Birthday, 11 November 2014. Ed. Clair Donavan. Bristol: Trouser Press, 2014, pp. 48-56.

Pringle, Denys. "A Medieval *Graffito* Representing a *Trébuchet* in an Etruscan Tomb in Corneto-Tarquinia." In, *Crusading Warfare in the Middle Ages: Realities and Representations*. Ed. Simon John and Nicholas Morton. Farnham: Ashgate, 2014, pp. 37-46.

Pringle, Denys. "Reconstructing the Castle of Safad." *Palestine Exploration Quarterly* 117 (1985), pp. 139-49.

Pringle, Denys. *The Red Tower (al-Burj al-Ahmar): Settlement in the Plain of Sharon at the Time of the Crusaders and Mamluks, A.D. 1099-1516*. London: British School of Archaeology in Jerusalem, 1986.

Pringle, Denys. "Richard I and the Walls of Ascalon." *Palestine Exploration Quarterly* 116 (1984), pp. 133-47.

Pringle, Denys. *Secular Buildings in the Crusader Kingdom of Jerusalem: An Archaeological Gazetteer*. Cambridge: Cambridge University Press, 1997.

Pringle, Denys. "Templar Castles between Jaffa and Jerusalem." *The Military Orders 2: Welfare and Warfare*. Ed. Helen Nicholson. Aldershot: Ashgate, 1998, pp. 89-109.

Pringle, Denys. "Templar Castles on the Road to the Jordan." In *The Military Orders: Fighting for the Faith and Caring for the Sick*. Ed. Malcolm Barber. Aldershot: Ashgate, 1994, pp. 148-66.

Pringle Denys. "A Thirteenth-Century Hall at Montfort Castle in Western Galilee." *Antiquaries Journal* 66.1 (1986), pp. 52-81.

Pringle, Denys. "Towers in Crusader Palestine." *Château Gaillard* 16 (1994), pp. 335-50.

Pringle, Denys. "Town Defences in the Crusader Kingdom of Jerusalem." In *The Medieval City Under Siege*. Ed. Ivy A. Corfis and Michael Wolfe. Woodbridge: Boydell, 1995, pp. 69-121.

Pringle, Denys. "The Walls of Ascalon in the Byzantine, Early Islamic and Crusader Periods: A Preliminary Report on Current Research." In, *Guerre et paix dans le Proche-Orient médiéval: Colloque international, 15–18 décembre 2011*. Ed. S. Denoix, M. Eychenne, S. Pradines and A. Zouache. Cairo: Institut français d'archéologie orientale du Caire, forthcoming.

Prouteau, Nicolas. "'Beneath the Battle'?: Miners and Engineers as 'Mercenaries' in the Holy Land (XII-XIII Siècles)." In *Mercenaries and Paid Men: The Mercenary Identity in the Middle Ages*. Ed. John France. Leiden: Brill, 2008, pp. 105-17.

Prutz, Hans. *Kulturgeschichte der Kreuzzuge*. Berlin: E.S. Mittler und Sohn, 1883.

Pryor, John and Michael Jeffreys. "Alexios, Bohemond, and Byzantium's Euphrates Frontier: A Tale of Two Cretans." *Crusades* 11 (2012), pp. 31-86.

Purton, Peter. *Early Medieval Siege c. 450-1200*. Woodbridge: Boydell, 2009.

Purton, Peter. "The Myth of the Mangonel: Torsion Artillery in the Middle Ages." *Arms and Armour* 3.1 (2006), pp. 79-90.

Rabbat, Nasser. "The Militarization of Taste in Medieval Bilad al-Sham." In *Muslim Military Architecture in Greater Syria: From the Coming of Islam to the Ottoman Period.* Ed. Hugh Kennedy. Leiden: Brill, 2006, pp. 84-105.

Raphael, Kate. "Mighty Towers and Feeble Walls: Ayyubid and Mamluk Fortifications in the Late Twelfth and Early Thirteenth Centuries in the Light of the Decline of Crusader Siege Warfare." *Crusades* 9 (2010), pp. 147-58.

Raphael, Kate. *Muslim Fortresses in the Levant: Between Crusaders and Mongols.* Abingdon: Routledge, 2011.

Raphael, Kate. "A Thousand Arrowheads from the Crusader Fortress at Vadum Iacob." In *In the Hill Country, and in the Shephelah, and in the Arabah.* Ed. R. Beeri, D. Bern Yosef, O. Cohen and M. Heltzer. Jerusalem: Ariel House, 2008, pp. 252-61.

Raphael, Kate and Yotam Tepper. "The Archaeological Evidence from the Mamluk Siege of Arsuf." *Mamluk Studies Review* 9.1 (2005), pp. 85-100.

Reich, Ronny and Eli Shukron. "Excavations in the Mamillah Area, Jerusalem: The Medieval Fortifications." *'Atiqot* 54 (2006), pp. 125-52.

Rey, E.-G. *Les colonies Franques de Syrie aux XII*[me] *et XIII*[me] *siècles.* Paris: Alph. Picard, 1883.

Rey, E.-G. *Étude sur les monuments de l'architecture militaire des croisés en Syrie et dans l'île de Chypre.* Collection de documents inédits sur l'histoire de France. Paris: Imprimerie Nationale, 1871.

Rice, Tamara Talbot. *Ancient Arts of Central Asia.* London: Thames and Hudson, 1965.

Richard, Jean. "Les Causes des Victoires Mongoles d'Apres les Historiens Occidentaux du XIIIe Siècle." *Central Asiatic Journal* 23 (1979), pp. 104-17.

Richard, Jean. *Le comté de Tripoli sous la dynastie Toulousaine (1102-1187).* Paris: Geuthner, 1945.

Richard, Jean. *Le royaume latin de Jérusalem.* Paris: Presses Universitaires de France, 1953. Trans. Janet Shirley as *The Latin Kingdom of Jerusalem.* Amsterdam: North-Holland Publishing Company, 1979.

Richards, D.S. "A Consideration of Two Sources for the Life of Saladin." *Journal of Semitic Studies* 25.1 (1980), pp. 46-65.

Richter-Bernburg, Lutz. "'Imad al-Din al-Isfahani." In *Medieval Muslim Historians and the Franks of the Levant.* Ed. Alex Mallet. Leiden: Brill, 2015, pp. 29-51.

Rihll, Tracey. *The Catapult: A History.* Yardley: Westholme, 2007.

Riley-Smith, Jonathan. *The Crusades: A History.* 2nd ed. New Haven: Yale University Press, 2005.

Rogers, Randall. *Latin Siege Warfare in the Twelfth Century.* Oxford: Oxford University Press, 1992.

Roll, Israel. "Introduction: History of the Site, Its Research and Excavations." In *Apollonia-Arsuf: Final Report of the Excavations, Vol. I: The Persian and Hellenistic Periods.* Ed.

Israel Roll and Oren Tal. Tel Aviv: Emery and Claire Yass Publications in Archaeology, 1999, pp. 1-62.

Runciman, Steven. *A History of the Crusades*. 3 vols. Cambridge: Cambridge University Press, 1951-54.

Saadé, Gabriel. "Histoire du Château de Saladin." *Studi medievali* 3.9 (1968), pp. 980-1,016.

Saimre, Tanel. "Trebuchet - A Gravity-Operated Siege Engine: A study in Experimental Archaeology." *Estonian Journal of Archaeology* 10.1 (2006), pp. 61-80.

Salame-Sarkis, Hassan. "Chronique Archéologique du Liban-Nord." *Bulletin de Musée de Beyrouth* 2 (1938), pp. 91-98.

Savage, Stephen H. and Donald R. Keller. "Archaeology in Jordan, 2005 Season." *American Journal of Archaeology*. 110.3 (Jul., 2006), pp. 471-91.

Sayers, William. "The Name of the Siege Engine *trebuchet*: Etymology and History in Medieval France and Britain." *Journal of Medieval Military History* 8 (2010), pp. 189-96.

Schiller, Ely, ed. *Jerusalem & the Holy Land in Old Engravings and Illustrations (1483-1800)*. Jerusalem: Ariel Publishing House, 1981.

Schiller, Ely and Dan Kyram, eds. *The First Photographs of the Holy Land*. Jerusalem: Ariel Publishing House, 1980.

Schneider, Rudolf. *Die Artillerie des Mittelalters*. Berlin: Weidmannsche Buchhandlung, 1910.

Schramm, Erwin. *Die antiken Geschütze der Saalburg: Bemerkungen zu ihrer Rekonstruktion*. Berlin: Weidmann, 1918.

Sauvaget, Jean. "La citadelle de Damas." *Syria* 11 (1930), pp. 59-90.

Setton, Kenneth M., gen. ed. *A History of the Crusades*. 6 vols. Madison: University of Wisconsin Press, 1969-89.

Shatzman, Israel. "Stone-Balls from Tel Dor and the Artillery of the Hellenistic World." *Scripta Classica Israelica* 14 (1995), pp. 52-72.

Shotten-Hallel, Vardit, Dorit Korngreen and Lydia Perelis Grossoicz. "The Stones of Montfort: Sources of Stone for Montfort Castle." In *Montfort*. Ed. Adrian Boas. Leiden: Brill, 2017, pp. 137-49.

Siano, Donald B. *Trebuchet Mechanics*. asme.usu.edu/wp-content/uploads/2013/09/trebmath35.pdf (3 Feb. 2006).

Smail, R.C. *Crusading Warfare (1097-1193)*. Cambridge: Cambridge University Press, 1956.

Smith, Caroline. *Crusading in the Age of Joinville*. Aldershot: Ashgate, 2006.

Soedel, Werner and Vernard Foley. "Ancient Catapults." *Scientific American* (Mar., 1979), pp. 150-60.

Solar, Giora. "Jerusalem, citadel moat." *Excavations and Surveys in Israel* 3 (1984), pp. 47-48.

Solomon, Anne. *The Last Siege of Dryslwyn Castle*. Carmarthen: V.G. Lodwick & Sons, 1982.

Stager, Lawrence and Douglas Esse. "Ashkelon, 1985-1986." *Israel Exploration Journal* 37 (1987), pp. 68-72.

Stepansky, Yosef. "The Crusader Castle of Tiberias." *Crusades* 3 (2004), pp. 179-81.

Stepansky, Yosef. "Das kreuzfahrerzeitliche Tiberias Neue Erkenntnisse." In *Burgen und Städte der Kreuzzugszeit*. Ed. Mathias Piana. Petersburg: Michael Imhof Verlag, 2008, pp. 384-95.

Stepansky, Yosef. "Tiberias, the Courtyard of the Jews. *Hadashot Arkheologiyot* 121 (2009), 24/8/2009.

Stepansky, Yosef. "Tiberias, Map Survey." *Hadashot Arkheologiyot* 117 (2005), 14/9/2005.

Stewart, Angus. "Qal'at al-Rum / Hromgla / Rumkale and the Mamluk Siege of 691AH/1292CE." In *Muslim Military Architecture in Greater Syria: From the Coming of Islam to the Ottoman Period*. Ed. Hugh Kennedy. Leiden: Brill, 2006, pp. 269-80.

Syon, Danny and Zvi Yavor. "Gamla 1997-2000." *'Atiqot* 50 (2005), pp. 37-71.

Tal, Oren. "Apollonia-Arsuf, 2006 and 2009." *Israel Exploration Journal* 60.1 (2010), pp. 107-14.

Tarver, W.T.S. "The Traction Trebuchet: A Reconstruction of an Early Medieval Siege Engine." *Technology and Culture* 36.1 (Jan., 1995), pp. 136-67.

Taylor, A.J. "Master James of St. George." *English Historical Review* 65 (1950), pp. 433-57.

Tepper, Yotam. "Stone Balls from 'Akko: Artillery in the Crusader and Mamluk Periods." In *'Akko III*. Ed. Danny Syon. IAA Reports. Jerusalem: Israel Antiquities Authority, forthcoming.

Thompson, A. Hamilton. *Military Architecture in England during the Middle Ages*. London: Oxford University Press, 1912.

Thorau, Peter. *Sultan Baibars I. von Ägypten: Ein Beitrag zur Geschichte des Vorderen Orients im 13. Jahrhundert*. Wiesbaden: Ludwig Reichert, 1987. Trans. P.M. Holt as *The Lion of Egypt: Sultan Baybars I & the Near East in the Thirteenth Century*. Harlow: Longman, 1992.

Tibble, Steven. *Monarchy and Lordships in the Latin Kingdom of Jerusalem, 1099-1291*. Oxford: Oxford University Press, 1989.

Tonghini, Cristina. "Die Burg Saizar." In *Burgen und Städte der Kreuzzugszeit*. Ed. Mathias Piana. Petersburg: Michael Imhof Verlag, 2008, pp. 234-40.

Tonghini, Cristina. *Shayzar I: The Fortification of the Citadel*. Leiden: Brill, 2012.

Tonghini, Cristina, et al. "The Evolution of Masonry Technique in Islamic Military Architecture, the Evidence from Shayzar." *Levant* 35 (2003), pp. 179-212.

Tonghini, Cristina and Nadia Montevecchi. "The Castle of Shayzar: The Fortification of the Access System." In *Muslim Military Architecture in Greater Syria: From the Coming of Islam to the Ottoman Period*. Ed. Hugh Kennedy. Leiden: Brill, 2006, pp. 2001-24.

Tonghini, Cristina and Nadia Montevecchi. "The Castle of Shayzar: The Results of Recent Archaeological Investigations." In *La fortification au temps des croisades*. Ed. Nicolas Faucherre, Jean Mesqui and Nicolas Prouteau. Rennes: Presses Universitaires des Rennes, 2004, pp. 137-50.

Toy, Sidney, *Castles: Their Construction and History*. London: W. Heinemann, 1939, reprinted New York: Dover Publications Inc., 1985.

Vachon, Véronique. "Les Châteaux Ismâʿîliens du Djabal Bahrâʿ." In *La fortification au temps des croisades*. Ed. Nicolas Faucherre, Jean Mesqui and Nicolas Prouteau. Rennes: Presses Universitaires des Rennes, 2004, pp. 219-41.

Van Berchem, Max and Edmond Fatio. *Voyages en Syrie*. 2 vols. Cairo: L'institut français d'archéologie orientale, 1914-15.

Vannini, Guido and Cristina Tonghini. "Medieval Petra. The Stratigraphic Evidence from Recent Archaeological Excavations at al-Wu'ayra." *Studies in the History and Archaeology of Jordan* 6 (1997), pp. 371-84.

Viollet-le-Duc, Eugene-Emmanuel. *Dictionnaire Raisonne de l'Architecture Française du XI au XVI Siècle*. 10 vols. Paris: Librairies-Imprimeries Réunies, 1860.

Viollet-le-Duc, M. *Description du Château de Coucy*. Paris: Bance, 1857.

Vryonis, Speros Jr. "The Evolution of Slavic Society and the Slavic Invasion in Greece, The First Major Slavic Attack on Thessaloniki, A.D. 597." *The Journal of the American School of Classical Studies at Athens* 50.4 (1981), pp. 378-90.

Wagner, Olaf and Heiko Lass, eds. *Beihefte zur Mediaevistik*. Vol. 7: *... wurfen hin in steine/ grôze und niht kleine ...* Gen. ed. Peter Dinzelbacher. Frankfurt am Main: Peter Lang GmbH, 2006.

Weksler-Bdolah, Shlomit. "The Fortification System in the Northwestern Part of Jerusalem from the Early Islamic to the Ottoman Periods." *'Atiqot* 65 (2011), pp. 105-30.

White, Lynn Jr. "The Crusades and the Technological Thrust of the West." In *War, Technology and Society in the Middle East*. Eds. V.J. Parry and M.E. Yapp. London: Oxford University Press, 1975, pp. 97-112.

White, Lynn Jr. *Medieval Technology and Social Change*. Oxford: Oxford University Press, 1962.

Whitney, George H. *Hand-Book of Bible Geography*. New York: Eaton & Mains, 1899.

Wightman, G.J. *The Damascus Gate, Jerusalem: Excavations by C.-M. Bennett and J. B. Hennessy at the Damascus Gate, Jerusalem, 1964-66*. Oxford: B.A.R., 1989.

Wilkins, Alan, Hans Barnard and Pamela J. Rose. "Roman Artillery Balls from Qasr Ibrim, Egypt." *Sudan & Nubia* 10 (2006), pp. 66-80.

Yadin, Y. "The Excavation of Masada - 1963/64: Preliminary Report" *Israel Exploration Journal* 15 (1965), pp. 1-120.

Yannopoulos, Panayotis. "La pénétration Slave à Argos." *Bulletin de Correspondance Hellénique, Etudes Argiennes* 6 (1980), pp. 323-72.

Yasmine, Jean. "Die Burg Beaufort (Qal'at Saqif 'Arnun) - Neue Bauaufnahme, neue Erkenntnisse." In *Burgen und Städte der Kreuzzugszeit*. Ed. Mathias Piana. Petersburg: Michael Imhof Verlag, 2008, pp. 274-84.

Yates, Robin D.S. "Siege Engines and Late Zhou Military Technology." In *Explorations in the History of Science and Technology in China*. Ed. Li Guohao, Zang Mehwen and Cao Tianqin. Shanghai: Shanghai Chinese Classics Publishing House, 1982, pp. 409-52.

Yovitchitch, Cyril. "Die Aiyubidische Burg 'Aglun." In *Burgen und Städte der Kreuzzugszeit*. Ed. Mathias Piana. Petersburg: Michael Imhof Verlag, 2008, pp. 118-25.

Yovitchitch, Cyril. "Die Befestigung der Tore Aiyubidischer Burgen - Herausbildung eines Standards." In *Burgen und Städte der Kreuzzugszeit*. Ed. Mathias Piana. Petersburg: Michael Imhof Verlag, 2008, pp. 110-17.

Yovitchitch, Cyril. "Bosra: Eine Zitadelle des Fürstentums Damaskus." In *Burgen und Städte der Kreuzzugszeit*. Ed. Mathias Piana. Petersburg: Michael Imhof Verlag, 2008, pp. 169-77.

Yovitchitch, Cyril. "La citadelle de Bosra." In *La fortification au temps des croisades*. Ed. Nicolas Faucherre, Jean Mesqui and Nicolas Prouteau. Rennes: Presses Universitaires des Rennes, 2004, pp. 205-17.

Yovitchitch, Cyril. *Forteresses du Proche-Orient l'architecture militaire des Ayyoubides*. Paris: Presses de l'Université Paris-Sorbonne, 2011.

Yovitchitch, Cyril. "The Tower of Aybak in 'Ajlun Castle: An Example of the Spread of an Architectural Concept in Early 13th Century Ayyubid Fortification." In *Muslim Military Architecture in Greater Syria: From the Coming of Islam to the Ottoman Period*. Ed. Hugh Kennedy. Leiden: Brill, 2006, pp. 225-42.

Zimmer, John and Werner Meyer. "Le Krak des Chevaliers, Travaux de Mesurage et Observation sur l'Histoire Architecturale." *Château Gaillard* 22 (2006), pp. 359-73.

Zimmer, John, Werner Meyer and Letizia Boscardin. *Krak des Chevaliers in Syrien: Archäologie und Bauforschung 2003 bis 2007*. Braubach: Europäisches Burgeninstitut, 2011.

Index

Places

People